Chemistry: A Contemporary Approach

Chemistry
A Contemporary Approach
Third Edition

G. TYLER MILLER, JR.

DAVID G. LYGRE
Central Washington University

With contributions by
Wesley D. Smith
Ricks College,
Coauthor of Second Edition

Wadsworth Publishing Company
Belmont, California
A Division of Wadsworth, Inc.

Chemistry Editor: Jack Carey
Editorial Assistant: Kathy Shea
Production Editor: Gary Mcdonald
Managing Designer: Andrew H. Ogus
Print Buyer: Karen Hunt
Art Editor: Donna Kalal
Permissions Editor: Robert M. Kauser
Copy Editor: Betty Duncan-Todd
Photo Researcher: Sarah Bendersky
New Illustrations: Victor Royer, Amy Hennig
Compositor: Syntax International
Cover: Andrew H. Ogus
Cover photo credits and Part Opening photo credits are found following the index.

Printed in the United States of America   19

    3 4 5 6 7 8 9 10—95 94 93 92

Library of Congress Cataloging-in-Publication Data

Miller, G. Tyler (George Tyler), 1931–
    Chemistry, a contemporary approach / G. Tyler Miller, Jr., David
Lygre, with contributions by Wesley D. Smith.—3rd ed.
        p. cm.
    Includes bibliographical references and index.
    ISBN 0-534-14280-X
    1. Chemistry.   I. Lygre, David G.   II. Smith, Wesley D., 1930–
    III. Title.
QD31.2.M48   1991
540—dc20   90-39904
                                                                    CIP

Contents In Brief

v

Contents

Preface

To the Student

WHY STUDY CHEMISTRY?   Chemistry is fun and interesting and is involved in almost everything you do. When you decide what to buy, use, or eat, you are making chemical decisions. Throughout your life you will encounter controversial issues such as acid rain, nuclear power, use of pesticides, and genetic engineering. A knowledge of chemistry can help you make wiser decisions in these and other matters that affect you, your loved ones, and society as a whole.

HOW THIS BOOK IS ORGANIZED   Look at the brief table of contents on p. v. Notice that this book is divided into five major parts.

Part I is a brief introduction to science and chemistry. Part II presents some basic principles of chemistry that are used throughout the rest of the book.

Parts III, IV, and V discuss applied areas of chemistry. Part III examines the air, water, soil, energy, and mineral resources on which we depend. You will learn how we use these resources and how abusing them can pollute the environment that keeps us and other animals and plants alive and healthy.

Part IV is devoted to the chemistry of everyday consumer products such as plastics, soap, detergents, toothpastes, deodorants, and skin and hair products. You will also learn about the chemistry of fertilizers, pesticides, and chemicals in the foods you eat.

Part V discusses drugs used in medicine and those that alter behavior and moods. You will also learn about toxic chemicals and how chemistry can be used to genetically change organisms, replace body parts, and improve athletic performance. To get a better picture of what you will be learning, take a few minutes to look at the detailed table of contents on pp. viii–xiv.

THIS BOOK IS FLEXIBLE Chemistry teachers differ in what they consider to be most useful and important in a chemistry course. Courses also differ in length, so many teachers have to decide which parts of this book to use.

We have designed this book to give your teacher great flexibility in making such choices. Once Parts I and II have been studied, the other three parts can be covered in any order; don't be concerned if your instructor skips around.

Chemistry teachers also differ over whether you should be introduced to chemical arithmetic in a course of this type. Some will concentrate only on chemical principles that involve only arithmetic. Others will use the brief sections in Chapters 1 and 5 and the two appendices at the end of this book to introduce you to chemical arithmetic. In these sections and appendices, the only math you will be using is simple addition, multiplication, and division.

Most chapters contain one or more "Chemistry Spotlights" highlighted in boxes. Some give interesting applications of chemistry; others describe how certain chemical discoveries were made. Some show that scientists are people who do not fit the stereotyped images you often see on TV and in the movies. We hope you enjoy reading these spotlights as much as we did writing them.

LEARNING OBJECTIVES AND CHAPTER SUMMARIES We have put a number of aids in this book to help you learn about chemistry. One way to help people learn is to give them an overview of what they will learn, present the necessary details, and then summarize what has been presented. This is our approach.

Each chapter begins with a few general questions, or learning objectives, written in easy-to-understand language (for example, see the questions on p. 24). They give you an idea of what you will be learning in each chapter. After you finish a chapter, you should try to answer these questions to review what you have learned.

At the end of each chapter, you will find a brief summary of the major ideas discussed in that chapter. Reading it will reinforce what you have learned and help you answer the questions posed in the general objectives. Don't read only the summary and skip reading the chapter. The summary is not complete, and it is written on the assumption that you have read the chapter.

CHEMICAL VOCABULARY Chemistry, like all subjects, introduces you to new terms with specialized meanings. We have used three ways to help you identify these key terms. When each term is introduced and defined, it is printed in **boldfaced type**. For review we have put a list of the key terms in each chapter at the end of each chapter. This list also shows the page number where each term is defined. Key terms are also included in the index.

You need to learn this chemical vocabulary and some of the shorthand symbols chemists use to represent chemical elements, compounds, and reactions. Think of chemical elements as the letters of chemistry, chemical compounds as its words or combinations of letters, and chemical reactions as its phrases or sentences.

One good way to memorize this information is to use flash cards. When you see a chemical term or formula that your instructor expects you to know, take a small card or piece of paper. Write the term or formula on one side and its meaning or chemical name on the other side. Take a few of these cards with you each day. Look at each term or formula and see if you can give its correct definition or name. Put the ones you get wrong in a separate place and keep studying them until you know them.

PRACTICE EXERCISES AND END-OF-CHAPTER REVIEW QUESTIONS In many chapters, you will find practice exercises to test what you have just studied. We suggest that you cover the answer given below each question. Next read the question and give your answer. Then uncover the answer and see if you got it right. If you can't answer the question, restudy the material or get help from your instructor.

At the end of each chapter, you will find other questions to test your knowledge. Answers to the odd-numbered questions are given in the back of the book. We suggest that you try to answer the question and then look up the answer to see if you got it right.

There are also several discussion questions that test your ability to think about and apply the chemical knowledge you have learned in each chapter. Because many of these questions involve opinions, value judgments, and different possible answers, we have not provided answers for them.

SIMPLIFYING COMPLEX MOLECULES If you look through this book, you will see some complex chemical formulas. In most cases, our purpose in giving these formulas is to show you key similarities and differences between important chemicals. Usually, the part of the structure that is similar (or different) is shown in **boldfaced color**. We also use boldfaced color to identify certain key parts of molecules known as *functional groups*. These groups of one or more atoms in a complex molecule are the key to that molecule's chemical behavior. So when you look at the formula of a molecule, focus on the key groups that are in color.

VISUAL AIDS We have developed a variety of diagrams to illustrate complex chemical ideas in a simple manner. We have also used many photos to give you a better picture of how chemistry occurs in the real world.

FURTHER READINGS   You may want to read other books and articles to get more information about some of the topics in this book. A list of suggested readings for each chapter is given in the back of this book beginning on p. 650.

IF YOU NEED MORE HELP   If you don't understand something, ask questions in class. You can also seek out your instructor and any available teaching assistants after class. Studying with other students also helps.

Don't wait until the last day or night before a test to get help. Most of the things you learn in one section or chapter are needed to understand what follows.

There is also a study guide that can help you learn the material in this book. You should be able to buy or order one from your bookstore. For each chapter section, this guide gives detailed learning objectives and key terms. It has many short-answer questions with answers that allow you to test your knowledge. A glossary covering the entire book is also provided.

HELP US IMPROVE THIS BOOK   We need your help in improving this book in future editions. Writing and publishing a book is an incredibly complex process. This means that any book will have some typographical errors and other minor problems. If you find what you believe is an error, write it down and send it to us.

We would also appreciate hearing about what you like and dislike about this book. This information helps us make the book better. Some of the things you will read in this edition were suggested by students like you.

Send any errors you find and any suggestions for improvement to Jack Carey, Wadsworth Publishing Company, 10 Davis Drive, Belmont, CA 94002. He will send them on to us.

AND NOW   Relax and enjoy yourself as you learn more about the exciting world of chemistry.

To the Instructor

OUR GOALS   We believe that chemistry can be presented in an accurate and meaningful way to students who have little or no prior interest or background in chemistry. We also want to help students discover that chemistry is fun, interesting, and important in their lives.

Recent studies have shown that the scientific literacy of the average citizen has declined. Yet such knowledge is more important than ever.

We wrote this book to provide students who don't intend to become scientists with a basic knowledge of chemistry and its applications. We believe that this information can help them participate more effectively in societal issues. It can also help them make more effective decisions about environmental and resource problems, their health, and the numerous products they buy.

READABILITY   Students often complain that their textbooks are difficult and boring to read. We have tried to overcome this problem by writing this book in a clear, interesting, and informal style.

We try to relate the chemical information in this book to the student's own life. We keep sentences and paragraphs short. We do not use long words when short ones can express an idea just as well.

Most chapters contain one or more "Chemistry Spotlights" highlighted in boxes. Some give further interesting applications of chemistry. Others describe how certain chemical discoveries were made. Some show that scientists are people who do not fit the stereotyped images we often see on TV and in the movies.

FLEXIBILITY   Teachers using this book have courses that last for different lengths of time. They also may disagree on the topics to be covered and their sequence.

With these problems in mind, we have designed this book to be flexible enough to meet your particular needs. This book is organized into five major parts. These parts are shown in the brief table of contents on p. v and described briefly in the preface to the student.

This organizational pattern gives you considerable flexibility. After you have covered all or most of the principles in Part II, you can cover the rest of the book in almost any order. We have included in Part II only those principles of chemistry that we apply and reinforce in other parts of the book.

One key to this flexibility is including a simple introduction to biochemistry (Chapter 9) in Part II. This material is useful for Chapter 15 and beyond and is especially useful for Chapters 18 to 22. Other books for this type of course either do not give enough of this biochemical information or scatter it throughout the book. This makes it hard to vary the order of certain chapters.

Some of you prefer to make no use of mathematics in this type of course. Others want to include some of the simple arithmetic used in making conversions between units of measurement and stoichiometric calculations.

We have allowed for both possibilities. Conversions between metric units are mentioned briefly in Section 1.4. This can be expanded by using any or all parts of the two appendices on exponential numbers, units, and unit conversions. The only part of the main text that includes stoichiometry is Section 5.5. Instructors not wishing to use any chemical arithmetic can omit these two sections and the two appendixes.

EMPHASIS ON LEARNING AIDS   To help students learn more effectively, we have included a comprehensive system of learning aids. These include the following:

- *General Objectives*   Each chapter begins with several simply worded questions (see p. 24). They give the student an overview of the chapter and can be used as review questions after the chapter is completed.
- *Chapter Summaries*   (See page 54.)
- *Key Terms*   When any new term is defined, the term is shown in **boldfaced type**. A list of these terms is given at the end of each chapter (see p. 55). Each term is followed by the page number where the term was defined. Key terms are also included in the index. We have not provided a glossary because we believe that it is better for students to review terms in the context where they are defined. A glossary is available in the optional *Student Study Guide.*
- *Practice Exercises and End-of-Chapter Questions*   Practice exercises with answers are given within some chapters to allow students to test their knowledge before proceeding to new material (see p. 32). More questions are given at the end of each chapter (see p. 56), with answers to odd-numbered questions given in the back of the book. Each chapter also has several discussion questions designed to have students think about and apply what they have learned (see p. 57).
- *Simplified Chemical Structures*   The structures of complex molecules have been simplified by showing key parts in **boldfaced color** (see pp. 230 and 231).
- *Visual Aids*   Many diagrams and photographs are used to illustrate complex chemical ideas in a simple manner and to show how chemistry occurs in the real world (see pp. 186, 187, and 188).
- *Further Readings*   A list of books and articles provides further information about the material in each chapter for students and instructors. These readings are listed by chapter at the end of the book (p. 650).

EXTENSIVE MANUSCRIPT REVIEW   This edition and the two earlier ones were reviewed by 61 chemistry teachers. Their names are given in the list that follows this preface. This extensive reviewing system has provided us with many helpful suggestions and minimizes errors. It also helps make each edition accurate and up-to-date.

MAJOR CHANGES IN THIS EDITION   We have devoted considerable effort to making this book even more readable. Sentences and paragraphs have been shortened. The wording has been simplified. We have also used a more personal writing style to help make the book more interesting.

The entire book has been updated and new material has been added. For example, we have added new material on enzymes, AIDS, indoor air pollution, drugs to improve athletic performance, and various consumer products. We have also deleted unnecessary material to keep the book from getting longer. The order of Chapters 14 and 15 has been reversed from the previous edition to improve clarity. Some topics within chapters have also been rearranged to improve flow and clarity.

Chapter summaries have been added. More practice exercises have been added to some chapters. We have substantially increased the number of questions at the end of each chapter.

SUPPLEMENTARY MATERIALS   A student study guide and an instructor's manual accompany this textbook. For each chapter section, the *Student Study Guide* gives detailed learning objectives and key terms. There are also many short-answer questions with answers that allow your students to test their knowledge of each chapter section. The study guide also contains a glossary for the entire book.

The *Instructor's Manual* has hundreds of test questions and answers to end-of-chapter questions not answered in the text. Master sheets for making *overhead transparencies* of many key diagrams are available from the publisher.

FEEDBACK   We need your help in improving this book in future editions. Writing and publishing a textbook is an extremely complex process. Any textbook is almost certain to have some typographical errors and other minor problems. Our extensive reviewing system helps minimize errors, but some will probably slip through. If you find any errors, please write them down and send them to us. Most errors can be corrected in subsequent printings of this edition, rather than our waiting for a new edition.

We would also appreciate you telling us how to improve this book. We all have the same goal of trying to find the best way to teach students about chemistry. Helping us helps you and your students. We also hope you will encourage your students to evaluate this book and send us their suggestions for improvement.

Send any errors you find and your suggestions for improvement to Jack Carey, Wadsworth Publishing Company, 10 Davis Drive, Belmont, CA 94002. He will send them on to us.

ACKNOWLEDGMENTS We want to thank the chemistry teachers who took time from their busy schedules to make detailed reviews of each edition of this book and those who corrected errors and sent in many helpful suggestions. Any remaining deficiencies and errors are ours, not theirs.

We are especially indebted to Wesley D. Smith, Ricks College, who served as a coauthor for the second edition of this book. His work on Chapters 15 ("Laundry Products") and 16 ("Personal Products") and on the study guide for the second edition is especially appreciated. For this edition, we give special thanks to Robert D. Gaines, Central Washington University, for revising the *Student Study Guide* and the *Instructor's Manual*.

Others have also made important contributions. They include production editor Gary Mcdonald, copy editor Betty Duncan-Todd, managing designer Andrew H. Ogus, art editor Donna Kalal, and artists Victor Royer and Amy Hennig. We also thank Sue Belmessieri, editorial assistant, for coordinating reviews and somehow juggling several hundred tasks at the same time with competence and good humor. Above all, we wish to thank Jack Carey, chemistry editor at Wadsworth, for help, encouragement, and an extremely useful reviewing system.

Manuscript Reviewers

David L. Adams, North Shore Community College
Ronald M. Backus, American River College
Virgil R. Baker, Arizona State University
Ian G. Barbour, Carleton College
Eunice Bonar, University of Texas at Arlington
Georg Borgstrom, Michigan State University
Arthur C. Borror, University of New Hampshire
Robert C. Brasted, University of Minnesota at Minneapolis
Fred Breitbeil, DePaul University
Richard A. Cellarius, Evergreen State College
Preston Cloud, University of California, Santa Barbara
Richard A. Cooley, University of California, Santa Cruz
Danette Dobyns, California State College at Dominguez Hills
W. T. Edmonson, University of Washington
Gordon Evans, Tufts University
Frank Fazio, Indiana University of Pennsylvania
Paul Feeny, Cornell University

Robert D. Gaines, Central Washington University
Patrick M. Garvey, Des Moines Area Community College
Ted L. Hanes, California State University at Fullerton
Richard Hanson, University of Arkansas
M. Lynne Hardin, University of Texas at Arlington
Keith Harper, North Texas State College
E. Lyndol Harris, McMurray College
James Heinrich, Southwestern College
Vincent Hoagland, California State College at Sonoma
C. S. Holling, University of British Columbia
David R. Inglis, University of Massachusetts
Floyd L. James, Miami University
Ray L. Johnson, Hillsdale College
Stanley N. Johnson, Orange Coast College
Lee Kalbus, California State College, San Bernardino
Donald King, Western Washington University
C. R. Kistner, University of Wisconsin
Edward J. Kormondy, Evergreen State College
William W. Murdoch, University of California, Santa Barbara
John E. Oliver, Indiana State University
Harry Perry, Legislative Reference Service, Library of Congress
William F. Pfeiffer, Utica College
Albert Plaush, Saginaw Valley State College
Grace L. Powell, University of Akron
John T. Riley, Western Kentucky University
Paul E. Robbins, Armstrong State College
Henry A. Schroeder, Dartmouth Medical School
Howard M. Smolkin, United States Environmental Protection Agency
John E. Stanley, University of Virginia
Max Taylor, Bradley University
Tinco E. A. van Hylckama, Texas Tech University
Donald E. Van Meter, Ball State University
Pothen Varughese, Indiana University of Pennsylvania
Kenneth E. F. Watt, University of California, Davis
Douglas West, Illinois State University
Ross W. Westover, Cañada College
Charles G. Wilber, Colorado State University
D. H. Williams, Hope College
Samuel J. Williamson, New York University
Archie S. Wilson, University of Minnesota
Byron J. Wilson, Brigham Young University
George M. Woodwell, Brookhaven National Laboratory
Noel S. Zaugg, Ricks College
Arden Zipp, State University of New York, Cortland

PART I

What Is Chemistry?

JACOB BRONOWSKI

Many people persuade themselves that they cannot understand mechanical things or that they have no head for science. These convictions make them feel enclosed and safe, and of course save them a great deal of trouble. But readers who have a head for anything are pretty sure to have a head for whatever they really want to put their minds to.

CHAPTER 1

Chemistry: Its Nature and Impact on Society

General Objectives

1. What is chemistry, and why is it important to you?

2. What is the nature of science? What are scientific hypotheses, theories, and laws? How does science differ from technology?

3. What is matter, and how do chemists classify matter?

4. What is the difference between a physical change and a chemical change?

5. What is energy, and what are some major forms of energy?

6. What are some important units that chemists use in their measurements of matter and energy?

Chemistry is in action everywhere: Plants use sunlight to make oxygen for you to breathe and food for you to eat, autumn leaves turn red or gold, and nerve impulses race to your brain as you read this book. Chemicals are in every form of matter—the material stuff of which the world is made.

You are a walking, talking, thinking bundle of chemicals undergoing thousands of reactions that keep you alive and healthy. You live in a chemical age. Chemistry is used to help satisfy your wants and needs and, ideally, to make the world a better place for all.

This chapter introduces you to the nature and importance of chemistry, science, and technology.

1.1 Our Chemical Age: Opportunities and Responsibilities

Signs of the chemical age are visible everywhere: plastics, antibiotics, gasoline, foods, detergents, and microchips. Scientists have solved the genetic code, defeated deadly diseases, found how to convert crude oil into gasoline and heating oil, and harnessed nuclear energy. Understanding these and many other discoveries is an important reason to study chemistry.

But there is another reason. Many scientific and technological advances bring harmful side effects along with their benefits. You must be able to assess whether the benefits of such advances outweigh the disadvantages.

For example, medicines that have saved the lives of millions of people are also partly responsible for the growing population we must feed, clothe, and house. Pesticides and fertilizers help us feed this population, but they can also pollute our food, water, and soil.

Fossil fuels—oil, natural gas, and coal—are used to propel vehicles, warm and cool buildings, produce electricity, and make plastics. But extracting, processing, and burning fossil fuels pollute air, water, and soil and may even be changing the climate. Nuclear energy can help support our energy-rich life-styles, but it also brings stockpiles of radioactive wastes and the risk of nuclear power plant accidents. Nothing in life is risk-free.

Chemistry is a key to reducing many of these harmful side effects. For example, chemists are working to develop cleaner-burning fuels from such things as coal, plants, and garbage. They are trying to develop safer food additives, better contraceptives, cheap solar cells that convert sunlight into electricity, and new genetic strains of rice and wheat to help feed the world's 5.2 billion people.

These examples should help convince you that understanding chemistry is important for everyone, not just for chemists.

1.2 Science and Technology

SCIENCE: THE SEARCH FOR ORDER   Which of the following statements do you believe?

1. Science emphasizes facts or data.
2. Science discovers absolute truth about the physical universe.
3. Science has a method, a "how-to-do-it" scheme for learning about the physical universe.
4. Science emphasizes logic, not creativity, imagination, or intuition.

Most people believe at least one of these ideas. Yet all are at least partly false.

Science is based on the assumption that the natural world behaves in an orderly, consistent, and predictable way. Scientists try to discover this order in nature. Then they use this knowledge to predict what will or will not happen in the natural world.

In this search for order, scientists first try to identify and describe *what* happens in nature over and over again. Then they try to invent ideas to explain *how* or *why* things happen in this way.

To find out what is happening, scientists collect **data**, or facts, by making observations and measurements in nature. Such data are color changes, temperature readings, and other numbers and facts collected in scientific experiments (Figure 1.1). These data are the stepping stones to scientific laws and theories. As the French scientist Henri Poincaré put it, "Science is built up of facts. But a collection of facts is no more science than a heap of stones is a house."

Figure 1.1 Science relies on making observations and measurements. The chemistry laboratory is one place to collect scientific data under controlled conditions. (Courtesy of Central Washington University)

Once facts are gathered, scientists try to describe what is happening by organizing the data into a generalization or scientific law. A **scientific law** describes the orderly behavior observed in nature. It is a summary of what we find repeatedly happening in nature. It summarizes past observations and predicts what will happen in new situations. The law of gravity, for example, describes the attraction of objects by the earth. It also predicts that flying birds must always land (Figure 1.2).

Scientists also want to explain how or why things happen a certain way. They do this by inventing a **scientific hypothesis**, an educated guess or preliminary idea that explains a scientific law or certain scientific facts. Scientists then test the hypothesis by making more observations and measurements.

If many experiments by different scientists support a hypothesis, it becomes a theory. Thus, a **scientific theory** is a well-tested and widely accepted scientific hypothesis.

In the early 1800s, for example, John Dalton tried to explain the behavior of matter by proposing that all matter is made of tiny building blocks called *atoms*. After being tested by many other scientists, his hypothesis was found to be useful in describing and predicting the properties of matter. It then became an accepted scientific theory, known as the *atomic theory* (discussed in Chapter 2).

Figure 1.2 The law of gravity predicts that even a soaring California condor must land. (U.S. Fish and Wildlife Service)

PRACTICE EXERCISE

Classify each of the following as a scientific law, hypothesis, or theory:

a. Gravity causes our skin to sag as we get older.
b. AIDS is caused by viruses.
c. Matter cannot be created or destroyed, but it can change forms.

a is a hypothesis, b is a theory, and c is a law.

SOLUTION

Scientific theories and hypotheses are dynamic ideas that are constantly challenged. Often they must be modified, or even discarded, because of new data or more useful explanations of the data.

Scientific theories are not viewed as absolute truth. They are simply the most useful ideas anyone has found *so far* to describe, explain, and predict what happens in nature. Some scientific laws may also have to be modified as scientists learn more about nature. Little by little, this quest for scientific knowledge has given us a better—but not a perfect—understanding of the physical world.

WHAT IS TECHNOLOGY?   **Technology** is the invention of new products and processes to improve our survival, comfort, and quality of life. Whereas science is most concerned with understanding *how* the physical world operates, technology emphasizes the development of something practical.

Technology is often defined as *applied science*, but science and technology are so intertwined that it's sometimes hard to tell them apart. In some cases, technology develops from known scientific laws and theories. The laser, for example, was invented by applying knowledge about the structure of atoms.

But it isn't always this way. Some technologies developed long before anyone understood the science on which they are based. For example, people have known how to make wine for thousands of years. But only in the last century have scientists come to understand how yeast can change grape sugar into ethyl alcohol. Cathedrals were built, aspirin was used to relieve pain, and photography was invented—all by people who had little or no knowledge of the underlying chemistry and physics.

Science and technology also differ in the way their results are shared. Ideally, scientists publish and pass around their observations and ideas as widely, openly, and quickly as possible. This allows these ideas to be tested, challenged, verified, or modified. In contrast, technological discoveries are usually kept secret until the individual or company can get a patent for the new process or product.

THE METHODS OF SCIENCE   The way scientists gather data and formulate scientific laws, hypotheses, and theories is called a **scientific method**. Some people mistakenly believe that scientists learn nature's secrets by following a simple "how-to" procedure, but the scientific method is not a recipe. It is a set of questions, with no particular rules for answering them. The questions are

— What is the question about nature to be answered?

— What relevant facts and data are already known, and what new data can be collected?

— After the data are obtained, can they be used to formulate a scientific law?

— How can a hypothesis be invented that explains the data and predicts new facts? Is this the simplest and only reasonable hypothesis?

— What new experiments can be done to test the hypothesis (and modify it if necessary) so it can become a scientific theory?

New discoveries happen in many ways. Some follow a data $\longrightarrow$ law $\longrightarrow$ hypothesis $\longrightarrow$ theory sequence. Other times, scientists simply follow a hunch, a bias, or, yes, a belief and then do experiments to test it. Other discoveries occur when an experiment gives totally unexpected results and the scientist insists on finding out what happened. So in reality, *there are many methods of science rather than one scientific method.*

Good scientists are curious, creative, and skeptical about new ideas until they can be tested. Devising a scientific experiment or formulating a hypothesis may require logical reasoning, but it also requires imagination and intuition. Albert Einstein, the famous theoretical physicist, once said, "Imagination is more important than knowledge and there is no completely logical way to a new idea." Indeed, we find that intuition, imagination, insight, and creativity are as important in science as in poetry, art, music, and other great adventures of the human spirit.

1.3 Chemistry: A Study of Matter and Energy

WHAT IS CHEMISTRY?   This book, your hand, the water you drink, and the air you breathe are all samples of **matter**—anything that has mass and occupies space. For example, this book has a certain amount of material, called its mass, and occupies a certain volume.

The light and heat streaming from a burning lump of coal and the force you must use to lift this book are examples of energy. **Energy** is the capacity to do work. Energy is needed or given off when matter moves or undergoes some type of change, such as boiling. Energy is also the heat that flows automatically from a hot object to a cold object. Touch a hot stove and you experience this energy flow in a painful way.

Chemists seek answers to certain key questions about matter and energy:

— What is each object made of?
— How does it differ from other types of matter?
— How can it change into another form of matter?
— How can these changes be controlled?
— How much energy is needed or given off when such changes occur?

Thus, **chemistry** can be defined as the study of matter, its properties, the changes matter undergoes, and the accompanying energy changes.

CLASSIFYING MATTER   There are millions of different chemicals. To help understand how all these chemicals are alike and how they differ, chemists classify matter according to its forms and properties. One way to do this

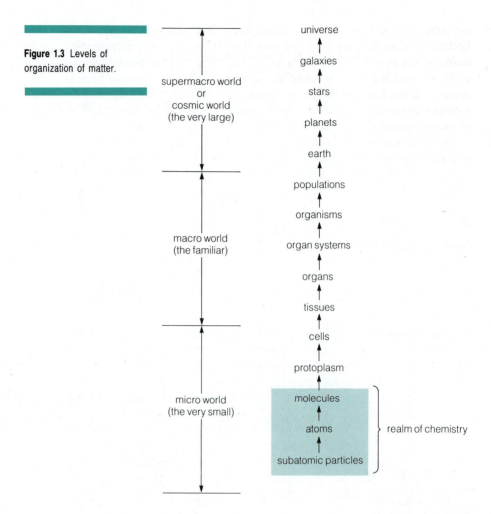

Figure 1.3 Levels of organization of matter.

is to classify matter into *levels of organization* from subatomic particles to the universe (Figure 1.3). Each higher level includes the types of matter found in all lower levels. But matter in each new level has matter from the lower levels organized in new patterns with new properties.

Chemistry is concerned mostly with three levels of organization of matter: (1) *subatomic particles* (electrons, protons, and neutrons), which make up the internal structure of atoms; (2) *atoms*, the distinct building blocks of all matter; and (3) *molecules*, combinations of two or more atoms held together by chemical bonds.

Another way chemists classify matter is by its *physical state* as a solid, a liquid, or a gas. A **solid** has a fixed volume (the amount of space it occupies) and a fixed shape. If you drop a sugar cube into a glass container, the cube just sits there. It cannot change its shape and volume to fit the bottom of the container. A **liquid** has a fixed volume but assumes the shape of the part of any container it occupies. If you pour milk into a glass, the milk will take the shape of the part of the glass it occupies. In contrast, a **gas** has neither a definite volume nor shape. If you put it in a closed container, a gas can rapidly expand or contract to assume the shape of the container and will fill its entire volume.

A third way chemists classify matter is according to its composition and properties (Figure 1.4). Imagine you are a chemist. Someone brings

Figure 1.4 A classification of matter according to its composition and properties.

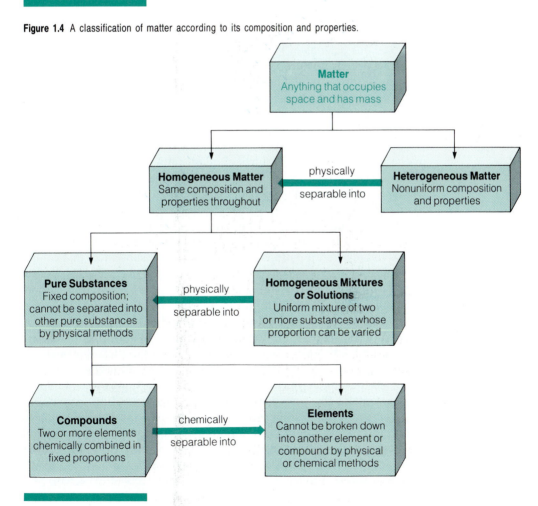

Figure 1.5 This reticulated python is heterogeneous matter. (Courtesy of Animals Animals)

you a sample of matter and asks you to find out what it is. What would you do? Your first step would be to look at the sample and run tests to decide whether it is homogeneous or heterogeneous. Materials with the same composition and properties throughout are **homogeneous matter**. Examples are pure water, copper, sugar, or salt dissolved in water. Matter having different parts with different properties and composition is **heterogeneous matter**. Examples are pizza, concrete, you, and a python (Figure 1.5).

If you find that the sample is homogeneous, your next step is to find out whether it is a pure substance or a homogeneous mixture (also called a solution). **Pure substances** have a fixed composition and cannot be separated into other pure substances by methods such as filtering, dissolving, and boiling. Examples are mercury, oxygen gas, sugar, and salt.

Homogeneous matter can contain two or more pure substances that can be mixed together in various proportions and be separated by physical methods such as evaporating or dissolving. Chemists call this kind of homogeneous mixture a **solution**. As you dissolve different amounts of instant coffee in hot water to make a cup of coffee with the desired strength, you vary the proportions in a solution. And you can separate out the solid instant coffee by evaporating the water.

You probably think of solutions as liquids (such as antifreeze) or solids (such as instant coffee) dissolved in liquids, especially in water. But the term also applies to mixtures of other states of matter. For example, air is a gaseous solution of at least five different pure substances. Brass (a homogeneous mixture of copper and zinc) is an example of a solid solution, called an alloy.

Suppose that your tests show that the unknown sample of matter is a pure substance. Your next step is to find out whether it is an element or a compound (see Figure 1.4). **Elements** are pure substances that cannot be broken down into simpler substances by ordinary chemical or physical means. All other materials form from these elements. Copper, iron, carbon, and sulfur are 4 of the 109 known elements.

A **compound** is a pure substance composed of two or more elements in fixed proportions. Water (containing the elements hydrogen and oxygen), sugar (containing carbon, hydrogen, and oxygen), and table salt (containing sodium and chlorine) are compounds. Like all compounds, pure water has the same chemical and physical properties throughout the world. It also contains the same fixed proportion of hydrogen and oxygen.

You can distinguish between elements and compounds by trying to break them down into simpler substances. If your pure substance cannot be decomposed into simpler substances, it is an element. If it can be broken down into simpler substances, it is a compound. For example, by passing electrical energy through a sample of water (a process called electrolysis) you can change liquid water into hydrogen gas and oxygen gas.

PHYSICAL AND CHEMICAL CHANGES   Matter can undergo physical and chemical changes. A **physical change** doesn't alter the chemical composition of the substance. Changing the size or shape of a substance, such as chipping off a small piece from a sugar cube, is one kind of physical change.

Another kind is altering the physical state of a substance (Figure 1.6). We can melt a solid and change it to its liquid state. Melting ice, for example, is a physical change. The liquid water produced by melting ice is in a different physical state. But it still has the same proportion of hydrogen and oxygen as when it was a solid. We could also boil liquid water and convert it into steam. This gaseous form of water has the same proportion of hydrogen and oxygen as do liquid water and ice.

A **chemical change** or **chemical reaction** occurs when the chemical composition of a substance changes. Decomposing water into hydrogen and oxygen gas is a chemical change. The burning of fuel in a rocket shown on page 1 is also a chemical change.

In a chemical reaction, one or more starting substances called **reactants** either combine or break down to form one or more different substances called **products**. Every chemical reaction requires or gives off energy. For example, when coal (which is mostly carbon) burns, it combines with oxygen gas in the air to form the gaseous compound carbon dioxide. Energy is given off by this reaction, which is why coal is a useful fuel. We can represent the reaction in the following manner:

$$Reactant(s) \longrightarrow Product(s)$$

carbon + oxygen $\longrightarrow$ carbon dioxide + **energy**

black solid   colorless gas   colorless gas

Figure 1.6 Changing the physical state of matter. A hot poker changes ice into water and steam. (Fritz Goro, *Life Magazine* © 1949 Time Inc.)

Figure 1.7 Energy is classified as potential energy or kinetic energy.

Type of Energy	Potential	Kinetic

mechanical — stone being held above ground | stone dropped, does work on experimenter's toe

electrical — charged battery | battery being discharged through a wire

light bulb

light energy is kinetic energy

wood stove

heat energy is kinetic energy

The arrow (⟶) used to separate reactants from products is a symbol that means "produces" or "yields." In English we would say that carbon reacts with oxygen to produce carbon dioxide and energy. Later you will learn how chemists use symbols and numbers to express a chemical reaction in a form of shorthand called a *chemical equation*.

CLASSIFYING ENERGY Like matter, energy affects every part of your life: You need it to keep your heart beating, to cook food, to travel, and to warm the buildings in which you live. You cannot touch or pick up energy, but you can use it to do work. You do work when you move matter, such as your arm or this book, or when you boil liquid water and change it into steam.

Energy comes in many forms: heat; light; electricity; mechanical energy to move matter; chemical energy stored in coal, sugar, and other materials; wind; moving water; and nuclear energy from certain elements such as uranium. Scientists usually classify energy as kinetic or potential (Figure 1.7).

Kinetic energy is the energy from motion. Wind, flowing rivers, heat, light, electricity, and moving trains have kinetic energy. The amount of kinetic energy a sample of matter has depends on both its mass and how fast it is moving. A bullet shot from a gun, for example, has more kinetic energy than the same bullet thrown by hand. Even objects at rest have some internal kinetic energy because tiny particles in their structure are constantly moving.

Potential energy is the energy stored in an object because of its position or the position of its parts. A rock in your hand has potential energy that changes into kinetic energy when you drop it (see Figure 1.7). Similarly, the energy stored in gasoline changes into heat, light, and mechanical (kinetic) energy that propels our cars when the gasoline burns.

Matter also has potential energy because of the forces holding its internal particles together; this is called **chemical energy**. You use the chemical energy in your food to move, breathe, think, and do all the other things that keep you alive.

1.4 Measuring Matter and Energy: The Importance of Numbers and Units

MEASUREMENTS AND UNITS Measurements of energy and matter are made relative to some *reference standard* and must include the *unit* used in the measuring device. For example, suppose you measure your height using a ruler marked off in units of length called meters. You are then comparing your height to an internationally agreed-on reference standard of length called the meter. As long as everyone uses the same units and agrees on their size, such a system can work. This allows us to make meaningful comparisons about the height, weight, volume, temperature, or other properties of samples of matter. For example, it would be confusing if your height were measured and found to be 1.7 meters, and your room-mate's height were reported as 6 feet. Who is taller?

METRIC SYSTEM OF UNITS   Today most scientists use an updated version of the *metric system of units* that was first developed in France in 1799. This version is called the **International System of Units** and is abbreviated **SI** (from the French *Système International d'Unités*). The system is simple and convenient. Converting most metric units to larger or smaller units involves multiplying or dividing by some multiple of 10.

By contrast, the English system has no systematic relationship among its units. This means you have to memorize such things as 5280 feet = 1 mile, 16 cups = 1 gallon, and 2000 pounds = 1 ton. You also have to deal with inches, pints, quarts, cubic feet, ounces, bushels, tablespoons, cubits, pecks, and acres.

Although American scientists use SI units, the United States is the only major nation not officially adopting the metric system. A gradual conversion is taking place in the United States, but it is happening slowly. The changes will affect such things as tools, manufacturing machinery, road signs, cookbooks, and thermometers. This conversion is expensive, but it makes it easier to sell American products abroad.

COMMON UNITS FOR MEASURING MATTER   The metric system uses a basic unit of length, mass, volume, and other properties and then uses prefixes to change their size by some factor of 10. Table 1.1 lists the common pre-fixes, their numerical values, and their abbreviations used in chemistry. (Appendix 1 discusses how to express numbers as exponents, or powers of 10. Appendix 2 shows how to convert one unit to another.)

The SI unit of length is the *meter* (m). A meter is slightly longer than a yard (1 m = 39.4 in.), so it takes a little longer to run 100 meters than 100 yards. Other common units of length are the *kilometer* (km), equal to 1000 m; the *centimeter* (cm), equal to 1/100 m; and the *millimeter* (mm), equal to 1/1000 m.

Volume is the amount of space an object occupies. The volume of a rectangular solid, for example, is simply its length times its width times its height. So the volume of a rectangular solid with sides of 2 m, 5 m, and 6 m is (2 m) × (5 m) × (6 m) = 60 m³.

Although the SI unit for volume is the *cubic meter* (m³), a smaller and more convenient unit for measuring volume is the *liter* (L). It is slightly

TABLE 1.1 PREFIXES FOR METRIC UNITS

| Prefix | Symbol | Multiply Base Unit by | |
		Decimal or Fraction Form	Exponential Form
mega-	M	1,000,000	10^6
kilo-	k	1000	10^3
deci-	d	0.1 or 1/10	10^{-1}
centi-	c	0.01 or 1/100	10^{-2}
milli-	m	0.001 or 1/1000	10^{-3}
micro-	μ	0.000 001 or 1/1,000,000	10^{-6}

larger than a quart (1 L = 1.06 qt). Another common unit, the *milliliter* (mL), is 1/1000 L and is equal to the *cubic centimeter* (cm³, or cc). So 1 mL = 1 cc = 1 cm³ = 1/1000 L. For example, a standard soft-drink can has a volume of about 340 mL (0.34 L).

Mass is the amount of matter in an object. We usually use the terms *mass* and *weight* interchangeably, but technically the two are different. Mass is a measure of the quantity of matter, whereas weight measures a force. On earth, weight is a measure of the gravitational force of attraction between the earth and an object. You have zero weight in space. Your weight on the moon (because of its weaker gravity) is about one-sixth as much as on earth. But you have the same mass on earth, on the moon, or in space.

The SI unit of mass is the *kilogram* (kg), which equals 1000 grams or 2.2 pounds. Other common mass units are the *gram* (g) and the *milligram* (mg), equal to 1/1000 g. A 110-pound ballerina, for example, weighs 50 kg (or 50,000 g), whereas a typical hamburger patty weighs about 100 g (or 100,000 mg).

CONVERTING BETWEEN UNITS   To convert between units in the English system, we have to memorize the appropriate conversion factors. For example, we could express 1600 inches in miles by

$$1600 \text{ inches} \times \frac{1 \text{ foot}}{12 \text{ inches}} \times \frac{1 \text{ mile}}{5280 \text{ feet}} = 0.025 \text{ miles}$$

(If you are not familiar with this method of converting units, see Appendix 2.)

Conversions between metric units are much easier because the units differ by multiples of 10. For example, we can convert 1600 millimeters (mm) into kilometers (km) by

$$1600 \text{ mm} \times \frac{1 \text{ m}}{1000 \text{ mm}} \times \frac{1 \text{ km}}{1000 \text{ m}} = 0.0016 \text{ km}$$

Notice that in going from millimeters to kilometers, we simply moved the decimal point six places to the left.

Most other metric conversions also involve simply moving the decimal point to the left or right. For example, we can express in milliliters (mL) the volume of a 2-liter (L) beverage bottle as follows:

$$2 \text{ L} \times \frac{1000 \text{ mL}}{1 \text{ L}} = 2000 \text{ mL}$$

PRACTICE EXERCISE

Suppose you weigh 68 kilograms (kg). What is your weight in grams (g)?

$$68 \text{ kg} \times \frac{1000 \text{ g}}{1 \text{ kg}} = 68,000 \text{ g}$$

SOLUTION

(Appendix 2 contains more practice exercises for converting from one unit to another.)

COMMON UNITS FOR MEASURING ENERGY AND TEMPERATURE   Most chemists use either the *joule* (J) or the *calorie* (cal) (1 cal = 4.18 J) for measuring heat or any form of energy. Larger units are the *kilojoule* (kJ), equal to 1000 J, and the *kilocalorie* (kcal), equal to 1000 cal.

The kilocalorie, also abbreviated Cal (with a capital *C*), is used in dietary tables to show the energy content of foods. For example, a piece of pie containing 500 Cal (500 kcal) actually has 500,000 cal or 2,090,000 J (2090 kJ) of energy.

Heat is a measurement of the *amount* of energy in a sample. **Temperature** is a measurement of the *intensity* (relative degree) of the heat energy.

For example, a cup of boiling water has a higher temperature, but less total heat energy, than the water in a waterbed at room temperature. And a small slice of pie has the same temperature but less total energy (Calories) than the rest of the pie.

The temperature of a sample of matter is usually measured with a *thermometer*. The *kelvin* (K) is the SI unit for temperature (Figure 1.8). However, scientists usually measure temperatures with the *Celsius* (C) scale, on which water freezes at 0°C and boils at 100°C. Adding 273 to the Celsius temperature converts the temperature into kelvins. The coldest temperature possible, called absolute zero, is −273°C, which has a value of 0 K.

Figure 1.8 Comparison of the Kelvin, Celsius, and Fahrenheit temperature scales.

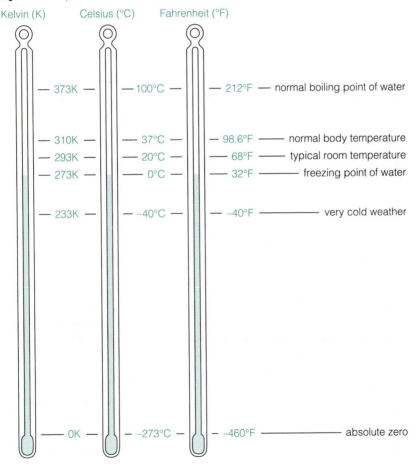

Kelvin (K)   Celsius (°C)   Fahrenheit (°F)

— 373K —   — 100°C —   — 212°F — normal boiling point of water

— 310K —   — 37°C —   — 98.6°F —— normal body temperature
— 293K —   — 20°C —   — 68°F —— typical room temperature
— 273K —   — 0°C —   — 32°F —— freezing point of water

— 233K —   — −40°C —   — −40°F —— very cold weather

— 0K —   − −273°C −   − −460°F —— absolute zero

In the United States, most people (except scientists) use the *Fahrenheit* (F) scale, on which water freezes at 32°F and boils at 212°F. All three scales are compared in Figure 1.8. (Appendix 2 gives a method to convert between the scales.)

CHEMISTRY AND THE FUTURE   Scientific knowledge itself is neither good nor bad; only the ways we choose to use such knowledge can be judged in this manner. Thus, the real question facing us is how to use scientific knowledge wisely.

To make wise and humane decisions, we need the insights of scientists and nonscientists—politicians, artists, lawyers, factory workers, theologians, businesspeople, and many others—who understand the nature of science and technology. Understanding the chemistry presented in this book will help you participate more effectively in resolving important issues—acid rain, use of food additives, genetic engineering, nuclear energy, and many others—that touch your life and the lives of others.

JACOB BRONOWSKI   Science is a great many things but in the end they all return to this: science is the acceptance of what works and the rejection of what does not. That needs more courage than we might think.

Summary

Chemistry is the study of matter, its properties, the changes matter undergoes, and the accompanying energy changes. Chemists and other scientists learn about the physical world by gathering data and formulating scientific laws, hypotheses, and theories. Each way of doing this is called a scientific method. Scientists also use their intuition, insight, imagination, and creativity in this process.

Matter is anything that has mass and occupies space. Matter can be classified by its level of organization (such as subatomic particles, atoms, and molecules), physical state (solid, liquid, or gas), or its composition and properties. All matter can be classified as homogeneous or heterogeneous. Homogeneous matter can be classified as pure substances or solutions (homogeneous mixtures). Pure substances can be classified as elements or compounds.

Matter can undergo physical changes that don't alter its chemical makeup. Matter can also undergo chemical changes, in which its chemical makeup changes.

Energy—the capacity to do work—is classified as potential or kinetic. Energy can be converted into different forms.

Scientists use the International System of Units (SI), an updated version of the metric system, in measuring matter and energy. Various prefixes, which change units by some factor of 10, are used in combination with units for mass (gram), length (meter), volume (cubic meter or liter), and energy (joules or calories). Temperatures are measured by the Celsius (C) or Kelvin (K) scale.

Terms for Review

After completing this chapter, you should know and understand the meaning of the following terms:

chemical change (p. 11)

chemical energy (p. 13)

chemical reaction (p. 11)

chemistry (p. 7)

compound (p. 10)

data (p. 4)

element (p. 10)

energy (p. 7)

gas (p. 9)

heat (p. 16)

heterogeneous matter (p. 10)

homogeneous matter (p. 10)

International System of Units (SI) (p. 14)

kinetic energy (p. 13)

liquid (p. 9)

mass (p. 15)

matter (p. 7)

physical change (p. 11)

potential energy (p. 13)

product (p. 11)

pure substance (p. 10)

reactant (p. 11)

scientific hypothesis (p. 5)

scientific law (p. 5)
scientific method (p. 6)
scientific theory (p. 5)
solid (p. 8)

solution (p. 10)
technology (p. 6)
temperature (p. 16)
volume (p. 14)

Questions

Odd-numbered questions are answered at the back of this book.

1. Classify each of the following as a law, a theory, or a hypothesis:
   a. All matter is composed of elements.
   b. $2 + 2 = 4$
   c. Black cats bring bad luck.
   d. The sun will set tonight.
   e. The gravitational pull of the sun and moon causes tides.
   f. It will rain tomorrow.
   g. Heat flows spontaneously from hot to cold areas.
   h. There will be a party tonight on campus.

2. Distinguish among a scientific law, a hypothesis, and a theory. Are scientific laws broken? Explain.

3. With further testing, hypotheses may become theories, but not laws. Explain.

4. What does the phrase "establish scientifically" mean? Why do scientists rarely use the phrase "it has been proven scientifically"?

5. Why are there many different scientific methods, not just one?

6. You are handed a sample of matter with a constant volume and an indefinite shape. What is its physical state?

7. Classify each of the following as an element, a compound, or a mixture: (a) wood, (b) cup of coffee, (c) toothpaste, (d) paint, (e) pure phosphorus, (f) pure table salt, (g) beer, (h) pure water, (i) muddy water, (j) iced tea, (k) pure ice, (l) pure mercury, (m) rubber.

8. Classify each item in Question 7 as (a) a pure substance, (b) a homogeneous mixture, or (c) heterogeneous matter.

9. Classify each of the following as a physical or a chemical change:
   a. A candle burns.
   b. Snowflakes form.
   c. Sugar dissolves in hot water.
   d. Deodorant spray escapes from a can.
   e. A ham is sliced.
   f. Strawberries ripen.
   g. Wax melts.
   h. A solvent removes paint.
   i. Paint dries.
   j. Grapes ferment into wine.

10. Convert the following:
    a. 27 cm = _____ m
    b. 81,000 mg = _____ g
    c. 2.5 kJ = _____ J
    d. 35 mL = _____ L

11. Convert the following:
    a. 78,000 g = _____ mg
    b. 12 L = _____ mL
    c. 725 cal = _____ kcal
    d. 41 m = _____ km

12. Hydrogen freezes at $-259°C$. What is this temperature on the Kelvin scale?

13. Liquid mercury boils at 630 K. What is this temperature on the Celsius scale?

14. Examine Figure 1.8. How many Fahrenheit degrees are there per 100 Celsius degrees? Is a change of 1°C a larger or smaller temperature change than 1°F?

15. Suppose the United State adopted another unit of currency, the millidollar, in which *milli* has the same numerical meaning as in the metric system. How many millidollars would you need to buy a carton of milk that costs $1.10?

16. For buying large amounts of an item, which rate would be the better buy?
    a. $1.29 per kg or 15¢ per gram
    b. 10 kg for $3.95 or 600 mg for 36¢

17. Your height is approximately 170 (a) m, (b) km, (c) cm, (d) dm, (e) mm.

18. Your weight is approximately 70,000 (a) g, (b) kg, (c) mg, (d) lb, (e) dg.

19. Suppose a football team needs to go 10 yards to make a first down. The metric equivalent is about how many meters? (a) 9140, (b) 9.14, (c) 1.09, (d) 10.9, (e) 0.0109.

20. Are measurements using SI units more exact than those using English units? Explain.

Topics for Discussion

1. Suppose you have been appointed to a technology-assessment board. List drawbacks and advantages for the following: the contraceptive pill, snowmobiles, videocassette recorders, nuclear power plants, and artificial hearts. In each case, would you recommend limits to the use of the technology or that it should be stopped? Defend your position.

2. To what extent are scientists responsible for the applications of knowledge they discover? Consider, for example, the following:
   a. A scientist discovers the chemical process for vision. Other scientists then use this discovery to develop blinding agents for warfare.
   b. A scientist discovers nuclear fission, which enables others to develop and use atomic bombs.

3. Is science always ethically neutral? Is science ever ethically neutral?

4. Do you think fraud is more or less likely to occur in science than in other areas? Explain.

5. Someone once said that the difference between a scientist and an artist is the difference between discovery and creation. For example, if Isaac Newton hadn't formulated the law of gravity, some other scientist would have; gravity was there to be discovered. On the other hand, only William Shakespeare could have written *Macbeth*. Do you think there are differences between creativity and imagination in science and in art or literature? Explain.

PART II

Some Chemical Principles

ALBERT EINSTEIN   The most incomprehensible thing about the universe is that it is comprehensible.

CHAPTER 2

Atoms as Building Blocks of Matter: Atomic Structure and the Periodic Table

General Objectives

1. What did early Greek philosophers think matter was made of?

2. How did John Dalton use experiments to propose and test the idea that matter consists of atoms?

3. What is the periodic table of the elements, and why is it so useful?

4. Why do chemists believe that atoms contain electrons, protons, and neutrons, with the protons and neutrons located in a tiny center called the nucleus?

5. How do atoms of different elements differ from one another?

6. How are electrons arranged around the nucleus of each atom? How does this explain the periodic table of the elements?

The foundation on which all chemistry rests is the *atomic theory of matter*. It is the idea that tiny particles, called atoms, are the basic building blocks of all matter. Greek philosophers proposed this idea more than 2400 years ago. But only in the last two centuries were scientists able to carry out experiments that converted this hypothesis into a scientific theory.

During the past 100 years, scientists have also done experiments that suggest that atoms have an internal structure. Atoms consist of subatomic particles known as electrons, protons, and neutrons. The story of how these ideas developed is an excellent example of how scientific theories develop and change.

How do atoms of different elements differ, and how are some of them alike? In 1869 Dmitri Mendeleev and Lothar Meyer proposed a way to arrange the elements into a table that grouped elements with similar chemical behavior. This classification, called the periodic table of the elements (see inside front cover), has become one of the most important and useful concepts in chemistry.

2.1 Early Greek Models: Imagining What Matter Is Made Of

Cut a piece of pure silver metal in half; then cut one of the halves in half. Suppose you had a "magic knife" that could repeat this chopping process again and again. Could you go on cutting the silver into smaller and smaller pieces indefinitely? Or would you eventually reach some basic structural unit of silver that cannot be subdivided further? In other words, 25 is an element such as silver continuous, as it appears to your naked eye? Or is it made up of a huge number of tiny building block units that are "uncuttable"?

About 2400 years ago, Greek philosophers tried to answer this question, but no one had a magic knife to actually do the experiment. Thus, they had to rely entirely on logic, intuition, and imagination. Different philosophers came up with different answers. Between 450 and 400 BC, Leucippus and his pupil Democritus (Figure 2.1) proposed that dividing matter again and again would eventually lead to tiny, hard, permanent, and indivisible bits (Figure 2.2).

Democritus used the word *atomos*—meaning *indivisible* or *uncuttable*—to describe these ultimate building blocks of matter. He proposed that all matter was composed of such **atoms**. He also suggested that various kinds of matter were made up of different kinds of atoms with different sizes and shapes.

Two other Greek philosophers, Plato and his student Aristotle, disagreed; they believed that matter was continuous rather than made up of atoms. Back then no one knew how to do experiments to help settle this question; all these philosophers could do was argue about who was right. Plato and Aristotle were so good at reasoning and debate that their view won out. Indeed, their fame and influence helped prevent any serious revival of the atom concept for more than 2000 years.

Figure 2.1 Democritus. More than 2400 years ago, this Greek philosopher proposed that all matter is made of atoms. (Courtesy National Library of Medicine)

Figure 2.2 Early Greek model of the atom. This model was based on logical thinking and imagination.

2.2 Dalton's Atomic Theory: A Model Based on Experiments

LAVOISIER: THE LAW OF CONSERVATION OF MASS   By the late 1700s, scientists had developed techniques and equipment for measuring the volumes and masses of solids, liquids, and gases. One scientist who carried out many measurements of matter during this period was Antoine Lavoisier (Figure 2.3). He was a French lawyer who became interested in science and devoted much of his life to research in chemistry.

Lavoisier measured the mass of all reactants before chemical reactions took place. Then he collected all products and measured their mass. He did this for many different reactions. In each case, he found that the total mass of the products was equal to the total mass of the reactants.

In 1783 he summarized his findings as the **law of conservation of mass**: *The total mass of the products of a chemical reaction is the same as the total mass of the reactants.* This scientific law is also stated in another way: *In any chemical reaction, matter is neither created nor destroyed but* *merely changes from one form into another.** Besides formulating this law, Lavoisier was one of the first to demonstrate clearly the nature of elements and compounds. He wrote what is considered to be the first chemistry textbook. Because of these achievements, he has been called the "father of chemistry."

PROUST: THE LAW OF CONSTANT COMPOSITION Early chemists carried out experiments to find out what various compounds were made of. They weighed samples and heated them to break them down into their elements and then weighed the amount of each element produced. They also weighed samples of two different elements, combined them in a chemical reaction, and weighed the resulting compound.

In 1799 French chemist Joseph Proust summarized the results of many such experiments in the **law of constant composition**: *The percentage by mass of the elements in a pure compound will always be the same.* For example, measurements show that any sample of pure water contains (by mass) 88.9 percent oxygen and 11.1 percent hydrogen. This is like saying that a carefully measured 1/4-lb hamburger sandwich will always contain a fixed percentage (by mass) of meat and a fixed percentage of bun.

DALTON'S ATOMIC THEORY In the early 1800s, the atom concept first proposed by Leucippus and Democritus was revived. It was restated in a more meaningful way by an English schoolmaster, John Dalton (Figure 2.4). According to him,

— Each element is made of extremely small particles called atoms.

— All atoms of a particular element are alike and have properties that differ from those of other elements.

— Atoms cannot be created, destroyed, subdivided, or converted into atoms of another element.

— A compound forms when atoms of different elements combine in a fixed, simple, whole-number ratio.

— A chemical change involves the joining, separation, or rearrangement of atoms in elements and compounds.

You might be thinking that this is about the same thing Democritus proposed 2200 years earlier. But there is a difference. Dalton was able to use his atomic hypothesis to explain scientific laws based on experiments. This elevated his hypothesis to the status of a theory.

Figure 2.3 Antoine Lavoisier (1743–1794) is considered the "father of chemistry." To get money for his research, he served as a tax collector—a very unpopular position—for King Louis XVI. After the king was overthrown by the French Revolution, Lavoisier was beheaded by the guillotine. (Burndy Library)

---

* We will see in Section 3.7 that the law of conservation of mass is not entirely correct because of the discovery—more than a century later—that mass can change into energy. This is an example of how scientific hypotheses, theories, and even laws may have to be modified as scientists learn more about nature.

Figure 2.4 John Dalton (1766–1844) lived a simple, unassuming life-style based on his Quaker beliefs. This meant he could not accept any honors for his important scientific work. (Smithsonian Institution)

Dalton used his theory to explain the law of conservation of mass and the law of constant composition. He reasoned that the total mass of the reactants and products in a chemical reaction must be equal because atoms cannot be created or destroyed. Instead, they rearrange during a chemical reaction.

Dalton also explained the law of constant composition. According to his ideas, a compound such as water consists of hydrogen and oxygen atoms combined in a specific, simple, whole-number ratio. This would explain why the mass ratio of hydrogen and oxygen atoms in any sample of pure water is always the same.

But Dalton went one step further. Based on his atomic theory, he proposed and tested a new scientific law, known as the **law of multiple proportions**: *If two elements can combine to form more than one compound, the masses of the elements that combine are in the ratio of small whole numbers such as 2 to 1 (2:1), 3 to 1 (3:1), and so on.*

For example, carbon and oxygen can react to form two different compounds, called carbon monoxide and carbon dioxide. Experimental measurements show that if we start with the same mass of carbon, the formation of carbon dioxide requires twice the mass of oxygen as does the formation of carbon monoxide. For example, 3 g of carbon combine with 4 g of oxygen to form carbon monoxide, or with 8 g of oxygen to form carbon dioxide.

According to Dalton's atomic theory, there is a simple explanation for this 2:1 mass ratio of oxygen in the two compounds. He reasoned that each molecule of carbon dioxide consists of *two* atoms of oxygen and one atom of carbon joined together, whereas each molecule of carbon monoxide consists of *one* atom each of oxygen and carbon joined together. (C is the symbol for an atom of carbon, and O represents an atom of oxygen; thus, carbon dioxide is CO_2, and carbon monoxide is CO.) This is like a double-hamburger sandwich consisting of two 1/4-lb patties and a bun (BP_2) having twice the mass of meat per bun as would a single-hamburger sandwich with one 1/4-lb patty (BP).

New and better measurements have led to a revision of some postulates of Dalton's atomic theory. However, his basic idea that matter consists of atoms remains intact today.

2.3 The Periodic Table: A Chemical Masterpiece

DEVELOPMENT OF THE PERIODIC TABLE   In the middle of the nineteenth century, chemists faced a difficult problem. The number of known elements had reached sixty-two, and it seemed likely more would be discovered. It was nearly impossible for anyone to memorize or comprehend the long list of chemical and physical properties of all these elements. And trying to keep track of the properties of the many thousands of compounds formed from these elements *was* impossible. What was needed was some kind of pattern to simplify matters.

In 1869 a Russian chemistry teacher, Dmitri Mendeleev ("Chemistry Spotlight" below), and a German physics teacher, Lothar Meyer (1830–1895), brought some order out of this chaos. Working independently and in different countries, they discovered that if they arranged the sixty-two known elements in order of increasing atomic masses, the elements would

CHEMISTRY SPOTLIGHT   DMITRI MENDELEEV: CHEMIST, BIGAMIST, AND ACTIVIST

Mendeleev was a professor of chemistry at the University of St. Petersburg in Russia. While preparing a chemistry textbook, he wrote the properties of each known element on a separate card. Then he began arranging the cards in different orders on his desk. He discovered that if the elements were arranged in order of increasing atomic mass, every eighth element had similar properties.

In 1869 he published his periodic table with six empty spaces. Instead of looking at these spaces as errors, he boldly predicted that six new elements would be discovered to fill these gaps. He also predicted their physical and chemical properties based on where they were in his periodic table.

Four of these elements were discovered before his death in 1907 and the other two after World War II. Their properties were remarkably close to those predicted by Mendeleev. For these reasons, and the fact that his table was published a few months before Lothar Meyer's, Mendeleev is usually given credit for discovering the periodic table.

In 1876 Mendeleev did something that attracted much more attention than did his scientific work. He disobeyed the laws of the Russian Orthodox Church by divorcing his first wife and marrying an art student. Officially, his new marriage made him a bigamist. But no one took legal action against him because of his fame and

importance as a scientist. As Czar Alexander II explained, "Mendeleev has two wives, but I have only one Mendeleev."

His protected status, however, did not save him when he began to implement his progressive ideas in education and to voice his concern about social problems in Russia. He got into political trouble by encouraging women to become scientists—a radical idea at that time. He also stood up for the rights of students to protest rules they considered unfair.

Mendeleev agreed to present a student petition, requesting that some rules be relaxed, to the minister of education. The minister refused to accept the petition. When more than 200 students tried to present the petition, the minister

had 175 of them arrested. Mendeleev protested this action by resigning. At his last lecture, his students gave him a standing ovation. For several years, he was unable to get another job.

In 1907, a few months before his death, Mendeleev missed receiving the Nobel Prize in chemistry by one vote. Perhaps he was passed over because of his marital and political behavior. Perhaps French chemist Henri Moissan (who won the prize for isolating fluorine) made a more important contribution. In recognition of his achievements, however, a new chemical element, made in a laboratory in 1955, was named *mendelevium*.

fall into eight vertical columns. Each column had elements with similar chemical properties. In other words, chemical properties of the elements repeated on a cycle, or period, of eight. This arrangement, known as a **periodic table of the elements**, has become one of the most useful generalizations in science.

THE MODERN PERIODIC TABLE   Because of new chemical knowledge, the original periodic table has been revised. But the basic idea of arranging elements in groups with similar chemical properties remains. The most widely used form of the periodic table is shown inside the front cover of this book. It contains 109 elements. About 90 are found in nature; the rest have been made in laboratories through various nuclear reactions (see Section 3.4).

Each of the seven horizontal rows in the table is called a **period**. Each vertical column lists elements with similar chemical properties and is called a family, or **group**.

Over the years, these groups have been labeled several different ways. The periodic table (inside front cover) shows two systems. One is the new, official international system in which the groups are numbered 1 through 18 going from left to right. The other system uses Roman numerals and the letter *A* or *B* to number groups. It is still widely used in the United States.

Several groups are also identified by common names:

— Group 1 or IA: **Alkali metals**
— Group 2 or IIA: **Alkaline earth metals**
— Group 17 or VIIA: **Halogens**
— Group 18 or VIIIA: **Noble gases**

You should be able to locate quickly any of these important groups in the periodic table.

Table 2.1 shows how elements in three of these groups have similar chemical properties. There is also a trend in some physical properties (such as melting point) as we move down each of these groups.

MAJOR TYPES OF ELEMENTS   Look at the periodic table inside the front cover. Notice that the elements can be classified as metals (eighty-four elements), nonmetals (seventeen elements), and metalloids (eight elements). The table also shows the shorthand symbol for each element. Remember that the elements in each vertical column or group have similar chemical properties.

More than 75 percent of the elements are metals, found to the left and at the bottom of the table. Examples are aluminum (Al), calcium (Ca),

TABLE 2.1 SOME PROPERTIES OF ELEMENTS
IN THREE MAJOR GROUPS IN THE PERIODIC TABLE

Element	Appearance and Physical State at 25°C	Melting Point (°C)	Boiling Point (°C)	Chemical Reactivity
Group 1 or IA, Alkali Metals				
Lithium (Li)	Soft, silvery metal	180	1336	Very high
Sodium (Na)	Soft, silvery metal	98	883	Very high
Potassium (K)	Soft, silvery metal	64	758	Extremely high
Rubidium (Rb)	Soft, silvery metal	39	700	Extremely high
Cesium (Cs)	Soft, silvery metal	29	670	Extremely high
Francium (Fr)	Available only in trace amounts	27	?	Extremely high
Group 17 or VIIA, Halogens				
Fluorine (F)	Light yellow gas	−220	−188	Extremely high
Chlorine (Cl)	Yellow-green gas	−101	−34	Very high
Bromine (Br)	Dark, red-brown liquid	−7	59	High
Iodine (I)	Purple solid	114	184	High
Group 18 or VIIIA, Noble Gases				
Helium (He)	Colorless gas	−270	−269	No known compounds
Neon (Ne)	Colorless gas	−249	−246	Forms a few compounds
Argon (Ar)	Colorless gas	−189	−186	Forms a few compounds
Krypton (Kr)	Colorless gas	−157	−153	Forms a few compounds
Xenon (Xe)	Colorless gas	−112	−108	Forms a few compounds
Radon (Rn)	Colorless gas	−71	−62	Forms a few compounds

copper (Cu), gold (Au), iron (Fe), magnesium (Mg), mercury (Hg), nickel (Ni), platinum (Pt), sodium (Na), tin (Sn), and zinc (Zn).

Except for mercury, the metallic elements are all solids at room temperature. **Metals** usually conduct electricity and heat, are shiny, and can be hammered into thin sheets or drawn into wires.

Sulfur (S), nitrogen (N), phosphorus (P), and chlorine (Cl) are common examples of nonmetal elements. They are found in the upper right of the table. Hydrogen (H), also a nonmetal, is placed by itself above the center of the table because it does not fit well into any of the groups. Even a masterpiece like the periodic table is not perfect. **Nonmetals** do not conduct electricity very well, are usually not shiny, and cannot be hammered into thin sheets or drawn into wires.

Some key properties of four nonmetallic halogens are given in Table 2.1. Group 18 (VIIIA) is a special group of nonmetals (He, Ne, Ar, Kr, Xe, and Rn) called the noble gases. These elements have little tendency to undergo chemical reactions (see Table 2.1).

The elements arranged in a diagonal staircase pattern between the metals and nonmetals are called **metalloids**, or semimetals. They have a mixture of metallic and nonmetallic properties.

PRACTICE EXERCISE

By looking at the periodic table, you should be able to classify an element as a metal, nonmetal, or metalloid. For example, five elements that you may not have heard of are beryllium (Be), xenon (Xe), rhenium (Re), polonium (Po), and selenium (Se). Use the periodic table to classify each as a metal, nonmetal, or metalloid.

Be and Re are metals; Xe and Se are nonmetals; Po is a metalloid.

SOLUTION

Table 2.2 lists fifteen elements that will play major roles in this book. It gives you a brief preview of the importance of these elements and their compounds.

The periodic table is a masterpiece based on experiments. But how does it work? Why do all elements in a group have similar properties? To find out, scientists needed to know more about atoms; we now return to that story.

TABLE 2.2 ROLES OF FIFTEEN IMPORTANT ELEMENTS AND SOME OF THEIR COMPOUNDS

Element	Symbol	Essential for Life and Good Health	Important as a Resource and in Consumer Products	Potential Pollutant
Hydrogen	H	Carbohydrates, proteins, fats, water	Fossil fuels, plastics, fibers, drugs	Hydrocarbon air pollutants, pesticides, water pollutants
Carbon	C	Carbohydrates, proteins, fats	Fossil fuels, plastics, fibers, drugs	Carbon monoxide, carbon dioxide, hydrocarbon air pollutants, pesticides
Oxygen	O	Carbohydrates, proteins, fats, water, nitrate and phosphate plant nutrients	Combustion, industrial chemicals, drugs, glass	Ozone, oxides of carbon, nitrogen and sulfur air pollutants, nitrate and phosphate water pollutants
Nitrogen	N	Proteins, nitrate plant nutrients	Fertilizers, industrial chemicals, plastics, fibers	Nitrogen oxides and smog air pollutants, nitrate water pollutants
Phosphorus	P	Proteins, bones, phosphate plant nutrients	Fertilizers, detergents, industrial chemicals, matches	Phosphate water pollutants
Sulfur	S	Proteins	Industrial chemicals	Sulfur oxides air pollutants
Fluorine	F	Fluorides	Glass etchant, Teflon, Freon refrigerant, toothpaste	Freon (CFC) air pollutants
Sodium	Na	Salt (sodium chloride)	Industrial chemicals, baking soda, glass	Irrigation runoff
Aluminum	Al	—	Key structural material, foil, antiperspirants	Particulate air pollutants
Silicon	Si	Trace element	Sand, glass, quartz, clay, asbestos, computer chips	Asbestos inhalation
Chlorine	Cl	Salt (sodium chloride)	Industrial chemicals	Chlorinated pesticides such as DDT
Potassium	K	Chief ion inside human cells	Photography, electroplating, gunpowder, food preservative	Irrigation runoff
Calcium	Ca	Bones	Glass, limestone, drying agent	Runoff of road deicing compounds
Iron	Fe	Blood (hemoglobin), photosynthesis	Steel, alloys	—
Iodine	I	Thyroid hormone	Iodized salt, disinfectants	Emissions from nuclear explosions or accidents

2.4 Atomic Particles: What's Inside Is Important

THE INTERNAL STRUCTURE OF ATOMS   In the late 1800s, scientists did experiments showing that atoms were made up of particles with positive and negative electrical charges. Later they discovered that most atoms also have particles without an electrical charge. In this section, we look at some of the experiments that led to these discoveries about the internal structure of atoms.

Why do you need to know about what atoms are made of? Because the properties of different elements are determined by subatomic particles their atoms contain. By understanding something about atomic structure, you can understand why the periodic table works. You can also understand why atoms of certain elements combine to form millions of different compounds.

DISCOVERY OF THE ELECTRON   Between 1875 and 1900, scientists carried out experiments with *gas-discharge* or *cathode-ray tubes* (Figure 2.5). These were the forerunners of today's TV picture tubes.

Metal discs called *electrodes* were sealed into the opposite ends of a glass tube. Most of the air in the tube was sucked out by a vacuum pump. An electric current was then passed between the electrodes. The experimenters saw rays (called cathode rays) streaming in straight lines from

Figure 2.5 A cathode-ray tube. When the electrical current is turned on, cathode rays (consisting of negatively charged electrons) are emitted by the cathode and travel in straight lines to the positively charged anode.

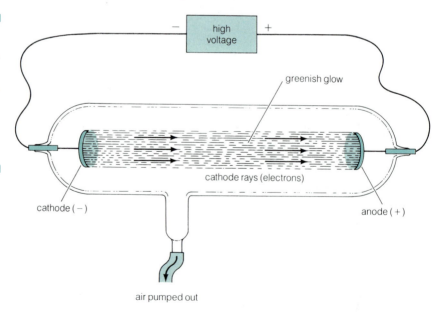

the negative electrode (the *cathode*) to the positive electrode (the *anode*). What could these rays be?

Some scientists believed they were streams of particles; others said they were light rays. British physicist J. J. Thomson (Figure 2.6) showed that the cathode rays could not be rays of light because they traveled at only a fraction of the speed of light. In 1897 he also showed that cathode rays were attracted to positively charged plates. He concluded that cathode rays are a stream of negatively charged particles.

Experiments showed that cathode rays could be produced using many different elements as cathodes. This led Thomson to propose that *negatively charged particles must be a fundamental component of all atoms.* Each particle had the same negative electrical charge. In 1891 George J. Stoney (1826–1911) named these negatively charged particles **electrons** (represented by the symbol *e*).

DISCOVERY OF THE PROTON   Experiments with cathode-ray tubes gave evidence that atoms contain positively charged units. In 1886 Eugene Goldstein (1850–1930) did experiments with a cathode-ray tube that had a cathode with holes in it (Figure 2.7). He saw the usual beam of electrons streaming from the cathode to the anode. But he also discovered that other rays formed in the tube and streamed in the opposite direction through the holes in the cathode. When these new rays passed through an electric field, they were all attracted to a negatively charged plate. This indicated that these new rays were made of positively charged particles.

Figure 2.6 Sir Joseph John Thomson (1856–1940) received the Nobel Prize in physics in 1906 for experiments leading to the discovery of the electron. He was clumsy in the laboratory, but he had an outstanding ability to design original experiments and to interpret their results. Seven of his laboratory assistants eventually won Nobel Prizes in science. (Courtesy Burndy Library)

Figure 2.7 Positively charged particles form in a cathode-ray tube when cathode rays (electrons) strike atoms of residual gas. These particles are attracted to the negatively charged cathode, where some pass through the holes and are deflected toward a negatively charged plate behind the cathode.

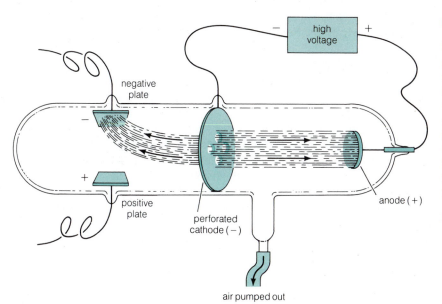

Figure 2.8 The Thomson model for a carbon atom.

Figure 2.9 Ernest Rutherford (1871–1937) left his native New Zealand to become Thomson's first laboratory assistant. He worked with Thomson on experiments that led to the discovery of the electron. Later he devised experiments that led to the discovery of alpha and beta particles, work for which he received the Nobel Prize in chemistry in 1908. (American Institute of Physics Niels Bohr Library)

Thomson had shown that cathode rays consisted of electrons, all with the same mass. Goldstein, however, observed that the positively charged particles passing through his perforated cathode had different masses. The mass varied depending on the type of gas left in the tube.

Goldstein reasoned that these positive particles seemed to be a basic component of all matter. Positive particles with the lowest mass were produced when a small amount of hydrogen gas, the lightest element, remained in the tube. Goldstein deduced that a positively charged hydrogen atom (H^+) represents the fundamental particle of positive charge in all atoms. We now call these positively charged particles **protons** (represented by the symbol p).

Later experiments showed that the proton is about 1840 times heavier than an electron. This is roughly the same as comparing the mass of a pickup truck to that of a golf ball.

THOMSON MODEL OF THE ATOM   In 1903 Thomson used the discoveries of electrons and protons to propose the first model of the internal structure of the atom. According to the *Thomson model*, the atom is a positively charged sphere with negatively charged electrons scattered throughout the sphere (Figure 2.8). The total positive charge equals the total negative charge. This means that the overall electrical charge of the atom is zero. In this model, the electrons would be like the chocolate chips in a cookie or the seeds in a watermelon.

DISCOVERY OF THE ATOMIC NUCLEUS: RUTHERFORD MODEL   How do protons and electrons fit together to make up atoms? The next step in answering this question came from a surprising discovery. Many scientific experiments give unexpected results. *Serendipity* is the name given to the chance discovery of one thing while looking for another. Instead of getting mad and throwing out such results, a good scientist asks "why?"

One example of serendipity occurred in 1896 when French physicist Henri Becquerel (1852–1908) accidentally discovered that certain substances emit highly energetic rays that can penetrate paper and expose photographic plates. He had discovered radioactivity.

Later experiments revealed three types of radioactivity: *alpha particles* (which have a positive charge), *beta particles* (which have a negative charge), and *gamma rays* (which have no electrical charge). We discuss radioactivity in more detail in Chapter 3.

Radioactivity gave scientists a useful tool for studying the internal structure of matter. One such experiment became a classic. In 1910 British physicist Ernest Rutherford (Figure 2.9) noticed a bright but inexperienced student named Ernest Marsden. Marsden was working as a janitor in Rutherford's laboratory just to be around eminent scientists. Rutherford gave him a research project to verify the Thomson model of the atom.

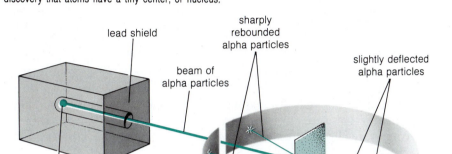

Figure 2.10 Rutherford and Mardsen's experiment with gold foil. This experiment led to the discovery that atoms have a tiny center, or nucleus.

He proposed that Marsden fire a beam of alpha particles (emitted by a radioactive element) at a thin piece of gold foil and observe how the alpha particles were scattered. Rutherford expected the alpha particles to pass through the gold foil virtually undisturbed—much like shooting machine gun bullets through a haystack. Why did he expect this? One reason is that the gold foil was very thin. Another was that—according to the Thomson model (see Figure 2.8)—the positive charge and the electrons were uniformly distributed throughout each gold atom.

In a darkened basement, Marsden carefully counted each alpha particle as it hit the detection screen and gave off a microscopic flash of light.* Each alpha particle had a mass 4 times that of a proton and more than 7300 times that of an electron. Marsden found that most of the speedy alpha particles whisked right through the foil undeflected or were only slightly deflected, just as Rutherford expected (Figure 2.10).

---

* The process was so tedious that Rutherford's colleague, Hans Geiger, later invented an automatic device for detecting radioactive emissions, the Geiger counter.

To Rutherford's great surprise, however, Marsden found that a few alpha particles—about 1 in 8000—rebounded off the foil and almost retraced their paths back to the radioactive material from which they came. Rutherford later remarked, "It was the most incredible event that ever happened in my life. It was as though you had fired a 15-inch artillery shell at a piece of tissue paper and it came back and hit you."

A year later, Rutherford proposed a new model of the atom to explain these results. He concluded that most of the atom must be empty space because most of the alpha particles passed through the gold foil with little or no deflection. He also reasoned that the few positively charged alpha particles that were reflected at sharp angles must have collided with or come very close to a tiny but concentrated mass of a positive electrical charge.

Thus, Rutherford proposed that most of the mass and positive charge must be concentrated in an extremely small space at the atom's center, which he called a **nucleus**. In other words, all of an atom's positively charged protons, each with a mass 1840 times that of a negatively charged electron, were packed into a dense speck at the center of each atom (Figure 2.11).

He also reasoned that, because the nucleus contained all the positive charge, the negatively charged electrons must be distributed somewhere outside the nucleus. Because atoms were known to be electrically neutral, the number of negatively charged electrons outside the nucleus would have to be equal to the number of positively charged protons inside the nucleus (see Figure 2.11).

DISCOVERY OF THE NEUTRON   When Rutherford used the number of protons in the nucleus to calculate the atomic masses of various elements, he could account for only about half of the mass. He suggested that the nuclei of atoms must contain another type of particle. The particle should have no electrical charge and about the same mass as a proton.

In 1932 James Chadwick (1891–1974), one of Rutherford's former students, discovered these particles. Because they are electrically neutral, they were called **neutrons** (represented by the symbol n). Chadwick then revised Rutherford's model and proposed that the nucleus of an atom contains both positively charged protons and uncharged neutrons (Figure 2.12).

Atoms are often represented as tiny hard spheres (see Figure 2.2), but the Rutherford–Chadwick model shows that atoms are mostly free space (see Figure 2.12). Since 1930 physicists have discovered more than 100 other fundamental particles that exist in the nucleus (such as *quarks* and *mesons*). But most chemists use the simplified electron–proton–neutron model of the atom because it helps account for nearly all the chemical behavior of matter.

Figure 2.11 Rutherford's model for a carbon atom. The nucleus is in the center.

Figure 2.12 Rutherford–Chadwick model for a carbon atom. This model shows that the nucleus of a carbon atom also contains neutrons.

TABLE 2.3 CHARACTERISTICS OF SUBATOMIC PARTICLES				
Particle	Symbol	Location in Atom	Electrical Charge	Mass (in g)
Proton	p	Inside nucleus	+	1.6726×10^{-24}
Neutron	n	Inside nucleus	0	1.6750×10^{-24}
Electron	e	Outside nucleus	−	9.1095×10^{-28}

SUBATOMIC PARTICLES: A SUMMARY   Table 2.3 summarizes the key properties of electrons, protons, and neutrons. Notice that protons and neutrons have about the same mass, which is much greater than that of an electron.

You can see from this why nearly all an atom's mass is in its nucleus, even though the nucleus occupies only a tiny amount of the atom's space. For example, if the nucleus of a hydrogen atom were the size of a softball, the single electron would be moving about in a spherical space extending out more than 3 km (2 miles) from the nucleus.

Put different numbers of protons, neutrons, and electrons together and you get different types of atoms. These atoms, however, are incredibly small. In fact, the period at the end of this sentence is big enough to hold about 1,000,000,000,000,000,000 atoms.

2.5 Atomic Number, Atomic Mass, and Isotopes: Some Ways to Tell One Atom from Another

ATOMIC NUMBER: A WAY TO TELL ONE ELEMENT FROM ANOTHER   We can now put together some of the pieces of the puzzle. In the original periodic table, Mendeleev arranged the elements in order of increasing *atomic mass*. In the modern periodic table, however, the elements are listed in order of increasing atomic number.

The **atomic number** is the number of protons in the nucleus of an atom. Atoms have no net electrical charge because they have the same number of positively charged protons and negatively charged electrons (see Figure 2.12). Thus, the atomic number is not only the number of protons in the nucleus but also the number of electrons outside the nucleus.

Look at the periodic table (inside front cover) and find the element carbon (C). Just above its symbol you will see its atomic number, which is 6 (Figure 2.13). This atomic number tells you the number of protons

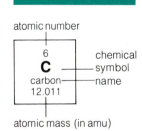

Figure 2.13 How to read the periodic table.

and the number of electrons in each atom of carbon. Compare this with the number of protons and electrons in the carbon atom in Figure 2.12.

The number of protons in the nucleus of an atom—the atomic number—is a unique characteristic that identifies each element. For example, any atom with one proton in its nucleus has an atomic number of 1 and must be an atom of hydrogen (H). Similarly, any atom with six protons in its nucleus must be an atom of carbon (C) with an atomic number of 6. If the atomic number is 7, the atom cannot be carbon; it must be nitrogen (N). Thus, we can redefine an **element** as a type of matter composed of atoms with the same atomic number.

PRACTICE EXERCISE

Use the periodic table to determine the number of protons and electrons in the atoms of each of the following elements: gold (Au), fluorine (F), calcium (Ca), and uranium (U).

79, 9, 20, and 92, respectively.

SOLUTION

NEUTRONS, ISOTOPES, AND MASS NUMBERS   The periodic table lists not only atomic numbers but also atomic masses (see Figure 2.13). Neutrons don't affect the atomic number, but they have about the same mass as protons (see Table 2.3) and thus are an important factor in an element's atomic mass.

Here matters get more complicated. Experiments have shown that atoms of the same element may contain different numbers of neutrons in their nuclei. For example, about 99.985 percent of the hydrogen atoms found on earth have one proton and no neutrons in their nuclei. But 0.015 percent (about 3 out of every 20,000 hydrogen atoms) have one proton and one neutron. An almost negligible percentage of earth's hydrogen atoms have one proton and two neutrons in their nuclei.

Atoms of the same element must have the same atomic number or number of protons in their nuclei, but they can have different numbers of neutrons. These different types of atoms of an element are called **isotopes**. Because neutrons have no electrical charge, the isotopes of a particular element differ only in the masses of their nuclei, not their charge. Table 2.4 shows the differences between the naturally occurring isotopes of hydrogen and nitrogen.

Isotopes of the same element may differ slightly in physical properties that depend on mass. But all of an element's isotopes have the same gen-

TABLE 2.4 ISOTOPES OF THE ELEMENTS HYDROGEN AND NITROGEN

Isotope	Symbol and Name	Percentage Abundance in Nature	Isotope	Symbol and Name	Percentage Abundance in Nature
1e 0n 1p	$^1_1$H, hydrogen-1	99.985	7e 7n 7p	$^{14}_7$N, nitrogen-14	99.63
1e 1n 1p	$^2_1$H, hydrogen-2 or deuterium (D)	0.015	7e 8n 7p	$^{15}_7$N, nitrogen-15	0.37
1e 2n 1p	$^3_1$H, hydrogen-3 or tritium (T)	Trace			

eral chemical properties. Why? Because chemical properties primarily depend on an atom's electrons, which are the same in all isotopes of an element.

Each isotope of an element has a different **mass number**. This is the number of protons plus neutrons ($p + n$) in the nucleus of each atom of that isotope. Table 2.4 shows that although the atomic number is the same for all isotopes of an element, the mass number differs.

Figure 2.14 shows some ways to represent the mass number and atomic number of an isotope. For example, the three isotopes of hydrogen can be represented as $^1_1$H, or hydrogen-1; $^2_1$H, or hydrogen-2 (also known as *deuterium*); and $^3_1$H, or hydrogen-3 (also known as *tritium*).

$$^{\text{mass number}}_{\text{atomic number}}X$$

mass number = number of protons plus neutrons
atomic number = number of protons
X = symbol of element

Examples:

$^{12}_6$C $\qquad$ $^{13}_6$C $\qquad$ $^{14}_6$C

Carbon-12 $\qquad$ Carbon-13 $\qquad$ Carbon-14
(98.89%) $\qquad$ (1.11%) $\qquad$ (trace amount)

Figure 2.14 Symbols for isotopes. Numbers shown in parentheses are the abundance by mass of each isotope in a natural sample of carbon.

If you know the mass number and atomic number of an isotope, you can find out how many protons, neutrons, and electrons this type of atom has. To do this, you need to know two things:

atomic number = number of protons = number of electrons

number of neutrons = mass number $(n + p)$ − atomic number (p)

For example, suppose you are asked to find the number of protons, electrons, and neutrons in atoms of the isotope uranium-235. The first step is to find this element's atomic number. How? Look it up in the periodic table (inside front cover).

The element U has an atomic number of 92. This tells you that all isotopes of uranium, including uranium-235, have ninety-two protons inside their nuclei and ninety-two electrons outside their nuclei. Now all you have to do is find the number of neutrons inside the nuclei of uranium-235. The 235 is this isotope's mass number.

number of neutrons = mass number − atomic number
$$= 235 - 92 = 143$$

PRACTICE EXERCISE

How many protons, electrons, and neutrons are in each atom of the following: uranium-238, carbon-14, and nitrogen-14?

$^{238}_{92}U$ $^{14}_{6}C$ $^{14}_{7}N$

number of protons = atomic number =   92   6   7
number of electrons = atomic number =   92   6   7
number of neutrons = mass number − atomic number =   146   8   7

SOLUTION

ATOMIC MASS OR WEIGHT OF AN ELEMENT   A single atom is so minute that scientists cannot determine its mass directly, even with the most sensitive weighing device. Fortunately, we only need to know the *relative masses* of the atoms of different isotopes and elements. One isotope is assigned a certain value for its mass. It is then used as a reference standard to compare the relative masses of all other isotopes.

The standard chosen is the most abundant isotope of carbon, carbon-12 (see Figure 2.14). Each atom of carbon-12 is assigned a relative mass of *exactly* 12 atomic mass units. Thus, one **atomic mass unit** (1 **amu**) is defined as a mass exactly equal to one-twelfth the mass of a carbon-12 atom:

$$1 \text{ amu} = \frac{\text{mass of one carbon-12 atom}}{12}$$

You can use this reference standard to find the relative mass of any isotope. For example, if an isotope such as magnesium-24 has twice the mass of carbon-12, then each magnesium-24 atom must have a relative mass of 24 amu. With this system, protons and neutrons each have a relative mass of approximately 1 amu, while an electron has approximately 0 amu. This means that the mass number of a particular isotope of an atom (its number of protons plus its number of neutrons) is approximately equal to its relative mass in atomic mass units.

Now look at the periodic table or Figure 2.13. Note that the atomic mass for carbon is 12.011 amu and not the 12 amu assigned to carbon-12. Why? Because the **atomic mass** (also called atomic weight) of an element is the *average* mass of all isotopes found in a natural sample of the element. This average mass is weighted to reflect the abundance and mass of each isotope in a natural sample of the element.

In nature carbon occurs as a mixture containing 98.89 percent carbon-12, with an isotopic mass of exactly 12 amu, and 1.11 percent carbon-13, with a mass of 13.003 amu. Thus, the atomic mass of carbon, listed in the periodic table, is $(0.9889 \times 12 \text{ amu}) + (0.0111 \times 13.003 \text{ amu}) = 12.011$ amu. Note that to multiply a number by a percentage you have to move the decimal point in the percentage value two places to the left.

PRACTICE EXERCISE

Suppose a natural sample of an element contains two isotopes. Two-thirds of the sample is an isotope with a mass of 150 amu; the other one-third is an isotope with a mass of 149 amu. What is the atomic mass for this element?

SOLUTION

We can find the answer in two ways. For every three atoms, two have a mass of 150 amu and one has a mass of 149 amu. The weighted average would be

$$\frac{150 \text{ amu} + 150 \text{ amu} + 149 \text{ amu}}{3} = 149.67 \text{ amu}$$

The second way is

$$(0.667 \times 150 \text{ amu}) + (0.333 \times 149 \text{ amu}) = 149.67 \text{ amu}$$

2.6 Models of Electronic Structure: How Are Electrons in Atoms Arranged?

WHERE ARE ELECTRONS?   We still have one more piece to fit into the puzzle. The periodic table groups elements that have similar properties. But *why* do the elements in the same group have similar properties? The answer lies in their electrons. Electrons are the parts of an atom outside its nucleus. So when atoms interact, it is their electrons that interact. For example, suppose you are riding in a bumper car at an amusement park. When you bump into another car, it is the rubber bumpers surrounding the car that interact. By stretching the example a bit, we can think of an atom's electrons as its bumpers.

But can all electrons in an atom interact with other atoms? The answer is no. It turns out that electrons outside the nucleus of an atom have a structure, or arrangement. Some are on average farther away from the nucleus than others. As you might expect, the electron or electrons farthest from the nucleus are the ones that mainly interact with other atoms. These *outermost electrons* are the primary bumpers of an atom.

In the early 1900s, scientists began trying to learn more about the arrangement of the electrons in atoms. Two major models of electronic structure were proposed: the *Bohr model* and the *quantum mechanical model*. We now examine these two models.

LINE SPECTRA: EMISSION OF LIGHT BY ELEMENTS   To understand how these models were developed, we need to learn about the types of light various elements emit when trace amounts are placed in a cathode-ray tube (see Figure 2.5). White light, like that emitted by the sun or an incandescent light bulb, consists of all visible colors. Each color represents different amounts of radiant energy, which decreases as we move from violet to blue, green, yellow, orange, and red.

How do we know this? The answer is to pass white light from the sun or an incandescent light bulb through a glass prism. The light then separates into a *continuous spectrum*, or rainbow, of the visible colors it contains (Figure 2.15). You have probably seen a rainbow after a rainstorm; it occurs when water droplets in the sky act as prisms to separate sunlight into its component colors.

Suppose you leave a small amount of hydrogen gas in a cathode-ray tube, turn on the electricity, and pass the light emitted by the hydrogen through a prism. Instead of getting a continuous spectrum of colors, you would see only four narrow lines. Each represents a specific color with a particular energy (Figure 2.16). This is called a *line spectrum*. Suppose you do the same experiment but replace hydrogen in the cathode-ray tube

Figure 2.15 A continuous spectrum of visible colors forms when white light passes through a prism.

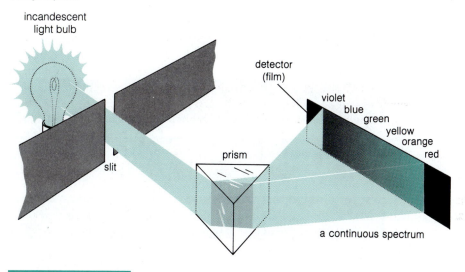

Figure 2.16 A discontinuous line spectrum of visible colors forms when the light emitted by hydrogen gas in a cathode-ray tube passes through a prism.

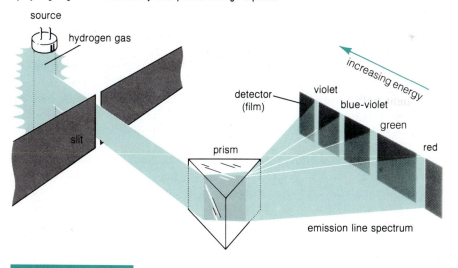

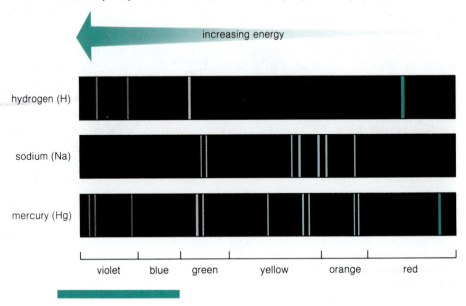

Figure 2.17 The discontinuous line spectra of visible colors emitted by hydrogen, sodium, and mercury. They show how each element has a unique spectral "fingerprint."

increasing energy

hydrogen (H)

sodium (Na)

mercury (Hg)

violet   blue   green   yellow   orange   red

with another element, such as sodium or mercury. Then you will get a different line spectrum (Figure 2.17).

Scientists discovered that each element has a unique line spectrum, which is like a distinctive "fingerprint" that shows the presence of a particular element. Indeed, some elements emit a particular color so strongly that you can see it if you heat some of it in a flame. For example, heat sodium (Na) compounds and you get a yellow-orange flame. Copper (Cu) compounds give a blue-green flame, potassium (K) compounds a fleeting violet flame, and lithium (Li) compounds a brilliant red flame. Such compounds are used to produce the colors in flares and fireworks. You also see strong emissions of certain colors in the yellow glow from sodium-vapor street lamps, the bluish white glow from mercury-vapor street lamps, and the red-orange glow from neon lights used for advertising.

THE BOHR MODEL   What do line spectra have to do with electron structure? Shortly after Rutherford proposed his nuclear model of the atom, scientists realized that any useful model of the atom should be able to explain and predict the specific colors or energies in the line spectra of hydrogen and other elements.

Rutherford envisioned electrons whizzing around, like tiny gnats, somewhere outside the nucleus. Because of their negative charge and small size, they were attracted toward the more massive, positively charged nucleus. The potential energy would vary depending on how close electrons were to the nucleus, but electrons must have enough kinetic energy to keep from being pulled into the nucleus.

In 1913 a young Danish physicist, Niels Bohr (Figure 2.18), suggested a way to relate the line spectrum of hydrogen (and presumably those of other elements) to its electronic structure. Until then, scientists had thought that an electron outside an atom could have any potential energy value. Bohr proposed, however, that the single electron in a hydrogen atom could have only certain potential energy values, called **major energy levels**. Furthermore, he said that the allowed values were spaced unequally, like the shelves of a bookcase (Figure 2.19). Unlike the book in Figure 2.19,

Figure 2.18 Niels Bohr (1885–1962) won the Nobel Prize in physics in 1922 for his investigations of atomic structure and radioactivity. (American Institute of Physics Niels Bohr Library)

Figure 2.19 Allowed major energy levels for the single electron in an atom of hydrogen (left). They are somewhat like the allowed values of potential energy that a book can have in a bookcase (right).

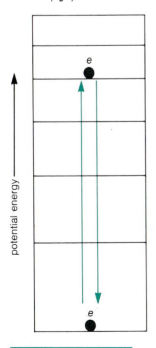

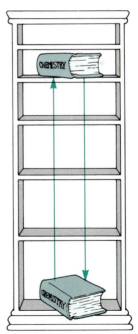

6th energy level
excited state

5th energy level
excited state

4th energy level
excited state

3rd energy level
excited state

2nd energy level
excited state

1st energy level
ground state

potential energy

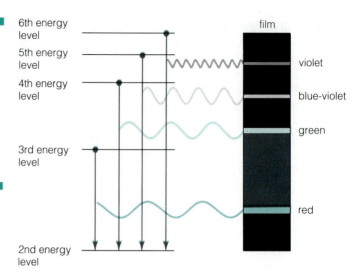

Figure 2.20 Explanation of the visible line spectrum of hydrogen due to electrons in higher excited states returning to the second energy level. Electrons returning to the first energy level (not shown) emit higher-energy, ultraviolet radiation.

however, the electron also has kinetic energy because it is in continuous motion.

Most scientists didn't accept this new idea, but they paid attention when Bohr used his ideas to explain the line spectrum of hydrogen and to calculate the four colors or energies in that spectrum. Bohr reasoned that an electron is normally found in its lowest possible major energy level, called its **ground state**. When a flame or electrical current supplies energy to a hydrogen atom, its electron absorbs a specific amount of energy. This moves it into one of the higher available major energy levels, called an **excited state**. When the electron loses this extra energy, it returns to a lower major energy level. When this occurs, a specific amount of energy is emitted. It is equal to the difference between the two energy levels.

Bohr calculated these values of energy and found that certain transitions from higher to lower major energy levels corresponded to the four colors in the line spectrum of hydrogen (Figure 2.20). Hydrogen also emits specific higher energies of ultraviolet light and lower energies of infrared light, which we can't see. They are emitted when electrons drop from other energy levels (not shown in Figure 2.20). Those spectral lines were also found exactly where Bohr predicted them to be.

You might be wondering, as Bohr did, just where the electron in a hydrogen atom is if it can only have certain allowed energy values. Bohr proposed that each energy level represented the electron in a hydrogen atom whirling about in a circular orbit at a fixed distance from the nucleus (Figure 2.21).

Figure 2.21 Bohr's planetary model of the hydrogen atom.

In its ground state, the electron would be in the innermost orbit, closest to the nucleus. In an excited state, the electron would be in one of the allowed circular orbits farther from the nucleus. This was known as the *planetary model* because it pictured the electron orbiting the nucleus, much as the planets revolve around the sun.

The Bohr model was easy to understand and visualize, but it had a fatal flaw: It was a one-element model. It worked beautifully for hydrogen. But the model could not predict the correct line spectra of other elements, all of which contain more than one electron. Scientists had to return to the drawing board.

THE QUANTUM MECHANICAL MODEL   Scientists tried to modify the Bohr model to make it work for other atoms. One approach was to have electrons travel in elliptical rather than circular orbits, but that didn't work either. In 1926 the Bohr model of electronic structure was replaced with an entirely new model. This currently accepted *quantum mechanical model* of electronic structure was developed primarily by Austrian physicist Erwin Schrödinger (Figure 2.22). (A *quantum* is the smallest amount, or "packet," of energy associated with light.)

This model is mathematically complex, so we will not discuss the details here. The important point to remember, however, is that scientists were able to use this model to account for the line spectra not only of hydrogen but also of elements with more than one electron.

This model, like the Bohr model, predicts that each electron in an atom can only have certain allowed energy values. These values correspond to the major energy levels shown earlier for the hydrogen atom (see Figure 2.19). But this new model also predicts that electrons can exist in various *energy sublevels* found within each major energy level, except the first.

The Bohr model and the quantum mechanical model differ sharply in their description of where each electron in an atom can be found. According to the Bohr model, each electron moves in a fixed, circular orbit around the nucleus (see Figure 2.21). According to the quantum mechanical model, however, it is impossible to determine both the exact location and speed of an atom's electrons at any given instant. Instead, we can only determine the mathematical *probability* (or likelihood) of finding a given electron in a particular space at a given instant. Thus, a particular electron may have a 5 percent probability of being at one point in space at a given moment, a 2 percent chance of being found at another point, and so on.

We can get a crude pictorial version of this purely mathematical model by plotting the probability values for finding each electron at various points in space outside an atom's nucleus. These diagrams are sometimes

Figure 2.22 Erwin Schrödinger (1887–1961) developed mathematical equations for the quantum mechanical model of electronic structure of atoms. In 1933 he won the Nobel Prize in physics for this work. (California Institute of Technology Archives)

Figure 2.23 Electron probability cloud for an atom of hydrogen. Regions with more dots have a higher probability of containing the electron at a given instant. (Courtesy John Douglas, Eastern Washington University)

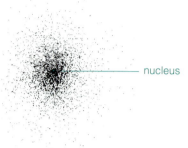

nucleus

called *probability clouds*. Figure 2.23 shows what this looks like for the single electron in a hydrogen atom. In three dimensions, the overall shape is spherical. A greater number of dots in a region indicates a higher probability of finding the electron there. Electrons in energy sublevels of higher energy levels have probability clouds with different and more complex shapes.

Figure 2.24 Number of electrons in each major energy level for elements with atomic numbers 1 through 20.

	1 IA	2 IIA		13 IIIA	14 IVA	15 VA	16 VIA	17 VIIA	18 VIIIA
1			1 **H** 1e						2 **He** 2e
2	3 **Li** 2e, 1e	4 **Be** 2e, 2e		5 **B** 2e, 3e	6 **C** 2e, 4e	7 **N** 2e, 5e	8 **O** 2e, 6e	9 **F** 2e, 7e	10 **Ne** 2e, 8e
3	11 **Na** 2e, 8e, 1e	12 **Mg** 2e, 8e, 2e		13 **Al** 2e, 8e, 3e	14 **Si** 2e, 8e, 4e	15 **P** 2e, 8e, 5e	16 **S** 2e, 8e, 6e	17 **Cl** 2e, 8e, 7e	18 **Ar** 2e, 8e, 8e
4	19 **K** 2e, 8e, 8e, 1e	20 **Ca** 2e, 8e, 8e, 2e							

ELECTRON ARRANGEMENTS According to both the Bohr and quantum mechanical models, each major energy level can contain some maximum number of electrons. The first energy level can hold a maximum of two electrons; the second, a maximum of eight; the third, a maximum of eighteen; and so on. Figure 2.24 shows the number of electrons in the major energy levels for the first twenty elements in the periodic table.

Sometimes chemists use the simplified Bohr model to show how electrons are arranged in energy levels or orbits. For example, the Bohr models for the elements hydrogen (H), lithium (Li), sodium (Na), and potassium (K) are

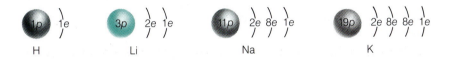

PRACTICE EXERCISE

Write the Bohr electronic structure diagrams for atoms of carbon, silicon, and phosphorus.

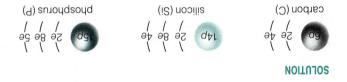

SOLUTION

VALENCE ELECTRONS Notice from Figure 2.24 and the Bohr diagrams just given that the elements in the same group have the same number of electrons in their highest energy level. These electrons, which are on the average farthest from the nucleus, are called **valence electrons**. They are the outermost electrons in an atom that "bump into" or interact with the outermost electrons in other atoms.

A simple way to represent the number of valence electrons in an atom is to give its **electron dot structure**. To do this, write the chemical symbol for an element and surround it with as many dots as there are valence electrons in each of its atoms. For example, the electron dot structure for hydrogen (with one valence electron) is H·. For chlorine (with seven valence electrons), it is $:\ddot{Cl}·$.

From the information in Figure 2.24, you can draw the electron dot structures for the first twenty elements in the periodic table. If you do this, you will notice a pattern. All Group 1 or IA elements have a single valence electron and thus a common electron dot structure. All Group 2 or IIA elements have two valence electrons and thus a common electron dot structure that is different from Group 1 or IA elements.

Table 2.5 shows this pattern of electron dots for several of the groups in the periodic table. The electron dot structures for groups in the middle of the periodic table and the two rows below the periodic table (inside front cover) are not given. These elements have more complex electronic structures that are beyond the scope of this book.

If you study Table 2.5, you can uncover two important generalizations:

— The maximum number of valence electrons for the atoms in Table 2.5 is eight.

— The number of valence electrons for atoms of any element in an A group of the periodic table is equal to its group number.

TABLE 2.5 VALENCE ELECTRONS AND ELECTRON DOT STRUCTURES OF ELEMENTS

Group	1/IA	2/IIA	13/IIIA	14/IVA	15/VA	16/VIA	17/VIIA	18/VIIIA
Number of Valence Electrons	1	2	3	4	5	6	7	8 (Except He)

Period	Electron Dot Structure							
1	H·							He:
2	Li·	Be:	·B·	·C·	·N·	:O·	:F·	:Ne:
3	Na·	Mg:	·Al·	·Si·	·P·	:S·	:Cl·	:Ar:
4	K·	Ca:	·Ga·	·Ge·	·As·	:Se·	:Br·	:Kr:
5	Rb·	Sr:	·In·	·Sn·	·Sb·	:Te·	:I·	:Xe:
6	Cs·	Ba:	·Tl·	·Pb·	·Bi·	:Po·	:At·	:Rn:
7	Fr·	Ra:						

For example, all atoms in Group IIA have two valence electrons; all those in Group VIA have six valence electrons.

ATOMIC STRUCTURE AND THE PERIODIC TABLE   We now have found the missing piece needed to explain how the periodic table works. Mendeleev and Meyer knew nothing about atomic structure. They grouped the elements in the periodic table according to common, experimentally observed, chemical properties.

Now we know that they also grouped the elements according to atomic structure—specifically the number of valence electrons in the atoms of each element. Beginning in Chapter 4, you will learn more about how valence electrons are the key factor in determining an element's chemical properties. The Bohr and quantum mechanical models show us *why* elements in the same group in the periodic table have similar chemical properties.

LOOKING BACK AT THE SEARCH FOR ORDER   This chapter demonstrated an excellent example of how science works. Observations of how matter consistently behaves led scientists to formulate the law of conservation of mass and the law of constant composition. Dalton then tried to explain these laws in terms of atoms. His hypothesis fit so well with the data that it became a widely accepted theory. Then scientists did experiments that suggested that atoms have an internal structure. This led to new hypotheses and theories about what an atom is like. As these ideas were tested, more new models and theories had to be invented to explain the new data (Figure 2.25).

You can see why science is a search for order that is always changing; new models and theories are developed that work better than old ones. Remember that science is a search for models or ideas that work best in describing or explaining something we observe in nature.

Figure 2.25 A Scanning Tunneling Microscope photo of xenon atoms. (Courtesy of International Business Machines Corporation)

In this chapter, you have seen how ideas about the periodic table, atomic number, atomic mass, and atomic structure all fit together to explain some of the key properties of matter. We hope you believe, as we do, that this search for order is exciting, beautiful, and useful.

LEONARDO DA VINCI   . . . those sciences are vain and full of errors which are not born from experiment, the mother of all certainty . . .

Summary

More than 2400 years ago, Leucippus and Democritus proposed that all elements are composed of atoms. In the early 1800s, Dalton revived this ancient idea. He strengthened this hypothesis by using it to explain the scientific laws of conservation of mass, constant composition, and multiple proportions. These laws summarized the behavior of matter.

Other experiments indicated that atoms have an internal structure. They contain positively charged protons and neutrons in a tiny center or nucleus and negatively charged electrons outside the nucleus. The protons and neutrons each have a relative mass of about 1 amu (atomic mass unit). They provide nearly all the mass of the atom because the electrons outside the nucleus have a relative mass close to 0 amu. Experiments also indicated that each atom has an equal number of positively charged protons and negatively charged electrons. Thus, an atom has no net electrical charge.

The atoms of a particular element have a unique atomic number that is equal to the number of protons (and electrons) in an atom of that element. All atoms of the same element have the same atomic number. Atoms of the same element, however, can have a different number of neutrons (and thus different masses) in their nuclei. These atoms are called isotopes.

More experiments suggested that the electrons found outside the nuclei of atoms are arranged in major energy levels. The electrons in the highest major level are called valence electrons and are represented by electron dot structures. The maximum number of valence electrons is eight.

Mendeleev and Meyer used data on the elements to develop the first periodic table of the elements. They placed the elements horizontally in order of increasing atomic mass. By choosing appropriate lengths for each row, they were able to group into vertical columns elements that had similar chemical properties.

Further experiments showed that the periodic table works better if elements are arranged horizontally by increasing atomic number. The Bohr and quantum mechanical models of electronic structure showed why the periodic table works. Elements in the same group in the periodic table have similar chemical properties because their atoms have the same number of valence electrons. These electrons, found farthest from the nucleus of an atom, are the ones that do most of the interacting with valence electrons in other atoms during chemical reactions.

Terms for Review

After completing this chapter, you should know and understand the meaning of the following terms:

alkali metal (p. 30)

alkaline earth metal (p. 30)

atom (p. 26)

atomic mass (p. 43)

atomic mass unit (amu) (p. 42)

atomic number (p. 39)

electron (p. 35)

electron dot structure (p. 51)

element (p. 40)

excited state (p. 48)

ground state (p. 48)

group (p. 30)

halogen (p. 30)

isotope (p. 40)

law of conservation of mass (p. 26)

law of constant composition (p. 27)

law of multiple proportions (p. 28)

major energy level (p. 47)

mass number (p. 41)

metal (p. 32)

metalloid (p. 32)

neutron (p. 38)

noble gas (p. 30)

nonmetal (p. 32)

nucleus (p. 38)

period (p. 30)

periodic table of the elements (p. 30)

proton (p. 36)

valence electrons (p. 51)

Questions

Odd-numbered questions are answered at the back of this book.

1. Why was the "law of conservation of mass" classified as a law instead of as a hypothesis or theory?

2. What was the major reason why the proposal by Democritus and Leucippus that matter consisted of atoms was not taken seriously for about 2200 years?

3. How did Dalton use his model of atomic structure to explain (a) the law of constant composition and (b) the law of conservation of mass?

4. In his atomic theory, Dalton proposed that all atoms of a particular element are identical and have properties that differ from those of other elements. Why did this idea have to be revised?

5. How can atoms have electrically charged particles and still have no overall electrical charge?

6. Explain how Rutherford used the results of Marsden's experiment to conclude that (a) the nucleus of an atom has a positive electrical charge and (b) the atom has a nucleus that contains most of the mass of an atom.

7. If X is a general symbol representing various elements, which ones of the following are isotopes of the same element?
   a. $^{32}_{16}X$ d. $^{15}_{8}X$
   b. $^{16}_{8}X$ e. $^{18}_{9}X$
   c. $^{30}_{16}X$ f. $^{32}_{15}X$

8. Which of the species in Question 7 have (a) the same number of protons and (b) the same mass number?

9. How many electrons, protons, and neutrons are in an atom of each of the following?

   a. $^{112}_{48}Cd$ c. Mercury-201
   b. $^{210}_{84}Po$ d. Barium-138

10. How are isotopes of the same element (a) identical to and (b) different from one another?

11. If 75 percent of an element were an isotope with a mass of 157 amu and 25 percent were an isotope with a mass of 155 amu, what would be the atomic mass listed on the periodic table for that element?

12. Why are atomic numbers in the periodic table whole numbers, but the atomic masses *not* whole numbers?

13. What does it mean to say that an electron in an atom is in (a) its ground state and (b) an excited state?

14. Explain how the Bohr and quantum mechanical models differ in their descriptions of the location of the single electron in a hydrogen atom.

15. Explain how both the Bohr and quantum mechanical models of electronic structure can account for the line spectrum of hydrogen. Why was the Bohr model abandoned?

16. Examine Tables 2.1 and 2.4. Explain how hydrogen (a) does and (b) does *not* fit as a Group 1/IA element.

17. Use the periodic table (inside front cover) to determine the symbol and name for (a) the third element in Group 12 going from top to bottom; (b) the fifth element in Period 3 going from left to right; (c) the fifth element in Group 17 or VIIA going from top to bottom; (d) the tenth element in Period 4 going from left to right; (e)

Period 5, Group 7; (f) Period 4, Group 17/VIIA; (g) two elements that behave chemically like cadmium; (h) three elements that behave chemically like phosphorus; and (i) the elements you would expect to have chemical properties similar to those of strontium (Sr).

18. Use the periodic table to classify each of the following elements as a metal, nonmetal, or metalloid: (a) francium, (b) tellurium, (c) As, (d) Mo, (e) Ra.

19. Give the group number and period in the periodic table for each of the elements in Question 18.

20. Which of the following sets of elements are members of the same group in the periodic table? (a) Fe, Os, Ru; (b) Al, Si, Ga; (c) Ar, He, Rn; (d) Pd, Ag, Cd.

21. List the names of two elements that are (a) noble gases, (b) alkali metals, and (c) halogens.

22. List the number of valence electrons for (a) phosphorus (P), (b) potassium (K), (c) iodine (I), (d) neon (Ne), and (e) aluminum (Al).

23. Write electron dot structures for (a) carbon (C), (b) strontium (Sr), (c) arsenic (As), and (d) krypton (Kr).

24. Which element listed in Table 2.4 does *not* have the same number of valence electrons as its group number (using the A group numbers)?

Topics for Discussion

1. Do you believe atoms really exist? Why or why not? What difference does it make?

2. Criticize the statement that, because smaller subatomic particles exist, the atom is not the fundamental building block of elements and compounds.

CHAPTER 3

Nuclear Chemistry: The Core of Matter

General Objectives

1. What are the three major types of radioactivity emitted by natural sources?

2. How can we use shorthand equations to represent nuclear reactions?

3. How can we measure the lifetimes of radioactive substances?

4. How can nuclear reactions be used to make new elements?

5. How can exposure to radioactive nuclei affect your body?

6. What are some of the beneficial uses of radioactivity?

7. What are nuclear fission and nuclear fusion, and how are they used in nuclear weapons?

You have learned that matter is made of atoms, which in turn are made of even smaller components. You also know that almost all of an atom's mass is crammed into its tiny nucleus, which contains protons and neutrons. Often these nuclear components stay together in a stable arrangement. But sometimes they part company, as you will learn in this chapter.

Unstable nuclei shoot out bits of mass and energy, an action called radioactivity. This and other changes that take place in the nuclei of atoms are *nuclear reactions*. We have learned to harness nuclear reactions to build atomic weapons that can take lives. We have also learned how to use such reactions to save lives and do many other useful things.

3.1 The Nature of Radioactivity: Spontaneous Emissions from Unstable Nuclei

DISCOVERY OF RADIOACTIVITY   One winter day in 1896, Henri Becquerêl, a French physicist, put a sample of uranium rock on top of an unexposed photographic plate in a desk drawer; then he left for a brief vacation. When he returned several days later, he was surprised to find that the film in the drawer had a faint image of the uranium rock. He did not expect this because the plate had been heavily wrapped to prevent accidental exposure.

Becquerel wondered why this happened. He concluded that somehow the uranium rock had emitted some highly energetic rays that penetrated the protective covering and exposed the photographic plate. (This is an example of a discovery by chance, or serendipity.) Marie Curie, a scientist working with Becquerel, suggested that this phenomenon be called **radioactivity**.

During the next few years, Marie Curie and her husband, Pierre, processed and purified several tons of pitchblende, an ore rich in uranium.   59

Working long hours in an abandoned shed, which was hot and smelly in summer and cold and damp in winter, they eventually obtained a few tenths of a gram of two new elements, both radioactive. One was the element polonium (Po), which Marie named for her homeland, Poland. The other was radium (Ra), named from the Latin word for *ray*.

TYPES OF RADIOACTIVITY   Becquerel and other scientists discovered three types of natural radioactive emissions. They were named after the first three letters in the Greek alphabet—*alpha* (α), *beta* (β), and *gamma* (γ). Alpha emissions are the least penetrating and have a positive electrical charge; beta emissions have a negative charge; and gamma rays, the most penetrating of all, have no charge.

These emissions come from the nuclei of atoms that have an unstable combination of protons and neutrons. Such nuclei change their internal structure by emitting one or more of these types of radiation (Figure 3.1). Some isotopes are unstable, and some are not. An isotope that spontaneously emits radiation is called a *radioactive isotope*, or **radioisotope**.

When it emits radiation, a radioisotope changes into a different isotope that is either stable and nonradioactive or one that is unstable. The latter type will emit radiation to form yet another isotope. This process continues until the original radioisotope changes completely into a stable, nonradioactive isotope.

Table 3.1 shows the important features of alpha, beta, and gamma emissions. An **alpha particle**, a package of two neutrons and two protons,

Figure 3.1 The three most common forms of natural radioactive emissions. Some radioisotopes emit only one of these types of radioactivity; others emit more than one type.

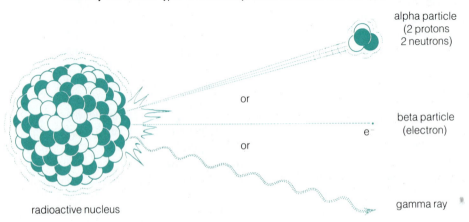

alpha particle
(2 protons
2 neutrons)

or

beta particle
(electron)

e⁻

or

gamma ray

radioactive nucleus

is identical to the nucleus of a helium-4 atom and has a positive charge
(2+) because of its two positively charged protons. It does not have the
two electrons found outside the nucleus of a helium atom. Because alpha
particles have a relatively large charge and size, they cannot travel very
far. They can be stopped by your skin, a thin sheet of aluminum, or several
sheets of paper (Figure 3.2).

TABLE 3.1 CHARACTERISTICS OF ALPHA, BETA, AND GAMMA EMISSIONS

Name and Symbols	Identity	Charge	Approximate Mass (amu)	Velocity	Penetrating Power
Alpha (α, $^4_2$He, $^4_2\alpha$)	Helium-4 nucleus	2+	4	5–10% of the speed of light	Low
Beta (β, $^0_{-1}\beta$, $^0_{-1}e$)	Electron	1−	0	Up to 90% of the speed of light	Low to moderate
Gamma (γ, $^0_0\gamma$)	High-energy radiation similar to X rays	0	0	Speed of light (3×10^{10} cm/s)	High

Figure 3.2 The three major types of natural radioactivity vary considerably in their penetrating power.

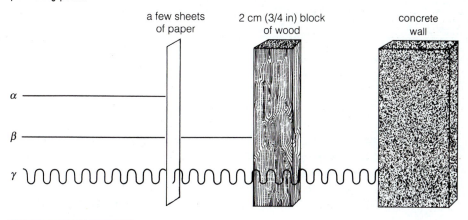

Beta particles are fast-moving electrons with a single negative charge. Because of their smaller charge and higher speeds, they are about 100 times more penetrating than alpha particles. It takes an aluminum plate 0.3 cm ($\frac{1}{8}$-in.) thick or a 2-cm ($\frac{3}{4}$-in.) block of wood to stop them.

Gamma rays are not particles but a form of high-energy radiation very much like X rays. Traveling at the speed of light, they are so penetrating that only thick layers of lead or concrete can stop them.

3.2 Nuclear Reactions:
Representing Nuclear Changes

NUCLEAR EQUATIONS Nuclear reactions are represented in shorthand form by nuclear equations. A *nuclear equation* shows the chemical symbol, mass number (as a superscript), and atomic number (as a subscript) for each reactant and product. The rule for balancing the equation is simple: *The sum of the atomic numbers must be the same on both sides of the equation, as must the sum of the mass numbers.*

ALPHA EMISSIONS An alpha particle's atomic number is 2 (due to the two protons), and its mass number is 4 (the sum of two protons plus two neutrons). Thus, an alpha particle has the symbol $^4_2\alpha$ or ^4_2He. We can represent the reaction in which an atom of uranium-238 emits an alpha particle to become an atom of thorium-234 (an isotope of a different element) by the following nuclear equation:

$$\text{\textit{total mass number}} = 238 \qquad \text{\textit{sum of mass numbers}} = 234 + 4 = 238$$
$$^{238}_{92}\text{U} \longrightarrow {}^{234}_{90}\text{Th} + {}^4_2\text{He}$$
$$\text{\textit{total atomic number}} = 92 \qquad \text{\textit{sum of atomic numbers}} = 90 + 2 = 92$$

Notice that the sum of the superscripts (mass numbers) is 238 on each side of the arrow, and the sum of the subscripts (atomic numbers) is 92 on each side.

BETA EMISSIONS Recall that electrons are found only outside the nuclei of atoms. How then can an unstable nucleus emit a beta particle, which is an electron? Scientists propose that the electron is produced in the nucleus when a neutron changes into a proton. The electron, or beta particle, is then ejected from the nucleus. This change doesn't alter the mass number (because protons and neutrons both count as one), but it increases the atomic number (number of protons) by one.

We can represent the beta particle by the symbol $_{-1}^{0}\beta$ or $_{-1}^{0}e$ (for an electron). We can then use this information to write the balanced nuclear equation for carbon-14 emitting a beta particle:

total mass number $= 14$ *sum of mass numbers* $= 14 + 0 = 14$

$$^{14}_{6}\text{C} \longrightarrow {}^{14}_{7}\text{N} + {}_{-1}^{0}e$$

total atomic number $= 6$ *sum of atomic numbers* $= 7 - 1 = 6$

Notice again that the subscripts and superscripts on both sides of the arrow are balanced. This also shows that emitting a beta particle changes a radioisotope into an element with the next higher atomic number.

GAMMA EMISSIONS   Gamma rays (γ) are not particles; they are a form of energy with no mass and no charge. When a radioisotope emits gamma rays, the isotope's atomic number or mass number does not change. We don't need to write equations for these reactions. Gamma rays also may be emitted along with beta or alpha particles, but we often omit them in the nuclear equations for these nuclear reactions. Gamma emissions remove excess energy from unstable nuclei.

You can also use a balanced nuclear equation to identify the element that forms from radioactive emissions (decay). For example, suppose you know that iodine-131 emits a beta particle (and a gamma ray) to form an unknown isotope. What is this unknown isotope? To find out, write the nuclear equation using the information you know:

$$^{131}_{53}\text{I} \longrightarrow \underline{\quad\quad} \text{(unknown isotope)} + {}_{-1}^{0}e$$

From the rule on balancing nuclear equations, fill in the values for the isotope's atomic number and its mass number:

total mass number $= 131$ *sum of mass numbers* $= 131 + 0 = 131$

$$^{131}_{53}\text{I} \longrightarrow {}^{131}_{54}\underline{\quad} + {}_{-1}^{0}e$$

total atomic number $= 53$ *sum of atomic numbers* $= 54 - 1 = 53$

Then using the periodic table (inside front cover), find the symbol for the unknown element with atomic number 54, which is Xe (xenon). Now write the complete nuclear equation:

$$^{131}_{53}\text{I} \longrightarrow {}^{131}_{54}\text{Xe} + {}_{-1}^{0}e$$

PRACTICE EXERCISE

Write nuclear equations for (a) alpha (and accompanying gamma) emission by plutonium-239, (b) alpha emission by polonium-204, and (c) beta emission by thorium-234.

$$\text{(a)} \quad {}^{94}_{239}\text{Pu} \longrightarrow {}^{2}_{4}\text{He} + {}^{235}_{92}\text{U}$$
$$\text{(b)} \quad {}^{84}_{204}\text{Po} \longrightarrow {}^{2}_{4}\text{He} + {}^{200}_{82}\text{Pb}$$
$$\text{(c)} \quad {}^{90}_{234}\text{Th} \longrightarrow {}^{-1}_{0}e + {}^{234}_{91}\text{Pa}$$

SOLUTION

3.3 Half-Life: How Long Does Radioactivity Last?

SOME RADIOISOTOPES DECAY FASTER THAN OTHERS   If you measure the radioactivity of a 1-g sample of iodine-131, you will find that the number of emissions per minute decreases steadily from day to day. If you make the same measurements on a sample of radioactive carbon-14 with the same number of atoms as your iodine sample, you will find two differences. First, carbon-14 gives off a different number of emissions per minute than does iodine-131. Second, the number of emissions per minute will not decrease as much in a few days as those from the iodine-131.

What accounts for these differences? Why do some radioisotopes give off their emissions—whether they are alpha, beta, or gamma—faster than others? Nuclei of some isotopes are more unstable than others, so they give off their emissions more quickly. Each radioisotope has its own rate of decay.

HALF-LIFE   We can express the rate of decay of a radioisotope in terms of **half-life**—the time needed for *one-half* of the nuclei in a radioisotope to emit its radiation. Each radioisotope has a characteristic half-life, which may range from a few millionths of a second to several billion years. For example, Table 3.2 shows that iodine-131 has a half-life of eight days, whereas the half-life of carbon-14 is 5730 years.

Figure 3.3 shows how a radioactive isotope decays over five half-life periods and changes into a new isotope. Notice that each half-life period reduces the amount of a radioisotope left by 50 percent.

TABLE 3.2 HALF-LIVES OF SELECTED RADIOISOTOPES

Isotope	Symbol	Half-Life	Radiation Emitted
Polonium-218	$^{218}_{84}Po$	3.05 minutes	α, β
Potassium-42	$^{42}_{19}K$	12.4 hours	β, γ
Iodine-131	$^{131}_{53}I$	8 days	β, γ
Cobalt-60	$^{60}_{27}Co$	5.27 years	β, γ
Hydrogen-3 (tritium)	$^{3}_{1}H$	12.5 years	β
Strontium-90	$^{90}_{38}Sr$	28 years	β
Carbon-14	$^{14}_{6}C$	5730 years	β
Plutonium-239	$^{239}_{94}Pu$	24,000 years	α, γ
Uranium-235	$^{235}_{92}U$	710 million years	α, γ
Uranium-238	$^{238}_{92}U$	4.5 billion years	α, γ

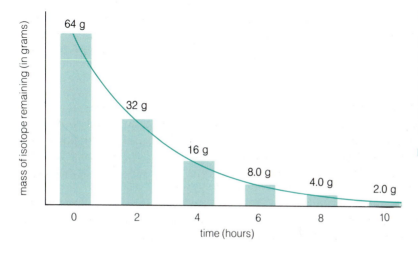

Figure 3.3 Pattern of decay for 64 g of any radioisotope with a half-life of two hours. With each half-life, the amount of the radioisotope decreases by one-half as it decays into a new isotope.

PRACTICE EXERCISE

Suppose you have 96 g of a radioisotope that has a half-life of thirty minutes. How many grams of that isotope would be left after two hours?

After one-half hour, 48 g remain; after one hour, 24 g remain; after one and one-half hours, 12 g remain; after two hours, 6 g remain.

SOLUTION

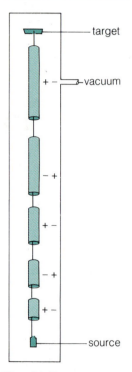

— target

— vacuum

— +

— +

— +

— +

— source

Figure 3.4 Model of a linear accelerator in which charged particles are accelerated as they move through cylinders having alternating electrical charges.

Half-life can be used to estimate how long a sample of a radioisotope must be stored in a safe enclosure before it decays to what is considered a safe level. The general rule is that this takes about ten half-lives. Thus, people would have to be protected from iodine-131, which has a half-life of eight days, for eighty days. Plutonium-239, which is in nuclear reactors and has a half-life of 24,000 years, would have to be stored for 240,000 years.

3.4 Artificial Radioactivity: Making New Isotopes and Elements

New elements and isotopes can be made by nuclear reactions. One way to do this is to use special machines: *cyclotrons* and *linear accelerators* (Figure 3.4). These machines accelerate charged particles such as electrons, protons, alpha particles, or even larger chunks of nuclear material. The accelerated particles are concentrated in a beam and then shot into the nuclei of target atoms. This causes nuclear interactions that produce new isotopes that would not form with slower-moving particles.

Until 1940 uranium was the heaviest known element. Then scientists started making new elements heavier than uranium by bombarding uranium and other heavy elements with nuclear material. For example, uranium-238 was made into uranium-239, which turned out to be a beta emitter:

$$^{239}_{92}\text{U} \longrightarrow {}^{\;\;0}_{-1}e + {}^{239}_{93}\text{Np}$$

And the new element Np was also a beta emitter:

$$^{239}_{93}\text{Np} \longrightarrow {}^{0}_{-1}e + {}^{239}_{94}\text{Pu}$$

The two new elements beyond uranium in the periodic table were named neptunium (Np) and plutonium (Pu) after the two planets in the solar system that lie beyond Uranus. Elements made in this way in a laboratory are called *synthetic elements* because they are not found in nature.

More recently, scientists made element 105. First they made a tiny amount of berkelium (Bk), element 97, and then they bombarded it with accelerated particles of oxygen-18 nuclei:

$$^{249}_{97}\text{Bk} + {}^{18}_{8}\text{O} \longrightarrow {}^{262}_{105}\text{Unp} + 5{}^{1}_{0}n$$

The new material didn't last long; its half-life was only 35 seconds. This new element still doesn't have an official name and symbol. For now, it is called unnilpentium (Unp).

3.5 Harmful Effects of Radioactivity: How Much Is Too Much?

EFFECTS OF RADIOACTIVITY Scientists have slowly learned, the hard way, how radioactivity affects living things. Little was known about this at the turn of this century when Marie Curie worked with radioactive materials in her Paris laboratory. Because of her exposure to radioactivity, she suffered from anemia, had a miscarriage, and eventually died of leukemia.

The dangers of radioactivity also weren't widely recognized in the 1920s when women worked in factories painting radium onto the dials of watches to make them glow in the dark. Many of those women died early from anemia or bone cancer.

The effects of radioactivity on your body depend on the amount and frequency of exposure, the type of radiation, and whether the radiation comes from outside or inside your body. From the outside, alpha emitters are the least dangerous because they cannot penetrate your skin (see Figure 3.2). Beta emissions don't penetrate the skin well, either. Gamma rays are the most dangerous because they easily pass through your skin and into your body.

But suppose you inhale air, drink water, or eat food that has a radioisotope that emits alpha or beta particles. Once inside your body, alpha- and beta-emitting isotopes can damage nearby cells. Alpha emitters are the most dangerous. Alpha particles don't travel as far as beta or gamma emissions, but they are heavy enough to cause considerable damage to nearby cells. The watch-dial painters, for example, ingested small amounts of radium, an alpha emitter, when they put the tips of their brushes on their tongues to make a finer point for painting.

Your tissues vary widely in their sensitivity to radiation. The most damage occurs in areas of rapidly dividing cells, such as bone marrow (where blood cells are made), spleen, gastrointestinal tract, reproductive organs, lymph glands, and developing embryos.

Several different units for measuring radioactivity are in use. In this book, we use a unit called the **rem**. It measures the amount of radioactivity to which 1 g of human tissue is exposed and the estimated amount of biological damage that exposure causes. A *millirem* (mrem) is one-thousandth of a rem.

NATURAL AND HUMAN-CAUSED EXPOSURE TO RADIOACTIVITY   You cannot completely avoid radioactivity. You eat, drink, and breathe small amounts of radioactive materials every day.

Several radioisotopes can concentrate in your body (Table 3.3). Two examples are long-lived strontium-90 and cesium-137. Radioisotopes accumulate in the same places in your body as other members of their group in the periodic table. Strontium-90, for example, can replace calcium in bones because both of these elements are in Group 2/IIA of the periodic table.

Some exposure to radioactivity is unavoidable because it comes from natural sources. This is called *natural* or *background radioactivity*. The typical U.S. resident is exposed to about 130 millirems (0.130 rem) of

TABLE 3.3 SOME RADIOISOTOPES THAT CAN CONCENTRATE IN YOUR BODY			
Radioisotope	Half-Life	Place of Concentration	Radiation Emitted
Iodine-131	8 days	Thyroid	β, γ
Zirconium-95	65 days	Bone	β, γ
Strontium-90	28 years	Bone	β
Cesium-137	30 years	All tissue	β, γ

background radioactivity from natural sources each year (Table 3.4). Your largest exposure to radioactivity comes from radon-222 (Rn-222) gas (see "Chemistry Spotlight").

Each year the average U.S. resident is exposed to about 100 mrem from human activities (see Table 3.4). The largest typical exposure in this category comes from dental and medical X rays. Another important source

TABLE 3.4 AVERAGE ANNUAL EXPOSURE TO RADIOACTIVITY FROM NATURAL SOURCES AND HUMAN ACTIVITIES IN THE UNITED STATES

Source of Exposure to Radioactivity	Approximate Dose in Millirems (1 mrem = 0.001 rem)
Radioactivity from Natural Sources	
Cosmic rays from space: add 1 mrem for each 30.5 m (100 ft) above sea level	40
Radioactive minerals in rocks and soil	55
Radioactivity in the human body from air, water, and food	25
Radioactivity from Human Activities	
Medical and dental X rays and tests: add 22 mrem for each X-ray film and 910 mrem for each whole-mouth dental X-ray film	76
Living or working in a stone or brick structure; add 40 mrem for living and an additional 40 mrem for working in such a structure	—
Smoking a pack of cigarettes a day: add 40 mrem	—
Nuclear weapons fallout	4
Air travel: add 2 mrem for each 2400 km, or 1500 miles, flown	—
TV or computer screens: add 4 mrem for each 2 hr of viewing a day	—
Occupational exposure: varies for uranium ore miners, nuclear power plant personnel, X-ray technicians, and jet plane crews	0.8
Living near a normally operating nuclear or coal-fired power plant: add 0.6–76 mrem	—
Normal operation of nuclear power plants, nuclear fuel reprocessing, and nuclear research facilities	0.06
Miscellaneous: luminous watch dials, smoke detectors, industrial wastes, and so on	2
Average annual exposure to radioactivity per person in the United States = 230 mrem (130 mrem from background radiation and 100 mrem from human activities)	

CHEMISTRY SPOTLIGHT   IS YOUR HOME CONTAMINATED WITH RADIOACTIVE RADON GAS?

Radon-222, a colorless, odorless, naturally occurring radioactive gas, is produced by the radioactive decay of uranium-238. Most soil and rock contain small amounts of uranium-238, but several types of rock have large amounts.

Radon gas produced in most of these deposits percolates up to the soil and is released outdoors. It then disperses quickly in the atmosphere and decays to harmless levels. However, the gas can seep through cracks and drains into houses and buildings built over or near such deposits. This allows the gas to build up to dangerously high levels inside such buildings.

The Environmental Protection Agency (EPA) considers radon-222 gas to be our most dangerous indoor air pollutant. Once inside, radon-222 quickly decays into solid particles of other radioactive elements that can be inhaled into the lungs. Once inhaled, the particles expose lung tissue to a large amount of radioactivity. Repeated exposure over 20–30 years can cause lung cancer. Smokers are especially vulnerable because the radioactive particles tend to adhere to tobacco-tar deposits in the lungs and upper respiratory tract.

A sampling of indoor radon-222 levels in thirty states indicates that at least one of every ten U.S. homes—perhaps as many as one of every five— may have harmful levels of this gas. According to the EPA and the National Research Council, prolonged exposure to radon-222 causes 6000 to 25,000 of the 136,000 lung cancer deaths each year in the United States.

Tests can measure indoor radon levels, and there are ways to deal with the problem if indoor levels are dangerously high. Has the building where you or your loved ones live been tested for radon-222?

is smoking cigarettes, which contain polonium-210 and other radioisotopes. Polonium-210 is an alpha emitter that can damage lung cells and may contribute to the higher risk of lung cancer in smokers.

HOW MUCH EXPOSURE TO RADIOACTIVITY IS HARMFUL?   How much harm does the typical exposure to 230 mrem of radioactivity a year cause? That is a hard question to answer. Scientists continue to study and argue about this issue, but we know what exposure to high-level radioactivity in a short time can cause (Table 3.5). It is very hard, however, to measure the effects of exposure to small amounts of radioactivity over a long time.

We do know that 230 mrem (0.230 rem) a year is a fairly small dose. Also, our bodies can repair some of the harm caused by exposure to radioactivity, which explains why many small doses of radioactivity over a long period of time causes less damage than the same total dosage given all at once. But most scientists agree that there is some risk from exposure to any level of radioactivity. The National Academy of Sciences estimates that the average annual exposure to 230 mrem of radioactivity causes about 1 percent of the fatal cancers and about 5 to 6 percent of the genetic defects in the United States.

TABLE 3.5 EFFECTS ON PEOPLE OF WHOLE-BODY EXPOSURE TO RADIOACTIVITY IN A SHORT PERIOD OF TIME

Dose (rems)	Effects
0–50	No consistent symptoms
50–200	Decreased white blood cells, nausea, vomiting; about 10% die within months at 200 rems
200–400	Loss of blood cells, fever, hemorrhage, hair loss, nausea, vomiting, diarrhea, fatigue, skin blotches; about 20% die within months
400–500	Same symptoms as 200–400 rems but more severe, increased infections due to lack of white blood cells; 50% death rate within months, at 450 rems
500–1000	Severe gastrointestinal damage, cardiovascular collapse, central nervous system damage; doses above 700 rems fatal within a few weeks
10,000	Death in hours

3.6 Useful Applications of Radioisotopes: Tracing Chemicals and Saving Lives

FINDING THE AGE OF OBJECTS   By measuring the relative amounts of a radioisotope and its decay product in an object, scientists can estimate the age of the object. For example, if equal amounts of the radioisotope and its product are present, the radioisotope has gone through one half-life. This means that the object's age is equal to the half-life. If most of the isotope has not yet decayed, the object is much younger than the half-life.

For this method to be accurate, the radioisotope needs to have a half-life not too different from the object's age. For example, if a 2000-year-old sample were dated using a radioisotope with a half-life of two days, not enough radioisotope would be left to measure accurately. Also, such measurements are accurate only if the decay product detected in the sample came entirely from the decay process and none of the decay product has leaked away.

One common method is *radiocarbon dating*. It uses radioactive carbon-14 to find the age of plants, wood, teeth, bone fossils, and other carbon-containing substances from plants or animals. Carbon-14 forms continuously in the atmosphere, and it is incorporated at a steady rate into plant and animal tissues as long as they are alive (Figure 3.5). But after a plant or animal dies, its carbon-14 slowly decays by beta emission, while the rest of its carbon (nonradioactive carbon-12) does not change.

Figure 3.5 How carbon-14 can be used to estimate the age of archaeological artifacts that were once living.

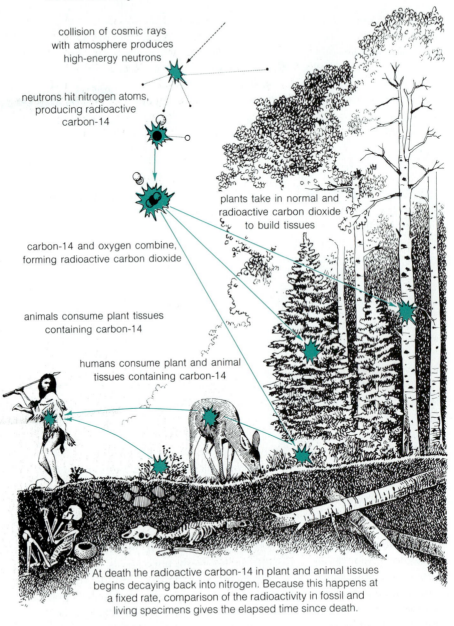

collision of cosmic rays with atmosphere produces high-energy neutrons

neutrons hit nitrogen atoms, producing radioactive carbon-14

carbon-14 and oxygen combine, forming radioactive carbon dioxide

plants take in normal and radioactive carbon dioxide to build tissues

animals consume plant tissues containing carbon-14

humans consume plant and animal tissues containing carbon-14

At death the radioactive carbon-14 in plant and animal tissues begins decaying back into nitrogen. Because this happens at a fixed rate, comparison of the radioactivity in fossil and living specimens gives the elapsed time since death.

Because the half-life of carbon-14 is 5730 years, the amount of carbon-14 will decrease by one-half during that time. Thus, measuring the ratio of radioactive carbon-14 to an object's total carbon gives an estimate of the object's age.

Because of its half-life, carbon-14 dating is limited to objects younger than 50,000 years. Several other methods have been used for older objects. One of them, *potassium-argon dating*, measures the ratio of argon-40 to potassium-40 in an object. Potassium-40, with a half-life of 1.3 billion years, is found in all organisms and decays to the stable isotope argon-40. Uranium-238, with a half-life of 4.5 billion years, can also be used to date very old objects. For example, uranium and potassium-argon dating of very old rocks indicate that the earth is about 4.6 billion years old.

USES OF RADIOACTIVITY IN INDUSTRY AND AGRICULTURE Radioisotopes are used as *tracers* in pollution detection, agriculture, industry, and the study of chemical reactions. Suppose, for example, a leak occurs somewhere in an underground pipeline. How can we locate the leak without digging up the pipeline until we find the leak? One way is to mix a radioisotope with the material being transported in the pipeline. Then using a Geiger counter to scan the ground above the pipeline and finding a place with high-level radioactivity, we can locate the leak.

In agriculture, scientists have exposed seeds to radioactivity to develop new and better genetic strains of wheat, corn, rice, and other crops. Exposure to radioactivity keeps crops from sprouting and can kill insects and bacteria. Such exposure allows some food to be stored for a long time. The U.S. Department of Agriculture has radiated millions of screwworm flies (an insect pest) in laboratories to produce sterile males. Once released into the natural population, these flies outnumber the fertile males. Matings then produce no offspring and help reduce the size of the pest population.

USE OF RADIOISOTOPES IN MEDICINE A special branch of medicine, called *nuclear medicine*, uses radioisotopes for both diagnosis and treatment. For example, physicians inject radioactive sodium-24 into the bloodstream as a salt solution. Then they measure radioactivity released by the radioisotope to trace the flow of blood and detect constrictions or obstructions in the blood vessels. Other radioisotopes are used to detect blockages to certain parts of the body.

Radioactive iodine-131 helps trace such things as liver activity, brain tumors, and possible cancer or malfunction of the thyroid gland. The thyroid gland removes iodine from the bloodstream and uses it to make a necessary hormone. Producing too much or too little of this hormone upsets important body functions. A patient drinks water containing iodine-131, a beta and gamma emitter. Then the physician scans the thyroid area

Figure 3.6 Scans showing the uptake of radioactive iodine-131 in a normal (top), enlarged (center), and cancerous (bottom) thyroid gland. (Courtesy Dr. Joseph Kriss, Stanford University Medical Center)

Figure 3.7 Deep-seated cancerous tumors can be treated with gamma rays emitted by cobalt-60. (Courtesy Varian Associates, Palo Alto, CA)

with a radioactivity detector to see if the iodine is accumulating there at the normal rate (Figure 3.6).

Because cancerous tumors grow very rapidly, they tend to accumulate materials faster than most tissues. Brain tumors, for example, rapidly absorb large amounts of iodine-131, copper-64, technetium-99, and indium-111. Physicians can use such radioisotopes to detect and precisely locate some types of cancer earlier than with other methods.

A risk occurs whenever a person ingests radioactive materials. This risk must be weighed against the benefits of doing the diagnosis. One way to reduce this risk is to use small doses. Another way is to use radioisotopes, with short half-lives, that are not alpha emitters.

Radioactivity can cause cancer, but it is also useful in treating cancer. Rapidly dividing cells, such as cancer cells, are especially vulnerable to radioactivity. Thus, exposing cancerous cells to radioactivity can kill them. However, rapidly growing normal cells—such as intestinal cells and blood cells—will also be killed when exposed to radioactivity. That is why people exposed to radioactivity to kill cancer cells often have nausea, diarrhea, and a low white blood cell count (which reduces their ability to fight infections).

Most treatments use an external source of radiation that can penetrate the skin and reach the tumor (Figure 3.7). The most common sources are gamma emitters, such as cobalt-60 and cesium-137, or X rays. The radiation beam is narrow and focused precisely on the tumor area to minimize the damage to other cells. Very high doses are used to kill the actively dividing cancer cells. Cancer patients typically receive 150 to 200 rems per treatment (a total of about 3000 rems in a month). Some

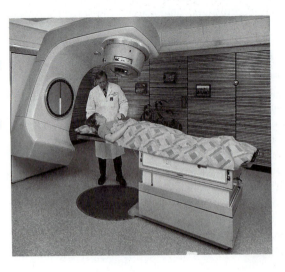

receive 6000 or more rems during a two- to three-month period. By comparing these doses with the information in Table 3.4, you can see why this treatment must be strictly confined to the tumor area.

3.7 Nuclear Fission and Fusion: Splitting and Combining Nuclei

SPLITTING HEAVY NUCLEI: NUCLEAR FISSION   In 1934 Italian scientists Enrico Fermi and Emilio Segrè bombarded uranium atoms with neutrons and were surprised to find several products, including four radioactive isotopes. They weren't able to identify all these products and couldn't explain what had happened.

In 1938, however, German radiochemists Otto Hahn and his student Fritz Strassman repeated this earlier experiment. They bombarded uranium with neutrons and found tiny amounts of barium (Ba), cerium (Ce), and lanthanum (La) in the reaction products. How could barium, which has an atomic number of 56, come from uranium, which has an atomic number of 92? It was very much as if the uranium nucleus had been cut nearly in half.

Hahn wrote a letter about this to Lise Meitner, his Jewish co-worker of thirty years who had fled Germany and was living in Sweden. During Christmas vacation she told her nephew, Otto Frisch, who was an undergraduate physics student at the University of Copenhagen, about the Hahn–Strassman experiment. They calculated that a nucleus would release a great amount of energy if it were to split. Frisch used the term **nuclear fission** to describe a large nucleus fragmenting into two or more smaller nuclei.

NUCLEAR FISSION BOMBS   Frisch told the news of Hahn's discovery to Niels Bohr, who was just leaving for New York. Bohr then passed the word to American scientists. In 1939 Albert Einstein wrote a letter to President Franklin D. Roosevelt saying that the energy released from nuclear fission could be released by extremely powerful bombs, which Germany was already trying to develop. So the United States launched the Manhattan Project, a secret mission to develop the first atomic bomb.

What would be the fuel for a nuclear fission bomb? It was known that uranium-235 and plutonium-239 nuclei could undergo fission when bombarded with neutrons moving at the right speed. A slow-moving neutron entering a uranium-235 nucleus can split the nucleus in several ways, one of which is shown in Figure 3.8. Notice that this nuclear reaction releases several neutrons. If they are not allowed to escape, they can be

Figure 3.8 Typical products of the fission of a uranium-235 nucleus by a slow-moving neutron.

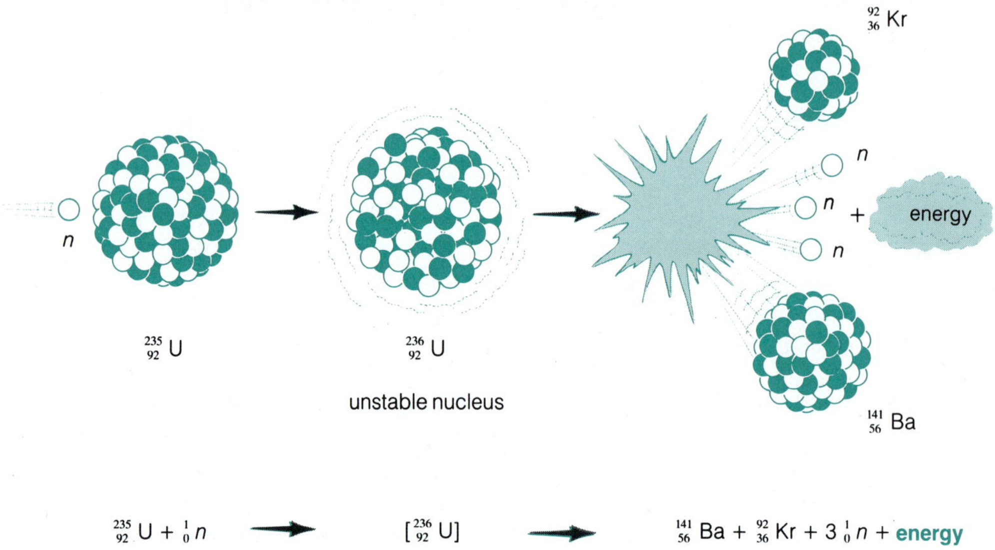

n
$^{235}_{92}U$
$^{236}_{92}U$
unstable nucleus
$^{92}_{36}Kr$
n
n
+ energy
n
$^{141}_{56}Ba$
$^{235}_{92}U + ^{1}_{0}n \longrightarrow [^{236}_{92}U] \longrightarrow ^{141}_{56}Ba + ^{92}_{36}Kr + 3\,^{1}_{0}n$ + energy

used to cause other uranium-235 nuclei to split apart. When this happens, a *chain reaction* rapidly occurs (Figure 3.9); a large number of uranium-235 nuclei split apart almost simultaneously. Together this chain of nuclear fission reactions releases a massive amount of energy. The trick is to figure out a way to start and keep a chain reaction going. Most of the neutrons released by each fission must be kept around so they can split other uranium-235 nuclei.

The minimum amount of fissionable fuel needed to keep a chain reaction going is called the **critical mass**. But getting a critical mass of uranium-235 for a bomb was a problem because natural uranium samples contain less than 1 percent of this isotope. The other 99 percent is uranium-238, which does not undergo fission.

Scientists decided that the solution was to find a way to separate the uranium-235 and uranium-238 in a natural sample of uranium. Then enough fissionable uranium-235 could be mixed with nonfissionable uranium-238 to get the critical mass. During World War II, a plant for separating and concentrating uranium-235 was built at Oak Ridge, Tennessee.

Scientists also knew that plutonium-239, an artificial isotope, could undergo fission and could be made by bombarding nuclei of uranium-238

Figure 3.9 A nuclear chain reaction started by one neutron triggering fission in a single uranium-235 nucleus. This series of nuclear fissions releases an enormous amount of energy in a fraction of a second.

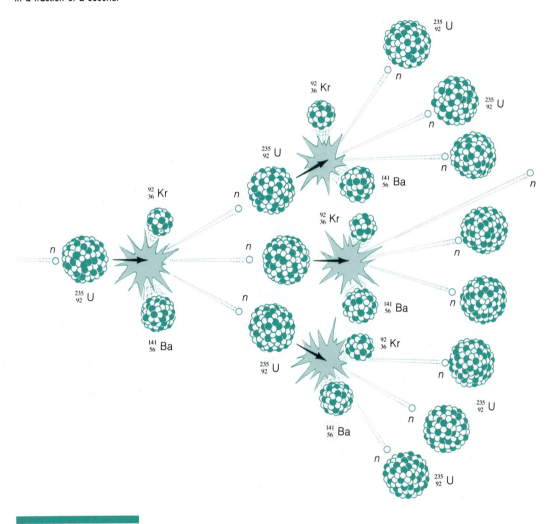

with neutrons. Another plant was built at Hanford, Washington, to produce plutonium-239 for nuclear fission bombs.

The first sustained chain reaction was achieved in 1942 in a test facility under the football stadium at the University of Chicago. The project was led by Enrico Fermi, an Italian physicist who fled from his country in 1938 because his wife was Jewish. The first nuclear fission bomb was tested successfully in New Mexico in July, 1945 (see "Chemistry Spotlight").

Harry Truman, who had just become president, decided to use this new bomb against Japan. Germany, the country that had instigated the research, had already surrendered. Some scientists and others argued that a demonstration of the bomb over an unpopulated area might persuade Japan to surrender. But Truman feared that a demonstration might not be successful. He believed that using this new bomb would make Japan surrender and thus save the lives of many U.S. and allied soldiers.

On August 6, 1945, a uranium-235 fission bomb was dropped on Hiroshima, Japan (Figure 3.10). On August 9, a plutonium-239 fission bomb was dropped on Nagasaki. These bombs killed 110,000 people instantly, and within several months, another 100,000 people died from injuries and radiation exposure. A few days after the bombs were dropped, Japan surrendered.

CHEMISTRY SPOTLIGHT   J. ROBERT OPPENHEIMER AND THE MANHATTAN PROJECT

At age 38, J. Robert Oppenheimer was one of the world's most brilliant physicists. During World War II, the U.S. government asked him to leave his teaching job at the University of California at Berkeley to direct the Los Alamos (New Mexico) Laboratory; his job was to develop the first atomic bomb. The code name for this top-secret operation was the Manhattan Project.

Under Oppenheimer's direction, the project was completed in less than three years. The first atomic bomb, detonated at a test site near Alamogordo, New Mexico, filled Oppenheimer and other onlookers with awe. The massive outburst of heat and light caused him to think of a line from the Hindu *Bhagavad-Gita*: "I am become death, the shatterer of worlds."

Less than a month later, the two atomic bombs that were dropped on Hiroshima and Nagasaki brought the war to an end. After the war, Oppenheimer became chairman of the General Advisory Committee to the newly formed Atomic Energy Commission (AEC).

He had reservations, however, about the United States embarking on another crash project to develop even more powerful nuclear fusion (hydrogen) bombs. This stance made him politically unpopular. Anticommunist fever was running high in the early 1950s when a security hearing was held to assess Oppenheimer's loyalty. His prewar associations with communists and communist sympathizers were linked to his reluctance to embrace new nuclear weapons. In 1954 the AEC voted four to one to deny him security clearance. The person most responsible for helping the United States develop the atomic bomb died in 1967, untrusted by his country.

Figure 3.10 A nuclear fission bomb of the type exploded over Hiroshima. The bomb is 3 m long, weighs 4000 kg, and has the explosive power equal to 17 million kg (20,000 tons) of TNT. (National Atomic Museum)

After World War II, scientists learned how to control the rate of nuclear fission. Controlled fission is now used to produce electricity at nuclear power plants. Here the uranium-235 nuclear fuel is not concentrated enough to have the critical mass needed for a nuclear explosion, as discussed in Section 11.4.

HOW DOES NUCLEAR FISSION RELEASE SO MUCH ENERGY?   Where does the energy come from when a nucleus splits? The answer is that some of the mass in the nucleus is converted into energy. This energy is released when an unstable nucleus, like that of uranium-235, splits apart and forms smaller nuclei that are more stable.

In 1905 Albert Einstein gave us the key to understanding what happens in nuclear fission. In his special theory of relativity, Einstein proposed that energy and mass are two aspects of the same thing. He expressed this relationship in the equation

$$E = mc^2$$

where E represents energy, m represents mass, and c represents the speed of light. Because c^2 is a very large number, even a small amount of mass (multiplied by c^2) changes into a very large amount of energy.* For example, the fission of 1 g of uranium produces about 10 million times as much energy as burning 1 g of coal.

---

* This discovery caused scientists to modify the *law of conservation of mass* (p. 26). It is now the *law of conservation of mass–energy*, which states that the total amount of mass and energy stays constant, though mass and energy may be interconverted. For ordinary chemical reactions, however, the law of conservation of mass still works.

Figure 3.11 Effect of nuclear size on nuclear energy and stability. Isotopes of certain elements release energy during nuclear fission or nuclear fusion when they form nuclei that are lower in energy and more stable.

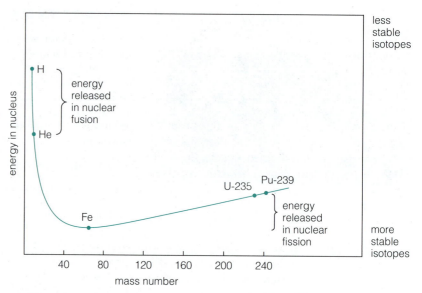

Scientists have discovered that the energy holding a nucleus together varies with the size of the nucleus (Figure 3.11). A nucleus with a lower internal energy is more stable than one with a higher internal energy. You can see from Figure 3.11 that nuclei about the size of iron have the lowest energy and are therefore the most stable. A large nucleus, such as uranium-235, is not as stable as nuclei of atoms about half its size and mass.

If the uranium-235 nucleus is struck by a neutron moving at the right speed, the nucleus can split apart and form more stable nuclei with lower internal energy (see Figure 3.11). When this happens, some of the mass of the particles in the uranium-235 nucleus is converted into energy and released. If this happens with a critical mass of uranium-235, a chain reaction occurs and releases a tremendous amount of energy in a short time (see Figure 3.9).

JOINING LIGHT NUCLEI: NUCLEAR FUSION   Scientists have also learned how to carry out another type of nuclear reaction, one in which two nuclei with low mass numbers unite, or fuse, to form a larger nucleus. It is called **nuclear fusion**. When this happens, an enormous amount of energy is released as some of the mass in the nuclei involved changes into energy (see Figure 3.11).

Because very high temperatures (10 to 100 million °C) are needed to get the positively charged nuclei to join together, nuclear fusion is harder to initiate than nuclear fission. But once initiated, nuclear fusion releases far more energy per gram of fuel than does nuclear fission.

Nuclear fusion is the major source of energy in the universe. The heat and light streaming out from the sun is produced as some of the sun's mass, which is mostly hydrogen, changes into energy during the nuclear fusion reaction:

$$\,^1_1H + \,^1_1H + \,^1_1H + \,^1_1H \longrightarrow \,^4_2He + \,^0_1e + \,^0_1e + \textbf{energy}$$

The two $\,^0_1e$ particles produced by this fusion reaction are *positrons*. Each has the same mass as an electron but has a positive charge.

Scientists have estimated that in the nuclear reaction above, 4 to 5 billion kg of the sun's mass becomes energy every second. Without this input of energy from the sun, we and all other earthly life could not exist.

NUCLEAR FUSION BOMBS After World War II, scientists developed extremely powerful *hydrogen*, or *thermonuclear*, *bombs* (Figure 3.12). The hydrogen bomb is based on the nuclear fusion of hydrogen-2 (*deuterium*) and hydrogen-3 (*tritium*):

Figure 3.12 Explosion of a hydrogen bomb involves nuclear fusion of two isotopes of hydrogen. (National Atomic Museum)

$$\,^2_1H \ + \ \,^3_1H \ \xrightarrow{\substack{\textit{extremely high} \\ \textit{temperature} \\ \textit{(millions of degrees)}}} \ \,^4_2He \ + \ \,^1_0n \ + \textbf{energy}$$

hydrogen-2   *hydrogen-3*   *helium-4*   *neutron*

In the hydrogen bomb, the fusion material is sandwiched between two layers of fissionable material (usually uranium-235), which on explosion generates the high temperature necessary for fusion. The fuel for fusion comes from water. Hydrogen atoms in natural water are mostly the isotope hydrogen-1, but a small amount is hydrogen-2 (deuterium) or hydrogen-3 (tritium). Scientists learned how to separate and concentrate these two isotopes from ordinary water to produce the fuel used in hydrogen bombs. Water enriched in these heavy isotopes of hydrogen is called "heavy water."

As hydrogen bombs were tested, however, their radioactive products (mostly from the accompanying nuclear fission) were released into the environment. Radioisotopes with fairly long half-lives, such as cesium-137 and strontium-90, became a health hazard.

In 1954 American chemist Linus Pauling (Figure 3.13) had received the Nobel Prize in chemistry for his work explaining how atoms in molecules are held together (chemical bonding). In the late 1950s, he became concerned about the harmful effects of radioactive fallout from testing nuclear weapons in the atmosphere. He and others led an international movement to ban the testing of these weapons above ground. In 1963 the world's major countries with nuclear power—except France, China, and India—signed a treaty banning above-ground tests of nuclear weapons. For this work, Pauling received the Nobel Peace Prize in 1962.

Figure 3.13 Linus Pauling, one of America's greatest chemists. (Linus Pauling Institute)

Now scientists are trying to develop nuclear fusion reactors to produce electricity for the coming decades, as discussed in Section 11.4.

NUCLEAR WAR   Today there are enough nuclear bombs and warheads to kill sixty-seven times the number of people on earth. According to the World Health Organization, a nuclear war involving about one-third of U.S. and Soviet nuclear arsenals would probably kill about 2.1 billion people. In the United States, at least 165 million people would be killed; another 60 million would suffer from some combination of trauma, burns, and radiation sickness.

Calculations made since 1982 indicate that even a limited nuclear war could kill 2 to 4 billion people—40 to 80 percent of the world's population. This could happen if less than 1 percent of the nuclear arsenals of the United States and the Soviet Union were detonated.

The explosions would immediately kill about 1 billion people in the Northern Hemisphere. Then the soot, dust, and smoke from these explosions could keep 50 to 90 percent of the sunlight from reaching most of the earth's surface for weeks or months. Temperatures would drop and plants—including food crops—would stop growing. This effect is called a *nuclear winter* or a *nuclear autumn*, depending on how much the temperatures would drop.

Experts disagree on how severe a nuclear winter or autumn would be and how much destruction it would cause. But they agree on one point: The risk of nuclear war is one of the most crucial issues that the human race must resolve.

GEORGE BERNARD SHAW   Indifference is the essence of inhumanity.

Summary

Unstable nuclei of isotopes called radioisotopes emit one or more types of radioactivity as alpha particles, beta particles, or gamma rays (see Table 3.1). Each radioisotope has a certain half-life: the time it takes for half of its nuclei to emit radioactivity and change into a different isotope.

These and other types of nuclear reactions can be represented by nuclear equations. Nuclear reactions can produce isotopes of new elements and new isotopes of natural elements. Some of these new isotopes may also be radioactive.

Everyone is exposed to small amounts of radioactivity from natural sources and human activities. Even this small amount of radioactivity damages cells, particularly those that divide rapidly. Gamma rays pass through the body and can damage cells. Alpha and beta particles cannot pass through skin. But if they enter the body in air, water, or food, they can damage internal cells and cause genetic defects and cancers.

Radioisotopes are used to estimate the age of objects, sterilize pests, produce new genetic strains of crops, and help trace various substances. In nuclear medicine, radioisotopes are used to diagnose diseases and treat cancer.

Nuclear fission is the fragmenting of large nuclei into smaller nuclei, with the release of energy. Atomic bombs and nuclear reactors are based on nuclear fission. Nuclear fusion is the joining of small nuclei (mostly hydrogen) into larger nuclei, with the release of energy. This is the basis for solar energy and the hydrogen bomb.

Terms for Review

After completing this chapter, you should know and understand the meaning of the following terms:

alpha particle (p. 60)

beta particle (p. 62)

critical mass (p. 76)

gamma ray (p. 62)

half-life (p. 64)

nuclear fission (p. 75)

nuclear fusion (p. 80)

radioactivity (p. 59)

radioisotope (p. 60)

rem (p. 68)

Questions

Odd-numbered questions are answered at the back of this book.

1. Gamma rays are more penetrating than alpha particles, yet alpha particles cause more biological damage. Explain.

2. How do alpha particles, beta particles, and gamma rays differ in their mass and electrical charge (sign and amount)?

3. Write or complete the following equations for nuclear reactions:
   a. $^{40}_{19}K \longrightarrow {}^{40}_{20}Ca + \underline{\quad}$
   b. $^{27}_{13}Al + \underline{\quad} \longrightarrow {}^{28}_{13}Al + {}^{1}_{1}H$
   c. Beta emission from uranium-239
   d. Alpha emission from uranium-233

e. $^{85}_{36}Kr \longrightarrow$ _____ $+ ^{0}_{-1}e$

f. _____ $\longrightarrow ^{4}_{2}He + ^{222}_{86}Rn$

g. $^{14}_{7}N + ^{4}_{2}He \longrightarrow$ _____ $+ ^{1}_{1}H$

h. Fusion of two deuterium nuclei to produce an alpha particle

4. Write the nuclear reaction for a radio-isotope that would produce gold-197 by emitting a beta particle.

5. If scientists tried to synthesize element 114 from californium (Cf), they would need to accelerate nuclei of which element to bombard the Cf?

6. Complete the following equations:

a. $^{4}_{2}He + ^{27}_{13}Al \longrightarrow ^{30}_{15}P +$ _____

b. _____ $\longrightarrow ^{0}_{-1}e + ^{131}_{54}Xe$

c. $^{237}_{95}Am \longrightarrow ^{4}_{2}He +$ _____

7. From the data in Table 3.4, compare the dose of radioactivity from (a) living near a normally operating nuclear power plant, (b) having a whole-mouth dental X ray, and (c) living at 1500 m (5000 ft) above sea level.

8. What is background radioactivity? List in decreasing order your probable exposures to natural and human-induced sources of radioactivity this year.

9. Explain why radioactivity is useful in treating cancer. Why isn't it just as damaging to normal cells as it is to cancer cells?

10. Explain the following:

a. Why can't a natural sample of uranium-235 sustain a chain reaction, but a sample in which the concentration of uranium-235 has been increased can?

b. How might radioactivity be used to determine whether a supposed Rembrandt painting is a forgery?

c. Why is plutonium-239 one of the most deadly substances on earth, but you can hold it for a brief period in a gloved hand without harm?

d. How might radioactivity be used to find a leak in an underground pipeline?

e. How might radioactivity be used to determine whether you have a blood clot in your leg?

f. Why can a cyclotron accelerate a proton, but not a neutron?

11. A radioactive substance you ingest for medical diagnosis should *not* emit _____, and it should have a _____ (short or long) half-life.

12. Would you use radiocarbon or potassium–argon dating to estimate the age of an object suspected of being 800,000 years old? Why?

13. Instruments have been developed to measure, or count, radioactive emissions. One gram of carbon in your body has enough radioactive carbon-14 to emit sixteen counts of radioactivity per minute while you are alive. If you died in the year 2050, in what year would your remains emit two counts per minute per gram of carbon?

14. Protactinium-234 has a half-life of 1 minute. How long would it take for an 8-g sample of this isotope to decay to 1 g of protactinium-234?

15. Look at the shape of the curve in Figure 3.3. Suppose that 50 percent of a radioisotope decays in two hours. How long would it take for 25 percent of that radioisotope to decay (a) less than one hour, (b) one hour, or (c) more than one hour?

16. What is nuclear fission? What two isotopes are usually used as fuel for nuclear fission? Why don't other isotopes work?

17. Why is energy released both in nuclear fission and in nuclear fusion?

18. Why is bone a common place for radio-active strontium-90 to accumulate?

19. What is nuclear fusion?

20. Compare nuclear fusion with nuclear fission in terms of (a) the fuel used and (b) amount of energy released per gram of fuel.

Topics for Discussion

1. Was President Harry Truman's decision to drop the atomic bombs on Hiroshima and Nagasaki justified? Explain.

2. What restrictions, if any, would you place on using radioactivity to sterilize food? Why? What additional information, if any, do you need to make a decision? (See the article by Weiss in "Further Readings" for more information.)

3. Should you be required to give informed consent before a radioactive material is put in your body for medical diagnosis? Why?

4. There are complex arguments between two opposing groups in the United States. One favors a freeze or reduction in nuclear weapons. The other says that this would put the nation at a strategic disadvantage and increase the risk of nuclear war. What do you think about this? Why?

CHAPTER 4

Chemical Bonding: Holding Matter Together

General Objectives

1. Why do atoms bond together to form compounds?

2. What is ionic bonding, and how do we write the chemical formulas and names for compounds held together by ionic bonds?

3. What is covalent bonding, and how do we write the formulas and names for simple covalent compounds?

4. What are the two major types of covalent compounds?

5. What is metallic bonding?

Look around. How many things can you see that are pure elements? Very few, if any. Most of the materials you see are made from combinations of elements. They are *compounds* such as sodium chloride and water.

When atoms of different elements combine to form compounds, something keeps the atoms together in the new arrangement. Chemists call those attractive forces **chemical bonds**. The compounds that form have properties very different from the original elements. For example, sodium, a soft, silvery metal that reacts violently with water, combines with chlorine, a greenish yellow, poisonous gas, to form sodium chloride, a white solid—salt, which you sprinkle on food.

Recall that chemistry is the study of matter, its properties, the changes matter undergoes, and the accompanying energy changes (Chapter 1). To understand the properties of matter and how matter changes, you need to learn how atoms bond together to form new compounds. This will help you understand what chemical formulas mean and why the formula for water is H_2O. It will also help explain the properties of compounds— why, for example, water is nothing like the colorless and odorless hydrogen and oxygen gases from which it forms.

In this chapter, we look at the different types of chemical bonding (called ionic, covalent, and metallic bonding). You will learn how to write chemical formulas and how to name compounds. In the next chapter, you will learn more about how to predict the properties of compounds.

4.1 Chemical Bonds and the Octet Rule: How Do Compounds Form?

THE MINIMUM POTENTIAL ENERGY PRINCIPLE Why do compounds such as sodium chloride and water form? The answer lies in understanding the 87 electron structure of atoms (Section 2.6) and the principle of minimum potential energy. Recall that potential energy is the energy of position (Section 1.3). One form, chemical energy, comes from the forces holding the internal particles of matter together. This is the type of potential energy we examine in this chapter.

According to the **minimum potential energy principle**, any object spontaneously tends toward the lowest state of potential energy available to it. Once an object reaches a lower potential energy state, it is less likely to change further unless it can achieve an even lower potential energy. In other words, a lower potential energy state is a more stable state.

When we apply this principle to atoms, the question becomes: How do atoms achieve a more stable state—that is, a state of lower potential energy?

THE OCTET RULE: TWO AND EIGHT AS MAGIC NUMBERS   How can we predict when two or more atoms will be able to lower their potential energy by interacting? When two atoms come close together, the regions that first meet are those containing (with a high probability) their outermost, valence electrons. It turns out that we can understand a great deal about bonding by focusing on the interactions of those valence electrons. The question then becomes: How can we know what valence-electron arrangement is the most stable?

The periodic table of the elements can help answer this question. Recall that elements with similar chemical properties are in the same vertical column (group) and have the same number of valence electrons (Section 2.6). Also remember that the number of valence electrons in each atom of an A group element is equal to its Roman-numeral group number in the periodic table. Those valence electrons are represented by electron dot formulas (see Table 2.5).

During the early 1900s, chemists observed that while atoms of most elements react and readily form compounds, atoms of the noble gas elements in Group VIIIA (18) of the periodic table, such as helium and neon, have little tendency to form compounds. Could it be that those atoms don't react because they already have stable electron configurations, with either two (He) or eight (Ne, Ar, Kr, Xe, Rn) valence electrons?

Using this information, several researchers proposed the octet rule to predict how atoms tend to form compounds (see the "Chemistry Spotlight"). According to the **octet rule**, *atoms of the A group elements tend to lose, gain, or share one or more of their valence electrons in order to achieve the electron configuration of the nearest noble gas in the periodic table.*

For most atoms, this rule means that they will tend to change their electron structure to end up with eight valence electrons. For a few very

CHEMISTRY SPOTLIGHT   THE LEWIS–LANGMUIR RIVALRY OVER THE OCTET RULE

Figure 4.1 American chemists G. N. Lewis (top) and Irving Langmuir (above) made great contributions to understanding the principles of atomic structure and chemical bonding. (Lewis photo courtesy of John Hagemeyer, Bancroft Library, courtesy of American Institute of Physics Niels Bohr Library; Langmuir photo courtesy of General Electric Company)

For any major scientific discovery, history usually singles out one or two people as responsible for the critical contribution, no matter how much they may have benefited from other peoples' work.

In the years following World War I, at least four scientists could claim credit for the octet rule: two Germans, Richard Abegg and Walter Kossel, and two Americans, Gilbert Newton Lewis and Irving Langmuir (Figure 4.1). A campaign of egos began. But postwar rivalries between nationalities, though seemingly natural, never materialized: It was the two Americans who squared off.

A professor of chemistry at the University of California, G. N. Lewis was a dynamic character. A friend said that "in any company he [was] the focus of the liveliest discussion and the center of the merriest group." While lecturing, he would "pace back and forth across the platform, covering miles at breakneck speed." Thus, when Lewis published his 1916 paper on the "rule of eight" and the notion of the chemical bond as a shared

pair of electrons, he let people know those were *his* ideas.

Irving Langmuir, a chemist at General Electric, was every bit as charismatic. One colleague described him as "the most convincing lecturer I ever heard." Writing and lecturing throughout the early 1920s, Langmuir extended, refined, and popularized Lewis's work. He had a talent for catchy phrases, and soon the scientific community was referring to Langmuir's "octet rule" instead of Lewis's "rule of eight" and speaking of "covalent bonds" (Section 4.3) rather than "shared-electron-pair bonds." Many were even calling the whole concept *Langmuir's* theory of valence.

Lewis was delighted with his theory's widespread acceptance but upset by the new terminology. He thought that connecting Langmuir's name to his brainchild was inexcusable. He wryly allowed that "sometimes parents show a singular infelicity in naming their children, but on the whole they seem to enjoy having the privilege." Langumir always acknowl-

edged his debt to Lewis, but he also felt he had added enough to deserve equal credit.

The controversy never became bitter, and in the long run it did not matter: History has remembered them both.

light elements—mainly, H, Li, and Be—the most stable structure is two valence electrons, like helium. Now let us see how atoms do this.

4.2 Ionic Bonds: Giving and Taking Electrons

THE IONIC BOND: ELECTRON TRANSFER   Certain combinations of atoms can become more stable by transferring electrons among themselves. This usually happens in reactions between metals and nonmetals.

Atoms of metallic elements, such as Na· and Mg:, tend to *lose* their few valence electrons to achieve the stable octet structure of the nearest noble gas atom (Figure 4.2). Sodium, for example, loses its one valence electron, whereas magnesium loses two.

When atoms lose electrons, they no longer have the same number of electrons as protons in the nucleus. They acquire a net positive electrical charge, equal to the number of negatively charged electrons they lost. Because they are no longer electrically neutral, they are no longer atoms; they become **ions**, charged particles that form when atoms gain or lose electrons.

Metallic elements in Groups IA, IIA, and IIIA (1, 2, and 13) tend to form stable ions with $1+$, $2+$, and $3+$ charges, respectively (Figure 4.3). Transition metal atoms, which are in the B groups (3 to 12), are more variable, but most can form ions with a charge of $2+$.

Atoms of nonmetal elements are mostly in Groups VA, VIA, VIIA, and VIIIA (15 to 18) and have a relatively large number of valence electrons. The noble gases don't react because they already have a stable valence-electron structure. The other nonmetal atoms, such as :F̈· and

Figure 4.2 The tendency of metal atoms to lose valence electrons to form positively charged ions having eight valence electrons, like the nearest noble gas atom.

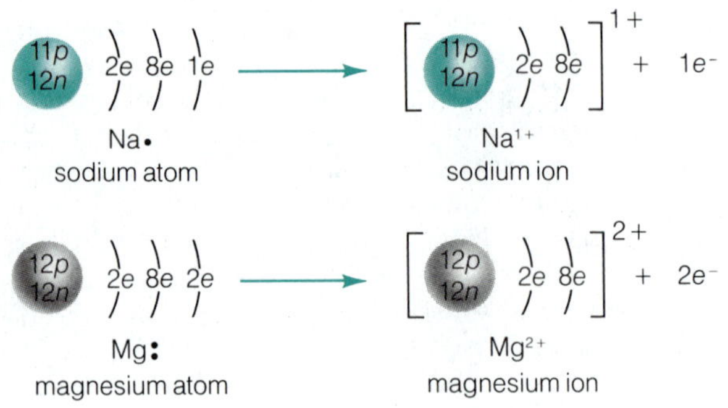

:Ö⁻, tend to *gain* electrons to form negatively charged ions with eight valence electrons (Figure 4.4). Atoms of elements in Groups VA, VIA, and VIIA (15, 16, and 17) tend to form stable ions with $3-$, $2-$, and $1-$ charges, respectively (see Figure 4.3). In addition, the names of these ions end in the suffix *-ide* (Table 4.1).

Figure 4.3 Common stable ions.

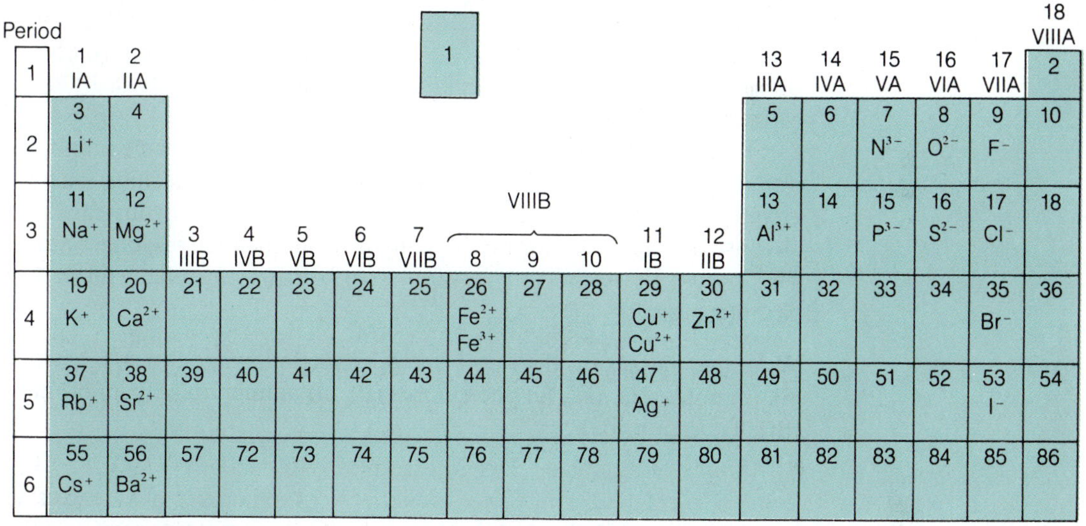

Period	1 IA	2 IIA						VIIIB			11 IB	12 IIB	13 IIIA	14 IVA	15 VA	16 VIA	17 VIIA	18 VIIIA
									1									2
1																		
2	3 Li⁺	4											5	6	7 N³⁻	8 O²⁻	9 F⁻	10
3	11 Na⁺	12 Mg²⁺	3 IIIB	4 IVB	5 VB	6 VIB	7 VIIB	8	9	10			13 Al³⁺	14	15 P³⁻	16 S²⁻	17 Cl⁻	18
4	19 K⁺	20 Ca²⁺	21	22	23	24	25	26 Fe²⁺ Fe³⁺	27	28	29 Cu⁺ Cu²⁺	30 Zn²⁺	31	32	33	34	35 Br⁻	36
5	37 Rb⁺	38 Sr²⁺	39	40	41	42	43	44	45	46	47 Ag⁺	48	49	50	51	52	53 I⁻	54
6	55 Cs⁺	56 Ba²⁺	57	72	73	74	75	76	77	78	79	80	81	82	83	84	85	86

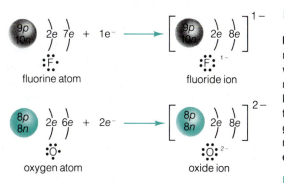

Figure 4.4 The tendency of nonmetal atoms to gain valence electrons to form negatively charged ions having eight valence electrons, like the nearest noble gas atom. Notice that the names of these negative ions end in *-ide* (see Table 4.1).

TABLE 4.1 NAME OF NEGATIVELY CHARGED IONS
OF SELECTED NONMETAL ELEMENTS

Name of Element with Stem in Boldface Type	Symbol of Element	Name of Ion (Add -ide to the Stem)	Symbol of Ion
Fluorine	F	Fluoride	F^-
Chlorine	Cl	Chloride	Cl^-
Bromine	Br	Bromide	Br^-
Iodine	I	Iodide	I^-
Oxygen	O	Oxide	O^{2-}
Sulfur	S	Sulfide	S^{2-}
Nitrogen	N	Nitride	N^{3-}
Phosphorus	P	Phosphide	P^{3-}

PRACTICE EXERCISE

Write the charges on the ions formed by the following atoms: (a) sulfur
(S), (b) potassium (K), (c) phosphorus (P), (d) aluminum (Al), (e) bromine
(Br), (f) barium (Ba).

SOLUTION

Consult the periodic table to obtain the group number of each ele-
ment or refer to Figure 4.3. (a) S^{2-}, (b) K^+, (c) P^{3-}, (d) Al^{3+}, (e) Br^-,
(f) Ba^{2+}.

When one or more electrons are transferred from a metal atom (electron donor) to a nonmetal atom (electron acceptor), a positively charged
metal ion and a negatively charged nonmetal ion form. These ions are
more stable than the original atoms, and the potential energy is lowered
further by the mutual attraction of the oppositely charged ions. This attractive force is known as an **ionic bond**, and the substances that form in
this way are **ionic compounds**.

A sodium atom (Na), for example, can transfer its valence electron to
a chlorine atom (Cl) to form a positively charged sodium ion (Na^+) and
a negatively charged chloride ion (Cl^-), enabling each to attain a stable

C H E M I S T R Y   SALT—A SPICE
S P O T L I G H T   OR VICE OF LIFE?

Sodium chloride (NaCl), often called salt, is a fascinating and vital compound. If you have too little salt in your body, you die. Indeed, getting enough salt to stay alive has been important in human history. In Roman times, soldiers received part of their pay in *salarium*—the Latin word for salt money and also the root for the word *salary*. This accounts for the phrase, "lazy people are not worth their salt."

To stay alive, you must have about 300 g (2/3 lb) of salt in your body. You need to take in about 0.5 g of salt a day to replace what you lose in blood, sweat, tears, and urine. The average American, however, takes in about 11 g of salt a day. Most of this excess comes from salting food, eating salty processed foods, drinking water, and using nonprescription drugs such as aspirin, antacids, laxatives, and sleeping aids. People taking in more than about 2 g of salt a day over a long period can suffer from hypertension (high blood pressure), which is a major cause of heart disease and strokes.

Salt is one of the most widely used chemicals in industry. It is involved in producing or processing most of the things we use, including dyes, cloth, drugs, plastics, tires, leather, soap, brass, steel, aluminum, detergents, varnish, toothpaste, wallpaper, wood, and many processed foods. For centuries salt has been used as a food preservative to prevent or slow the growth of undesirable bacteria. Salt is added to other foods—such as sauerkraut, pickles, fermented cheeses, and sausages—to promote the growth of bacteria in the fermentation process. Salt is also spread on highways and sidewalks to help melt ice and snow by forming a salt and water mixture that has a lower freezing point than pure water.

Thus, we see that salt can be a beneficial spice of life if used properly and a harmful vice of life if used improperly.

noble gas electron structure:

metal atom + nonmetal atom ⟶ positive metal ion + negative nonmetal ion

$$\text{Na·} \quad + \quad \text{:C̈l·} \quad \longrightarrow \quad \text{Na}^+ \quad + \quad \text{:C̈l:}^- \ (\text{or NaCl})$$

The oppositely charged ions formed by this *electron transfer reaction* are then attracted to each other by an ionic bond. They form a **formula unit**, which consists of the simplest ratio of oppositely charged ions that has no net electrical charge. For the compound of sodium and chlorine, the simplest electrically neutral formula unit consists of one sodium ion (Na^+) and one chloride ion (Cl^-). The formula is written as Na_1Cl_1, or more simply as NaCl. (To simplify matters, chemists do not write subscripts when they are 1.)

The oppositely charged ions in each formula unit attract each other by ionic bonds. Large numbers of formula units link together to form a highly ordered, three-dimensional array called an **ionic crystal lattice**, as shown for NaCl in Figure 4.5. In such an ionic crystal lattice, each ion attracts all the oppositely charged ions around it.

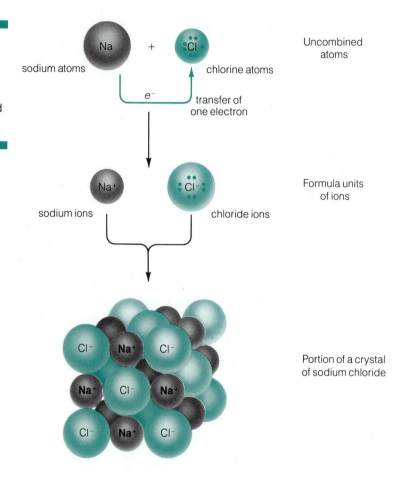

Figure 4.5 A solid ionic crystal lattice of sodium chloride consisting of a three-dimensional array of oppositely charged ions held together by ionic bonds.

So the next time you look at a crystal of table salt (sodium chloride), imagine its underlying structure—a gigantic array of sodium and chloride ions, all packed around each other and held together by ionic bonds.

Because of their strong electrical forces of attraction, ionic compounds are solids at room temperature and have relatively high melting points. In the liquid (but not solid) state, they are good conductors of electricity because their positive and negative ions are free to move and transfer electrical charges.

WRITING FORMULAS FOR SIMPLE IONIC COMPOUNDS   The formula for an ionic compound shows the number of each ion in a formula unit. The numbers are written as subscripts, omitting subscripts where the number is 1. In the formula, the symbol for the metallic element comes first.

Because an ionic compound has no net charge, you can predict its formula by adding the appropriate number of + and − ions to give a

TABLE 4.2 FORMULAS AND FORMULA UNITS FOR SEVERAL IONIC COMPOUNDS, ILLUSTRATING THE RULE OF ELECTRICAL NEUTRALITY			
Name	**Formula**	**Formula Unit**	**Balancing of Charges**
Sodium chloride	NaCl	Na^+, Cl^-	$1+$ and $1- = 0$
Calcium chloride	$CaCl_2$	$Ca^{2+}, 2Cl^-$	$2+$ and $2- = 0$
Sodium oxide	Na_2O	$2Na^+, O^{2-}$	$2+$ and $2- = 0$
Calcium oxide	CaO	Ca^{2+}, O^{2-}	$2+$ and $2- = 0$
Aluminum oxide	Al_2O_3	$2Al^{3+}, 3O^{2-}$	$6+$ and $6- = 0$

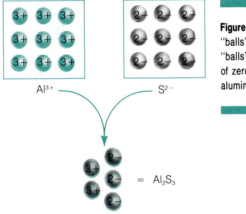

Al³⁺ — S²⁻

= Al₂S₃

Figure 4.6 Two aluminum "balls" and three sulfide "balls" have a net charge of zero. The formula for aluminum sulfide is Al_2S_3.

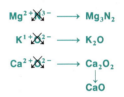

$Mg^{2+} \quad N^{3-} \longrightarrow Mg_3N_2$

$K^{1+} \quad O^{2-} \longrightarrow K_2O$

$Ca^{2+} \quad O^{2-} \longrightarrow Ca_2O_2$
$\downarrow$
CaO

Figure 4.7 The crisscross method for predicting formulas of ionic compounds.

net charge of 0, as shown in Table 4.2. This is known as the *rule of electrical neutrality*.

One way to do this is shown in Figure 4.6. Think of each type of ion as a ball with the appropriate electrical charge. Imagine two boxes, each holding one type of ball. To predict the formula, you need to figure out the smallest number of balls from each box that can be combined to have a net charge of 0; that combination is the formula. For example, it takes two aluminum ions (each having a 3+ charge) and three sulfide ions (each having a charge of 2−) to have a net charge of 0. The formula, then, is Al_2S_3.

You also can predict formulas by crisscrossing the numerical values of the two charges so that each becomes a subscript for the other ion (Figure 4.7). This is another way to make the net charge on the formula unit 0. For example, the formula for the compound between magnesium (Mg) and nitrogen (N) is Mg_3N_2; between potassium (K) and oxygen (O), it is K_2O. If both subscripts come out the same—as with calcium (Ca) and oxygen (O), for example—the formula consists of one ion of each.

PRACTICE EXERCISE

Predict the correct formulas for the ionic compounds between (a) barium (Ba) and fluorine (F), (b) strontium (Sr) and sulfur (S), and (c) aluminum (Al) and oxygen (O).

formulas are (b) SrS and (c) Al_2O_3.
[$2x(-1) + 1x(+2)$] = 0. Thus, the formula becomes BaF_2. The other trality, we must match two fluoride ions with one barium ion so that to the octet rule and fluorine forms a 1− ion. To obtain electrical neu-
Consulting Figure 4.3, we find that barium forms a 2+ ion according

SOLUTION

NAMING SIMPLE TWO-ELEMENT IONIC COMPOUNDS   To name an ionic compound containing ions of two different elements, first name the metallic element and follow this with the name of the nonmetallic element shown in Table 4.1 (using the suffix *-ide*), as shown in Figure 4.8.

Figure 4.8 Naming simple two-element ionic compounds.

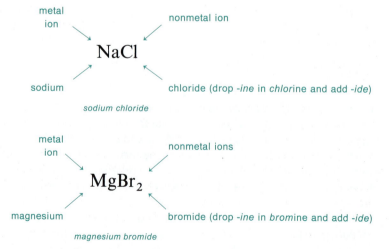

PRACTICE EXERCISE

Name the following ionic compounds: (a) CaO, (b) $SrCl_2$, (c) Na_2S, (d) Mg_3N_2.

SOLUTION

(a) calcium oxide, (b) strontium chloride (c) sodium sulfide (d) magnesium nitride.

PRACTICE EXERCISE

Predict the correct *formulas* and *names* for the ionic compounds that form by reactions between (a) K and S, (b) Ca and I, (c) Zn and O, (d) Al and N, (e) Li and O, (f) Mg and F.

SOLUTION

(a) K_2S, potassium sulfide; (b) CaI_2, calcium iodide; (c) ZnO, zinc oxide; (d) AlN, aluminum nitride; (e) Li_2O, lithium oxide; (f) MgF_2, magnesium fluoride.

Note in Figure 4.3 that the transition metal elements iron (Fe) and copper (Cu) can each form two stable positive ions. These ions are named by placing a Roman numeral corresponding to their positive charge after the name of the element:

— Fe^{2+} is called iron(II) ion.

— Fe^{3+} is called iron(III) ion.

— Cu^+ is called copper(I) ion.

— Cu^{2+} is called copper(II) ion.

Thus, the compound $FeCl_2$, formed by a combination of Fe^{2+} and Cl^- ions, is called iron(II) chloride, and the compound $FeCl_3$, formed by a combination of Fe^{3+} and Cl^- ions, is called iron(III) chloride. Similarly, the compound Cu_2O, consisting of Cu^+ and O^{2-} ions, is called copper(I) oxide, and the compound CuO, consisting of Cu^{2+} and O^{2-} ions, is called copper(II) oxide.

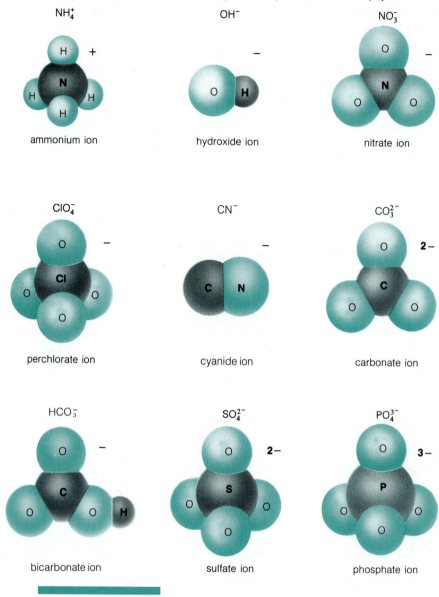

Figure 4.9 Formulas, names, electrical charges, and shapes of some common polyatomic ions.

NH_4^+

H +

H N H

H

ammonium ion

OH^-

−

O H

hydroxide ion

NO_3^-

O −

N

O   O

nitrate ion

ClO_4^-

O −

Cl

O   O

O

perchlorate ion

CN^-

−

C N

cyanide ion

CO_3^{2-}

O 2−

C

O   O

carbonate ion

HCO_3^-

O −

C

O   O H

bicarbonate ion

SO_4^{2-}

O 2−

S

O   O

O

sulfate ion

PO_4^{3-}

O 3−

P

O   O

O

phosphate ion

FORMULAS AND NAMES OF IONIC COMPOUNDS CONTAINING POLYATOMIC IONS Atoms of two or more different elements (usually nonmetals) can combine chemically by covalent bonds (Section 4.3) to form an electrically charged species called a **polyatomic ion**. Nitrogen and hydrogen, for example, can form a polyatomic ion called an *ammonium ion* with a positive charge and the formula NH_4^+ (read as "N-H-four-plus"). Figure 4.9 gives the formulas, names, electrical charges, and shapes of some common polyatomic ions.

To write the formulas for compounds containing polyatomic ions, again use the principle of electrical neutrality. Figure 4.10 shows how to do this using the crisscross method. Notice that when more than one of a particular polyatomic ion is in the formula, the polyatomic ion is enclosed in parentheses and a subscript is placed after the parentheses to indicate the number of such ions in the formula. When the subscript is 1, write neither the subscript nor the parentheses.

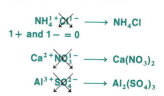

Figure 4.10 The crisscross method for predicting formulas of compounds containing polyatomic ions.

Compounds with polyatomic ions are named just as simple ionic compounds are, except that the full name of the polyatomic ion is included. Thus, the names for the three compounds shown in Figure 4.10 from top to bottom are ammonium chloride, calcium nitrate, and aluminum sulfate. Table 4.3 shows some common ionic compounds and their uses.

TABLE 4.3 SOME IONIC COMPOUNDS AND SOME OF THEIR USES		
Name	Formula	Uses
Aluminum chloride	$AlCl_3$	Deodorant
Aluminum phosphate	$AlPO_4$	Antacid
Ammonium carbonate	$(NH_4)_2CO_3$	Smelling salts
Ammonium chloride	NH_4Cl	Diuretic (helps urine form), expectorant (liquefies bronchial secretion)
Barium sulfate	$BaSO_4$	X-ray examination of internal organs (barium enema)
Calcium carbonate	$CaCO_3$	Antacid, chalk
Iron(II) sulfate	$FeSO_4$	Treatment of iron deficiency
Magnesium chloride	$MgCl_2$	Laxative
Magnesium sulfate	$MgSO_4$	Laxative
Potassium nitrate	KNO_3	Diuretic, antiseptic
Silver nitrate	$AgNO_3$	Antiseptic, prevents eye infection in newborn
Sodium bicarbonate	$NaHCO_3$	Antacid, baking soda
Sodium bromide	$NaBr$	Sedative
Sodium fluoride	NaF	Prevention of tooth decay
Sodium iodide	NaI	Treatment of iodine deficiency
Zinc oxide	ZnO	Skin ointments, paint pigments
Zinc sulfate	$ZnSO_4$	Eye wash, skin ointments

4.3 Covalent Bonds: Sharing Electrons

COVALENT BONDING Certain combinations of atoms can become more stable by sharing—instead of transferring—electrons. Atoms of nonmetal elements tend to bond together in this way.

When a nonmetal atom such as hydrogen ($H\cdot$) combines with another nonmetal atom (such as another hydrogen), it can match its unpaired electron to an unpaired electron from the other atom so that both achieve a stable electron structure, like the noble gas helium. A chemical bond formed by sharing a pair of electrons is known as a **covalent bond**. Substances held together by covalent bonds are **covalent compounds**. The simplest unit of a covalent compound, a group of atoms held together by covalent bonds, is called a **molecule**:

$$H\cdot \quad + \quad \cdot H \quad \longrightarrow \quad H\!:\!H \text{ or } H\!-\!H$$

| hydrogen | hydrogen | hydrogen molecule (H_2) |
| atom | atom | |

Often the dots representing a shared pair of electrons (a covalent bond) are represented by a dash. In this abbreviated form, H_2 ($H\!:\!H$) would be represented as $H\!-\!H$.

As the two hydrogen atoms come close together, their positively charged nuclei repel each other but attract the electrons; the electrons repel each other but are attracted to the nuclei (Figure 4.11). The attractive forces exceed the repulsive forces at a certain distance, and the more stable valence-electron structures of both atoms result in the molecule having a lower chemical potential energy than the atoms. So a sample of hydrogen gas actually consists of H_2 molecules that formed from pairs of hydrogen atoms in order to become more stable.

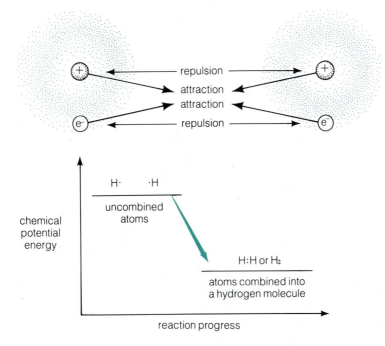

Figure 4.11 Potential energy diagram for the formation of a hydrogen molecule.

Atoms of other nonmetal elements behave in a similar way. For example, two chlorine atoms would tend to share a pair of electrons (form a covalent bond) with each other to achieve a stable octet of valence electrons, like the noble gas argon:

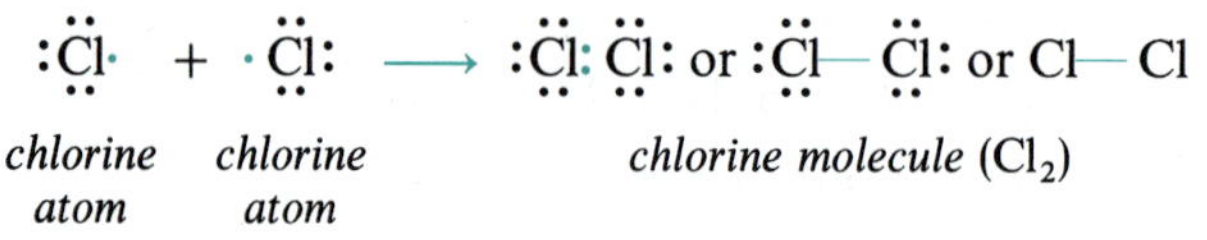

$$\ddot{\ddot{:Cl}}\cdot \; + \; \cdot \ddot{\ddot{Cl}}: \; \longrightarrow \; \ddot{\ddot{:Cl}}:\ddot{\ddot{Cl}}: \text{ or } \ddot{\ddot{:Cl}} - \ddot{\ddot{Cl}}: \text{ or } Cl-Cl$$

| chlorine atom | chlorine atom | chlorine molecule (Cl$_2$) |

Because the noble gas structure represents a lower state of potential energy, each nonmetal atom tends to share enough electrons to satisfy the octet rule. Oxygen, having six valence electrons, needs two more electrons to complete its octet; thus, oxygen shares two pairs of electrons and forms two covalent bonds. If oxygen and hydrogen atoms combine, the formula of the resulting molecule is H$_2$O:

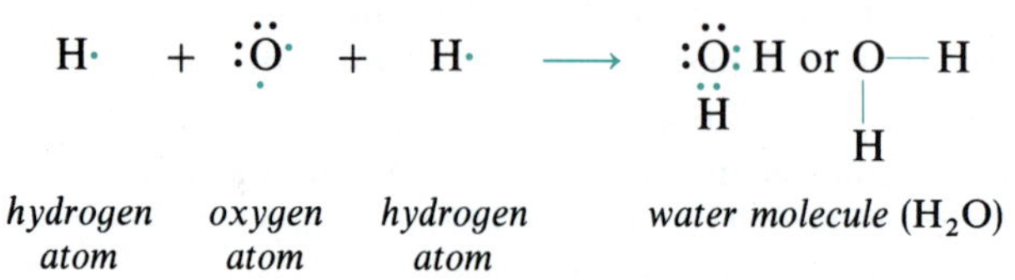

$$H\cdot \; + \; :\ddot{O}: \; + \; H\cdot \; \longrightarrow \; \begin{matrix} :\ddot{O}: H \\ \ddot{H} \end{matrix} \text{ or } \begin{matrix} O-H \\ | \\ H \end{matrix}$$

| hydrogen atom | oxygen atom | hydrogen atom | water molecule (H$_2$O) |

Carbon needs four valence electrons to complete its octet, so it will tend to form four covalent bonds. Chlorine, however, needs just one elec-

---

Figure 4.12 Number of covalent bonds formed by atoms of selected key elements.

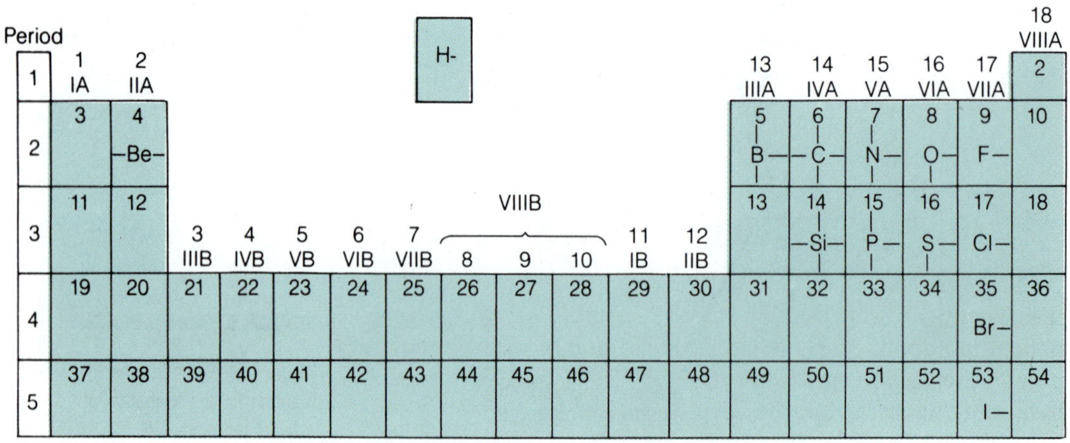

tron to attain an octet structure and thus will form one covalent bond. If carbon reacts with chlorine, the predicted result is a covalent molecule with the formula CCl_4:

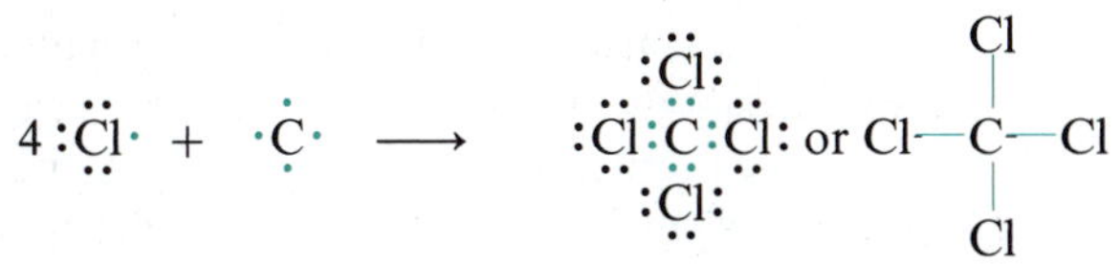

$$4 \,:\!\ddot{\underset{..}{C}l}\!\cdot \; + \; \cdot\!\overset{.}{\underset{.}{C}}\!\cdot \; \longrightarrow \; :\!\ddot{\underset{..}{C}l}\!:\!\overset{:\ddot{\underset{..}{C}l}:}{\underset{:\ddot{\underset{..}{C}l}:}{C}}\!:\!\ddot{\underset{..}{C}l}\!: \; \text{or} \; \overset{\text{Cl}}{\underset{\text{Cl}}{\text{Cl}\!-\!\text{C}\!-\!\text{Cl}}}$$

chlorine carbon   carbon tetrachloride molecule (CCl_4)
atoms   atom

An easy way to predict formulas of covalent compounds between two different elements is to first determine how many covalent bonds each nonmetal atom needs to complete its octet (Figure 4.12). Then write the simplest arrangement that provides the necessary number of covalent bonds to each atom. After doing this, write the formula, using subscripts to specify how many atoms of each element are in a molecule of the substance. The element on the left in the periodic table usually (a few exceptions occur, especially with H compounds) comes first in the formula.

PRACTICE EXERCISE

Using the appropriate number of covalent bonds (see Figure 4.12), write structures and then formulas for the covalent compounds made from (a) H and Br, (b) N and Cl, (c) O and F.

$$\text{(a) H—Br, HBr; (b) } \overset{\text{Cl}}{\underset{\text{Cl}}{\text{Cl—N—Cl}}}, \text{NCl}_3; \text{ (c) } \overset{\text{F}}{\underset{}{\text{O—F}}}, \text{OF}_2.$$

SOLUTION

Some molecules, especially those containing carbon, are more complex. Figure 4.13 shows a few examples. But notice that these same rules of bonding apply: Carbon atoms have four covalent bonds, hydrogen atoms have one, oxygens have two, and nitrogens have three. You will learn more about carbon compounds in Chapter 8.

Although the octet rule is not all-powerful—it cannot generate or explain every molecular substance—it does predict an enormous number

$$\overset{\text{H H H H}}{\underset{\text{H H H H}}{\text{H—C—C—C—C—H}}}$$

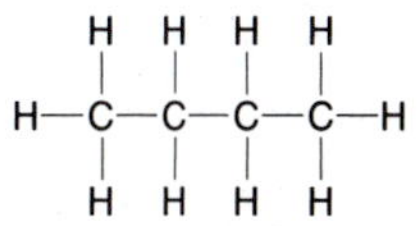

butane (C_4H_{10})
(a light fuel)

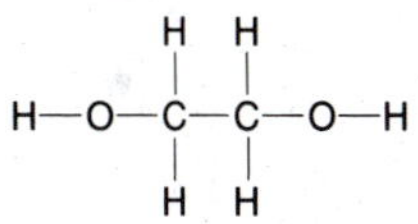

$$\overset{\text{H H}}{\underset{\text{H H}}{\text{H—O—C—C—O—H}}}$$

ethylene glycol ($C_2H_6O_2$)
(in antifreeze)

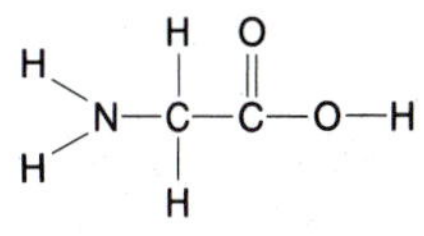

$$\overset{}{\underset{\text{H}}{\text{H}}}\!\!\!\!\!\!\text{N—}\overset{\text{H O}}{\underset{\text{H}}{\text{C—C}}}\text{—O—H}$$

glycine ($C_2H_5O_2N$)
(an amino acid in proteins)

Figure 4.13 Some carbon-containing molecules. Notice the number of bonds each type of atom forms.

of them. And it explains, for instance, why water's formula is H_2O rather than HO, or HO_2, or something else.

MULTIPLE COVALENT BONDS   Sharing a single pair of electrons is not the only way two atoms can achieve a state of minimum chemical potential energy. Nitrogen, for example, needs three electrons to complete its octet. It cannot attain a stable noble gas structure by sharing just one pair of electrons, yet it exists in nature as N_2 molecules. The N_2 molecules become stable because pairs of nitrogen atoms share *three* pairs of electrons with each other; this gives each nitrogen atom a stable octet of valence electrons:

$$\cdot \overset{\cdot\cdot}{\underset{\cdot}{N}} \cdot \; + \; \cdot \overset{\cdot\cdot}{\underset{\cdot}{N}} \cdot \; \longrightarrow \; \overset{\cdot\cdot}{N} ::: \overset{\cdot\cdot}{N} \text{ or } N \equiv N$$

nitrogen   nitrogen   nitrogen molecule (N_2)
atom   atom

In this case, a stable molecule results from the formation of a *triple bond*.

Similarly, atoms may share two pairs of electrons and form *double bonds*. Carbon dioxide (CO_2), for example, has two double bonds:

$$: \overset{\cdot\cdot}{O} :: \overset{\cdot\cdot}{C} :: \overset{\cdot\cdot}{O} : \text{ or } O = C = O$$

Notice also in Figure 4.13 that a double bond occurs in the amino acid glycine.

TABLE 4.4 NUMBER PREFIXES FOR NAMING NONMETAL-NONMETAL COMPOUNDS	
Number of Atoms of Each Element in the Compound	Number Prefix
1	mono- (optional)
2	di-
3	tri-
4	tetra-
5	penta-
6	hexa-
7	hepta-
8	octa-
9	nona-
10	deca-

NAMING COVALENT COMPOUNDS   To name covalent compounds, name the first element in the formula and then the second, using the *-ide* suffix. The names of covalent compounds (unlike those of ionic compounds) also include a prefix (Table 4.4) to indicate the number of atoms of the element in a molecule. The prefix *mono-* is optional; the absence of a prefix implies that there is only one atom of that element in the formula. Figure 4.14 shows the formulas, shapes, and names of several covalent substances.

PRACTICE EXERCISE

What are the names of (a) N_2O, (b) PCl_3, (c) N_2O_3, (d) P_4S_7, (e) CCl_4?

(a) dinitrogen oxide, (b) phosphorus trichloride, (c) dinitrogen trioxide, (d) tetraphosphorus heptasulfide, (e) carbon tetrachloride.

SOLUTION

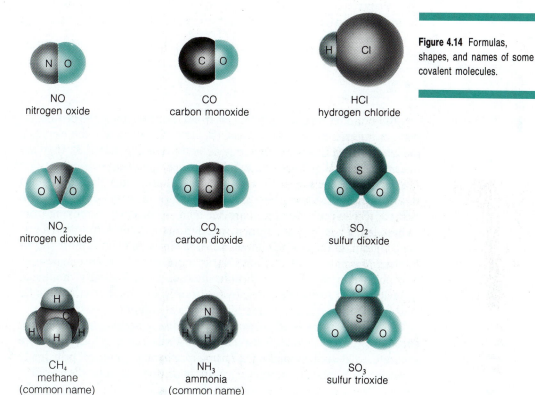

NO
nitrogen oxide

CO
carbon monoxide

HCl
hydrogen chloride

Figure 4.14 Formulas, shapes, and names of some covalent molecules.

NO_2
nitrogen dioxide

CO_2
carbon dioxide

SO_2
sulfur dioxide

CH_4
methane
(common name)

NH_3
ammonia
(common name)

SO_3
sulfur trioxide

4.4 Polar and Nonpolar Molecules: The Tug-of-War for Electrons

PARTIAL IONIC AND PARTIAL COVALENT CHARACTER   We have classified bonding between nonmetal atoms as covalent and that between metal and nonmetal ions as ionic. Using these rules, we can predict, for example, that nitrogen (N), a nonmetal, will form covalent bonds with oxygen (O), another nonmetal, and that it will form ionic bonds with magnesium (Mg), a metal.

PRACTICE EXERCISE

Identify the bond type within each of the following substances: (a) Cl_2, (b) CsCl, (c) PCl_3, (d) $CaBr_2$, (e) SO_2, (f) O_2.

CsCl and $CaBr_2$ are the only two substances on the list that contain metals, so they would be ionic. The other compounds are covalent.

SOLUTION

But bonding isn't actually that clear-cut. For one thing, the distinction between metals and nonmetals isn't definite. Recall that some elements are classified neither as metals nor as nonmetals (Section 2.3); they are intermediate in their properties and are called *metalloids*.

Table 4.5 summarizes some of the general properties that distinguish metals from nonmetals. These properties, however, vary somewhat from element to element. Metallic character tends to increase in moving from right to left in a given horizontal row and from top to bottom within a vertical column, in the periodic table. Thus, magnesium (Mg) is more metallic than aluminum (Al); sodium (Na) is more metallic than magnesium; and potassium (K) is more metallic than sodium. Similarly, nonmetal behavior tends to increase in moving from left to right in a horizontal row and from bottom to top within a vertical column.

Although examples of pure covalent and ionic bonding do exist, most substances fall somewhere between these two extremes. Thus, it is more accurate to describe bonding as being principally ionic or principally covalent. In nonmetal–nonmetal compounds, for example, the covalent

TABLE 4.5 SOME GENERAL PHYSICAL AND CHEMICAL PROPERTIES OF METALS AND NONMETALS	
Metals	**Nonmetals**
Physical Properties	
1. Most have a shiny or lustrous surface when clean.	**1.** Most have a dull surface with no metallic luster in their solid state.
2. Some are malleable and can be hammered or rolled into flat sheets without breaking.	**2.** Most are brittle and thus not malleable in their solid state.
3. Most are ductile and can be pulled into wires without breaking.	**3.** Most are brittle and thus not ductile in their solid state.
4. All are fair to good conductors of heat and electricity.	**4.** All, except carbon, are poor conductors of heat and electricity.
Chemical Properties	
1. Most do not readily combine with each other to form compounds.	**1.** Most, except the noble gases, can gain electrons by sharing electrons with nonmetals to form covalent compounds.
2. Most tend to lose electrons and combine with nonmetals to form ionic compounds.	**2.** Most also can gain electrons by combining with metals to form ionic compounds.

tendency to share electrons usually outweighs the ionic tendency to transfer electrons from one atom to another. This partial ionic, partial covalent character can be described using the concepts of *bond polarity* and *electronegativity*.

NONPOLAR AND POLAR COVALENT BONDS   When two identical atoms share one or more pairs of electrons (such as H—H, Cl—Cl, or N≡N), each atom exerts the same attraction for the shared electrons. Thus, the electrons are distributed or shared equally between the two atoms, and the bond is a pure or **nonpolar covalent bond**. Molecules with equal distribution of electrons are **nonpolar molecules**.

When two different atoms share electrons, however, the situation is quite different. Nuclei of different atoms have different capacities for attracting an electron pair. The tendency of each atom to attract electrons is known as its **electronegativity**. In covalent bonds, the shared electrons are pulled toward the atom with the greater electronegativity or electron-attracting power. As Figure 4.15 shows, nonmetals tend to have higher

Figure 4.15 Relative electro-negativity values (in color) for A group elements.

electronegativity values than do metals. In fact, the pattern of electro-negativity (increasing from bottom to top and from left to right, in the periodic table) is like the pattern for nonmetal behavior.

In a molecule with different atoms, such as H—Cl, the shared-electron pair is pulled toward the more electronegative chlorine and away from the hydrogen. This unequal distribution of shared electrons produces a slightly negative partial charge on the more electronegative atom (chlorine) and a slightly positive charge on the other atom (hydrogen), although the overall charge of the molecule is 0. Unequal sharing produces partial positive and partial negative poles within the molecule; thus, such bonds are **polar covalent bonds**.

The amount of electrical charge at each pole is less than 1; it is a *partial* charge because it is not as large as the full charge of an electron or proton. Indeed, an electron would have to be transferred completely (as occurs in pure ionic bonding) to produce an electrical charge as large as 1. Chemists use the Greek letter delta (δ) to represent the partial positive (δ^+) and negative (δ^-) charges at the poles. HCl, for example, is represented as

partial positive charge
partial negative charge
δ^+H—Cl$^{\delta^-}$

The greater the difference in electronegativity between the two atoms forming the bond, the more polar the bond is and the more ionic its character. Figure 4.16 depicts the range of bonding types.

NET MOLECULAR POLARITY   Most molecules that have polar covalent bonds are **polar molecules** (also called **dipoles**); they have overall partial positive and partial negative poles within the molecule.

But a few molecules with more than one polar covalent bond are nonpolar because their polar bonds are equal and in opposite directions in space; this leaves the molecule with no *net* polarity. To understand how this works, you need to know a bit about the shapes of molecules.

In a molecule, the atoms typically have eight valence electrons arranged as four pairs; some pairs are involved in covalent bonds, and some are not. One simple way to predict a molecule's shape is to envision each negatively charged pair of electrons repelling the other pairs. As a result, the four pairs (whether involved in a covalent bond or not) will be as far away from each other as possible. In three dimensions, the greatest separation produces angles about 109° apart (Figure 4.17).

Keep this idea (called the *valence-shell electron-pair repulsion*, or *VSEPR*, theory) in mind as you look at Table 4.6. Around each central atom (A) are four pairs of electrons, though not all may be used for bonding. Only the atoms bonded to A are shown. In the tetrahedral shapes, you can see the effect of all four pairs bonded to atoms. In the bent shape, such as in water, two pairs are bonded, and two are not. The nonbonded pairs, however, still repel the bonded pairs, causing the water molecule to be bent. (The actual angle between the two H atoms and the O is about 105°, close to the predicted value of 109°.) The shapes of some other molecules are shown in Figure 4.14.

Now let's return to the question of which molecules are nonpolar despite having polar covalent bonds. The right side of Table 4.6 shows the main two examples of this. One example is molecules such as carbon dioxide (CO_2), which is arranged in a straight line because of its two double bonds between carbon and oxygen. Even though both double

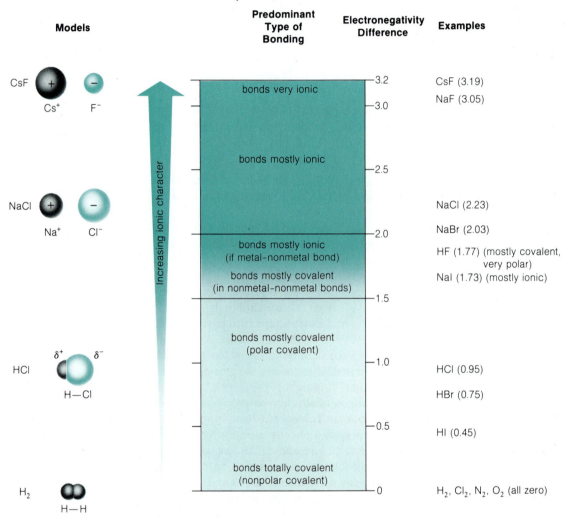

Figure 4.16 The transition from molecules whose bonds are totally covalent (nonpolar molecules) to molecules in which they are mostly covalent (polar molecules) to compounds in which the bonds are mostly ionic.

TABLE 4.6 EFFECT OF SHAPE ON THE NET POLARITY OF MOLECULES

Polar Covalent Molecules (Dipoles Do Not Cancel Out)		Nonpolar Covalent Molecules (No Dipoles or Dipoles Cancel Out)	
Type and Shape	Examples	Type and Shape	Examples
A B linear	$\overset{\delta^+}{H}\!\!-\!\!\overset{\delta^-}{Cl}$ $\overset{\delta^+}{H}\!\!-\!\!\overset{\delta^-}{F}$	A A linear	H—H N≡N
B A B bent	$\overset{\delta^-}{O}$ $\underset{\delta^+}{H}$ $\underset{\delta^+}{H}$ $\overset{\delta^-}{S}$ $\underset{\delta^+}{H}$ $\underset{\delta^+}{H}$	B A B linear	$\overset{\delta^-}{O}\!\!=\!\!\overset{\delta^+}{C}\!\!=\!\!\overset{\delta^-}{O}$
C B A B B tetrahedral	$\overset{\delta^+}{H}$ $\overset{\delta^+}{C}$ $\underset{\delta^-}{Cl}$ $\underset{\delta^-}{Cl}$ $\underset{\delta^-}{Cl}$	B A B B tetrahedral	$\overset{\delta^-}{Cl}$ $\overset{\delta^+}{C}$ $\underset{\delta^-}{Cl}$ $\underset{\delta^-}{Cl}$ $\underset{\delta^-}{Cl}$

Figure 4.17 An atom has four pairs of valence electrons arranged about 109° apart, according to the VSEPR theory.

bonds are polar, the CO_2 molecule is nonpolar because the double bonds are equal and pull symmetrically in opposite directions (just as in a tug-of-war between two equally matched teams). As a result, the two partial negative poles on the oxygen atoms cancel each other, leaving no net polarity. The same effect occurs in molecules such as CCl_4, in which the four polar covalent bonds between carbon and chlorine are equal and pull symmetrically in opposite directions, canceling one another.

For water (H_2O), the situation is quite different. Because this molecule is bent, the two polar bonds do not cancel each other in space (see Table 4.6).

What practical difference does it make whether a molecule is polar or nonpolar? Literally, all the difference in the world. If water molecules weren't bent, making water polar, water would be a gas, and life as we know it couldn't exist on earth. You will learn more in Chapter 5 about how polarity helps make substances solids, liquids, or gases and how it affects what materials can dissolve in each other.

4.5 Metallic Bonding

METALLIC BONDS AND METALLIC PROPERTIES   We have discussed what happens when metals combine with nonmetals (principally ionic bonding) and when nonmetals combine with nonmetals (principally covalent bonding, either polar or nonpolar). But what happens when all combining atoms are of metal elements? Here each atom, in order to become stable, needs to *lose* its one or two or three valence electrons. And none of the other atoms tend to accept additional electrons.

A pure metal consists only of atoms of that same metal element. The metal is a solid (except for mercury) consisting of an orderly three-dimensional array of structural particles called a **metal crystal lattice** (Figure

Figure 4.18 Model of a typical solid metal as a gigantic number of spherical atoms packed together in a regular, repeating pattern. Models of the atoms shown in this figure are about 43 million times larger than the actual size of a typical metal atom.

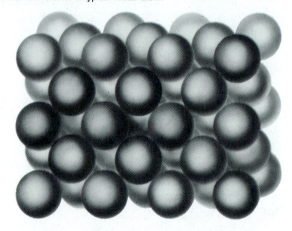

individual metal atom   array of metal atoms packed together

4.18). According to one theory, each lattice point is a positively charged ion, formed when the metal atom loses its valence electrons to achieve an electron configuration like a noble gas. The valence electrons lost by all the metal atoms then remain in the crystal and freely move throughout the lattice of positively charged ions, as shown for Na and Ca in Figure 4.19. Thus, **metallic bonding** consists of the forces of attraction between the positively charged ions in the crystal lattice and the mobile "sea" of negatively charged valence electrons. In other words, the valence electrons act as a kind of mobile "glue" that helps hold together the lattice of metal ions.

This arrangement is similar regardless of whether the ions are of the same or different metal elements. A mixture of metals is called an **alloy**. Some familiar examples are brass (which contains copper, zinc, and small amounts of tin, lead, and iron) and various coins. There are no formulas to write for alloys because alloys are mixtures (Section 1.3) that can contain varying proportions of metals.

Pure metals and alloys generally conduct heat and electricity (see Table 4.5). This can be explained in terms of their valence electrons moving

Figure 4.19 Crude model of metallic bonding. The solid crystals of Na and Ca are held together by the attractive forces between their positively charged metal ions and their mobile "sea" of negatively charged valence electrons lost by the metal atoms when they formed ions.

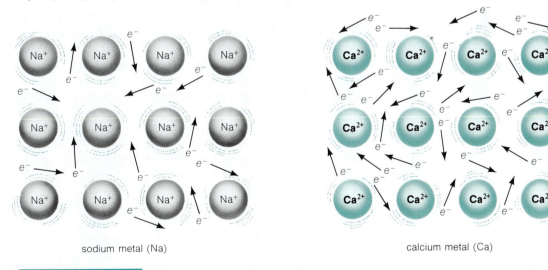

sodium metal (Na)   calcium metal (Ca)

about in the crystal (see Figure 4.19) and thus conducting electricity and heat from one point to another. Metals are generally pliable rather than brittle. They can be flattened into thin sheets or foils or drawn into long wires. No matter what shape the lattice is distorted into, the sea of mobile valence electrons still holds the ions together. The arrangement of electrons at the surface also tends to reflect any light that falls on them, which makes metal surfaces shiny.

JOHN DALTON

All changes we produce consist in separating particles that are in a state of cohesion or combination, and joining those that were previously at a distance.

Summary

Atoms react with each other by forming chemical bonds to produce new substances. Bonding enables atoms to achieve a state of lower potential energy, which is more stable. Atoms typically form bonds so as to achieve a stable octet of valence electrons.

Metals and nonmetals typically react to form ionic compounds. Atoms of metal elements transfer their valence electrons to atoms of nonmetal elements, producing electrically charged ions that attract each other by ionic bonds. The charge of the ion can be predicted by the group number of the element. Some ionic compounds also contain polyatomic ions. For all ionic compounds, the simplest ratio of ions that has no net electrical charge is the formula unit and is written as the formula for the compound. These compounds exist as an ionic crystal lattice.

Nonmetals typically react with each other to form covalent compounds. Atoms of nonmetal elements share valence electrons to become more stable. A group of atoms held together by covalent bonds is a molecule. The formula for covalent compounds consists of the number of atoms of each element in a molecule. Covalent bonds are polar or nonpolar, depending on whether the electrons are shared equally. Polar bonds result from atoms of different electronegativities sharing electrons. Polar covalent bonds result in polar molecules, except for symmetric molecules in which the polar bonds cancel each other in space.

Metal atoms bond and form a metal crystal lattice. In the lattice are metal ions and a "sea" of valence electrons that the atoms lost to become ions. A mixture of metals is an alloy.

Terms for Review

After completing this chapter, you should know and understand the meaning of the following terms:

alloy (p. 113)	metal crystal lattice (p. 112)
chemical bond (p. 87)	metallic bonding (p. 113)
covalent bond (p. 101)	minimum potential energy
covalent compound (p. 101)	principle (p. 88)
dipole (p. 109)	molecule (p. 101)
electronegativity (p. 107)	nonpolar covalent bond (p. 107)
formula unit (p. 93)	nonpolar molecule (p. 107)
ion (p. 90)	octet rule (p. 88)
ionic bond (p. 92)	polar covalent bond (p. 108)
ionic compound (p. 92)	polar molecule (p. 109)
ionic crystal lattice (p. 93)	polyatomic ion (p. 99)

Questions

Odd-numbered questions are answered at the back of this book.

1. Why do chemical bonds form?

2. What is the octet rule?

3. Which of the following has a lower potential energy? (a) Na^+ ion or Na atom, (b) H_2 molecule or H atom.

4. How many electrons would each of the following atoms need to *gain* to achieve an octet of valence electrons? (a) H, (b) S, (c) C, (d) Ne, (e) Br, (f) P.

5. How many electrons would each of the following atoms need to *lose* to achieve an octet of valence electrons (a) Na, (b) Mg, (c) Ar, (d) Sr, (e) K, (f) Al?

6. Would the elements listed in Question 5 (if they reacted at all) be more likely to form ionic or covalent compounds? Why?

7. Would the elements listed in Question 4 (if they reacted at all) be more likely to form ionic or covalent compounds? Why?

8. If an alkaline earth element combined with a halogen, would the resulting compound be predominantly covalent or ionic?

9. Use the periodic table to predict the major type of bond (metallic, ionic, or covalent) that might be expected to form

between atoms of (a) sodium and fluorine, (b) phosphorus and oxygen, (c) C and H, (d) K and Cl, (e) S and O, (f) Na and Na.

10. Complete the following table:

Formula	Name	Principal Bond Type (Ionic or Covalent)
_____	Sulfur dioxide	_____
CO_2	_____	_____
_____	Carbon monoxide	_____
KCl	_____	_____
O_3	Ozone	_____
_____	Hydrogen chloride	_____
$CaCl_2$	_____	_____
NH_3	Ammonia	_____

11. Identify the ion (if any) that each of the following atoms might be expected to form: (a) Ra, (b) Rn, (c) Sb, (d) Al, (e) Sr, (f) I, (g) As.

12. Identify the ion (if any) that each of the following atoms might be expected to form: (a) K, (b) P, (c) O, (d) Ca, (e) F, (f) He, (g) Se.

13. List the total number of electrons in each of the following ions: (a) Br^-, (b) Te^{2-}, (c) Cs^+, (d) Mg^{2+}, (e) O^{2-}, (f) Cd^{2+}.

14. What is the difference between Mg and Mg^{2+}? Why should Mg^{2+} form? Explain why you would not expect Mg^{3+} to form.

15. Predict the formula unit for an ionic compound formed from atoms of (a) Cs and F, (b) gallium and chlorine, (c) Sr and I, (d) strontium and oxygen, (e) K and O, (f) Al and I, (g) Ca and P.

16. Why is the formula for potassium bromide KBr and not K_2Br, KBr_2, or something else?

17. Name the compounds that form by each of the combinations of atoms in Question 15.

18. Name the following compounds: (a) KCN, (b) Li_2CO_3, (c) NaOH, (d) $Mg(NO_3)_2$, (e) $Al_2(SO_4)_3$.

19. Write formulas for the following compounds: (a) sodium phosphate, (b) calcium hydroxide, (c) ammonium oxide, (d) sodium bicarbonate (bicarbonate of soda.)

20. Predict how many (if any) covalent bonds each of the following form: (a) C, (b) S, (c) Cl, (d) N, (e) O.

21. Using dashes to represent covalent bonds, write structures of the molecules for each of the compounds formed from atoms of (a) phosphorus and chlorine, (b) H and Br, (c) C and F, (d) H and S.

22. In the compounds that form from the combinations listed in Question 21, how many valence electrons does each atom have?

23. Predict the molecular formula for each of the compounds formed by the combinations in Question 21.

24. Which of the compounds formed by the combinations in Question 21 have polar covalent bonds?

25. Which of the compounds formed by the combinations in Question 21 have polar covalent bonds but are nonpolar molecules?

26. Why does aluminum foil exist but not sulfur foil?

27. Where does the mobile "sea" of electrons come from in a metal?

Topics for Discussion

1. The octet rule states that some atoms tend to form bonds such that some of them obtain eight valence electrons and others obtain two. Because there are many exceptions, what good is such a rule?

2. List as many examples as you can of pure elements that you use or see in your daily activities. Why are there so many more compounds than elements in the world?

3. What would the world be like if water were nonpolar?

CHAPTER 5

Physical and Chemical Changes: Rearranging Matter

General Objectives

1. What are the characteristic properties of the solid, liquid, and gaseous states of matter, and how can we account for these properties?

2. What happens when a substance dissolves in another substance? How can we predict what will dissolve in a given substance?

3. How can we write and balance chemical equations that represent chemical reactions?

4. How do chemists calculate the amounts of reactants needed or products formed in a chemical reaction?

Matter can change in two ways: physically and chemically. In a physical change, all molecules, atoms, and ions retain their identities; they still have the same composition and can be represented by the same formulas. For example, when water freezes into ice or evaporates from your skin, all molecules remain H_2O. And when you stir some table sugar ($C_{12}H_{22}O_{11}$) into water to dissolve it, the water and sugar molecules still remain.

But in chemical changes, matter rearranges to form new substances with different formulas. When hydrogen gas (H_2) and oxygen gas (O_2) combine to form water (H_2O), the change is chemical. Chemists represent chemical changes by writing chemical equations. From a balanced equation, they can calculate the amounts of reactants and products that participate in a chemical reaction.

5.1 Physical States: Solids, Liquids, and Gases

STATES OF MATTER You can classify any material you see as a solid, liquid, or gas. These three physical states of matter (Section 1.3) are properties of large collections of atoms, ions, or molecules. It is meaningless to say that a single atom, ion, or molecule is a solid, liquid, or gas. Instead, the physical state depends on the interactions between the individual atoms, ions, or molecules in a substance.

THE KINETIC MOLECULAR THEORY Why is a particular substance a solid, liquid, or gas? One of the most useful explanations is the **kinetic molecular theory**, in which scientists visualize how molecules, ions, and atoms move.

To illustrate the theory, imagine that molecules, ions, or atoms are like a large group of students under three different conditions. The solid state is approximated by several hundred students aligned in rows of seats during a lecture. There is only a small amount of disorder as they fidget (vibrate) in their seats. The more boring the lecture, the more rigid and ordered the "student solid" is, as more of them slump to their desks, sound asleep. If the lecturer gives the students a 10-minute break but asks them to stay in the building, their movement approximates the liquid state. The disorder increases as students move about and cluster in small groups here and there. The gaseous state is approximated when class is over, and the students spread out all over campus.

Although hardly scientific, this analogy captures the essence of the kinetic molecular theory. Atoms, molecules, or ions of any substance are in continuous, rapid, random motion. The higher the temperature, the

Figure 5.1 Relative order and disorder of solids, liquids, and gases.

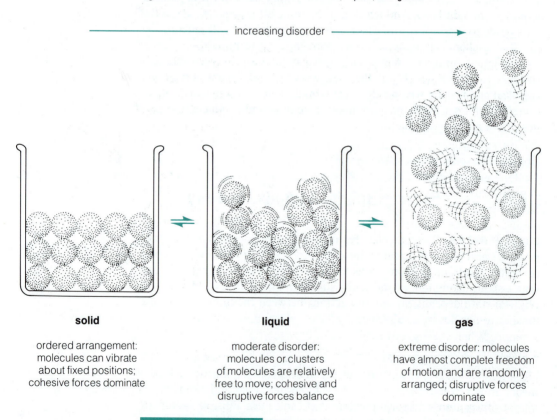

increasing disorder

solid

ordered arrangement:
molecules can vibrate
about fixed positions;
cohesive forces dominate

liquid

moderate disorder:
molecules or clusters
of molecules are relatively
free to move; cohesive and
disruptive forces balance

gas

extreme disorder: molecules
have almost complete freedom
of motion and are randomly
arranged; disruptive forces
dominate

more energetic this motion is. Their movement constitutes a disruptive or chaos-causing force that competes directly with attractive or cohesive forces holding atoms, molecules, or ions together.

When attractive forces are stronger than disruptive ones, the substance is a solid (Figure 5.1). Here atoms, ions, or molecules remain in orderly, fixed positions, never changing neighbors. If disruptive and attractive forces are balanced, the substance is a liquid; particles move more freely and jostle around each other, but they still stay close together. When disruptive forces dominate, the substance is a gas. In this state, particles move freely and randomly; apart from occasional collisions, they never touch each other.

The kinetic molecular theory can account for the properties of gases, liquids, and solids, as summarized in Table 5.1.

SOME PHYSICAL PROPERTIES OF GASES Besides having a changeable shape and volume, a gas is unique in several ways. One liter of a gas weighs next to nothing compared to one liter of a liquid or solid. It is almost impossible to compress a liquid or a solid, but you can easily squeeze a gas into a smaller volume. And unlike a liquid or a solid, a gas exerts a steady outward push, or *pressure*, on every part of its container.

No matter how a gas may behave chemically, it acts physically like all other gases. Its pressure, volume, and temperature are all interrelated in a predictable way. For example, as you drive along the road, the air in the tires gets hotter, and the pressure of the gas in the tire rises. When you stop, the pressure and temperature of the gas both drop; you can check this with a tire gauge. At a constant volume then, the pressure of a gas is directly related to the temperature.

One way to inflate a balloon with little effort (but a lot of trouble) is to get it started, tie it off, and then put it in a closed container (such as a bell jar) from which the air can be withdrawn to lower the air pressure. As the pressure falls outside the balloon, the gas pressure inside the balloon causes it to expand. In fact, the balloon keeps expanding until the internal pressure matches the external pressure (Figure 5.2). If air is let back into the container around the balloon, the balloon returns to its original, smaller size. The volume increases as the external pressure decreases (and vice versa) if the temperature stays the same.

If you put a balloon in the refrigerator, the balloon will shrink as its air cools. When you take the balloon out of the refrigerator, the air inside it gradually expands to its original volume as it warms to room temperature. Volume, like pressure, is directly related to temperature.

FORCES BETWEEN MOLECULES: WHY EVERYTHING IS NOT A GAS Using the kinetic molecular theory, you might predict that substances with strong forces

TABLE 5.1 THE KINETIC MOLECULAR THEORY AND CHARACTERISTIC PROPERTIES OF GASES, LIQUIDS, AND SOLIDS

Property	Kinetic Molecular Theory Explanation
Gases	
Gases have an indefinite shape that allows them to fill the entire volume of any closed container.	The particles are moving rapidly in all directions; with little force of attraction between them, they can move freely to fill the container.
Gases have a low density (mass per unit volume).	Particles are widely separated, so that there are relatively few of them in a given volume.
Gases are easily compressed.	Because the particles are so small and so far apart (relative to their own size), they can easily be pushed closer together by an outside force.
Gases exert pressure equally in all directions on any surface they contact.	The gas particles are moving rapidly and randomly in all directions, and as a result they continually bombard the container walls to exert a uniform force on each unit area of surface.
Gases mix spontaneously with one another.	Because the rapidly moving gas particles are relatively far apart, there is ample room for them to mix completely.
Liquids	
Liquids have a definite volume and assume the shape only of the portion of the container occupied by their definite volume.	The molecules are held in a definite volume by their forces of attraction; these intermolecular forces are weak enough to allow the molecules to move freely and assume the shape of the container occupied by the liquid, but not so weak that they can leave the surface of the liquid freely.
Liquids have a higher density than gases.	The stronger intermolecular attractive forces in liquids hold the molecules relatively close together; this results in a relatively high density, or mass per unit volume.
Liquids are very difficult to compress.	Because the molecules are fairly close to one another, an increase in pressure can move them only slightly closer together.
Liquids flow readily.	The intermolecular forces between the molecules are weak enough to allow them to flow freely.
Liquids mix with other liquids slowly.	The molecules in liquids are so close to one another that they collide very frequently, thus slowing their ability to mix freely.
Solids	
Solids have a definite volume, a definite, rigid shape, and normally do not flow.	The forces of attraction are so strong that the particles are held in fixed positions and cannot move freely.
Solids normally have a higher density than gases and liquids.	The particles are so close together that the density is high.
Solids are extremely difficult to compress.	The particles are so close to one another that even a large increase in pressure cannot move them much closer together.
Solids have very little tendency to mix with other solids.	The particles are held together so closely and tightly that they cannot move freely.

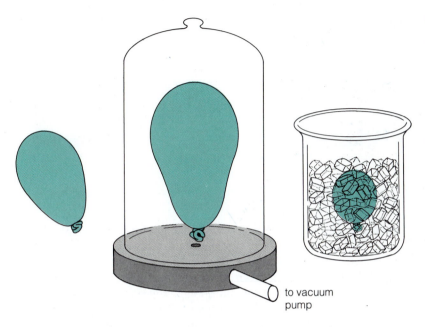

to vacuum
pump

Figure 5.2 The volume of a balloon will increase if the external pressure is lowered (center) and decrease if the temperature is lowered (right).

of attraction between their particles (atoms, ions, or molecules) would be solids, whereas those with the weakest attractions would be gases; liquids would have attractions of intermediate strength.

Let's examine this idea. In ionic compounds, recall that strong ionic bonds hold the formula units together in a crystal lattice (Section 4.2). Indeed, ionic compounds are solids. Metals, too, have strong electrical attractions; the positively charged metal ions attract the negatively charged "sea" of electrons (Section 4.5). This fits with our observation that metals (except mercury) are solids. At the other extreme are the noble gases. With a stable valence-electron structure, these atoms have very little interaction with each other. No wonder they are gases.

With covalent compounds, the situation is more complicated. The forces between molecules are not as strong as in metals and ionic compounds, but they tend to be stronger than the forces between noble gas atoms. There are three types of forces between molecules: (1) London forces, (2) dipole–dipole interactions, and (3) hydrogen bonds.

Temporary internal shifts in electron distribution cause **London forces**. Because electrons are in constant motion, for fleeting instants more electrons may be at one side of an atom, ion, or molecule than at the other. These temporary unsymmetrical distributions of charge set up a temporary dipole. This in turn can induce a similar dipole in a nearby atom, ion,

Figure 5.3 London forces of attraction between two nonpolar molecules of Cl_2.

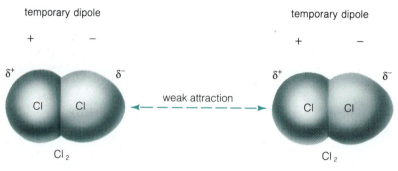

Figure 5.4 Increasing strength of London forces with the weight of molecules affects the physical state at room temperature.

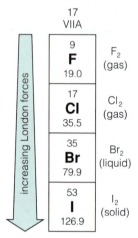

or molecule. The attraction between these temporary dipoles (Figure 5.3) is only about one-thousandth as strong as the covalent bonds holding the atoms together in a typical molecule.

London forces exist between all atoms, ions, and molecules, but they are the *only* attractions between nonpolar molecules such as H_2, CH_4, CO_2, and Cl_2 (Figure 5.3), and between noble gas atoms. As these atoms and molecules get larger, they have stronger London forces because their large electron clouds more readily form temporary dipoles. Figure 5.4 shows how this increase in London forces affects the physical state of the halogens, which all exist as nonpolar molecules.

In addition to London forces, polar molecules have somewhat stronger attractive forces between their permanent dipoles. These are the **dipole–dipole forces** of attraction between the negative pole of one molecule and the positive pole of another (Figure 5.5). Because these attractions are between *partial* electrical charges (δ^- and δ^+), dipole–dipole forces are much weaker than the ionic bonds holding ionic compounds together (Section 4.2).

A **hydrogen bond** is the relatively strong dipole–dipole force between a hydrogen atom covalently bonded to an atom and a more electronegative atom in the same or a nearby molecule. Despite being the strongest of the intermolecular forces, hydrogen bonds are still much weaker than the covalent bonds within molecules.

The strongest and most important hydrogen bonds are between or within molecules having a F—H, —O—H, or —N—H group. Hydrogen bonded to a highly electronegative F, O, or N atom in a polar covalent molecule is strongly attracted to another F, O, or N atom in the same or another molecule. Water (H_2O), ammonia (NH_3), and hydrogen fluoride (HF) have hydrogen bonds between their molecules (Figure 5.6). Hydrogen bonds also occur between and within many other molecules, including such vital molecules of life as proteins, carbohydrates, and DNA (Chapter 9).

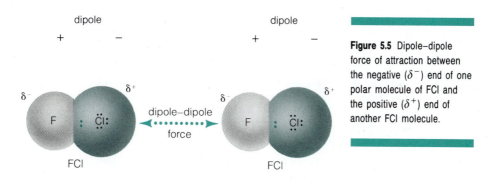

Figure 5.5 Dipole–dipole force of attraction between the negative (δ^-) end of one polar molecule of FCl and the positive (δ^+) end of another FCl molecule.

Figure 5.6 Hydrogen bonding (indicated by dotted lines) between polar molecules of HF, H_2O, and NH_3.

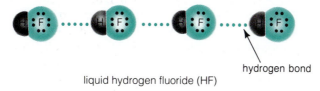

liquid hydrogen fluoride (HF)

hydrogen bond

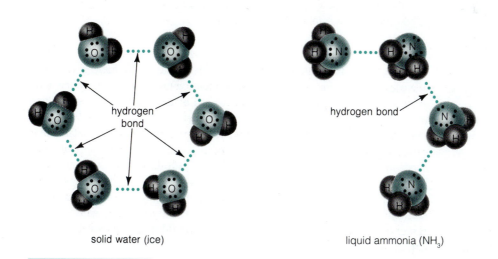

solid water (ice)

liquid ammonia (NH_3)

5.2 Physical Changes: Melting and Boiling

MELTING AND FREEZING   When we call something a solid, liquid, or gas, we are usually describing their state at normal pressure and at normal (room) temperature. But substances can readily change physical states under different conditions. You can now use the kinetic molecular theory and your understanding of bonding to picture what happens when substances change their physical state.

The atoms, ions, or molecules in a solid are very close together and their motion is restricted primarily to quivering or vibrating in place. When a solid is heated, it absorbs heat energy, and its particles begin to vibrate more violently. Eventually, the energy increases enough to overcome the forces of attraction holding the particles together in the solid crystal. At this point, the solid undergoes **melting**; it changes to a liquid. For a pure substance, this transformation occurs at a single temperature known as the *melting point.*

Because the strength of attractive forces among particles of any substance is unique, the temperature at which those forces are overcome is also unique. Thus, one way to identify a sample of a pure substance is to measure its melting point and compare that value with those of known substances.

The reverse change—from a liquid to a solid—is **freezing** or *solidification*. This change occurs at the *freezing point*, which for every pure substance is the same temperature as the melting point. Above the freezing point, the material remains liquid, but at the freezing point it solidifies.

A few substances can change directly from a solid into a gas, a process called **sublimation**. You may have seen dry ice (solid CO_2) do this. Another substance that sublimes is iodine (I_2).

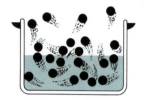

Figure 5.7 Vaporization and vapor pressure. At a given temperature, some particles in a liquid in an open container (top) have enough energy to overcome the attractive forces and escape to the gaseous state. In a closed container (bottom) at a fixed temperature, a dynamic equilibrium occurs between particles escaping to the gaseous state and those returning to the liquid.

VAPORIZATION AND VAPOR PRESSURE   Particles in a liquid have a range of energies and speeds, but they have a fixed average speed at a given temperature. They are somewhat like a group of people milling around and bumping into one another at a crowded party. Although attractive forces tend to keep particles relatively close together, a few near the surface have enough energy to escape into the gaseous state, just as a few people near the door at a crowded party may escape outdoors for fresh air or to go home.

Changing from the liquid state to the gaseous or vapor state is **vaporization** (Figure 5.7). This is what happens when you leave an open container of a liquid out long enough; the liquid may vaporize (dry up) completely.

If you put a liquid in a covered container, its particles escape into the vapor state as before, but the cover prevents them from leaving the vicinity of the liquid's surface. As they ricochet around in the confined space, some particles return to the liquid state because they are attracted to the particles of the liquid. This change from vapor back to liquid is **condensation**.

In a closed container, the number of particles escaping to the gaseous state over a given period of time is eventually balanced by an equal number condensing into the liquid. When this occurs, the total number of particles in the vapor state becomes constant; it neither increases nor decreases. The balance in the rates of two opposing processes is known as a **dynamic equilibrium**. *Dynamic* implies motion rather than a static situation. In this case, vaporization and condensation both continue steadily, but at equilibrium the total number of particles in each state stays constant.

The pressure of a gas or vapor over a liquid at dynamic equilibrium in a closed container is the **vapor pressure** of the liquid (see Figure 5.7). The weaker the forces between particles, the more easily a liquid evaporates at a given temperature, and the higher its vapor pressure is. Thus, vapor pressure is another physical property that can be used to identify a substance.

BOILING POINT   As the temperature of a liquid increases, its vapor pressure rises, and its rate of vaporization increases. At some temperature, bubbles of vapor appear throughout the liquid and escape through the surface. This process, known as **boiling**, differs from vaporization in two ways. First, molecules must be converted to the gaseous state throughout the entire liquid instead of just at the surface. Second, the vapor bubbles must escape from the liquid with sufficient force to displace the atmosphere above the liquid. The *boiling point* is the temperature at which the vapor pressure of the liquid equals the pressure of the atmosphere above the liquid.

Because the pressure of the atmosphere varies, so does a substance's boiling point. For example, the boiling point of water in a sea-level city like Miami Beach is 100°C. But in mile-high Denver, where the atmospheric pressure is consistently less, the boiling point is about 95°C (Figure 5.8). The *normal boiling point* of a substance is its boiling temperature under typical sea-level atmospheric conditions.

By increasing the external pressure on a liquid, we can raise its boiling point. The pressure cap on a modern water-cooled car radiator uses this principle to allow radiator water to heat up to about 120°C without boiling. Similarly, a pressure cooker allows food to cook at a temperature greater than 100°C without the water boiling away; this shortens the cooking time.

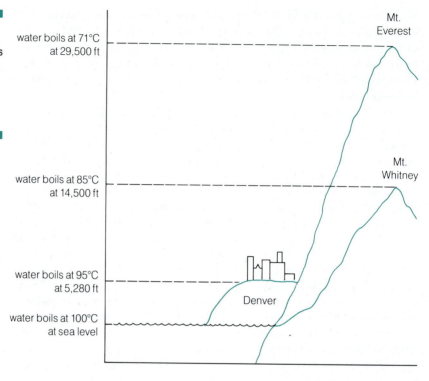

Figure 5.8 The boiling point of water at various elevations above sea level. The boiling point is lower at higher elevations because the atmospheric pressure is lower.

Substance (Formula)	Type of Bonding Between Particles	Melting Point (°C)	Boiling Point (°C)
Fluorine (F₂)	London forces only	−220	−188
Nitric oxide (NO)	Dipole–dipole forces	−164	−152
Methanol (CH₄O)	Hydrogen bonds	−94	65
Calcium (Ca)	Metallic bonding	893	1484
Sodium fluoride (NaF)	Ionic bonding	993	1695

TABLE 5.2 EFFECT OF INCREASING BONDING STRENGTH BETWEEN PARTICLES ON THE MELTING AND BOILING POINTS OF SUBSTANCES OF SIMILAR WEIGHTS

MELTING POINT, BOILING POINT, AND INTERMOLECULAR FORCES   The forces exerted by particles on each other help determine a substance's melting and boiling point. The stronger those forces, the more heat energy it takes to overcome them, and the higher the melting and boiling temperatures become. Table 5.2 shows how, for substances of a similar weight, the type of bonding affects their melting and boiling points.

5.3 Physical Changes: Forming Solutions

When two different substances occupy the same container, they form a *mixture*. The more finely divided each substance is, the more uniformly it can be distributed throughout the container, and the better the mixture. The best mixture occurs when the individual atoms, molecules, or ions of one substance—its tiniest particles—are evenly interspersed with those of the other substance. Such a homogeneous mixture is a **solution**. The dissolved substance, usually present in the lesser amount, is the **solute**, and the substance in which it is dissolved is the **solvent**. The actual amount of the solute dissolved in a specific amount (volume or mass) of the solvent is known as the *concentration* of the solution.

Although solutions can be made by dissolving a gas in a gas (air), a solid in a solid (metal alloys), or some other combinations, most common solutions involve solids in liquids (such as salt in water), liquids in liquids (such as alcohol in water), or gases in liquids (such as carbon dioxide in water).

THE SOLUTION PROCESS   For a solute to dissolve in a solvent, the particles of each must become interspersed. Thus, the solution process first involves overcoming the forces of attraction between the individual solute particles. Solvent particles must then be separated so that solute particles can be distributed uniformly among them. Both of these separations require energy. However, new forces of attraction between the mixed particles release energy. The successful formation of a solution depends, in large measure, on whether this exchange leads to a state of minimum potential energy. If the attractions before and after are similar, a solution forms; if they are dissimilar, the substances will not mix.

A rough rule of thumb for predicting whether one substance will be soluble in another is: *Likes dissolve one another; unlikes do not.* The two categories are (1) substances that have fully or partially charged particles (ionic and polar covalent compounds) and (2) substances that do not have partially or fully charged particles (nonpolar covalent compounds). Table 5.3 summarizes how this rule works.

| TABLE 5.3 COMBINATIONS ILLUSTRATING THE RULE THAT LIKES DISSOLVE ONE ANOTHER; UNLIKES DO NOT |||
Type of Solute	Type of Solvent	Is Solution Likely?
Polar covalent or ionic	Polar covalent	Yes
Nonpolar covalent	Nonpolar covalent	Yes
Polar covalent or ionic	Nonpolar covalent	No
Nonpolar covalent	Polar covalent	No

Consider, for example, dissolving an ionic compound such as sodium chloride in water (Figure 5.9). First, the strong ionic bonds between the Na^+ and Cl^- ions must be overcome before the ions can be separated. This occurs when the Na^+ and Cl^- ions on the surface of the crystal strongly attract (and are attracted by) the positive (H) and negative (O) ends of the polar water molecules. These forces are strong enough to pull the ions away from corners, edges, and surfaces of the NaCl lattice. Once the ions are free, they are surrounded by water molecules (see Figure 5.9). Thus, salt dissolves in water.

WHY OIL AND WATER DON'T MIX   The rule that like dissolves like can be reversed to predict that solvents will not dissolve unlike solutes. A nonpolar solute such as oil should not dissolve in a polar solvent such as water (Figure 5.10). Nonpolar molecules have no charged ions or partially charged ends, so positive and negative ends of water molecules are not attracted to them. Instead, water molecules attract each other by hydrogen bonds. The oil and the water cannot reach a favorable energy state by dissolving, and the two do not mix.

OTHER APPLICATIONS OF SOLUBILITY   If you spill some pancake syrup on a fabric place mat or if a Popsicle drips on your clothes, you can usually clean up the mess with water because sugary substances, being polar, dissolve well in water. If you spill bacon grease or butter instead, water will probably not work satisfactorily because these, like other oily stains, will not go into water solution. You would be much more successful using cleaning fluid in place of water. Cleaning fluids are nonpolar solvents, usually made from petroleum, that will dissolve nonpolar stains.

Dry cleaning uses the same like-dissolves-like principle. Rather than washing clothes in water, dry cleaners use a nonpolar liquid such as perchloroethylene (C_2Cl_4) to remove most of the soil from the fabric. Then

Figure 5.9 Dissolving an ionic solid (sodium chloride) in a polar solvent (water). Notice that the negative (O) poles of water molecules attract the Na^+ ions, while the positive (H) poles of water attract the Cl^- ions.

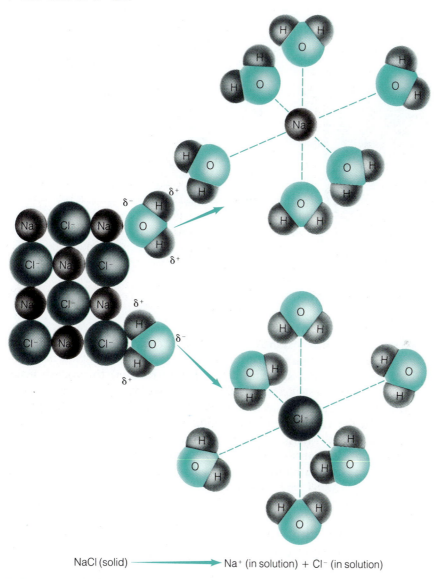

NaCl (solid) $\longrightarrow$ Na$^+$ (in solution) + Cl$^-$ (in solution)

Figure 5.10 Oil and water don't mix because the hydrogen bonds between water molecules are stronger than the weak attractive forces between nonpolar oil molecules and polar water molecules.

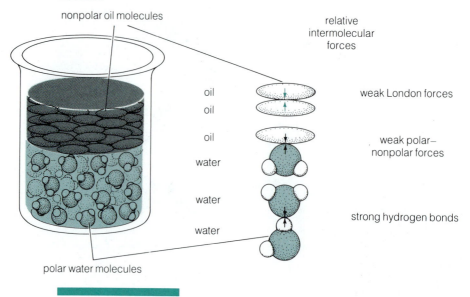

nonpolar oil molecules

relative
intermolecular
forces

oil
oil

weak London forces

oil

water

weak polar–
nonpolar forces

water

water

strong hydrogen bonds

polar water molecules

a specially trained person known as a spotter tackles any remaining stains with a whole arsenal of other solvents.

Applications of solubility have allowed industry to produce many everyday articles such as cavity-fighting fluoride toothpaste. Fluoride ions (F^-) help prevent your teeth from decaying, but they have to be in solution to do so. Until the 1950s, no toothpaste containing fluoride was effective because F^- ions would not stay dissolved in the paste. Now that compatible substances have been discovered, dozens of brands of fluoride toothpaste can truthfully advertise their effectiveness (see Section 16.1).

5.4 Chemical Changes: What Is a Chemical Reaction?

CHEMICAL REACTIONS: MAKING AND BREAKING BONDS   Recall that a chemical change occurs when the starting substances (*reactants*) either break down or combine chemically to form one or more different substances (*products*)

TABLE 5.4 EXAMPLES OF CHEMICAL CHANGES AND THEIR ACCOMPANYING PHYSICAL CHANGES

Chemical Change	Accompanying Physical Changes
Mix vinegar and baking soda.	Mixture "fizzes" and bubbles of gas are given off.
Burn sulfur in air.	Yellow solid sulfur changes to sharp-smelling sulfur dioxide gas.
Burn gasoline.	Liquid gasoline burns, emits heat and light, and is converted to a mixture of carbon monoxide gas, carbon dioxide gas, and water vapor.
Dissolve silver nitrate in water, dissolve sodium chloride in water, and then mix the two solutions.	When the two colorless solutions are mixed, a white precipitate or insoluble compound called *silver chloride* forms and settles out of the solution.

(Section 1.3). Because a chemical change is almost always accompanied by one or more physical changes, telling the difference between a physical and chemical change is not always easy. You can usually conclude that a chemical change has taken place when you observe one or more of the following physical changes:

— Bubbles of gas form without heating.

— Odors and/or colors change.

— The solution turns cloudy, or a solid material called a **precipitate** settles out of solution.

— The material burns.

— A relatively large amount of heat, light, or both is given off.

Table 5.4 lists a few examples of chemical reactions that are accompanied by one or more of these physical changes.

CHEMICAL EQUATIONS   We could describe the change of reactants into products either in words or in pictures (Figure 5.11). But for convenience, chemists express a chemical reaction by arranging chemical symbols in what is known as a **chemical equation**. The formulas of all reactants are listed, separated by + signs, to the left of an arrow ($\longrightarrow$); the formulas of all products, also separated by + signs, are listed to the right of the arrow. Each + sign can be translated as "and," and the arrow denotes "reacts to yield." A chemical equation is like a sentence in which words

Figure 5.11 Different ways to describe a chemical reaction.

Word description:

Solid carbon reacts with oxygen gas to form carbon dioxide gas

Atomic-molecular description:

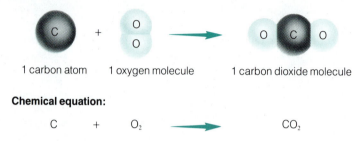

1 carbon atom   1 oxygen molecule   1 carbon dioxide molecule

Chemical equation:

$$C \quad + \quad O_2 \quad \longrightarrow \quad CO_2$$

(in the form of the chemical formulas of the substances involved) are combined in a particular way.

BALANCING CHEMICAL EQUATIONS   A chemical equation is more than just a list of reactants and products. It is also a form of chemical bookkeeping by which chemists can keep track of the atoms involved, as required by the law of conservation of mass (Section 2.2). This law implies that, if a reaction begins with a particular number of atoms of an element among the reactants, the products must contain the same number of atoms of the element—no more and no less. Chemists use chemical equations to represent this balance. When each side of the equation has the same number of atoms for each element, the law of conservation of mass is upheld, and a *balanced* chemical equation exists.

To balance a chemical equation, first write the correct formula for each reactant and product and separate the reactants and products with a reaction arrow. Then place the smallest possible set of whole-number coefficients in front of some or all of the formulas to make the number of atoms of each element the same on each side of the equation. *Do not change the subscripts in the formulas of the reactants and products to balance an equation because this would change the chemicals involved in the reaction.*

For example, when exposed to an electrical spark, hydrogen gas (H_2) reacts explosively with oxygen gas (O_2) to form water vapor (H_2O). To balance this equation, first write the correct formula for each reactant and product:

$$H_2 + O_2 \xrightarrow{\text{spark}} H_2O \qquad \textbf{Unbalanced equation}$$

This equation is not balanced because there are two atoms of oxygen on the left side and only one atom of oxygen on the right.

Now put numbers (coefficients) in front of the formulas to balance this equation, but do not change the formulas themselves. If you place a 2 in front of the formula for water, the oxygen atoms will be in balance (two atoms of O on each side):

$$H_2 + O_2 \xrightarrow{\text{spark}} 2H_2O \qquad \textbf{Unbalanced equation}$$

But this creates a hydrogen imbalance (four H atoms on the right and two H atoms on the left). To balance the hydrogens, you need to place a 2 in front of the H_2:

$$2H_2 + O_2 \xrightarrow{\text{spark}} 2H_2O \qquad \textbf{Balanced equation}$$

colorless gas   *colorless gas*   *colorless liquid*

Figure 5.12 illustrates what this balanced equation represents. Note that each subscript indicates the number of atoms of the element immediately preceding it in the formula and is fixed for a given substance. The coefficients, however, indicate the number of molecules needed to balance the equation so that the law of conservation of mass is not violated. (If no coefficient is written, it is understood to be 1.)

$$2H_2$$

coefficient 2
indicates two molecules   *subscript 2*
                            indicates two atoms per molecule

Figure 5.12 Meaning of a balanced chemical equation.

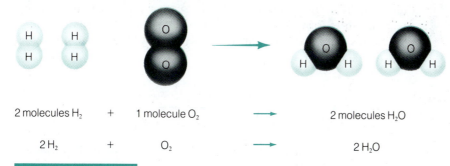

2 molecules H_2 +   1 molecule O_2 ⟶   2 molecules H_2O

2 H_2 + O_2 ⟶   2 H_2O

Figure 5.13 Explosion of the German dirigible *Hindenberg* during landing at Lakehurst, New Jersey, in 1937. (The Bettmann Archive)

PRACTICE EXERCISE

How many atoms of each element are represented by the following:
(a) NH_3, (b) $3O_2$, (c) $2CH_4$, (d) $3Mg(OH)_2$, (e) $2Ca(NO_3)_2$,
(f) $6H_2O$, (g) $2C_6H_{12}O_6$.

SOLUTION

(a) one N, three H; (b) six O; (c) two C, eight H; (d) three Mg, six O, six H; (e) two Ca, four N, twelve O; (f) twelve H, six O; (g) twelve C, twenty-four H, twelve O.

Hydrogen gas has the lowest density of any gas and is fifteen times less dense than air. Dirigibles used to be filled with H_2 gas to give them buoyancy in the denser air. After 1937, however, when the hydrogen-filled dirigible *Hindenberg* exploded and killed a number of people (Figure 5.13), this practice was discontinued. The tragedy occurred because a spark caused the hydrogen gas in the dirigible to react explosively with oxygen gas in the air to form water. Today dirigibles, such as the familiar Goodyear blimps, are filled with unreactive helium gas to eliminate the possibility of a similar disaster.

Figure 5.14 Chemical reaction taking place in a photographic flashbulb. (GTE Products Corporation)

PRACTICE EXERCISE

Figure 5.14 shows the reaction taking place in a photographic flashbulb. An electrical current produced by a battery causes a thin filament of magnesium metal (Mg) to react with oxygen gas (O_2) inside the bulb to produce magnesium oxide (MgO), a white powder, and energy in the form of heat and light. The unbalanced equation is $Mg + O_2 \longrightarrow MgO$. Balance this equation.

$$2Mg + O_2 \longrightarrow 2MgO$$

SOLUTION

NINE IMPORTANT CHEMICAL REACTIONS   The equations that follow are important reactions in air pollution, chemical manufacturing, and biological processes. They are not necessarily balanced as presented. Try to balance each equation. (The balanced equations are given in the back of the book as the answer to Question 21 in this chapter.)

The first two reactions represent the thousands of reactions involved in burning fossil fuels. One shows the burning of carbon (C), the main component of coal, and the other represents the combustion of octane (C_8H_{18}), an ingredient of gasoline:

carbon + *oxygen* $\longrightarrow$ *carbon dioxide*

Reaction 1 _____ C + _____ O_2 $\longrightarrow$ _____ CO_2

octane + *oxygen* $\longrightarrow$ *carbon dioxide* + *water*

Reaction 2 _____ C_8H_{18} + _____ O_2 $\longrightarrow$ _____ CO_2 + _____ H_2O

At ordinary temperatures, nitrogen gas (N_2) and oxygen gas (O_2), which make up most of the air around us, do not react with each other. At the high temperatures inside automobile engines, however, the two gases combine to produce colorless nitric oxide (NO). Although NO is relatively harmless itself, it reacts further with oxygen to form nitrogen dioxide (NO_2). A yellow-brown gas with a pungent odor, NO_2 contributes to the brown haze of polluted air:

nitrogen + *oxygen* $\longrightarrow$ *nitric oxide*

Reaction 3 _____ N_2 + _____ O_2 $\longrightarrow$ _____ NO

nitric oxide + *oxygen* $\longrightarrow$ *nitrogen dioxide*

Reaction 4 _____ NO + _____ O_2 $\longrightarrow$ _____ NO_2

Sulfur dioxide forms from the reaction of sulfur (S) with oxygen (O_2). Sulfur is either mined specifically to undergo this reaction or comes as an impurity in coal or oil that is burned. Once in the air, sulfur dioxide can react with oxygen to produce sulfur trioxide (SO_3), which then reacts with water to form sulfuric acid (H_2SO_4). All are air pollutants, but these compounds and reactions are also useful. In the United States alone, sulfur dioxide and compounds made from it generate more than \$3 billion in business each year. H_2SO_4, the workhorse of the chemical industry, plays a major role in the preparation of fertilizers and paints, in the refining of ores and petroleum, and in the manufacture of steel and many other products:

sulfur + *oxygen* $\longrightarrow$ *sulfur dioxide*

Reaction 5   ____ S + ____ O_2 $\longrightarrow$ ____ SO_2

sulfur dioxide + *oxygen* $\longrightarrow$ *sulfur trioxide*

Reaction 6   ____ SO_2 + ____ O_2 $\longrightarrow$ ____ SO_3

sulfur trioxide + *water* $\longrightarrow$ *sulfuric acid*

Reaction 7   ____ SO_3 + ____ H_2O $\longrightarrow$ ____ H_2SO_4

Ammonia (NH_3) is a vital compound. Huge quantities are manufactured for use as fertilizer and as a raw material to synthesize other chemicals. During World War I, the German scientist Fritz Haber developed the industrial process for combining nitrogen gas (N_2) and hydrogen gas (H_2) directly into ammonia. The Haber process is still used throughout industry today:

nitrogen + *hydrogen* $\longrightarrow$ *ammonia*

Reaction 8   ____ N_2 + ____ H_2 $\longrightarrow$ ____ NH_3

In *photosynthesis*, green plants use energy from sunlight to power a complex series of reactions. The series can be summarized in a rather simple net equation, in which H_2O and CO_2 from the atmosphere react to produce glucose ($C_6H_{12}O_6$), a simple sugar, plus O_2, which goes into the air:

carbon dioxide + *water* $\longrightarrow$ *glucose* + *oxygen*

Reaction 9   ____ CO_2 + ____ H_2O $\longrightarrow$ ____ $C_6H_{12}O_6$ + ____ O_2

LIMITATIONS OF BALANCED EQUATIONS   It is important to note that a balanced equation does not tell us everything. *A balanced chemical equation does not indicate how fast a reaction will go, or even whether it happens at all.* In practice the reactants may not react, or they may form different products from those shown in the equation. One example is

He + 2Ne $\longrightarrow$ HeNe$_2$ **Does not take place**

Although this equation is balanced, Ne and He are inert gases that refuse to combine with each other under any known laboratory conditions.

A balanced chemical equation does not indicate how the reaction occurs. The equation is merely a bookkeeping device, showing the initial reactants and the possible final products, without indicating exactly how the reactants are converted to products. The balanced equation for the photosynthesis reaction, for example, does *not* mean that six molecules of H_2O and six molecules of CO_2 collide to form one molecule of glucose and six molecules of O_2. Photosynthesis and many other reactions take place in a series of steps that when added together give the *net* or *overall reaction* represented by the balanced equation.

5.5 The Mole Concept:
Some Simple Arithmetic of Chemical Reactions

AVOGADRO'S NUMBER: COUNTING PARTICLES OF ELEMENTS AND COMPOUNDS Any piece of matter you can see contains an unbelievably large number of atoms, molecules, or formula units. These particles are so small compared to everyday objects that you can see them or weigh them only in extremely large numbers. For example, the smallest number of typical atoms you can see is about 1 quadrillion (1,000,000,000,000,000, or 10^{15}).

Chemists use a standard bunch that can be seen and weighed to represent an extremely large number of particles that make up elements and compounds. This number is called **Avogadro's number**, named in honor of Italian physicist and chemist Amadeo Avogadro. Avogadro's number is a counting unit, just like one dozen, except that it represents approximately 602,000,000,000,000,000,000,000 particles of an element or compound. It is more convenient to write this number as 6.02×10^{23} (see Appendix 1).

You can hardly comprehend how big this number is. If you tried to spend 6.02×10^{23} dollars at the rate of $1 billion a day, you would need 1.6 trillion years (400 times longer than the earth has existed). Or if you spread out 6.02×10^{23} grains of sand, they would cover the United States with a layer 7.5 cm (3 in.) deep.

AVOGADRO'S NUMBER AND THE MOLE CONCEPT Chemists use Avogadro's number to define the **mole** (mol) as the amount of substance represented by 6.02×10^{23} particles of an element or compound:

$$1 \text{ mol Fe atoms} = 6.02 \times 10^{23} \text{ Fe atoms}$$

$$1 \text{ mol } H_2O \text{ molecules} = 6.02 \times 10^{23} \text{ } H_2O \text{ molecules}$$

$$1 \text{ mol NaCl formula units} = 6.02 \times 10^{23} \text{ NaCl formula units}$$

$$3 \text{ mol } O_2 \text{ molecules} = 3 \times 6.02 \times 10^{23} \text{ } O_2 \text{ molecules}$$

Figure 5.15 The molar mass of an element is the mass in grams of 1 mol (6.02×10^{23}) of atoms in a natural sample of the element.

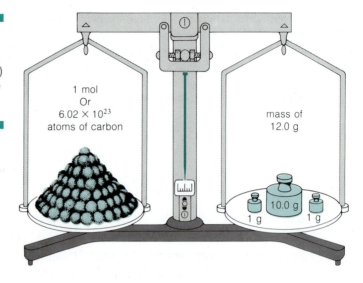

1 mol
Or
6.02×10^{23}
atoms of carbon

mass of
12.0 g

10.0 g

1 g   1 g

THE MOLE CONCEPT AND THE MOLAR MASS OF ELEMENTS AND COMPOUNDS   Atoms of different elements have different masses or weights. Although we cannot weigh a single atom or molecule, we can compare their relative masses by weighing the same large number (such as a mole) of each. Those masses will be in the same proportion as the masses of the individual atoms or molecules.

The **molar mass** is the mass in grams of 1 mol, or Avogadro's number (6.02×10^{23}), of the particles (atoms, molecules, or formula units) in a substance. To identify the molar mass of an element, look up the element's atomic mass (see the periodic table, inside front cover, or the table, inside back cover); that number of grams is the molar mass of the element. For example, the element carbon has a molar mass of 12.0 g. This means that 1 mol (Avogadro's number) of carbon atoms in a natural sample of carbon has a mass of 12.0 g (Figure 5.15). Similarly, the molar mass of iron is 55.8 g.

PRACTICE EXERCISE

What is the molar mass of (a) N and (b) copper?

(a) 14.0 g, (b) 63.5 g.

SOLUTION

To calculate the molar mass of a compound, add the molar masses of all atoms or ions in the formula of the compound. For example, carbon dioxide (CO_2) has 1 mol of carbon atoms and 2 mol of oxygen atoms in every mole of its molecules. Thus, the calculation is

$$\begin{aligned}\text{molar mass} &= (1 \times \text{molar mass of C}) + (2 \times \text{molar mass of O})\\ &= (1 \times 12.0\ \text{g}) + (2 \times 16.0\ \text{g})\\ &= 12.0\ \text{g} + 32.0\ \text{g} = \mathbf{44.0\ g}\end{aligned}$$

You can calculate the molar mass of the ionic compound Al_2O_3, which has 2 mol of Al atoms and 3 mol of O atoms per mole of formula units, in a similar way:

$$\begin{aligned}\text{molar mass} &= (2 \times \text{molar mass of Al}) + (3 \times \text{molar mass of O})\\ &= (2 \times 27.0\ \text{g}) + (3 \times 16.0\ \text{g})\\ &= 54.0\ \text{g} + 48.0\ \text{g} = \mathbf{102.0\ g}\end{aligned}$$

PRACTICE EXERCISE

What is the molar mass of (a) H_2, (b) H_2O, (c) Na_3PO_4?

SOLUTION

(a) $\text{molar mass of } H_2 = (2 \times \text{molar mass of H})$
$= (2 \times 1.0\ \text{g}) = \mathbf{2.0\ g}$

(b) $\text{molar mass of } H_2O = (2 \times \text{molar mass of H}) + (1 \times \text{molar mass of O})$
$= (2 \times 1.0\ \text{g}) + (1 \times 16.0\ \text{g})$
$= 2.0\ \text{g} + 16.0\ \text{g} = \mathbf{18.0\ g}$

(c) $\text{molar mass of } Na_3PO_4 = (3 \times \text{molar mass of Na})$
$+ (1 \times \text{molar mass of P})$
$+ (4 \times \text{molar mass of O})$
$= (3 \times 23.0) + (1 \times 31.0) + (4 \times 16.0)$
$= \mathbf{164.0\ g}$

USING THE MOLE AND MOLAR MASS CONCEPTS TO MAKE CALCULATIONS FROM BALANCED CHEMICAL EQUATIONS We can use the mole and molar mass concepts to interpret a balanced equation in several ways. The coefficients in a balanced equation give the mole-to-mole ratios for the substances

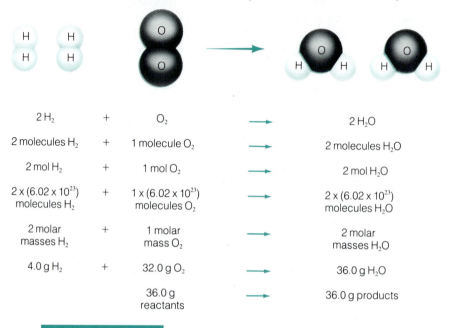

Figure 5.16 Relationships between reactants and products in a balanced chemical equation.

$2 H_2$	$+$	O_2	$\longrightarrow$	$2 H_2O$
2 molecules H_2	$+$	1 molecule O_2	$\longrightarrow$	2 molecules H_2O
2 mol H_2	$+$	1 mol O_2	$\longrightarrow$	2 mol H_2O
$2 \times (6.02 \times 10^{23})$ molecules H_2	$+$	$1 \times (6.02 \times 10^{23})$ molecules O_2	$\longrightarrow$	$2 \times (6.02 \times 10^{23})$ molecules H_2O
2 molar masses H_2	$+$	1 molar mass O_2	$\longrightarrow$	2 molar masses H_2O
4.0 g H_2	$+$	32.0 g O_2	$\longrightarrow$	36.0 g H_2O
		36.0 g reactants	$\longrightarrow$	36.0 g products

involved in a chemical reaction. In the reaction between H_2 and O_2 (Figure 5.16), for example, the balanced equation tells us that the reaction involves 2 mol of H_2 per 1 mol of O_2. We can express this as the following mole-to-mole ratios:

$$\frac{2 \text{ mol } H_2}{1 \text{ mol } O_2} \textit{ or its inverse } \frac{1 \text{ mol } O_2}{2 \text{ mol } H_2}$$

Other mole-to-mole ratios from the same balanced equation are

$$\frac{2 \text{ mol } H_2}{2 \text{ mol } H_2O} \textit{ or its inverse } \frac{2 \text{ mol } H_2O}{2 \text{ mol } H_2}$$

$$\frac{1 \text{ mol } O_2}{2 \text{ mol } H_2O} \textit{ or its inverse } \frac{2 \text{ mol } H_2O}{1 \text{ mol } O_2}$$

By using such simple mole-to-mole conversion factors, you can make many important calculations for chemical reactions. For example, you can determine how many moles or grams of reactant are required to produce a certain number of moles or grams of product. You can also calculate how much product would be formed by the complete reaction of a specified amount of the reactants.

MOLE-TO-MOLE CALCULATIONS The simplest of these calculations uses the mole-to-mole ratios alone.

PRACTICE EXAMPLE 5.1

How many moles of O_2 react with H_2 to produce 80 mol of H_2O?

SOLUTION

Step 1: Balance the equation (Section 5.4) and identify the unknown and known substances and their units:

$$2H_2 + O_2 \longrightarrow 2H_2O$$

$$\underset{\substack{\text{unknown}\\(\text{mol } O_2)}}{} \qquad \underset{\substack{\text{known}\\(\text{mol } H_2O)}}{}$$

Step 2: Perform the mole-to-mole conversion:

Unknown: **mol O_2**

Known: 80 mol H_2O; $\dfrac{1 \text{ mol } O_2}{2 \text{ mol } H_2O}$ (from balanced equation)

Plan: Convert **mol H_2O** (known) to **mol O_2** (unknown)

Result: Start with the known quantity (mol H_2O) and multiply it by the mole-to-mole ratio between the unknown (O_2) and the known (H_2O) so that the units cancel:

$$\underset{known}{80 \text{ mol } H_2O} \times \underset{\substack{\text{mole-to-mole}\\ \text{ratio}}}{\frac{1 \text{ mol } O_2}{2 \text{ mol } H_2O}} = \underset{unknown}{40 \text{ mol } O_2}$$

PRACTICE EXAMPLE 5.2

How many moles of O_2 are required for the complete combustion of 38 mol of methane (CH_4) to produce CO_2 and H_2O?

SOLUTION

Step 1: Balance the equation and identify the unknown and known substances and their units:

$$CH_4 + 2O_2 \longrightarrow CO_2 + 2H_2O$$

$$\underset{\substack{known \\ (mol\ CH_4)}}{} \qquad \underset{\substack{unknown \\ (mol\ O_2)}}{}$$

Step 2: Perform the mole-to-mole conversion:

Unknown: **mol O_2**

Known: **38 mol CH_4**; $\dfrac{2 \text{ mol } O_2}{1 \text{ mol } CH_4}$ (from balanced equation)

Plan: Convert **mol CH_4** (known) **to mol O_2** (unknown)

Result: $38 \text{ mol } CH_4 \times \dfrac{2 \text{ mol } O_2}{1 \text{ mol } CH_4} = $ **76 mol O_2**

Notice that in order for the units to cancel, we had to use the inverse of the mole-to-mole ratio shown as a known.

MOLE-TO-MASS CALCULATIONS Suppose you want to find out how many grams of H_2O are produced by the complete reaction of H_2 with 14 mol of O_2. In this case, the number of moles of H_2O produced from 14 mol O_2 (the known) is determined in a manner similar to the one used in Practice Example 5.1. The result is then multiplied by the molar mass of H_2O to convert from moles of H_2O into grams of H_2O (the unknown).

Step 1: Balanced equation:

$$2H_2 + O_2 \longrightarrow 2H_2O$$

$$\underset{\substack{known \\ (mol\ O_2)}}{} \qquad \underset{\substack{unknown \\ (g\ H_2O)}}{}$$

Step 2: Mole-to-mass conversion:

Unknown: **g H_2O**

Known: 14 mol O_2

Plan: mol $O_2 \longrightarrow$ mol $H_2O \longrightarrow$ **g H_2O**

         known   *unknown*

Result: $14 \text{ mol } O_2 \times \dfrac{2 \text{ mol } H_2O}{1 \text{ mol } O_2} \times \dfrac{18.0 \text{ g } H_2O}{1 \text{ mol } H_2O}$

      known   *mole-to-mole*   *molar mass*
                 ratio   *of unknown*

            = 504 g H_2O

PRACTICE EXAMPLE 5.3

How many moles of water are produced by the complete combustion of 600 g of methane?

SOLUTION

Step 1: Balanced equation:

$$CH_4 + 2O_2 \longrightarrow CO_2 + 2H_2O$$

    known   *unknown*
    (g CH_4)   (mol H_2O)

Step 2: Mass-to-mole conversion:

    Unknown: **mol H_2O**

    Known: 600 g CH_4; molar mass of known (CH_4)

    $= (1 \times 12.0 \text{ g C}) + (4 \times 1.0 \text{ g H})$

    $= 16.0 \text{ g } CH_4;\ \dfrac{2 \text{ mol } H_2O}{1 \text{ mol } CH_4}$ (from balanced equation)

Plan: **g $CH_4 \longrightarrow$** mol $CH_4 \longrightarrow$ **mol H_2O**

       known   *unknown*

Result: $600 \text{ g } CH_4 \times \dfrac{1 \text{ mol } CH_4}{16.0 \text{ g } CH_4} \times \dfrac{2 \text{ mol } H_2O}{1 \text{ mol } CH_4}$

          known   *molar mass*   *mole-to-mole*
                  of known   *ratio*

          = 75.0 mol H_2O

MASS-TO-MASS CALCULATIONS   Let's take these simple arithmetic calculations a step further by assuming that you want to find out how many grams of water will be produced by the complete reaction of 78.0 g of H_2 with O_2.

Step 1: Balanced equation:

$$2H_2 + O_2 \longrightarrow 2H_2O$$

known   *unknown*
(g H_2)   (g H_2O)

Step 2: Mass-to-mass conversion:

Unknown: **g H_2O**

Known: 78.0 g H_2; molar mass of known (H_2) = 2 × 1.0 g H

$$= 2.0 \text{ g } H_2; \frac{2 \text{ mol } H_2O}{2 \text{ mol } H_2} \text{ (from balanced equation);}$$

molar mass of unknown (H_2O) = (2 × 1.0 g H)

$$+ (1 \times 16.0 \text{ g O})$$

$$= 18.0 \text{ g } H_2O$$

Plan: g $H_2 \longrightarrow$ mol $H_2 \longrightarrow$ mol $H_2O \longrightarrow$ **g H_2O**

known   *unknown*

Result: $78.0 \text{ g } H_2 \times \dfrac{1 \text{ mol } H_2}{2.0 \text{ g } H_2} \times \dfrac{2 \text{ mol } H_2O}{2 \text{ mol } H_2} \times \dfrac{18.0 \text{ g } H_2O}{1 \text{ mol } H_2O}$

known   *molar mass*   *mole-to-mole*   *molar mass*
of known   *ratio*   *of unknown*

$$= \textbf{702 g } H_2O$$

PRACTICE EXAMPLE 5.4

How many grams of H_2 are needed to react completely with N_2 to produce 85 g of NH_3?

Solution continued on page 147.

(g NH_3)   (g H_2)
known   *unknown*

$$N_2 + 3H_2 \longrightarrow 2NH_3$$

Step 1: Balanced equation:

SOLUTION

Step 2: Mass-to-mass conversion:

Unknown: **g H_2**

Known: 85 g NH_3; molar mass of known (NH_3)

= $(1 \times 14.0 \text{ g N}) + (3 \times 1.0 \text{ g H})$

= 17.0 g NH_3; $\dfrac{2 \text{ mol } NH_3}{3 \text{ mol } H_2}$ (from balanced equation);

molar mass of unknown (H_2) = $(2 \times 1.0 \text{ g H})$

= 2.0 g H_2

Plan: g $NH_3 \longrightarrow$ mol $NH_3 \longrightarrow$ mol $H_2 \longrightarrow$ **g H_2**

    known   *unknown*

Result: 85 g $NH_3 \times \dfrac{1 \text{ mol } NH_3}{17.0 \text{ g } NH_3} \times \dfrac{3 \text{ mol } H_2}{2 \text{ mol } NH_3} \times \dfrac{2.0 \text{ g } H_2}{1 \text{ mol } H_2}$

    known   *known molar mass*   *mole-to-mole ratio*   *molar mass of unknown*

= **15.0 g H_2**

Now you know how to make the key arithmetic calculations involving reactants and products in balanced equations for chemical reactions.

ANTOINE LAVOISIER   We may lay it down as an incontestable axiom that, in all the operations of art and nature, nothing is created; an equal quantity of matter exists both before and after the experiment . . .

Summary

The kinetic molecular theory helps account for properties of the three physical states of matter—solids, liquids, and gases. The main forces between particles are ionic bonds, metal bonds, hydrogen bonds, dipole–dipole forces, and London forces.

The physical state of a substance at room temperature depends on the strength of the bonds between its particles. The temperature at which a substance changes states (melting point, boiling point, or sublimation temperature) also depends on the strength of the bonds between particles.

The ability of materials to form solutions depends on the type of interactions between their particles. A general rule is that like dissolves like.

Chemical changes occur when new substances form. A chemical change is represented by a chemical equation. A balanced chemical equation has coefficients in front of each formula as necessary so that equal numbers of atoms of each element occur in products and in reactants.

A mole is Avogadro's number (6.02×10^{23}) of atoms, molecules, or formula units of a substance. One mole of a substance equals the molar masses of the atoms in the formula. In a balanced chemical equation, coefficients are the ratio of moles of each substance that participate in the reaction. Using this information, you can calculate the amount (in moles or grams) of reactants or products that participate in chemical reactions.

Terms for Review

After completing this chapter, you should know and understand the meaning of the following terms:

Avogadro's number (p. 139)

boiling (p. 127)

chemical equation (p. 133)

condensation (p. 127)

dipole–dipole forces (p. 124)

dynamic equilibrium (p. 127)

freezing (p. 126)

hydrogen bond (p. 124)

kinetic molecular theory (p. 119)

London forces (p. 123)

melting (p. 126)

molar mass (p. 140)

mole (p. 139)

precipitate (p. 133)

solute (p. 129)

solution (p. 129)

solvent (p. 129)

sublimation (p. 126)

vaporization (p. 126)

vapor pressure (p. 127)

Questions

Odd-numbered questions are answered at the back of this book.

1. Classify each of the following as a physical change or a chemical reaction:
   a. A candle burns.
   b. Dirty clothes become clean in a washing machine.
   c. Homemade bread becomes moldy after several days.
   d. The filament of a light bulb glows when it is connected to electricity.
   e. Honey becomes thick and hard to pour in the cold.
   f. Iron rusts.
   g. A steak gets charred in all the places it touches on the grill.

2. Use the kinetic molecular theory to explain each of the following facts:
   a. Beer fizzes when the top is removed from the container.
   b. The volume of a balloon increases as it rises in the atmosphere.
   c. The volume of a balloon decreases when it is put into a refrigerator and increases when the balloon is removed.
   d. An aerosol can blows up when it is thrown into a fire.
   e. When an automobile is driven, the air in its tires becomes hot.
   f. The boiling point of water is 95°C in Denver (altitude 1624 m) and 120°C in a pressure cooker.

3. Distinguish between a solid, a liquid, and a gas (a) in terms of physical properties and (b) in terms of the kinetic molecular theory.

4. Describe what happens to the volume of a gas if you (a) increase its temperature but leave its pressure the same, (b) decrease its pressure but leave its temperature the same, and (c) increase its pressure and decrease its temperature simultaneously.

5. Explain what you could do to a sample of nitrogen gas (a) to increase its pressure, (b) to decrease its volume, and (c) to increase the average speed of its molecules.

6. Wood alcohol (CH_4O) has a higher vapor pressure than rubbing alcohol (C_3H_8O) under the same conditions. Which compound has stronger forces of attraction between its molecules? Explain.

7. If you set out equal amounts of the liquids butyl alcohol ($C_4H_{10}O$) and ethyl ether ($C_4H_{10}O$), you would observe that the ether vaporized faster. Which compound has stronger forces of attraction between its molecules?

8. Identify which of the following have dipole–dipole interactions between their molecules: (a) I_2, (b) NO, (c) CCl_4, (d) BCl_3, (e) SO_3.

9. Identify which of the following have hydrogen bonds between their molecules: (a) N_2, (b) HF, (c) H_2, (d) CH_4, (e) H_2O_2 (or H—O—O—H), (f) NH_3.

10. Describe how (a) an ionic compound such as CsCl and (b) a polar substance such as hydrogen chloride gas (HCl) dissolve in water.

11. Carbon tetrachloride (CCl_4) is nonpolar (Table 4.6). Is carbon tetrachloride soluble in water? Explain.

12. Explain why the following "business ventures" should or should not work:
    a. Use CCl_4 to clean up rust stains (Fe_2O_3).

b. Use CCl_4 to remove reddish purple stains from iodine (I_2).

c. Market a carbonated cleaning fluid by dissolving carbon dioxide (CO_2) in carbon tetrachloride (CCl_4).

d. Wash with water some $100 potassium bromide (KBr) plates for an infrared spectrophotometer.

e. Market "instant nitrogen" prepared by dissolving N_2 gas in water and bottling the solution.

f. Clean oil-base paint off brushes with water.

g. Clean water-soluble paint off brushes with CCl_4.

13. What is a chemical reaction? How could you detect whether a chemical reaction has occurred?

14. What information does a balanced chemical equation provide? What information does it *not* provide?

15. What is the relationship between a balanced chemical equation and the law of conservation of mass?

16. Identify which of the following equations are balanced:
a. $Na + 2H_2O \longrightarrow NaOH + H_2$
b. $2KClO_3 \longrightarrow 2KCl + 3O_2$
c. $C_3H_8 + 5O_2 \longrightarrow 3CO_2 + 4H_2O$
d. $HCl + 2Mg(OH)_2 \longrightarrow 2MgCl_2 + H_2O$
e. $Al + Fe_2O_3 \longrightarrow Fe + Al_2O_3$

17. Balance the following equations:
a. $Na + Cl_2 \longrightarrow NaCl$
b. $KNO_3 \longrightarrow KNO_2 + O_2$
c. $NaHCO_3 \longrightarrow$
$\qquad Na_2CO_3 + H_2O + CO_2$
d. $K + H_2O \longrightarrow KOH + H_2$
e. $NaCl + Pb(NO_3)_2 \longrightarrow PbCl_2 + NaNO_3$

18. Using simple circles to represent the atoms, draw a picture (similar to Figures

5.11 or 5.12) representing the following reaction:

$$CH_4 + 2O_2 \longrightarrow CO_2 + 2H_2O$$

19. Calculate the molar mass of (a) NaCl, (b) Na_2CO_3, (c) C_8H_{18}, (d) Ar, (e) N_2.

20. Calculate the molar mass of (a) I_2, (b) $Mg(HCO_3)_2$, (c) $CaSO_4$, (d) As, (e) cholesterol ($C_{27}H_{55}O$).

21. Balance the nine equations near the end of Section 5.4 and use these balanced equations to answer the questions below.

22. How many moles of ammonia will be produced by the complete reaction of 45 mol of H_2 with N_2?

23. How many moles of N_2 would be needed to produce 600 mol of ammonia by complete reaction with H_2?

24. How many grams of N_2 are needed to produce 285 mol of ammonia by complete reaction with H_2?

25. How many grams of O_2 will react completely with NO to produce 150 mol of NO_2?

26. How many moles of SO_3 will be produced by the complete reaction of 95.0 g of SO_2 with O_2?

27. How many grams of SO_3 will be produced by the complete reaction of 95.0 g of SO_2 with O_2?

28. In photosynthesis, how many moles of O_2 are produced by the complete reaction of 25 mol of CO_2 with H_2O?

29. If a plant consumes 125 g of CO_2 during photosynthesis, how many grams of glucose ($C_6H_{12}O_6$) and how many grams of O_2 will it produce?

Topics for Discussion

1. Water is sometimes called the *universal solvent*, implying that anything left in water long enough will dissolve. Is this true? Explain.

2. One encyclopedia stated that H_2O is the only common substance that exists in all three physical states: gas (steam), liquid (water), and solid (ice). Do you agree with that statement? Why?

3. Some ionic compounds, such as silver chloride (AgCl), do *not* dissolve in water. So what good is the rule *like dissolves like*? What kind of solvent would you try to use to dissolve silver chloride?

4. What makes the simple reaction in which sulfur (S) and oxygen (O_2) combine to produce sulfur dioxide (SO_2) so important to the chemical industry? Does *important* mean good or bad?

CHAPTER 6

Acid–Base and Oxidation–Reduction Reactions: Transferring Protons and Electrons

General Objectives

1. What are acids, bases, and acid–base reactions?

2. What are some practical examples of acid–base reactions in your daily life?

3. What are oxidation, reduction, and oxidation–reduction reactions?

4. What are some practical uses of oxidation–reduction reactions?

$\mathbf{O}$f the many chemical reactions that occur in your daily activities, two types stand out in importance. Both involve the transfer of simple particles—either protons or electrons—from one substance to another. The first type, acid–base reactions, occurs when you take an antacid to relieve an upset stomach, when someone bakes a cake, or when rain lands on the ground. The second type, oxidation–reduction reactions, helps release energy from such things as food, fossil fuels, and batteries. We also use them to kill bacteria, remove stains, and take pictures. In this chapter, you learn how to recognize these reactions and some of their uses.

6.1 Acids, Bases, and Acid–Base Reactions: Pass the Protons

ACIDS You are familiar with many substances that are acids. Hydrochloric acid in your stomach helps digest food. Lemons, grapefruits, and limes taste sour because they contain ascorbic acid and citric acid. Vinegar used in cooking or in salad dressing is a solution containing about 95 percent water and 5 percent acetic acid. The active ingredient in aspirin is acetyl-salicyclic acid. The liquid inside a car battery is a solution of water and sulfuric acid.

An **acid** is any substance that in water donates hydrogen ions (H^+) and thus produces **hydronium ions** (H_3O^+). The H^+ ion is often called a *proton* because it is a bare hydrogen nucleus containing one positively charged proton. Thus, we can also define an acid as a proton donor.

An example of an acid is hydrogen chloride (HCl), a gas consisting of polar covalent molecules (see Figure 4.16). When HCl gas is bubbled into water, its molecules react with water to form hydronium ions (H_3O^+)

and chloride ions (Cl^-):

$$HCl + H_2O \longrightarrow H_3O^+ + Cl^-$$

Notice that in this reaction each molecule of HCl donates a proton (H^+, a hydrogen atom that has lost its single electron) to water to form a hydronium ion.

Besides donating protons and producing H_3O^+ ions in water, acids have several other properties in common (Table 6.1):

1. Their chemical formulas generally start with H. From the formulas alone, you can recognize that HNO_3, H_2SO_4, and $HC_2H_3O_2$ are acids, and NaCl, NH_3, CH_4, NaOH, and I_2 are not acids.

2. They have a sour or tart taste. Only fools and TV detectives try to identify unknown and potentially dangerous substances with their tongues. Nevertheless, when you taste something sour or tart, it is almost always an acid.

3. They turn a vegetable dye called *litmus* from blue to red. Litmus paper contains this dye. If you dip a strip of litmus paper into the solution and the litmus paper turns red, the dissolved substance is an acid. This is the simplest and safest way to test for an acid.

Acids are classified as strong or weak. Acids that react completely, or nearly so, with water to produce hydronium ions are **strong acids**. For example, hydrochloric acid is a strong acid because HCl reacts completely with water to produce hydronium ions and chloride ions:

Strong acid: $HCl + H_2O \longrightarrow H_3O^+ + Cl^-$

TABLE 6.1 SOME CHARACTERISTIC PROPERTIES OF ACIDS AND BASES

Property	Acids	Bases
Ions produced in water solution	H_3O^+	OH^-
Proton (H^+) transfer	Proton donor	Proton acceptor
Formula	Usually starts with H	Usually ends with OH
Taste in foods	Sour or tart	Bitter
Feel on skin	—	Slippery
Litmus paper test	Red	Blue

Weak acids provide relatively few of the potential hydronium ions when dissolved in water. For example, the highly poisonous gas hydrogen cyanide (HCN) produces relatively few hydronium ions in water. This is indicated by placing a double reaction arrow ($\rightleftharpoons$) between the reactants and products, with the long arrow pointing to the left to indicate that most of the reactants remain unchanged:

Weak acid: $HCN + H_2O \rightleftharpoons H_3O^+ + CN^-$

The reactions going in both directions eventually produce a state of dynamic equilibrium (Section 5.2), with a constant but higher concentration of HCN and H_2O than of H_3O^+ and CN^-.

Table 6.2 lists some common strong and weak acids. Many people think that all acids eat holes in clothes and burn the skin. Some strong acids like hydrochloric acid (HCl), nitric acid (HNO_3), and sulfuric acid (H_2SO_4) are indeed dangerous if used improperly; if any of these acids

TABLE 6.2 SOME COMMON ACIDS

Name	Formula	Important Uses
Strong Acids		
Hydrochloric acid	HCl	Digesting food; cleaning hard water deposits from fixtures; removing excess mortar from bricks; removing scale or rust from metals
Nitric acid	HNO_3	Manufacturing ammonium nitrate fertilizer, explosives, dyes, plastics, and synthetic fibers; refining steel
Sulfuric acid	H_2SO_4	Producing hydrochloric acid; cleaning steel: refining petroleum; manufacturing plastics, dyes, fertilizers, and paper; liquid in car batteries; drain cleaner
Weak Acids		
Acetic acid	$HC_2H_3O_2$	Component of vinegar
Benzoic acid	$HC_7H_5O_2$	Food preservative
Ascorbic acid	$HC_6H_7O_6$	Vitamin C
Oxalic acid	$H_2C_2O_4$	Stain remover
Hydrocyanic acid	HCN	Fumigant
Carbonic acid	H_2CO_3	Carbonated soft drinks
Boric acid	H_3BO_3	Eyewash
Palmitic acid	$HC_{16}H_{31}O_2$	Making soap

gets in your eyes or spills on your skin or clothes, you should wash it off immediately with plenty of water.

Because they produce a much lower concentration of corrosive hydronium ions, many weak acids are harmless and do not deserve their bad reputation. Ascorbic acid ($HC_6H_7O_6$), for example, is vitamic C, and boric acid (H_3BO_3) is mild enough to be used as an eyewash.

BASES   You also are familiar with substances called bases. Most drain cleaners contain the base sodium hydroxide (NaOH), or lye. The main ingredient in milk of magnesia and certain other antacids is the base magnesium hydroxide, $Mg(OH)_2$. You also may have used ammonia gas (NH_3) dissolved in water, a household cleaner that is a base.

A **base** is any substance that produces hydroxide ions (OH^-) in water or accepts protons (H^+). The two definitions are similar because hydroxide ions (OH^-) accept protons from hydronium ions (H_3O^+) or other sources:

$$OH^- + H_3O^+ \rightleftharpoons 2H_2O$$

Most common bases are ionic compounds that contain hydroxide ions. KOH (potassium hydroxide), NaOH (sodium hydroxide), and $Ca(OH)_2$ (calcium hydroxide) are examples. Ammonia gas (NH_3), however, does not contain hydroxide ions but is classified as a base because when placed in water it accepts a proton (H^+ ion) from water to produce hydroxide ions:

$$NH_3 + H_2O \rightleftharpoons NH_4^+ + OH^-$$

Bases are classified as strong or weak depending on how many hydroxide ions they release or produce in water. From the direction of the arrows in the reaction above, you can see that NH_3 is a **weak base** because it produces relatively few hydroxide ions (OH^-) in water. Another example of a weak base is aluminum hydroxide:

Weak base: $Al(OH)_3 \rightleftharpoons Al^{3+} + 3OH^-$

A **strong base** such as sodium hydroxide (NaOH) releases all of its hydroxide ions in water:

Strong base: $NaOH \longrightarrow Na^+ + OH^-$

Table 6.3 lists some strong and weak bases and their uses. Strong bases such as sodium hydroxide and potassium hydroxide are used in drain cleaners because they can dissolve hair, grease, and various organic materials clogging a drain. One name-brand drain cleaner, for example, is simply a high-priced water solution of two very cheap (and very strong) chemicals, NaOH and KOH. Strong bases can destroy hair, skin, other

TABLE 6.3 SOME COMMON BASES

Name	Formula	Important Uses
Strong Bases		
Sodium hydroxide	NaOH	Refining petroleum and vegetable oils; producing aluminum, rayon, cellophane, pulp and paper, soaps, and detergents; etching; drain and oven cleaners
Potassium hydroxide	KOH	Producing liquid soaps, detergents, and fertilizers; fuel cells
Weak		
Ammonia	NH_3	Producing fertilizer, explosives, plastics, synthetic fibers, paper and pulp, rubber, refrigerants, detergents, insecticides, and food additives; cleaning agents
Calcium hydroxide	$Ca(OH)_2$	Producing mortar (used in laying bricks and cinder blocks), bleaching powder, and paper and pulp; purifying sugar; tanning hides; softening water; reducing soil acidity
Magnesium hydroxide	$Mg(OH)_2$	Antacids and laxatives (milk of magnesia)
Aluminum hydroxide	$Al(OH)_3$	Fixing dyes to fabrics; purifying water; antacids

body tissue, and many fabrics. If you spill a strong base on your skin or clothing, wash it off immediately with plenty of water.

Most bases share a number of characteristic properties (see Table 6.1):

1. Their ionic formulas often end in OH. You can easily recognize that potassium hydroxide, KOH, and magnesium hydroxide, $Mg(OH)_2$, are bases.

2. They taste bitter (a dangerous and inconclusive test). Unsweetened chocolate, herbs, cod-liver oil, and antacids all contain bitter-tasting bases.

3. They have a slippery feel, like soap. Bases break down fatty materials in the skin, making skin cells flat and thus able to slip over each other. Obviously, touching a substance that can decompose your skin is an unsafe way to test for a base.

4. A water solution of a base turns red litmus paper blue—the simplest and best way to identify a base.

THE pH SCALE How strong is an acid or a base solution? The sulfuric acid in batteries certainly is stronger than the acetic acid in vinegar, which in turn is stronger than the boric acid in eyewashes. And the lye, or sodium hydroxide, used to unclog drains and clean ovens is stronger than the ammonia used to clean windows. But *how much* stronger or weaker are these solutions?

Figure 6.1 The pH scale used to measure acidity and alkalinity of water solutions. Values shown are approximate.

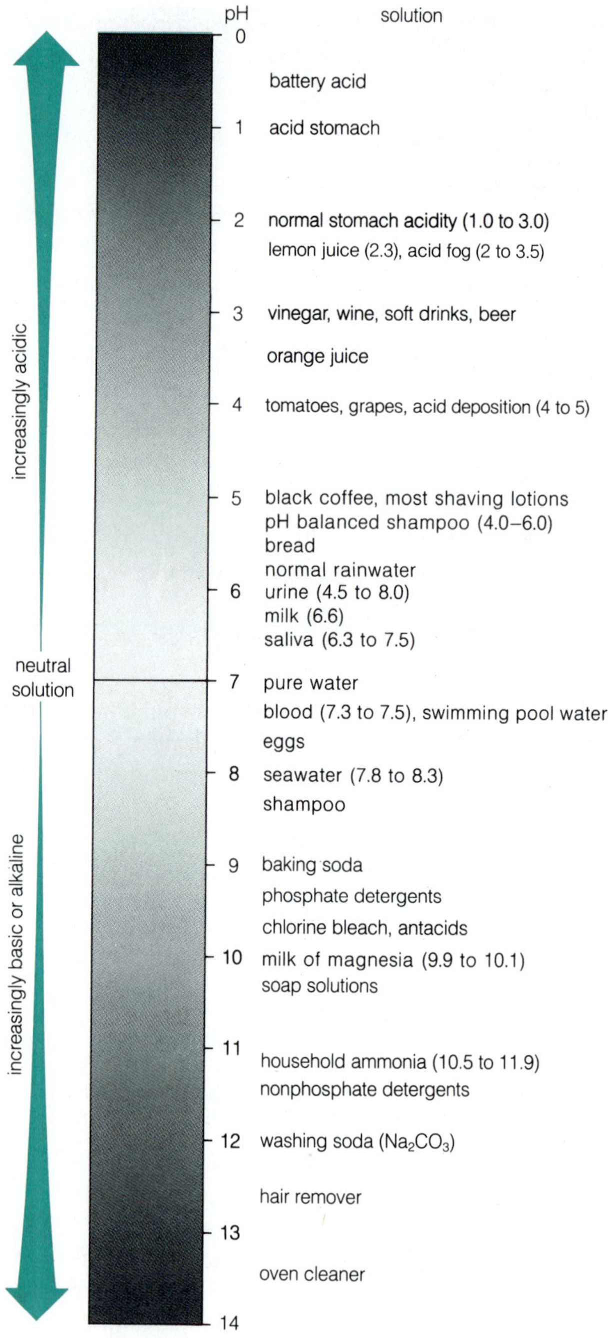

increasingly acidic

neutral solution

increasingly basic or alkaline

pH	solution
0	
	battery acid
1	acid stomach
2	normal stomach acidity (1.0 to 3.0)
	lemon juice (2.3), acid fog (2 to 3.5)
3	vinegar, wine, soft drinks, beer
	orange juice
4	tomatoes, grapes, acid deposition (4 to 5)
5	black coffee, most shaving lotions
	pH balanced shampoo (4.0–6.0)
	bread
	normal rainwater
6	urine (4.5 to 8.0)
	milk (6.6)
	saliva (6.3 to 7.5)
7	pure water
	blood (7.3 to 7.5), swimming pool water
	eggs
8	seawater (7.8 to 8.3)
	shampoo
9	baking soda
	phosphate detergents
	chlorine bleach, antacids
10	milk of magnesia (9.9 to 10.1)
	soap solutions
11	household ammonia (10.5 to 11.9)
	nonphosphate detergents
12	washing soda (Na_2CO_3)
	hair remover
13	
	oven cleaner
14	

To measure the acid or base strength of various water solutions, chemists have set up the *pH scale*. **pH** is a measure of the number of hydronium ions in 1 L of solution. For most acid or base solutions, the scale of pH values ranges from 0 to 14, as shown in Figure 6.1. A scientific instrument known as a *pH meter* (Figure 6.2) can be used to measure the pH value of a water solution.

Pure water has a pH value of 7. A tiny fraction of water molecules exist as hydronium and hydroxide ions:

$$H_2O + H_2O \rightleftharpoons H_3O^+ + OH^-$$

Figure 6.2 A pH meter. (Beckman Instruments, Inc.)

In pure water, the amount of hydronium ions equals the amount of hydroxide ions. Thus, water is **neutral**; it is neither acidic nor basic. Solutions having an excess of hydronium ions (H_3O^+) or hydroxide ions (OH^-) are *acidic* or *basic*, respectively. Another name for basic is **alkaline**.

Any solution with a pH value below 7 is acidic. pH is inversely related to acid strength. The lower the pH, the higher the number of hydronium ions per liter of solution and the greater the solution's acid strength.

The pH scale is set up mathematically so that each change of 1 pH unit corresponds to a tenfold change in the number of hydronium ions per liter of solution. Thus, a liter of vinegar whose pH is 3 (see Figure 6.1) has 10 times as many hydronium ions as a liter of grape juice whose pH is 4. A liter of lemon juice (pH 2) has 100 times as many hydronium ions as a liter of tomato juice whose pH is 4.

Any water solution with a pH above 7 is basic, or alkaline. The higher the value of pH above 7, the higher the number of hydroxide ions per liter of solution and the greater the solution's base strength. Thus a liter of a water solution of $Mg(OH)_2$ (milk of magnesia) with a pH of 10 (see Figure 6.1) has one-tenth the number of hydroxide ions as a liter of household ammonia with a pH of 11.

PRACTICE EXERCISE

Using the information in Figure 6.1, classify the following as acids or bases and compare their strengths: (a) normal stomach acidity and beer, (b) baking soda and washing soda, (c) black coffee and vinegar.

(a) normal stomach acidity (pH 2) is 10 times more acidic than beer (pH 3); (b) washing soda (pH 12) is 1000 times ($10 \times 10 \times 10$) more basic than baking soda (pH 9); (c) vinegar (pH 3) is 100 times (10×10) more acidic than black coffee (pH 5).

SOLUTION

ACID–BASE REACTIONS   An **acid–base reaction** is one in which a proton (H^+) is transferred from an acid (proton donor) to a base (proton acceptor). The general reaction is

$$\text{HX} + \text{MOH} \longrightarrow \text{H}_2\text{O} + \quad \text{MX}$$

acid   base   ionic
compound

Because the resulting solution is neutral, being neither acidic nor basic, this is also called a **neutralization reaction**.

When water solutions of hydrochloric acid (a strong acid) and sodium hydroxide (a strong base) are mixed, a proton is transferred from the acid to the base to form sodium chloride and water:

proton transfer

$$\text{HCl} \quad + \quad \text{NaOH} \quad \longrightarrow \text{NaCl} + \text{H}_2\text{O}$$

acid   base
(proton donor)   (proton acceptor)

Another example is the reaction between acetic acid ($HC_2H_3O_2$), a weak acid, and the strong base sodium hydroxide to form sodium acetate ($NaC_2H_3O_2$) and water.

proton transfer

$$\text{HC}_2\text{H}_3\text{O}_2 + \quad \text{NaOH} \quad \longrightarrow \text{NaC}_2\text{H}_3\text{O}_2 + \text{H}_2\text{O}$$

acid   base
(proton donor)   (proton acceptor)

This acid–base reaction is used to make oven cleaning safer and easier. Most oven cleaners contain a gel of NaOH that converts the grease spattered on oven walls into soap, which can be easily washed away. But wiping away the soap and excess oven cleaner with a damp rag soaks the rag with NaOH, which can attack unprotected hands, producing painful burns. Worse, repeated scrubbing with the wet rag is necessary because the excess gel doesn't mix well with water. This problem can be solved by spraying the oven with vinegar before wiping; the acetic acid in vinegar converts the mess into a neutral mixture of water and sodium acetate, which can easily be wiped away with a rag.

6.2 Some Useful Acid–Base Reactions: Reducing Excess Stomach Acidity, Improving Lakes and Soils, and Baking Cakes

ANTACIDS: REDUCING EXCESS STOMACH ACIDITY Your stomach produces hydrochloric acid to help digest food. Stress, overeating, or other factors can cause the stomach to produce too much HCl, lowering the pH of gastric juice from about 2 (see Figure 6.1) to less than 1. If this condition persists, the excess acid can destroy the tissue lining the stomach or intestine, causing ulcers. But you can often get temporary relief by taking an antacid preparation containing one or more alkaline substances to neutralize the excess HCl (Table 6.4).

For example, the magnesium hydroxide in milk of magnesia reduces stomach acidity by the following acid-base reaction:

$$2HCl \quad + \quad Mg(OH)_2 \quad \longrightarrow \quad MgCl_2 + 2H_2O$$

$\quad\quad$ *acid* $\quad\quad\quad\quad$ *base*

$\quad$ (*proton donor*) $\quad$ (*proton acceptor*)

TABLE 6.4 COMMON ANTACIDS AND THEIR MAIN INGREDIENT(S)	
Brand Name	**Ingredient(s)**
Alka-Seltzer	$NaHCO_3$
BiSoDol	$NaHCO_3$
Di-Gel	$Al(OH)_3$, $Mg(OH)_2$
Gaviscon	$Al(OH)_3$, $NaHCO_3$
Gelusil	$Al(OH)_3$, $Mg(OH)_2$
Maalox	$Al(OH)_3$, $Mg(OH)_2$
Milk of magnesia	$Mg(OH)_2$
Mylanta	$Al(OH)_3$, $Mg(OH)_2$
Riopan	$AlMg(OH)_5$
Rolaids	$AlNa(OH)_2CO_3$
Tums	$CaCO_3$

When dissolved in water, the sodium bicarbonate in antacids such as baking soda and Alka-Seltzer releases bicarbonate ions (HCO_3^-). These ions act as a base by accepting a proton from the HCl in gastric juice to form a water solution of sodium chloride and carbonic acid, a weak acid:

$$HCl \quad + \quad NaHCO_3 \quad \longrightarrow \quad NaCl + H_2CO_3$$

acid   *base*

(proton donor)   *(proton acceptor)*

Carbonic acid (H_2CO_3) then decomposes at body temperature to form carbon dioxide gas and water:

$$H_2CO_3 \longrightarrow CO_2 + H_2O$$

The overall reaction is

$$HCl \quad + \quad NaHCO_3 \quad \longrightarrow \quad NaCl + CO_2 + H_2O$$

acid   *base*

(proton donor)   *(proton acceptor)*

PRACTICE EXERCISE

The reaction of $CaCO_3$ in Tums (Table 6.4) with stomach acid (HCl) is similar to the reaction above with $NaHCO_3$. Write a balanced chemical equation for the neutralization of HCl with $CaCO_3$.

SOLUTION

$$2HCl \quad + \quad CaCO_3 \quad \longrightarrow \quad CaCl_2 + CO_2 + H_2O$$

acid   *base*

(proton donor)   *(proton acceptor)*

Ideally, an antacid should

— Decrease stomach acidity rapidly (instant relief).

— Maintain normal stomach acidity for at least 30 minutes to 1 hour (lasting relief).

— Not decrease stomach acidity too much (this would impair digestion and could cause *acid rebound*, in which the body responds by secreting extra HCl and thus aggravating the problem).

— Cause no side effects such as constipation or diarrhea.

Each antacid preparation has certain advantages and disadvantages. Antacids containing $Al(OH)_3$, for example, tend to do the best job of meeting the first three conditions but tend to cause constipation. Sodium bicarbonate ($NaHCO_3$) reacts quickly but tends to cause acid rebound and typically provides relief for only a few minutes. Calcium carbonate ($CaCO_3$) provides relief for about 30 minutes but also tends to cause acid rebound.

REDUCING EXCESS ACIDITY IN LAKES AND SOILS   Acids in rain, snow, and fog can cause considerable damage to trees, crops, aquatic life, and building and other materials (Section 12.2). Sulfur dioxide (SO_2) and nitrogen oxides emitted by coal-burning power plants and nitrogen oxides emitted by cars can be converted in the atmosphere into droplets of sulfuric and nitric acids (Section 5.4; Figure 6.3). When these acids fall to earth in rain or snow, they harm the land and water unless they are neutralized.

Normal rainwater is slightly acidic (pH = 5.6) because a weak solution of carbonic acid (H_2CO_3) forms when rainwater dissolves carbon dioxide in the air:

$$CO_2 + H_2O \rightleftharpoons H_2CO_3$$

Figure 6.3 Acid deposition.

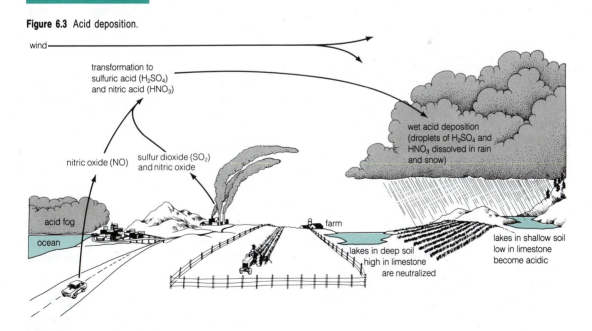

Figure 6.4 A farmer can spread lime (Ca(OH)$_2$), a base, to neutralize soil acidity from acid rain and from the use of synthetic fertilizers. (G. E. Lee/U.S. Department of Agriculture Soil Conservation Service)

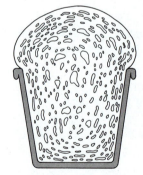

before rising

after rising because of bubbles of CO$_2$ gas

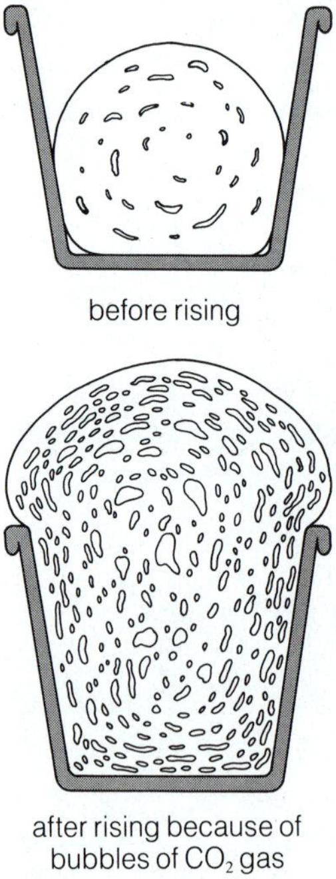

Figure 6.5 Dough rises when baking soda (NaHCO$_3$) reacts with a weak acid in the dough to produce tiny bubbles of CO$_2$.

Acid rain often has a pH of about 4, which is forty times more acidic than normal rainwater.

Soils in the Midwest states typically contain CaCO$_3$ (limestone), CaMg(CO$_3$)$_2$ (dolomite), and other basic substances that neutralize acids before they enter nearby lakes and streams. For example, limestone neutralizes sulfuric acid (H$_2$SO$_4$) by the reaction

$$H_2SO_4 + CaCO_3 \longrightarrow CaSO_4 + H_2O + CO_2$$

In other areas, especially the northeastern United States, southeastern Canada, and Sweden, many soils are naturally acidic and thus have little ability to neutralize additional acids. As a result, some lakes and streams in these areas have become so acidic that they no longer contain fish.

Most experts believe that the solution is to reduce the emissions of sulfur dioxide and nitrogen oxides. In the meantime, excess acidity in lakes or soil (Figure 6.4) can be neutralized by adding lime, Ca(OH)$_2$. The acid–base reaction is

$$H_2SO_4 \quad + \quad Ca(OH)_2 \quad \longrightarrow \quad CaSO_4 \quad + 2H_2O$$

sulfuric acid   *calcium hydroxide*   *calcium sulfate*
  (acid)   *(base)*

However, this procedure must be done frequently, is quite expensive, and adds to the mineral content of lakes and soil.

BAKING CAKES AND PUTTING OUT FIRES   The dough used to make cakes, biscuits, and many other bakery items rises because of an acid–base reaction involving sodium bicarbonate, commonly called *baking soda*. Bicarbonate ions (HCO$_3^-$) from the baking soda react with a weak acid in moist dough to produce carbonic acid (H$_2$CO$_3$), which decomposes into carbon dioxide gas and water when the dough is heated in the oven:

$$H_3O^+ + HCO_3^- \longrightarrow H_2CO_3 + H_2O$$

$$H_2CO_3 \longrightarrow H_2O + \mathbf{CO_2}$$

The tiny bubbles of CO$_2$ then cause the dough to rise (Figure 6.5). Other baked goods, such as bread, rise because yeast produces CO$_2$ by fermenting sugars in the dough.

Some fire extinguishers use this same acid–base reaction between sodium bicarbonate (NaHCO$_3$) and an acid. The reaction generates CO$_2$ inside the extinguisher, and the increase in pressure propels chemicals out to combat the fire.

6.3 Oxidation, Reduction, and Oxidation–Reduction Reactions: Pass the Electrons

OXIDATION AND REDUCTION: THE LOSS OR GAIN OF ELECTRONS Oxygen, the most abundant element on earth and in the human body (Figures 10.3 and 10.11) is also one of the most reactive elements. It can combine with nearly every other element. Reactions involving oxygen predominate in any list of important chemical reactions. For example, eight of the nine important chemical reactions discussed in Section 5.4 involve oxygen.

It is not surprising then that early chemists classified chemical reactions around oxygen. Any element or compound that gained oxygen was said to be *oxidized*. For example, carbon burned in oxygen gas to form carbon dioxide is oxidized because it gains two oxygen atoms:

$$C \quad + \quad O_2 \quad \longrightarrow \quad CO_2$$

black solid   *colorless,*   *colorless gas*
(oxidized)   *odorless gas*

Because oxygen is very electronegative (Figure 4.15) and thus strongly attracts electrons, substances that react with oxygen lose electrons partially or completely. Thus, **oxidation** is commonly defined as the loss of one or more electrons by an element in a chemical reaction. In the reaction above, each carbon atom ($\cdot \overset{\cdot}{C} \cdot$) begins with four valence electrons. Once it reacts with oxygen, a carbon atom in CO_2 shares a total of eight valence electrons:

$$:\overset{..}{O}::C::\overset{..}{O}:$$

But because of oxygen's much greater electronegativity, a carbon atom in CO_2 actually has an average of *less* than four valence electrons in the vicinity of its nucleus; carbon, in effect, has lost electrons and thus is *oxidized*.

Oxidation is more apparent when electrons are lost directly. Zinc (Zn), for example, is oxidized in any reaction in which it loses two electrons to form a zinc ion (Zn^{2+}):

Oxidation: $Zn \longrightarrow Zn^{2+} + 2e^-$

But this is only half of a reaction and cannot take place by itself. The zinc metal must donate its electrons to some other substance that accepts electrons. For example, copper(II) ions (Cu^{2+}) can accept two electrons from zinc and be converted to copper metal:

$$Cu^{2+} + 2e^- \longrightarrow Cu$$

This process, by which an element gains one or more electrons, is called **reduction**. Whenever an element gains electrons and thus has a more negative charge, it is said to be *reduced*.

Another sign of oxidation is in reactions involving hydrogen. Because hydrogen is the least electronegative of reactive nonmetals, nonmetals that react with hydrogen essentially gain the valence electron in each hydrogen atom. Hydrogen, which loses its electron in the process, thus is oxidized while the element that gains (combines with) hydrogen is reduced. One example we examined earlier (Section 5.4) is the synthesis of ammonia (NH_3):

$$N_2 \quad + \quad 3H_2 \quad \longrightarrow \quad 2NH_3$$
$$\textit{reduced} \quad \textit{oxidized}$$

TABLE 6.5 THREE SIGNS OF OXIDATION AND REDUCTION	
Oxidation	Reduction
Gain O	Lose O
Lose e^-	Gain e^-
Lose H	Gain H

Table 6.5 summarizes three signs by which you can recognize that oxidation and reduction have occurred.

OXIDATION–REDUCTION REACTIONS   Oxidation cannot take place without reduction. According to the law of conservation of mass, we can neither create nor destroy matter, including electrons, in a chemical reaction. Thus, in any reaction the number of electrons (or oxygen atoms or hydrogen atoms) lost by one substance must be equal to the number of electrons (or oxygen atoms or hydrogen atoms) gained by another substance.

In an **oxidation–reduction reaction** at least one element undergoes oxidation and at least one element undergoes reduction. For example, we can combine the oxidation half-reaction involving Zn with the reduction half-reaction involving Cu^{2+} to get an overall oxidation–reduction reaction:

Oxidation:	$Zn \longrightarrow Zn^{2+} + 2e^-$
Reduction:	$Cu^{2+} + 2e^- \longrightarrow Cu$
Oxidation–reduction reaction:	$Zn \quad + \quad Cu^{2+} \longrightarrow Zn^{2+} + Cu$
	$\textit{oxidized} \quad \textit{reduced}$

The net effect is that two electrons are transferred from Zn to Cu^{2+}

You could see that an oxidation–reduction reaction takes place between Zn and Cu^{2+} if you immersed a strip of zinc metal in a solution

Figure 6.6 A spontaneous oxidation–reduction reaction.

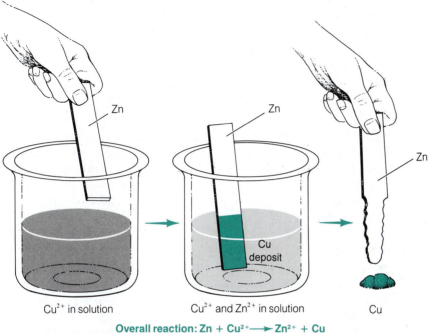

Zn

Zn

Zn

Cu deposit

Cu²⁺ in solution

Cu²⁺ and Zn²⁺ in solution

Cu

Overall reaction: Zn + Cu²⁺ ⟶ Zn²⁺ + Cu

containing copper(II) (Cu^{2+}) ions (Figure 6.6). The zinc would become covered with a reddish deposit that could be identified as copper metal (Cu). In addition, the blue color of the solution (due to the presence of Cu^{2+} ions) would fade, indicating a decrease in the number of Cu^{2+} ions.

In this reaction, Zn is oxidized, and Cu^{2+} is reduced. In an oxidation–reduction reaction, the substance that brings about oxidation by accepting electrons (and in the process is reduced) is called the **oxidizing agent**. Similarly, the substance that brings about reduction by giving up electrons (and in the process is oxidized) is called the **reducing agent**. Thus, Zn is oxidized and acts as reducing agent, and Cu^{2+} is reduced and acts as an oxidizing agent:

$$\text{Zn} + \text{Cu}^{2+} \longrightarrow \text{Cu} + \text{Zn}^{2+}$$

oxidized-   *reduced-*
reducing agent *oxidizing agent*

PRACTICE EXERCISE

Seawater contains small amounts of bromide ions (Br^-). These ions can be converted into Br_2 by bubbling chlorine gas through the seawater:

$$2Br^- + Cl_2 \longrightarrow Br_2 + 2Cl^-$$

Identify the substance oxidized, the substance reduced, the oxidizing agent, and the reducing agent.

SOLUTION

An element in its natural state, such as Cl_2 or Br_2, has an electrical charge of 0. In this reaction, Br^- increases its electrical charge from $1-$ to 0 and thus means that it acts as the reducing agent. Cl_2 is reduced because it undergoes a decrease in electrical charge from 0 to $1-$, and in the process it acts as an oxidizing agent.

When the changes in an electrical charge are not shown, it is harder to identify which elements have been oxidized and which have been reduced. Here you can look for the other two signs listed in Table 6.5.

For example, consider the reaction that takes place when phosphorus in an artillery shell explodes and reacts with oxygen gas to form a white smoke containing diphosphorus pentoxide (Figure 6.7). Because it combines with oxygen, phosphorus is oxidized and acts as a reducing agent; simultaneously, oxygen is reduced and is the oxidizing agent:

$$\underset{\substack{oxidized\text{-} \\ reducing\ agent}}{4P} \quad + \quad \underset{\substack{reduced\text{-} \\ oxidizing\ agent}}{5O_2} \quad \longrightarrow \quad 2P_2O_5$$

Figure 6.7 When an artillery shell explodes, phosphorus reacts with oxygen gas to produce a white smoke containing particles of diphosphorus pentoxide (P_2O_5). (U.S. Department of the Army)

PRACTICE EXERCISE

When iron rusts, it undergoes the following oxidation–reduction reaction:

$$4Fe + 3O_2 \longrightarrow 2Fe_2O_3$$

Identify the substance oxidized, the substance reduced, the oxidizing agent, and the reducing agent.

Fe is oxidized because it combines with oxygen; it also acts as the reducing agent. This means that O_2 is reduced and is the oxidizing agent.

SOLUTION

Another example is reacting a typical vegetable oil ($C_{57}H_{98}O_6$) with hydrogen gas to make shortening. In this reaction, you can recognize that the vegetable oil is reduced (and thus is the oxidizing agent) because it gains hydrogen atoms:

$$C_{57}H_{98}O_6 \quad + \quad 5H_2 \quad \longrightarrow \quad C_{57}H_{108}O_6$$

reduced-	*oxidized-*	*(shortening)*
oxidizing agent	*reducing agent*	

6.4 Some Useful Oxidation–Reduction Reactions

BLEACHING Common household liquid bleach is made by dissolving a small amount of sodium hypochlorite ($NaOCl$) in water. The hypochlorite ions (OCl^-) act as an oxidizing agent to add oxygen to a colored molecule in a fabric and thereby convert it into a colorless molecule. The general equation is

$$\text{colored molecule} + \quad OCl^- \quad \longrightarrow \quad Cl^- + \text{colorless molecule}$$

oxidized-	*reduced-*
reducing agent	*oxidizing agent*

A similar reaction occurs when people use oxidizing agents such as hydrogen peroxide (H_2O_2) on their hair. Hair pigments such as melanin become colorless, leaving the hair blond:

$$\text{colored hair} + \text{H}_2\text{O}_2 \longrightarrow \text{colorless hair} + \text{H}_2\text{O}$$
pigments   "pigment"

oxidized-   *reduced-*
reducing agent   *oxidizing agent*

DISINFECTANTS   Solid sodium hypochlorite and bleaching powder, which is calcium hypochlorite, $Ca(OCl)_2$, are also used to kill germs. The hypochlorite ions added to a swimming pool oxidize chemicals in the germ cells to disinfect the water:

$$\text{live germ} + \text{OCl}^- \longrightarrow \text{Cl}^- + \text{dead germ}$$

oxidized-   *reduced-*
reducing agent   *oxidizing agent*

A similar process occurs when wastewater is chlorinated to kill bacteria (Section 13.7).

Disinfectants are also put on the skin to prevent infections of cuts and scrapes. Two such oxidizing agents are tincture of iodine (a solution of I_2 in ethyl alcohol) and a 3 percent solution of hydrogen peroxide (H_2O_2) in water. The iodine works like this:

$$\text{live germ} + \text{I}_2 \longrightarrow 2\text{I}^- + \text{dead germ}$$

oxidized-   *reduced-*
reducing agent   *oxidizing agent*

REMOVING STAINS   Stain removers work in a variety of ways. Some adsorb the molecules that make up the stain. Cornstarch, for example, removes grease stains in this way. Other substances dissolve or wash out the stain. This is how acetone removes ballpoint-ink stains and detergents remove mustard stains. Certain stains can also be removed by using oxidizing and reducing agents, as shown in Table 6.6.

For instance, a solution of potassium permanganate—which has a purple color because of the presence of permanganate ions (MnO_4^-) in the solution—can remove most stains on white cotton fabric. The purple color then is removed by treatment with a reducing agent, oxalic acid $(H_2C_2O_4)$:

$$5\text{H}_2\text{C}_2\text{O}_4 + 2\text{MnO}_4^- + 6\text{H}_2\text{O} \longrightarrow 10\text{CO}_2 + 2\text{Mn}^{2+} + 14\text{H}_2\text{O}$$

oxidized-   *purple*   *practically*
reducing agent   *reduced*   *colorless*
                     oxidizing agent

TABLE 6.6 SOME REDUCING AND OXIDIZING AGENTS USED AS STAIN REMOVERS		
Name	**Formula**	**Use**
Reducing Agents		
Oxalic acid	$H_2C_2O_4$	Rust spots and potassium permanganate ($KMnO_4$) stains
Sodium thiosulfate	$Na_2S_2O_3$	Iodine and silver stains
Oxidizing Agents		
Sodium hypochlorite	NaOCl	Effective on most stains on cotton and linen (should not be used on wool or silk)
Hydrogen peroxide	H_2O_2	Blood stains on cotton or linen
Potassium permanganate	$KMnO_4$	Most stains on white fabrics except rayon (purple $KMnO_4$ stain must be removed, usually with oxalic acid)

Sodium thiosulfate (Table 6.6) removes iodine stains by reducing I_2 to the colorless iodide ion (I^-):

$$2Na_2S_2O_3 \ + \ \underset{\substack{\text{purple-brown} \\ \text{reduced} \\ \text{oxidizing agent}}}{I_2} \ \longrightarrow \ \underset{\text{colorless}}{2NaI} \ + Na_2S_4O_6$$

oxidized-
reducing agent

PHOTOGRAPHY A typical black-and-white photographic film is made by taking a plastic base and covering it with a gelatin in which millions of tiny crystals of a silver compound, such as silver bromide (AgBr), are suspended. When the film is exposed to a brief flash of light in a camera, an image is recorded on the film through an oxidation–reduction reaction in which silver ions (Ag^+) and bromide ions (Br^-) in the formula units of the ionic compound AgBr are converted into their respective elements:

$$\underset{\substack{\text{reduced-} \\ \text{oxidizing agent}}}{2Ag^+} \ + \ \underset{\substack{\text{oxidized-} \\ \text{reducing agent}}}{2Br^-} \ \xrightarrow{\text{light}} \ \underset{\substack{\text{shows as dark} \\ \text{clusters on film}}}{2Ag} \ + Br_2$$

Intense light striking the film converts large numbers of Ag^+ ions to Ag atoms, which show up as dark areas once the negative is developed. Less light produces lighter shades of gray because fewer Ag atoms are formed. When no light strikes an area of the film, no Ag atoms form, and the developed negative remains transparent.

In a darkroom, the photographer develops the negative by placing the exposed film in a solution of a reducing agent. Here the AgBr crystals that had formed some metallic silver by exposure to light will react further to form more silver:

$$C_6H_6O_2 \quad + \quad 2AgBr \quad \longrightarrow \quad 2Ag + 2HBr + C_6H_4O_2$$

hydroquinone oxidized	reduced	quinone
reducing agent	*oxidizing agent*	

Another solution, called a *fixer*, then removes the remaining, unexposed AgBr from the film before the light is turned on. The resulting *negative* is the inverse of the final black-and-white photograph. The *print* is made by shining light through the negative and onto light-sensitive photographic paper.

Instant films work on the same basic principle, except that the image on the negative is made into a positive print by chemicals known as transfer agents rather than by shining light through the developed negative.

BATTERIES   In Figure 6.6, we saw that if a strip of zinc is dipped into a solution containing Cu^{2+} ions, the following oxidation–reduction reaction occurs:

$$Zn + Cu^{2+} \longrightarrow Zn^{2+} + Cu$$

This transfer of electrons converts chemical energy into electrical energy. If we could devise a way to let these electrons travel through a wire from the zinc metal to the copper metal, we would have a source of electrical current. A device that converts chemical energy into electrical energy is an **electrochemical cell**.

An electrochemical cell can be made by placing each metal (Zn and Cu) in separate compartments, as shown in Figure 6.8. This allows oxidation and reduction to take place in separate compartments. To complete the electrical circuit, we must do two things: (1) run a wire between the two solid strips of metals, the *electrodes*, to let electrons flow from the oxidation compartment to the reduction compartment; and (2) connect the solutions in the two compartments with a tube containing a porous barrier that prevents mixing but lets ions flow from one compartment to the other. Such a cell can produce about 1.10 volts (V) of electricity, enough to run a tiny light bulb, until the Zn or Cu^{2+} is depleted.

One or a series of electrochemical cells is a **battery**. As Figure 6.9 shows, common flashlight batteries, smaller silver oxide batteries (which power devices such as hearing aids, electronic calculators, and watches), and automotive lead-storage batteries operate on similar oxidation–reduction reactions.

Figure 6.8 A simple battery. The electron flow is created by the oxidation–reduction reaction between zinc and copper(II) ions.

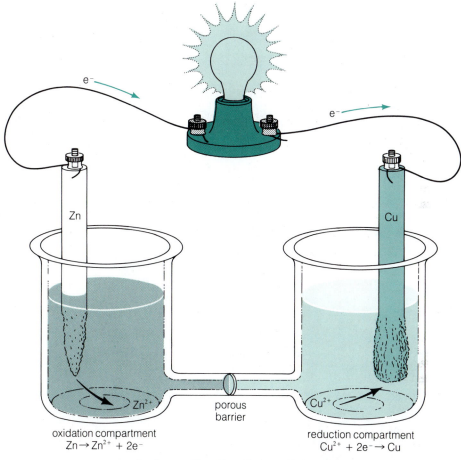

oxidation compartment
$Zn \rightarrow Zn^{2+} + 2e^-$

reduction compartment
$Cu^{2+} + 2e^- \rightarrow Cu$

net redox reaction: Zn + Cu²⁺ ⟶ Zn²⁺ + Cu

The longer-life alkaline batteries typically use manganese dioxide (MnO_2) and potassium hydroxide (KOH) as the moist paste. The net reaction is

$$2Zn + 3MnO_2 + 2H_2O \rightleftharpoons 2Zn(OH)_2 + Mn_3O_4 + \text{electrical energy}$$

oxidized   *reduced*

Figure 6.9 Some common batteries and their oxidation–reduction reactions.

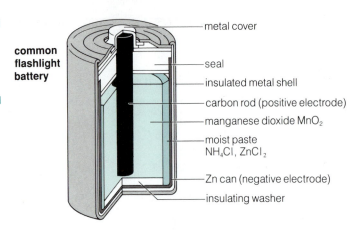

common flashlight battery

metal cover

seal

insulated metal shell

carbon rod (positive electrode)

manganese dioxide MnO_2

moist paste NH_4Cl, $ZnCl_2$

Zn can (negative electrode)

insulating washer

$$Zn + 2\,MnO_2 + 2\,NH_4Cl \longrightarrow ZnCl_2 + Mn_2O_3 + 2\,NH_3 + H_2O + \text{electrical energy}$$
oxidized *reduced*

metal cap

gasket

silver oxide cell

Zn

Ag_2O

separator

metal cup

$$Ag_2O + H_2O + Zn \longrightarrow 2\,Ag + Zn(OH)_2 + \text{electrical energy}$$
reduced   *oxidized*

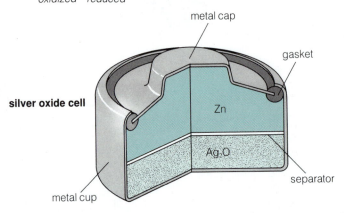

oxidation electrode (during discharge)

2 V

reduction electrode (during discharge)

12 V

single cell

spongy Pb

PbO_2

Pb plates

storage battery with 6 cells

H_2SO_4 solution

$$Pb + PbO_2 + 2\,H_2SO_4 \rightleftharpoons 2\,PbSO_4 + 2\,H_2O + \text{electrical energy}$$
oxidized *reduced*

For nickel–cadmium (Nicad) batteries, the reaction is

$$Cd + NiO_2 + 2H_2O \rightleftharpoons Cd(OH)_2 + Ni(OH)_2 + electrical\ energy$$

PRACTICE EXERCISE

In the reaction for nickel–cadmium batteries, which substance is oxidized, and which is reduced?

Cd is oxidized, and NiO₂ is reduced.

SOLUTION

Notice that the two reactions for the alkaline and nickel–cadmium batteries have arrows going in both directions, as does the reaction for the automobile battery (Figure 6.9, bottom). The double arrows mean these reactions go in both directions and are reversible. As a result, these batteries can be recharged by putting electricity back into them. A generator or alternator does this to an automobile battery when the engine is running. Other batteries can be recharged by plugging them into an appropriate source of electricity.

In a *fuel cell*, substances undergo oxidation and reduction continuously as they enter the cell as fuel, much as in a natural gas or oil furnace. Electricity is produced as long as the reactants keep entering the fuel cells and reacting.

One of the most promising candidates is the hydrogen–oxygen fuel cell (Figure 6.10) used on space flights. The reactants, H_2 and O_2, are stored in tanks that continuously supply the fuel cell. The oxidation–reduction reaction of hydrogen gas and oxygen gas produces water and generates electricity. Scientists expect various types of improved fuel cells to run cars and possibly to heat homes and buildings in the future.

STAYING ALIVE   Your body has to generate enough energy for you to grow new cells, breathe, pump blood, scream, dance, and do everything else that makes you alive. That energy comes from oxidizing food. When you metabolize food, such as the carbohydrate glucose ($C_6H_{12}O_6$), you produce energy:

$$C_6H_{12}O_6 + 6O_2 \longrightarrow 6CO_2 + 6H_2O + energy$$

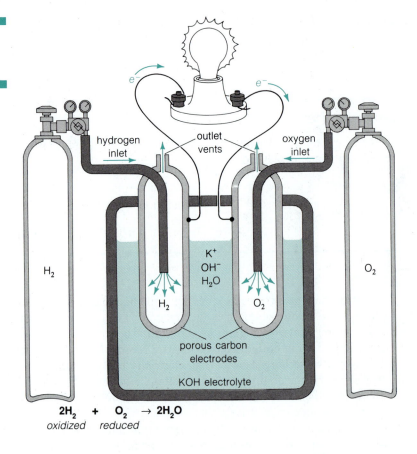

Figure 6.10 A hydrogen–oxygen fuel cell.

$$2H_2 + O_2 \rightarrow 2H_2O$$
oxidized   reduced

Look at this equation carefully. Notice that the carbon in glucose becomes enriched in oxygen when it becomes CO_2; thus, glucose is oxidized. Some of the O_2 gains hydrogen to become water, so O_2 is reduced.

Oxygen is the grand oxidizing agent in your body. It oxidizes carbohydrates, fats, and proteins to generate the energy you need to stay alive. Indeed, your body is full of oxidation–reduction reactions.

MICHAEL COLLINS,

APOLLO 11

ASTRONAUT

Those fuel cells are funny things. They are like human beings; they have their little ups and downs. They put out lots of electricity, but they do it only bitterly and with much complaining and groaning, and you have to worry about them and sort of talk to them sweetly.

Summary

Acids are proton donors. In water solution, they release hydrogen ions, H^+ (protons), which become hydronium ions (H_3O^+). Bases are proton (H^+) acceptors. Most bases produce hydroxide ions (OH^-) in water solution.

The strength of an acid or base depends on how much of its potential ions (H_3O^+ or OH^-, respectively) actually form in water. The pH scale is a measure of the acidity or basicity (alkalinity) of a solution. pH values above 7 are alkaline, those below 7 are acidic, and a value of 7 is neutral. In acid–base reactions, hydronium ions (H_3O^+) from the acid react with hydroxide ions (OH^-) from the base to produce water.

Oxidation is the losing of electrons in a reaction; this may involve gaining oxygen atoms or losing hydrogen atoms. Reduction has the opposite characteristics. In an oxidation–reduction reaction, the substance reduced is called the oxidizing agent, and the substance oxidized is the reducing agent.

Oxidation–reduction and acid–base reactions have many useful applications.

Terms for Review

After completing this chapter, you should know and understand the meaning of the following terms:

acid (p. 153)

acid–base reaction (p. 160)

alkaline (p. 159)

base (p. 156)

battery (p. 172)

electrochemical cell (p. 172)

hydronium ion (p. 153)

neutral (p. 159)

neutralization reaction (p. 160)

oxidation (p. 165)

oxidation–reduction reaction (p. 166)

oxidizing agent (p. 167)

pH (p. 159)

reduction (p. 166)

reducing agent (p. 167)

strong acid (p. 154)

strong base (p. 156)

weak acid (p. 155)

weak base (p. 156)

Questions

Odd-numbered questions are answered at the back of this book.

1. What is an acid? What is a base?

2. How do acids and bases differ in (a) formula, (b) taste in foods, and (c) litmus test?

3. Classify each of the following as an acid or a base:
   a. A bitter-tasting pill
   b. A clear liquid with pH = 10
   c. $HC_2H_3O_2$
   d. A liquid that turns litmus blue
   e. A substance that produces H_3O^+ ions in water
   f. Sour candy
   g. $Al(OH)_3$

4. Identify each of the following as an acid, a base, or neither: (a) $MgCl_2$, (b) H_2SO_4, (c) CH_4, (d) H_3PO_4, (e) LiOH, (f) $Mg(OH)_2$.

5. Distinguish between a weak acid and a strong acid.

6. What is an acid–base reaction? Why is it called neutralization?

7. Write a balanced equation for the reaction of hydrochloric acid, HCl, with calcium hydroxide, $Ca(OH)_2$, to form water and calcium chloride, $CaCl_2$.

8. Identify the acid (proton donor) and base (proton acceptor) in each of the following reactions:
   a. $H_2SO_4 + 2NaOH \longrightarrow$
   $\qquad Na_2SO_4 + 2H_2O$
   b. $H_2O + HCO_3^- \longrightarrow H_2CO_3 + OH^-$
   c. $H_2O + NH_4^+ \longrightarrow NH_3 + H_3O^+$
   d. $HS^- + HCl \longrightarrow H_2S + Cl^-$
   e. $K_2HPO_4 + HNO_3 \longrightarrow$
   $\qquad KH_2PO_4 + KNO_3$

9. Supply the missing product in the following acid–base reactions. Balance the equations where necessary.
   a. $NaOH + HCl \longrightarrow NaCl +$ _____
   b. $Mg(OH)_2 + HCl \longrightarrow$
   $\qquad MgCl_2 +$ _____
   c. $NaOH + HCN \longrightarrow$
   $\qquad$ _____ $+ H_2O$
   d. $HNO_3 + Ca(OH)_2 \longrightarrow$
   $\qquad$ _____ $+ H_2O$
   e. $H_2SO_4 + KOH \longrightarrow$
   $\qquad$ _____ $+$ _____

10. What is the pH scale? In what pH range is a solution considered acidic? Alkaline? Neutral?

11. Classify the following solutions as acidic, basic, or neutral: (a) pH = 4, (b) pH = 12, (c) pH = 7, (d) pH = 9.2.

12. Compare a glass of milk (pH = 6.6) to a glass of soda pop (pH = 3.3). Which is more acidic? Is it twice as acidic? Why?

13. Suppose your gastric juices have a pH of 2 and you take an antacid that makes your gastric juices only one-tenth as acidic as before. What would the pH value now be?

14. Suppose you have a solution of lye (sodium hydroxide) that has a pH value of 12 and you add enough water to make the solution only one-hundredth as alkaline as before. What would the pH value now be?

15. Identify which of the following could be used safely as an antacid: (a) HCl, (b) $Mg(OH)_2$, (c) $NaHCO_3$, (d) NaOH, (e) KCl.

16. What is oxidation? What is reduction?

17. What is an oxidizing agent? What is a reducing agent?

18. In each of the following reactions, identify the substance that is oxidized and the substance that is reduced:
    a. $N_2 + 3H_2 \longrightarrow 2NH_3$
    b. $WO_3 + 3H_2 \longrightarrow W + 3H_2O$
    c. $Fe_2O_3 + 2Al \longrightarrow 2Fe + Al_2O_3$
    d. $Pb + PbO_2 + 2H_2SO_4 \longrightarrow$
    $\qquad\qquad\qquad 2PbSO_4 + 2H_2O$
    e. $6CO_2 + 6H_2O \longrightarrow C_6H_{12}O_6 + 6O_2$
    f. $Al + 6HCl \longrightarrow 2AlCl_3 + 3H_2$

19. In each of the reactions in Question 18, identify the oxidizing agent and the reducing agent.

20. In a smelter, certain metals are produced from their oxides by heating them in the presence of carbon. In the following reactions, is carbon an oxidizing agent or a reducing agent?
    a. $CuO + C \longrightarrow Cu + CO$
    b. $Fe_2O_3 + 3C \longrightarrow 2Fe + 3CO$
    c. $SnO_2 + 2C \longrightarrow Sn + 2CO$

21. In the following reactions, identify which reactant is oxidized:
    a. $SiO_2 + 2Mg \longrightarrow Si + 2MgO$
    b. $2BCl_3 + 3H_2 \longrightarrow 2B + 6HCl$

c. $PbS + Fe \longrightarrow Pb + FeS$
d. $2Cr + 3H_2O + 3HgO \longrightarrow$
$\qquad\qquad\qquad 2Cr(OH)_3 + 3Hg$
e. $CH_2O + Cu(OH)_2 \longrightarrow$
$\qquad\qquad\qquad CH_2O_2 + Cu + H_2O$
f. $2Al + 3Cu^{2+} \longrightarrow 2Al^{3+} + 3Cu$
g. $C_{55}H_{100}O_6 + 77O_2 \longrightarrow$
$\qquad\qquad\qquad 55CO_2 + 50H_2O$

22. What is "alkaline" about an alkaline battery?

23. Write a balanced equation for the reaction of hydrogen gas with oxygen gas to produce water. In this reaction, which substance is oxidized, and which is reduced?

24. When an apple is peeled and sits out in the air, it becomes brown because a substance in it is oxidized. What do you think is the oxidizing agent?

25. When butter sits out in the air for days, it slowly becomes rancid as its ingredients become oxidized. What do you think is the oxidizing agent?

26. What major oxidizing agent do you use to produce energy for your body?

Topics for Discussion

1. Suppose a politician proposed (a) to ban the sale of all foods containing acids or (b) to require warning labels on all such products. Which proposal, if either, would you support? Why?

2. Calcium has become a popular mineral supplement to the diet, and the makers of calcium-containing antacids recommend their product as an important source of this mineral. Do you favor regarding antacids as food supplements? Why?

3. Does the electronegativity of its elements affect the strength of an acid or base? Does it affect a substance's tendency to be an oxidizing agent or a reducing agent? Why?

CHAPTER 7

Energy and Speed of Chemical Reactions: Influencing Chemical Changes

General Objectives

Energy Questions About Chemical Reactions

1. Does the reaction occur spontaneously? (Will it go of its own accord, or does it require energy and therefore money to make it go?)

2. Is energy released or absorbed during the reaction?

Rate Questions About Chemical Reactions

3. What is the rate (speed) of the reaction?

4. How can the rate be altered? (If it is too fast or explosive, how can it go slower? If it is too slow, how can it go faster? Either alteration will probably cost money.)

Yield Questions About Chemical Reactions

5. How far will the reaction go? (What is the expected yield? How efficient is the conversion of raw materials or reactants to the desired products?)

6. How can we make the reaction go farther? (How can the yield be increased?)

Imagine that you have graduated from college and are now a multi-millionaire. You want to make more money, but in a way that improves humanity's lot. A business entrepreneur approaches you with the idea of investing in a chemical process for making automobile fuel from water. What questions should you ask about the chemical reaction involved in this process before sinking your fortune into the venture?

Looking at your chemistry notes, which you have carefully preserved, you find that you should ask six fundamental questions—two about the energy involved in the reaction, two about the speed (rate) of the reaction, and two about how far the reaction goes to the right (yield). If the answers are not favorable, you would be throwing your money away.

7.1 The First Law of Thermodynamics: You Can't Get Something for Nothing

ENERGY CHANGES IN NATURAL PROCESSES   The physical and chemical changes that we observe in nature involve energy because *energy* is the capacity to do work or to cause action.

As discussed in Chapter 1, energy can take a number of forms. When water flows or apples fall, they possess energy of motion, or *kinetic energy*. Still water behind a dam and an apple attached to a branch of its tree possess *potential energy*, or energy of position. The power from a battery or a wall plug is electrical energy. The movement of springs, levers, pulleys, and other machines involve mechanical energy. The warmth from a fire is heat energy, the illumination from a lamp is light energy, and the energy that holds atoms together in molecules is chemical potential energy.

Energy can change from one form to another. For example, burning natural gas (mostly CH_4) changes the chemical potential energy in covalent molecules of CH_4 into heat and light. The generators at Hoover Dam transform the gravitational potential energy of the water stored in Lake Mead into electrical energy as the water flows through turbines to a lower elevation.

THE FIRST LAW OF THERMODYNAMICS   *Thermodynamics* is the relationship of heat to other forms of energy. In studying millions of physical and chemical processes, scientists have observed energy being transformed from one form to another, but they have never been able to create or destroy energy.* These observations are summarized in the **first law of thermodynamics**, commonly known as the **law of conservation of energy**: *In all chemical and physical changes, energy is neither created nor destroyed but merely transformed from one form to another.* In other words, the total energy of the observable universe remains constant.

This law is one of the basic rules governing what we can and cannot do. It states, for example, that we can harness energy from whatever source—solar, geothermal, natural gas, and so on—if we are clever enough to transform it usefully. But it also tells us that, no matter how ingenious we become, we will never be able to synthesize new energy to enlarge the observable universe's total energy supply.

Conservation of energy applies to the universe as a whole. No energy ever goes out or comes in. But if we focus on some small part of it, then we must allow for the possibility of energy flow. A **system** (the collection of matter under study) might gain energy from its **surroundings** (the rest of the accessible universe), or it might lose energy to its surroundings. The first law of thermodynamics says only that the total energy of the system plus its surroundings must stay the same.

SPONTANEOUS PROCESSES   Many natural processes occur of their own accord without any help from us. Water spontaneously flows downhill, and apples, books, or other objects drop spontaneously to the ground. In contrast, other processes are not spontaneous and don't occur without the input of energy. Water doesn't flow spontaneously uphill, nor do books leap spontaneously from the floor to the shelf.

Why are some processes spontaneous while others are not? What is the driving force for a natural spontaneous process? The feature common to all spontaneous, mechanical processes is that the system loses potential energy and goes to a lower potential energy state. A book on the shelf,

---

* Recall (Section 3.7) that in nuclear reactions, according to Einstein's famous equation $E = mc^2$, mass is considered just another form of energy.

Figure 7.1 A book falls spontaneously to a state of minimum potential energy consistent with its surroundings.

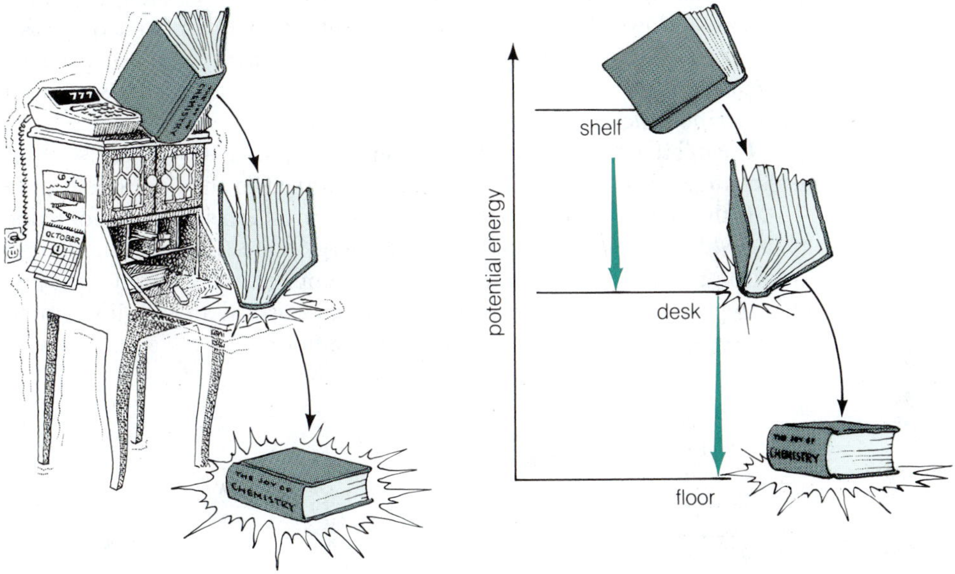

for example, has higher potential energy than a book that has fallen to the floor (Figure 7.1).

The driving force for spontaneous processes in mechanical systems appears very simple: *Mechanical systems tend to move spontaneously to the state of minimum potential energy consistent with their surroundings.* To understand the necessity for adding the phrase "consistent with their surroundings," consider a book falling to the top of a desk rather than directly to the floor (see Figure 7.1). The book reaches a minimum potential energy state consistent with its accessible universe (the desk top). But if new surroundings are provided—if the desk were tilted or the book pushed—the book would fall to the floor and achieve a new state of minimum potential energy consistent with its new surroundings.

THE TENDENCY TOWARD MINIMUM POTENTIAL ENERGY The minimum potential energy principle applies to spontaneous mechanical processes. Can we, by analogy, apply it to all physical and chemical changes? Recall (Section 4.1) that substances form bonds to achieve a lower chemical potential energy. Can we postulate that any chemical reaction will be spontaneous if the potential energy of the products is lower than that of the reactants? In other words, should reactions that "go downhill" in terms of potential energy be spontaneous?

A physical or chemical change that gives off energy to its surroundings is called **exothermic** (Figure 7.2). One that absorbs or requires heat from its surroundings is called **endothermic** (Figure 7.2). In any spontaneous process, energy should be released to the surroundings as the system goes to a state of lower potential energy. In summary, we can tentatively propose a minimum energy hypothesis: *Chemical and physical changes tend spontaneously toward a state of minimum potential energy consistent with their surroundings.*

The exothermic reaction for burning coal (mostly C) is like the physical change when a book falls to a desk and then to the floor (see Figure 7.1). If the coal is only partially burned, the product is carbon monoxide (CO), which exists at an intermediate energy state (Figure 7.3), like the book on the desk. If additional oxygen is supplied, the CO can react further to produce carbon dioxide (CO_2), which exists at an even lower state of energy (Figure 7.3), like the book on the floor.

Figure 7.2 Exothermic and endothermic reactions.

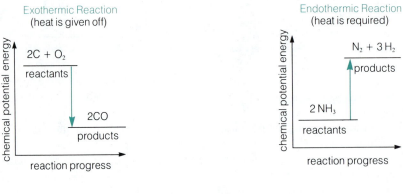

Figure 7.3 The complete combustion of coal can occur in two exothermic steps.

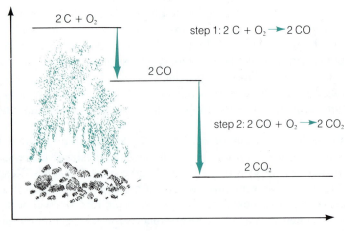

Have we found the driving force for spontaneous changes so quickly and easily? One test of any scientific hypothesis is whether it can account for known experimental data. When we go to the laboratory, do we find that all exothermic processes are spontaneous and all endothermic ones are nonspontaneous?

At room temperature, most spontaneous changes are indeed exothermic. A piece of sodium (Na) metal dropped in water instantly reacts to produce H_2 gas and so much heat that the hydrogen gas can ignite:

$$2Na + 2H_2O \longrightarrow 2NaOH + H_2 + \textbf{energy}$$

But some spontaneous processes—both physical and chemical—are dramatically endothermic. An ice cube melts on its own, taking heat from its surroundings as it does so. Barium hydroxide [$Ba(OH)_2 \cdot 8H_2O$ (which has eight water molecules per formula unit trapped in its lattice)] reacts spontaneously with ammonium chloride (NH_4Cl):

$$Ba(OH)_2 \cdot 8H_2O + 2NH_4Cl + \textbf{energy} \longrightarrow BaCl_2 + 2NH_3 + 10H_2O$$

Liquid water forms, the aroma of ammonia (NH_3) gas appears, and the temperature inside the flask drops more than 35°C (Figure 7.4). This is an endothermic but spontaneous chemical change.

So what can we conclude? Either the minimum potential energy principle is altogether incorrect, or it is not the sole driving force of spontaneous changes. As you learn in the next section, the latter conclusion is valid; there is a second driving force.

7.2 The Second Law of Thermodynamics: You Can't Even Break Even

THE SECOND LAW OF THERMODYNAMICS AND ENERGY QUALITY Because the first law states that energy can neither be created nor destroyed, you might think that there will always be enough energy. Yet when you fill a car's tank with gasoline and drive around, something is lost. If it isn't energy, what is it?

Energy varies in its quality, or ability to do useful work. For useful work to occur, high-quality (more concentrated) energy changes to lower-quality (less concentrated) energy. The chemical energy concentrated in a tank of gasoline is high-quality energy that can perform useful work in moving the car. But in the process, much of the high-quality energy in gasoline changes into less concentrated heat energy (in the engine and exhaust) that has little ability to do useful work.

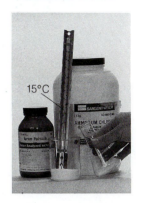

Figure 7.4 The endothermic reaction of $Ba(OH)_2 \cdot 8H_2O$ powder with NH_4Cl powder.

According to the **second law of thermodynamics**, *in any conversion of high-quality energy to do useful work, some energy is always degraded to a less useful form,* usually as heat given off at a low temperature to the surroundings. The supply of concentrated, high-quality energy continually decreases, and the supply of low-quality energy continually increases, so that the total energy remains the same.

When you drive a car, only about 10 percent of the high-quality chemical energy available in the gasoline changes into mechanical energy to propel the vehicle (Figure 7.5). The remaining 90 percent is degraded

Figure 7.5 Efficiency of some common energy converters.

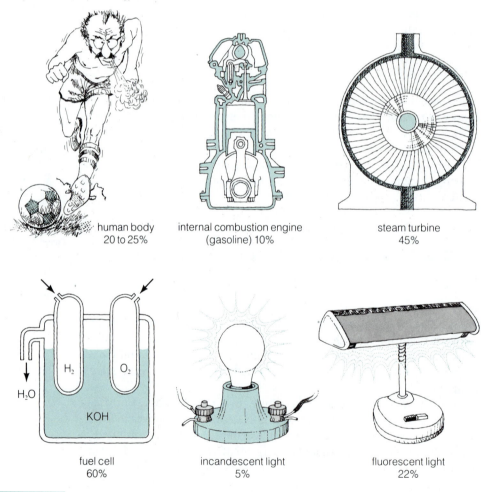

to low-quality heat that spreads out into the environment. Electrical energy flowing through the filament wires of an ordinary light bulb changes to a mixture of useful light (5 percent) and waste heat (95 percent) (Figure 7.5). Photosynthesis in plants converts solar energy (sunlight) into chemical energy (stored in sugar and starch molecules). After eating plant foods, your body converts their chemical energy into mechanical energy to move your muscles and perform other life processes. But each conversion produces some low-quality heat (Figure 7.6). In no case does 100 percent of the useful energy in one form change into useful energy in another form.

According to the first law of thermodynamics, we will never run out of energy; but according to the second law, we may run out of high-quality, or useful, energy. Thus, we can never recycle high-quality energy as we can recycle matter. Once a tank of gasoline has been burned or a meal has been metabolized, its high-quality potential energy is lost forever.

THE SECOND LAW AND INCREASING DISORDER   The second law can be stated in several ways. For example, because energy tends to change spontaneously from a concentrated and ordered form to a more dispersed and disordered

Figure 7.6 The second law of thermodynamics. When energy changes to another form, some is always degraded to low-quality heat, which is added to the environment.

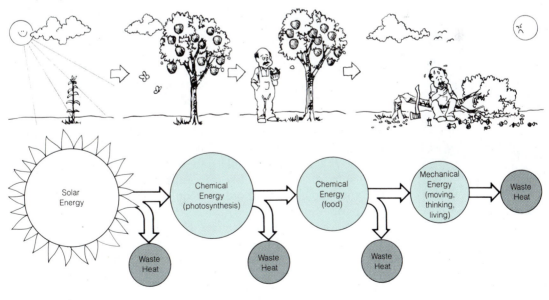

Figure 7.7 The spontaneous tendency toward increasing disorder (entropy) of a system and its surroundings.

form, the second law can also be stated: *Heat energy flows spontaneously from hot (high-quality energy) to cold (low-quality energy).*

This is hardly news. You learned about the flow of heat energy the first time you touched a hot stove. A cold sample of matter such as air has its heat energy spread out in the random motion of its molecules. This is why heat energy at a low temperature can do little, if any, useful work.

The second law gives other characteristics of spontaneous processes and their tendency toward increasing disorder. Consider, for example, your room. Is the most probable state of your room spontaneously increasing order or spontaneously increasing chaos (Figure 7.7)? Doesn't keeping your room neat and orderly require a continual input of energy on your part?

We can see many other examples. Houses tend spontaneously to fall into disrepair. If you drop a vase to the floor and it shatters, under no circumstances would you expect the fragments to leap back together to reform the vase. From the air, you can immediately see where humans have been by the presence of straight lines in fields, roads, and rows of housing (Figure 7.8). These lines represent an ordering process against the apparently natural tendency for the more random, or disorderly, growth

Figure 7.8 People tend to convert the world into straight lines by using large inputs of energy, as shown by this crop field in California. (EPA Documerica photo by Charles O'Rear)

of vegetation. These human ordering activities require enormous inputs of energy.

If you dropped a highly ordered crystal of a water-soluble dye into a glass of water, you could see the colored dye spontaneously spread throughout the solution; the relative disorder of the dye molecules increases. If a woman wearing Zapglow perfume or a man wearing Kung Gu aftershave walks into a room, molecules of the scent will spread spontaneously throughout the room in a few minutes, even if the air appears to be perfectly still. And when we dump chemical wastes in a lake or send car exhaust into the air, we expect them to spontaneously disperse into a more disordered (and less harmful) state. Indeed, we instinctively trust the natural tendency of disorder to increase.

One statement of the second law affirms our common sense about disorder: *For any process, there is always a net increase in disorder when both the system and its surroundings are considered.* For example, living organisms, with their highly ordered structures, are allowed by the second law; but when the system and its surroundings are taken together, disorder always increases. To form and keep the orderly array of molecules and the organized network of chemical reactions in your body, you must continually get high-quality energy and raw materials from your surroundings in the form of food; producing that food requires energy. In addition, metabolizing that food creates disorder in the environment, mostly as low-quality heat. Maintaining order in your body then costs a greater amount of disorder in your surroundings.

Scientists frequently use **entropy** as a measure of relative disorder, or randomness. A disorderly system has high entropy, and an orderly system has low entropy. We can use this idea to distinguish among the physical states of matter (Section 5.1). Solids have low entropy (that is, low disorder or high order), gases have high entropy (disorder), and the entropy of liquids is between the two extremes.

SPONTANEOUS PROCESSES REVISITED   Now let's return to the question: Why are some processes spontaneous while others are not? We have seen that two principles help make a change spontaneous: (1) The system goes to a lower potential energy, and (2) the system increases in entropy. If both occur, the process is spontaneous. If neither occurs, the change is not spontaneous. If one occurs but not the other, the process could be spontaneous or not, depending on which factor predominates.

Why, for example, does ice spontaneously melt at room temperature despite being an endothermic process? The answer is that the increase in entropy (the liquid state has higher entropy than the solid state) is the prevailing factor. Why, then, doesn't *everything* melt at room temperature? The answer is that for many substances the endothermic process of melting (going to a state of higher potential energy) predominates over the increase in entropy.

The reaction shown in Figure 7.4 is

$$Ba(OH)_2 \cdot 8H_2O + 2NH_4Cl + energy \longrightarrow BaCl_2 + 2NH_3 + 10H_2O$$

Why is this reaction spontaneous despite being endothermic?

over the increase in potential energy.
forming a liquid (H_2O) and a gas (NH_3) from two solids, predominates
The reaction is spontaneous because the increase in entropy, such as

SOLUTION

7.3 Reaction Rate: What Makes a Reaction Go Faster or Slower?

WHAT IS REACTION RATE?   The **reaction rate** is the quantity of reactants that are converted to products during a specified period of time. To measure rate, we need to follow the changes in the amount or concentration of one of the reactants or products over time. If we follow one of the reactants, its concentration (or number of molecules in some unit of volume) will decrease with time. Similarly, monitoring one of the products should show an increase in concentration as the reaction proceeds.

How can we measure the concentration changes as the reaction proceeds? In some reactions, in which a reactant or product is colored, we can measure the change in color. One example is Reaction 4 in Section 5.4 involving oxides of nitrogen:

$$2NO \quad + \quad O_2 \quad \longrightarrow \quad 2NO_2$$
colorless gas   colorless gas   brownish yellow gas

We can measure the rate of this reaction by measuring the speed at which the brownish yellow color (due to formation of the product, NO_2) deepens.

WHAT DETERMINES REACTION RATE?   Some reactions are very fast. All it takes is a spark to ignite the explosive reaction of hydrogen (H_2) and oxygen (O_2) to form water:

Fast rate: $2H_2 + O_2 \longrightarrow 2H_2O + $ **energy**

This reaction, under control, can be used to propel vehicles. Out of control, it caused the explosion of the hydrogen-and-oxygen-filled main fuel tank of the space shuttle *Challenger*.

Other reactions are slow. A piece of iron may take years to be oxidized to iron rust (Fe_2O_3):

Slow rate: $4Fe + 3O_2 \longrightarrow 2Fe_2O_3 + $ **energy**

Why do reactions differ in their rates? One reason is that atoms, ions, or molecules in a reaction must collide before they can change chemically. The more frequently they collide, the faster the reaction is. Thus, the rate of reaction should be determined by the collision frequency—the number of collisions in a given period of time.

But not all collisions produce reactions. Oxygen molecules (O_2) and nitrogen molecules (N_2) in the air collide with each other all the time, but they rarely react. For a reaction to occur, two other conditions must be met: the collisions must have the right orientation and energy.

First, let's consider the orientation of collisions. Figure 7.9 shows two ways that CO and NO_2 molecules can collide to form CO_2 and NO. Even if each type of collision takes place with enough energy, the reaction can only occur when the molecules are lined up or oriented correctly.

Figure 7.9 For a reaction to occur, molecules must collide with the proper orientation so that the atoms directly involved come into contact with each other.

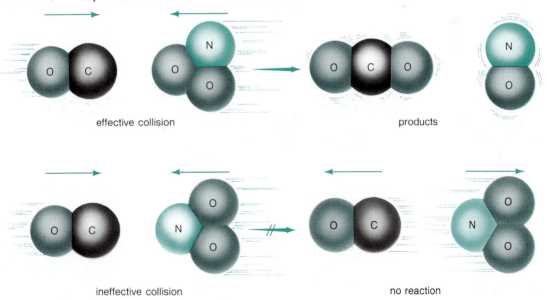

Second, collisions also have to occur with enough energy to bring about a reaction. By way of analogy, imagine the following "reaction":

black belt's fist + log $\longrightarrow$ sore fist + split log

A dainty slap doesn't break the log or injure the fist. The fist and the log will remain "unreacted" unless they come together with all the power of the martial arts. Even then, the blow must be properly aligned with the grain of the wood, or the "reaction" does not happen. The same is true in chemical reactions. The collisions must be effective ones—molecules must collide with sufficient energy and with the right orientation to bring about reaction.

The minimum energy required to cause a reaction is known as the **activation energy**. An energy barrier or hill must be surmounted before any reaction can proceed, just as a boulder may have to be rolled over a small hill before it can spontaneously roll down a mountain (Figure 7.10). The height of the initial energy barrier determines how fast the change will occur. If the barrier is high, not many molecules will have enough energy to cross it; with few effective collisions, the reaction is slow. With low activation energy, however, frequent effective collisions occur, and the reaction is rapid.

The reaction of H_2 and O_2 illustrates this idea (Figure 7.11). At room temperature in the absence of impurities or flames, hydrogen and oxygen gases can remain mixed indefinitely without reacting. The activation energy barrier is high enough that, at room temperature, almost none of

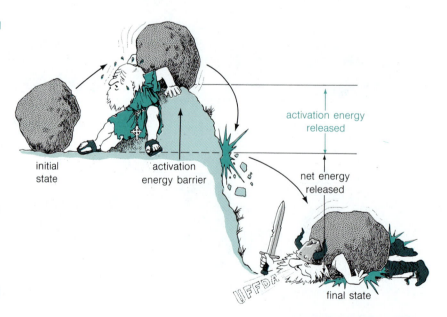

Figure 7.10 An analogy of the activation energy for a chemical reaction. If enough energy is supplied to the boulder to push it over the hill (activation energy barrier), it will spontaneously roll down the mountain, releasing energy as it moves to a state of lower potential energy. The rate at which boulders are pushed over the cliff depends on the height of the activation energy barrier (if all other factors remain the same).

initial state

activation energy barrier

activation energy released

net energy released

final state

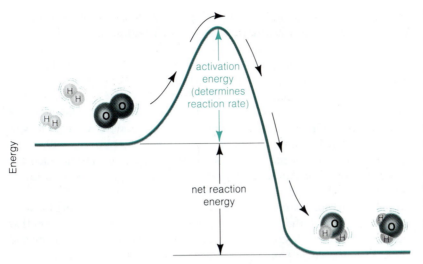

Reaction progress

Figure 7.11 The activation energy for the reaction between H_2 and O_2 to form H_2O is so high that the reaction is extremely slow at 25°C. This reaction actually takes place in a series of steps, each with its own activation energy barrier; this diagram is a simplified summary of the net energy changes.

the colliding molecules has enough energy to bring about an effective collision. Yet the tiniest spark or flames cause H_2 and O_2 to react explosively.

PRACTICE EXERCISE

Figure 7.11 shows an energy diagram for an exothermic reaction. What does an energy diagram for an endothermic reaction look like?

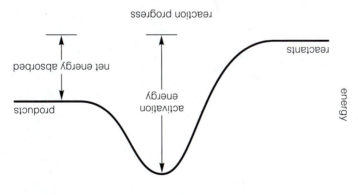

SOLUTION

HOW CAN WE ALTER THE RATE? Many reactions are too slow or too fast for our convenience, and much effort is spent in finding practical ways to change their rates. Four factors can affect the rate of a given chemical reaction: (1) concentration of the reactants, (2) temperature, (3) surface area of the reactants, and (4) catalysts or inhibitors.

Putting more reactants in a closed container makes it more crowded, just as adding dancers makes a dance floor more crowded. The increased concentration of molecules (or dancers) means that more will collide with each other. And among the increased number of collisions will be an increased number of effective ones. Thus, reactions go faster when the concentration of reactants increases and slower if the concentration is reduced.

Higher temperatures normally cause faster reactions. Raising the temperature increases the average kinetic energy of the molecules so that more of them move faster. This increases both the collision frequency and the collision energy. Thus, a roast will cook faster in an oven at 375°F than at 300°F. Slow-cooker appliances, which don't get very hot, may take all day to simmer a stew. Unwanted reactions, such as food spoiling,

Figure 7.12 A catalyst increases the rate of a reaction by lowering the activation energy so that more reactant molecules collide with enough energy to surmount the lower energy barrier.

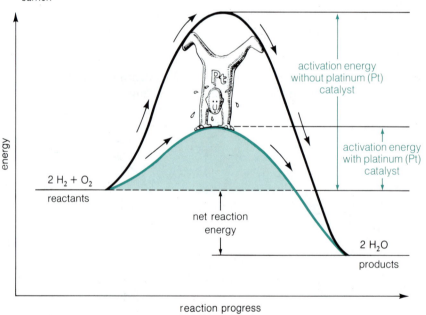

are much slower at cold temperatures; thus, freezing can keep food fresh for months.

If a reaction is highly exothermic, the heat evolved can accelerate the reaction to make it go faster still. These runaway reactions, once started, usually result in a ball of flame or an explosion. For example, the heat from a spark gets some molecules of H_2 and O_2 reacting, and the net reaction energy (see Figure 7.11) is enough to get the rest of them going. The result is a brilliant fireball and a deafening boom.

If reactants are in two different states—for example, a log (solid) reacting with oxygen (gas) while burning in a fireplace—we say that the reaction is heterogeneous. For heterogeneous reactions, the surface area of the reactants has an important effect on reaction rate. In the fireplace, the only part of the log that can burn is its surface; nothing else is exposed to the oxygen. Splitting the log into pieces exposes more surface, and the fire burns faster. Chopping it into splinters makes it burn completely in just a few seconds.

The more surface area available in a heterogenous reaction, the faster the reaction is. Dried kernels of grain in a silo do not pose much of a fire hazard, but those same kernels ground into flour have so much more surface area that they are a significant fire danger. Dust explosions at grain elevators are a constant threat. Surface area makes a difference in rates of heterogeneous physical changes, too. A salt block for livestock to lick can be left in a pasture for months without completely dissolving in the rain. The same amount of granulated salt would wash away in the first drizzle.

Many slow reactions go faster when certain outside substances are introduced. Substances that accelerate a reaction without themselves being consumed are called **catalysts**. They differ from the reactants in that they are not used up.

For example, if we add finely divided platinum (Pt) metal to a hydrogen–oxygen mixture at room temperature, the reaction goes very rapidly:

$$2H_2 + O_2 \xrightarrow{\;Pt\;} 2H_2O + \text{energy}$$

On analysis, however, we find that the platinum is no different after the reaction than before. Platinum is a catalyst for this reaction.

Catalysts increase the reaction rate by lowering the energy of activation for the reaction (Figure 7.12). Platinum catalyzes the hydrogen–oxygen reaction by immobilizing H_2 molecules on its surface and enabling O_2 molecules to interact with them more effectively (Figure 7.13).

Practically all major chemical industries need cheap and effective catalysts. For example, the petroleum industry depends heavily on catalysts for refining petroleum into gasoline and other products (see Section 8.3). The automobile industry relies on catalysts in exhaust systems to

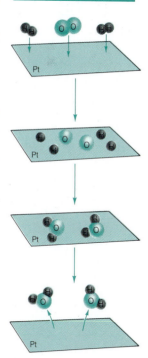

Figure 7.13 Platinum (Pt) works as a catalyst by providing a surface for hydrogen (H_2) and oxygen (O_2) molecules to separate into individual atoms, which then recombine to form water molecules (H_2O).

TABLE 7.1 RELATIVE ACTIVATION ENERGIES FOR THE REACTION
$2H_2O_2 \longrightarrow 2H_2O + O_2$

Conditions	Relative Energy of Activation
Uncatalyzed	18
Pt catalyst	13
Enzyme catalyst	7

convert air-polluting substances into relatively harmless wastes. Self-cleaning ovens contain catalysts that eliminate smoke by accelerating the breakdown of smoke particles.

One group of catalysts deserves special mention. These are *enzymes*, the catalysts of life that keep reactions going fast enough to meet the needs of living cells (see Section 9.4). Like other catalysts, enzymes lower the activation energy of a reaction (Table 7.1). As a result, some enzymes cause their particular reactions to go as much as a million times faster.

When reactions such as food spoilage go too fast for our convenience, the opposite of a catalyst is required. Substances that decrease reaction rates without being consumed (or otherwise altered) are called *inhibitors*. Forgotten homemade bread left in a drawer for several days will grow green mold. But if a trace of calcium propionate inhibitor is added to the dough (as in many commercial breads), the loaf can be forgotten until it is as dry as a cracker, without getting moldy.

7.4 Dynamic Chemical Equilibrium and Le Châtelier's Principle: How Far Will a Reaction Go?

DYNAMIC EQUILIBRIUM   Of the six questions posed at the beginning of this chapter, we have yet to answer the two concerning the yields of chemical reactions. How far will a reaction go, and how can it be made to go farther?

In a closed container at constant temperature and pressure, most reactions do not go to 100 percent completion; that is, not all reactants change into products. Instead, they reach a state of *dynamic equilibrium* (Section 5.2): Reactants form products and products revert to reactants at the same rate, leaving the net amounts of the two constant.

For example, suppose we put a mixture of sulfur dioxide (SO_2) gas and oxygen (O_2) gas in a closed container at room temperature and pressure. If we analyze the contents later, we will find that the container holds not only SO_2 and O_2 gases but also sulfur trioxide (SO_3) gas. SO_3 was formed by the (forward) reaction between SO_2 and O_2:

$$2SO_2 + O_2 \longrightarrow 2SO_3$$

SO_2 and O_2 are formed by the reverse reaction of SO_3:

$$2SO_3 \longrightarrow 2SO_2 + O_2$$

Once the forward and reverse reactions occur at the same rate, the net amounts of SO_2, O_2, and SO_3 remain constant. This dynamic equilibrium is symbolized by a double arrow ($\rightleftharpoons$) between the reactants and the products:

$$2SO_2 + O_2 \rightleftharpoons 2SO_3$$

Notice that a chemical equilibrium is a *dynamic* state, with constant movement. It is like a football game in a large stadium with 50,000 spectators. If 500 people are leaving every minute to go behind the stands for refreshments and if 500 people are returning every minute, then the concentration, or number, of people in the stands (the system) remains constant at 49,500 people. But this sameness is maintained by a continuous turnover; it is a dynamic equilibrium.

In the reaction above, billions and billions of reactant molecules (SO_2 and O_2) form product molecules (SO_3) each second; at the same time, billions and billions of product molecules are breaking up and forming the original reactant molecules. If we begin with reactants only (SO_2 and O_2), the forward reaction rate is initially greater than that of the reverse. Thus, the reactants decrease in number at first, and the amount of product (SO_3) increases. But eventually the rates of the two opposing reactions become equal, and the system reaches a state of dynamic chemical equilibrium.

MAKING A REACTION GO FARTHER: LE CHÂTELIER'S PRINCIPLE   When they synthesize chemicals, scientists want to get as high a yield of products as possible. Two approaches are used to increase the yield. One is to change the initial concentration of reactants and adjust the temperature so that more products form at equilibrium. (The trick here is to figure out whether raising or lowering the temperature will improve production.) The second approach is to remove products as they form. This prevents the reverse reaction and keeps the system from reaching equilibrium.

Suppose you are attending a rock concert in a large auditorium, and the audience has settled into a blissful state of equilibrium. A hundred or so people are leaving the hall every minute for refreshments, and a hundred or so are returning from the lobby. Suddenly, an amplifier burns out and fills the hall with an acrid stench. This dramatic intrusion, or stress, upsets the state of dynamic equilibrium. Now the number of people exiting for fresh air exceeds the number returning. After a while, when the people who could not stand the smell have left the scene entirely, a new state of dynamic equilibrium is eventually established. But the new equilibrium concentration of people in the room is lower than before. This adjustment to a stress imposed on a system in a state of dynamic equilibrium provides the basis for understanding how we can make a reaction go farther.

Figure 7.14 Henri Le Châtelier (1850–1936) proposed a principle to predict how the dynamic equilibrium in a reaction will shift when a stress is applied to the system. (Smithsonian Institution)

French chemist Henri Le Châtelier (Figure 7.14) proposed a way to predict the shift that will occur in an equilibrium mixture when the concentration of reactants or products is altered. **Le Châtelier's principle** states that *if a system in dynamic equilibrium is subjected to a stress, the system will change, if possible, to relieve the stress.* We can use this principle to predict whether a new equilibrium mixture will contain more or less of the products.

Suppose the reaction between SO_2 and O_2 to form SO_3 has reached a state of dynamic equilibrium in a closed vessel at a constant pressure. What happens if we upset the equilibrium by adding more O_2? According to Le Châtelier's principle, the system will shift to relieve this stress. In this case, the system will tend to use up some of the excess O_2. This can only occur if the reaction proceeds in the forward direction, where SO_2 combines with the excess O_2 to form more SO_3. Eventually, a new dynamic equilibrium position will be established, but one that contains more SO_3 than the original. Thus, by increasing the concentration of one reactant, we have shifted the equilibrium equation (as written) to the right.

If instead of adding excess O_2 we added excess SO_2, the equilibrium would still shift to the right to relieve the stress. But if we increased the concentration of the product (SO_3) instead, the equilibrium would shift to the left to remove the excess SO_3. In effect, adding more of any reactant or product to a system in equilibrium is like pushing up on one side of a seesaw; the equilibrium mixture will shift in the opposite direction.

Another way to shift the equilibrium is by temporarily opening the system and removing reactant or product molecules. For example, we can shift the equilibrium position to the right by removing SO_3, which causes more SO_2 and O_2 to react to replenish the SO_3 in the system.

PRACTICE EXERCISE

For the reaction $2SO_2 + O_2 \rightleftharpoons 2SO_3$, what is the effect on the equilibrium position if either SO_2 or O_2 molecules are removed?

The equilibrium position shifts to the left to replace SO_2 or O_2.

SOLUTION

We can also use Le Châtelier's principle to predict how a change in reaction temperature affects the equilibrium position. Let's consider two reactions, one endothermic (requiring heat) and one exothermic (giving off heat):

Endothermic: $N_2 + O_2 + $ **heat** $\rightleftharpoons 2NO$

Exothermic: $2SO_2 + O_2 \rightleftharpoons 2SO_3 + $ **heat**

If we heat a reaction mixture in a closed container, the reaction will shift in the direction that uses up the excess heat. For instance, in the endothermic reaction of N_2 and O_2, the addition of heat will shift the position of the equilibrium to the right in order to use up some (but not necessarily all) of the excess heat. Similarly, in the exothermic reaction between SO_2 and O_2, heating the mixture will shift the equilibrium to the left in order to use up the excess heat.

PRACTICE EXERCISE

What is the effect of cooling on the position of equilibrium in the two reactions above?

Cooling removes energy. The equilibrium in the endothermic reaction shifts to the left to replace the energy; the equilibrium in the exothermic reaction goes toward the right to replenish the energy removed.

SOLUTION

CHEMICAL FRAUD   At the beginning of the chapter, you were asked to imagine yourself as a potential investor in a venture involving chemical reactions. We have now explored six questions that you would need to ask to find out if the idea is based on sound chemical principles.

In the late 1970s, a prominent businessman actually had such an opportunity. Had he known what you know, perhaps he would not have been swindled. The scheme involved using hydrogen (H_2) as a fuel for automobiles. The businessman knew that, with surprisingly little modification, ordinary internal combustion engines could use hydrogen in place of gasoline. He was approached by an "inventor" who had a prototype fuel tank, reportedly lined with a "catalytic metal," that would produce H_2 from tap water. The businessman had the tank tested, and it did indeed generate hydrogen from water for several days. Satisfied, he invested his money—and lost it.

Evidently, he did not know to ask the six questions about the reaction. The energy questions should have aroused suspicion from the beginning. The first law of thermodynamics implies that the energy required to decompose water into its elements

$$2H_2O + \textbf{energy} \longrightarrow 2H_2 + O_2$$

is exactly the same as the energy returned by burning the hydrogen in air

$$2H_2 + O_2 \longrightarrow 2H_2O + \textbf{energy}$$

At best, this would be a break-even situation, with absolutely no energy left over to power the car. But the second law of thermodynamics says that it cannot work "at best"; it would really end up as a net loss of high-quality energy.

The questions about reaction rate should also have tipped him off. A catalyst cannot run a nonspontaneous reaction into a spontaneous one; it can only speed a reaction that is already spontaneous. Thus, the so-called catalytic metal in the tank was probably a reactant instead. Any alkali metal or alkaline earth metal reacts with water to give hydrogen. For example,

$$Ca + 2H_2O \longrightarrow Ca(OH)_2 + H_2$$

Finally, the questions about the yield of the reaction should have convinced him of the fraud. To shift the reaction toward the production of more H_2, you would have to add more reactants (alkali metals and alkaline earths are much more expensive than the hydrogen they produce) or change the temperature (heating or cooling costs energy).

With your knowledge about energy and chemical reactions, you have an edge this businessman lacked. Perhaps one day, you'll be able to measure its value in actual dollars and cents.

ARTHUR S.

EDDINGTON

The law that entropy increases—the second law of thermodynamics—holds, I think, the supreme position among laws of nature. . . . If your theory is found to be against the second law of thermodynamics, I can give you no hope; there is nothing to do but collapse in deepest humiliation.

Summary

Processes that take up heat from their surroundings are endothermic; those that give off heat are exothermic. The first law of thermodynamics states that energy cannot be created or destroyed. The second law states that in energy changes some energy changes into less useful, dispersed forms, often heat. Furthermore, in any process there is a net increase in the disorder (entropy) of a system and its surroundings. Spontaneous physical and chemical changes result in a state of lower potential energy, or greater entropy, or both.

Reaction rates depend on the frequency, orientation, and energy of collisions between reactants. The minimum energy required for a reaction to occur is the activation energy. Reaction rates can be altered by changing the concentration of reactants, the reaction temperature, or the surface area of reactants (for heterogeneous reactions) and using catalysts or inhibitors.

According to Le Châtelier's principle, if a system in dynamic equilibrium is subjected to a stress, the system will change, if possible, to relieve the stress. Therefore, increasing the concentration of the reactant(s) or removing the product(s) are both ways to increase the amount of product formed in a chemical reaction. Changes in temperature will also change the equilibrium position.

Terms for Review

After completing this chapter, you should know and understand the meaning of the following terms:

activation energy (p. 192)

catalyst (p. 195)

endothermic reaction (p. 184)

entropy (p. 189)

exothermic reaction (p. 184)

first law of thermodynamics (p. 182)

law of conservation of energy (p. 182)

Le Châtelier's principle (p. 198)

reaction rate (p. 190)

second law of thermodynamics (p. 186)

surroundings (p. 182)

system (p. 182)

Questions

Odd-numbered questions are answered at the back of this book.

1. Give two different statements of (a) the first law of thermodynamics and (b) the second law of thermodynamics.

2. At each step in the following sequence, identify the form and type (kinetic or potential) of energy involved: burning coal to generate electricity to power an electric saw to cut wood.

3. Distinguish between (a) an exothermic and (b) an endothermic reaction.

4. Which of the following systems has the higher entropy? (a) A solid sugar cube or the sugar cube dissolved in hot coffee, (b) a gas or a liquid at the same temperature, and (c) students sitting in a classroom listening to a lecture or the same students 30 seconds after the end of the period.

5. Is each of the following spontaneous primarily because of a decrease in potential energy or an increase in entropy? (a) Sand added to a glass of water settles on the bottom, (b) a sugar cube dissolves in a glass of water, and (c) individual nitrogen atoms bond to form a molecule of N_2.

6. Rationalize why it is possible for the following reaction to occur spontaneously:

$$C + H_2O + energy \longrightarrow CO + H_2$$

7. Criticize the statement: "Any spontaneous process results in an increase in the entropy of the system."

8. Criticize the statement: "Life is an ordering process, and because it goes against the natural tendency for increasing disorder, it constitutes a violation of the second law of thermodynamics."

9. Explain how the following nursery rhyme is a statement of the second law of thermodynamics:

Humpty Dumpty sat on a wall
Humpty Dumpty had a great fall!
All the King's horses
And all the King's men
Couldn't put Humpty Dumpty together again.

10. Why is the conversion of energy from one form to another never 100 percent efficient?

11. Use the second law of thermodynamics to explain why the ultimate pollutant on earth is heat.

12. How do scientists measure the rate of a reaction?

13. When reactants collide with each other, what three characteristics of those collisions determine the reaction rate?

14. Which one of the following is most likely to increase the percentage of collisions that have the right orientation to produce a reaction? (a) Increased concentration of reactant(s), (b) catalyst, or (c) increased temperature.

15. Which of the factors in Question 14 changes the activation energy of a reaction?

16. Draw an energy diagram (like the one in Figure 7.11) showing each of the three situations described in Table 7.1.

17. A lump of sugar warmed to body temperature will not decompose. Yet, if you swallow the sugar, it is oxidized at body temperature. Draw two net energy diagrams illustrating the difference between these two situations.

18. Why do wood shavings burn faster than logs?

19. Carbon monoxide gas (CO) reacts with chlorine gas (Cl_2) to form phosgene gas ($COCl_2$):

$$CO + Cl_2 \longrightarrow COCl_2$$

If this reaction takes place in a closed container at room temperature, which of the following are true about its rate?

a. The reaction rate decreases if the initial number of CO molecules in the container is doubled.

b. The reaction rate increases if the initial number of Cl_2 molecules in the container is increased.

c. The reaction rate increases if the temperature is raised.

d. The reaction rate increases if a catalyst is added.

20. Define dynamic chemical equilibrium. Give an example.

21. What is Le Châtelier's principle? Give two examples.

22. One way to make nylon is to position the two reactants (one is a liquid and the other is in a water solution) as two liquid layers, one on top of the other, in a container. Solid nylon forms only at the surface boundary where the two liquids meet. If you used tweezers to pull out the nylon as it formed at the boundary, how would this affect the amount of nylon formed? Explain.

23. Given the reaction of ammonia gas with oxygen to form nitrogen and water vapor, carried out at constant temperature in a closed container,

$$4NH_3 + 3O_2 \rightleftharpoons 2N_2 + 6H_2O + heat$$

predict whether the following concentration changes (stresses) will shift the equilibrium position to the right or to the left: (a) removal of oxygen, (b) removal of water, (c) increasing the concentration of ammonia, (d) decreasing the concentration of both ammonia and oxygen, (e) increasing the concentration of nitrogen, and (f) cooling the mixture.

24. Nitrogen gas (N_2) and oxygen gas (O_2) in the air can react to form nitric oxide (NO). The balanced equation is

$$N_2 + O_2 + heat \rightleftharpoons 2NO$$

What effect does the high temperature in an internal combustion engine have on this equilibrium? Why?

25. Nitric oxide (NO) in the air reacts with O_2 to form NO_2:

$$2NO + O_2 \rightleftharpoons 2NO_2$$

How does this reaction affect the equilibrium of the reaction in Question 24?

26. In which of the following reactions will a decrease in temperature shift the equilibrium to the left?
 a. $H_2 + I_2 + heat \rightleftharpoons 2HI$
 b. $CH_4 + 2O_2 \rightleftharpoons CO_2 + 2H_2O + heat$

27. How would an increase in the concentration of O_2 alter the equilibrium of Reaction b in Question 26?

Topics for Discussion

1. The U.S. Patent Office refuses even to review patent applications for so-called perpetual motion machines that produce more energy than they use. Is this a wise policy? Could the possible benefits of some technological breakthrough be kept from us in this way?

2. The theory of chemical evolution says that life arose from nonlife as chemicals combined, step-by-step, to form more complex substances, finally assembling into something that was alive. The theory of biological evolution says that humans evolved, step-by-step, from simpler organisms. Does the second law of thermodynamics allow for these processes?

3. Does your body ever reach a state of dynamic equilibrium? Why?

CHAPTER 8

Organic Chemistry: Some Important Carbon Compounds

General Objectives

1. Why does carbon form so many more compounds than other elements?

2. What are some important hydrocarbons (compounds containing carbon and hydrogen)?

3. What are the key components of coal, natural gas, and petroleum, and how are these fossil fuels used?

4. How does attaching other groups of atoms (called functional groups) to carbon compounds change their chemical behavior?

5. What important carbon-containing molecules have functional groups containing chlorine, fluorine, oxygen, and nitrogen?

Skin, food, gasoline, cotton, detergents, aspirin, and many other materials important to you and your life-style have one thing in common: They all contain *carbon*. Carbon is such an important element that an entire branch of chemistry is devoted to the study of carbon-containing compounds. This takes in a lot of territory, for more than 7 million carbon compounds are known, and an average of 100 or more new ones are made each week. In fact, there seems to be almost no limit to the number of organic compounds that can exist. In this chapter, we examine some of the important ones; in the next chapter, we study those that are essential in living things.

8.1 Carbon: A Unique Element

WHY DOES CARBON FORM SO MANY COMPOUNDS? What's so special about carbon? Why is there a separate branch of chemistry, called **organic chemistry**, devoted to carbon-containing compounds while another branch (called *inorganic chemistry*) deals with all the 108 other elements? And why are there so many different organic compounds?

Carbon is in Group 14 (IVA) in the periodic table, so its atoms contain four valence electrons and thus become stable by forming four covalent bonds (Section 4.3), as summarized in Table 8.1. Carbon has just the right combination of size and electronegativity to form long chains with stable covalent bonds between like (carbon) atoms. No other element can do this, not even silicon (Si) and germanium (Ge), which are in the same group in the periodic table as carbon. Thus, carbon alone can form molecules of almost all sizes and shapes. This helps explain the astonishing number of organic compounds.

TABLE 8.1 POSSIBLE COVALENT BONDING ARRANGEMENTS

Arrangement	Type of Covalent Bonds	Total Covalent Bonds
$-\overset{\displaystyle \mid}{\underset{\displaystyle \mid}{C}}-$	4 single bonds	4
$\overset{\displaystyle \mid}{\underset{\displaystyle \mid}{C}}{=}$	2 single bonds, 1 double bond	4
${=}C{=}$	2 double bonds	4
$-C{\equiv}$	1 single bond, 1 triple bond	4

Carbon atoms can link together like a long freight train of identical cars to form chains containing from two to thousands of carbon atoms. They can form straight chains

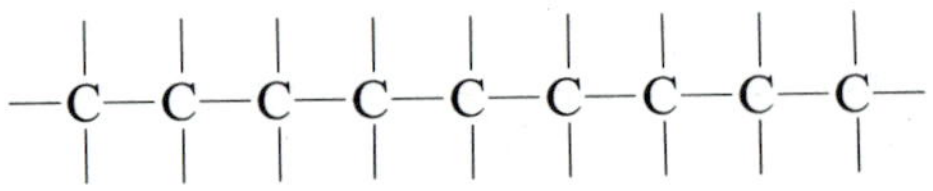

$$-\overset{\mid}{\underset{\mid}{C}}-\overset{\mid}{\underset{\mid}{C}}-\overset{\mid}{\underset{\mid}{C}}-\overset{\mid}{\underset{\mid}{C}}-\overset{\mid}{\underset{\mid}{C}}-\overset{\mid}{\underset{\mid}{C}}-\overset{\mid}{\underset{\mid}{C}}-\overset{\mid}{\underset{\mid}{C}}-\overset{\mid}{\underset{\mid}{C}}-$$

or branched chains

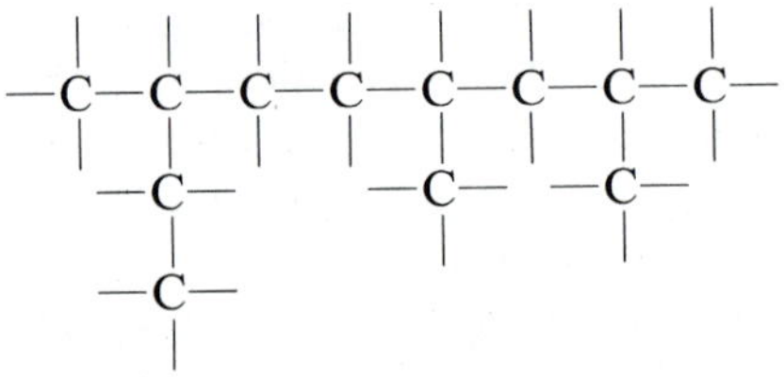

or even rings

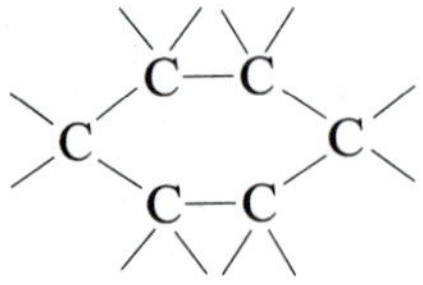

The length and variety of straight and branched carbon chains are almost limitless, but carbon rings usually have three to eight atoms, with six being the most common. In three dimensions, chains of three or more carbon atoms are staggered (see Table 8.3) because the four valence electrons carbon uses for bonding are not arranged at 90° angles but in a tetrahedral shape (Figure 8.1). To simplify drawings and save space, however, straight two-dimensional (flat) chains are used in most textbooks, including this one. As we will see, beginning in Table 8.3, carbon compounds can also be represented by ball-and-stick or space-filling models that show how the carbon atoms (shown as dark spheres) and their other bonded atoms (light spheres) are arranged in space.

Carbon can form double and triple covalent bonds with carbon and other atoms, which further increases its versatility. Examples include

$$-C-C-C=C-C-$$

double bond
(two shared pairs of electrons)

and

$$-C-C-C-C-C\equiv C-$$

triple bond
(three shared pairs of electrons)

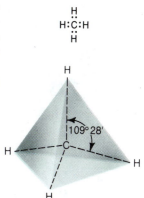

Figure 8.1 Electron dot structure of a carbon atom bonded to four hydrogen atoms (top). The tetrahedral structure of the molecule, with an angle between carbon and any two bonded atoms of 109°28′(bottom).

Remember, though, that each carbon atom still forms a total of four covalent bonds (see Table 8.1).

Carbon also forms covalent bonds with other nonmetals. Although elements such as Br, I, P, S, Mg, Ca, Al, Na, K, B, and Si are involved in organic chemistry, most organic compounds are combinations of carbon and five elements—hydrogen (H), oxygen (O), nitrogen (N), fluorine (F), and chlorine (Cl). This simplifies matters a bit.

Building organic molecules is like taking hundreds, or even thousands, of Tinker Toys and putting them together in all kinds of chains and rings of varying size and shape. Several other kinds of pieces (elements) can then attach to different places on each ring or chain. It's hardly surprising then that millions of different organic compounds exist.

FORMS OF CARBON   Carbon itself exists in more than one form, so we can see how the pattern of covalent bonding affects its properties. One form is *graphite* (Figure 8.2), a slippery black material used in lubricants, "lead" pencils, and the shafts of golf clubs, tennis rackets, and fishing rods. In

Figure 8.2 Two different forms of the element carbon, graphite (top) and diamond (bottom). (Photo courtesy of General Electric)

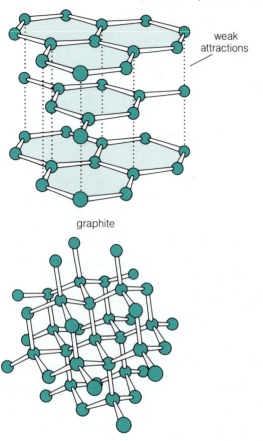

weak
attractions

graphite

diamond

graphite an array of hexagonal (six-sided) carbon rings joins together, with individual carbon atoms bonded to three—not four—other carbon atoms. Clusters of joined rings act like tiny sheets that slide past each other, making the graphite slippery. And the unusual electron-bonding arrangement makes graphite, alone among the nonmetal elements, a good conductor of electricity.

Another form of carbon—diamond—could hardly be more different (Figure 8.2). Here each carbon atom bonds to four neighboring carbon atoms in a huge, interconnected network of carbon atoms joined in a tetrahedral arrangement. Diamonds are hard, shiny, and clear, and they do not conduct electricity. In fact, their stable electron structure makes diamonds good electrical insulators.

8.2 Hydrocarbons: Chains and Rings

SATURATED HYDROCARBONS: ALKANES   Hydrogen is bound to carbon in virtually all organic compounds. An organic compound containing only carbon and hydrogen is known as a **hydrocarbon**. The four families of hydrocarbons are alkanes, alkenes, alkynes, and aromatic hydrocarbons.

If all bonds between carbon atoms are single bonds, the hydrocarbon is known as an **alkane**, or **saturated hydrocarbon**. Adding carbon atoms one at a time, and enough hydrogens to give each carbon four bonds, produces a series of straight-chain alkanes (Table 8.2). We name them by using the suffix -*ane* and a prefix that indicates the number of carbon atoms in the chain.

Table 8.3 shows formulas and models for methane (CH_4), the major component of natural gas; ethane (C_2H_6); and propane (C_3H_8), which is used in lighters and bottled gas. Included in the table are *condensed structural formulas*, a shorthand that chemists use to show the structures of organic compounds. With straight-chain and branched compounds, these condensed formulas show each carbon atom together with its bonded hydrogen (or other) atoms. Individual bonds in the chain aren't shown, however, except for double and triple bonds and where branching occurs; the other bonds are "understood" to be there. You need to get used to these condensed structures because you will see them used frequently in later chapters.

PRACTICE EXERCISE

Write condensed structural formulas for (a) butane and (b) decane.

SOLUTION

(a) $CH_3CH_2CH_2CH_3$ [also, $CH_3(CH_2)_2CH_3$]

(b) $CH_3CH_2CH_2CH_2CH_2CH_2CH_2CH_2CH_2CH_3$

[also, $CH_3(CH_2)_8CH_3$]

TABLE 8.2 THE FIRST TEN STRAIGHT-CHAIN SATURATED HYDROCARBONS

Name	Molecular Formula	State at 25°C	Structural Formula
Methane*	CH_4	Gas	
Ethane	C_2H_6	Gas	
Propane	C_3H_8	Gas	
n-Butant†	C_4H_{10}	Gas	
n-Pentane	C_5H_{12}	Liquid	
n-Hexane	C_6H_{14}	Liquid	
n-Heptane	C_7H_{16}	Liquid	
n-Octane	C_8H_{18}	Liquid	

* The prefix in color indicates the number of carbon atoms in the molecule.

† n- signifies a normal (straight-chain) molecule of a hydrocarbon.

TABLE 8.2 CONTINUED

Name	Molecular Formula	State at 25°C	Structural Formula
n-**Non**ane	C_9H_{20}	Liquid	
n-**Dec**ane	$C_{10}H_{22}$	Liquid	

TABLE 8.3 THE FIRST THREE MEMBERS OF THE ALKANE SERIES

Name	Methane	Ethane	Propane
Molecular formula	CH_4	C_2H_6	C_3H_8
Condensed structural formula	CH_4	CH_3CH_3	$CH_3CH_2CH_3$
Expanded structural formula			
Ball-and-stick model			
Space-filling model			

TABLE 8.4 THE TWO STRUCTURAL ISOMERS OF BUTANE

Name	n-Butane	Isobutane
Molecular formula	C_4H_{10}	C_4H_{10}
Condensed structural formula	$CH_3CH_2CH_2CH_3$	CH_3CHCH_3 CH_3
	straight chain	*branched chain*
Expanded structural formula		
Ball-and-stick model		
Space-filling model		
Physical properties:		
Melting point	−138.3°C	−160.0°C
Normal boiling point	−0.5°C	−12.0°C

STRUCTURAL ISOMERS   When two or more compounds have the same molecular formula but different structural formulas—different arrangements of atoms—they are called **structural isomers**. Notice in Table 8.3 that the name of each alkane with four or more carbon atoms is identified by the italic prefix *n-*. This prefix designates an alkane as the straight-chain (normal) isomer, for all alkanes containing four or more carbon atoms exist also as branched-chain isomers.

Butane (C_4H_{10}), for example, can have two possible structures (Table 8.4). Both isomers have four carbon atoms and ten hydrogen atoms, but they differ in what is bonded to what. Each isomer has slightly different properties because of its different structure.

PRACTICE EXERCISE

Write the expanded and condensed structural formulas for the three isomers of pentane, C_5H_{12}.

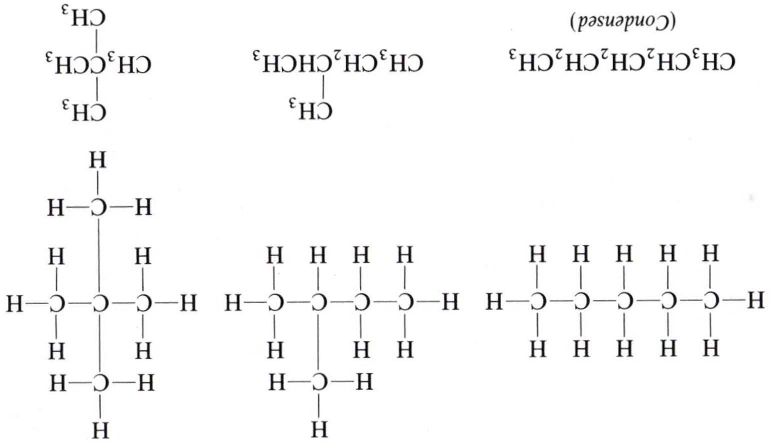

(Condensed)

$CH_3CH_2CH_2CH_2CH_3$

$CH_3CH_2CHCH_3$ with CH_3 branch

$(CH_3)_4C$

SOLUTION

As the size of the molecule increases, the number of structural isomers grows rapidly. For example, octane (C_8H_{18}) has 18 possible isomers; $C_{20}H_{42}$ has 366,319 structural isomers; and $C_{40}H_{82}$ has more than 62 trillion isomers, each with a different name. Structural isomerism is another reason why so many different organic compounds occur.

PRACTICE EXERCISE

Organic compounds that contain elements besides hydrogen and carbon also can form isomers. Write structural formulas for the two isomers that have the formula C_3H_7F.

SOLUTION

$$H-\overset{\overset{\displaystyle H}{|}}{\underset{\underset{\displaystyle H}{|}}{C}}-\overset{\overset{\displaystyle H}{|}}{\underset{\underset{\displaystyle H}{|}}{C}}-\overset{\overset{\displaystyle H}{|}}{\underset{\underset{\displaystyle H}{|}}{C}}-F \quad (CH_3CH_2CH_2F) \quad \text{and}$$

$$H-\overset{\overset{\displaystyle H}{|}}{\underset{\underset{\displaystyle H}{|}}{C}}-\overset{\overset{\displaystyle F}{|}}{\underset{\underset{\displaystyle H}{|}}{C}}-\overset{\overset{\displaystyle H}{|}}{\underset{\underset{\displaystyle H}{|}}{C}}-H \quad (CH_3CHFCH_3)$$

(In the first isomer, F is bonded to an end carbon. In the second isomer, F is bonded to the interior carbon. All other structures with the formula C_3H_7F are the same as one of these two isomers in terms of what is bonded to what.)

Alkanes have many practical uses. We use them, for example, as fuels for camp stoves, automobiles, and furnaces. Alkanes burn by combining with oxygen to form carbon dioxide and water in heat-producing oxidation–reductions. A typical combustion reaction is

$$C_3H_8 + 5O_2 \longrightarrow 3CO_2 + 4H_2O + \text{energy}$$
propane

Other alkanes are used as waxes, solvents, lubricants, and road materials such as asphalt. We discuss some of these uses further in Section 8.3.

UNSATURATED HYDROCARBONS   Hydrocarbons with double or triple covalent bonds between any two carbon atoms are called **unsaturated hydrocarbons**. We say they are "unsaturated" with hydrogen because, compared with alkanes, these compounds cannot bond as many hydrogen atoms per carbon atom. Compounds containing more than one multiple bond are called **polyunsaturated**. The most familiar examples are plant oils (used for cooking), which have long carbon chains (usually twelve to twenty carbon atoms) with two to four double bonds in each chain.

TABLE 8.5 TWO MEMBERS OF THE ALKENE SERIES

IUPAC name*	Ethene	Propene
Common name	Ethylene	Propylene
Molecular formula	C_2H_4	C_3H_6
Condensed structural formula	$CH_2{=}CH_2$	$CH_3CH{=}CH_2$

Expanded structural formula

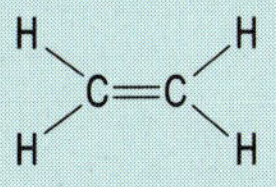

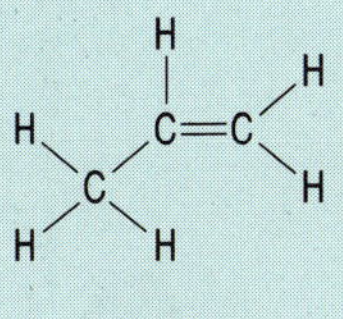

Ball-and-stick model

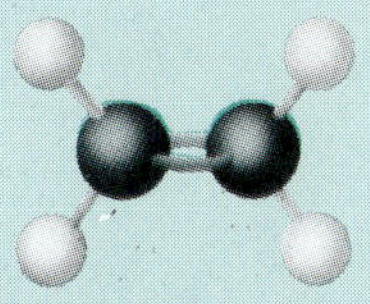

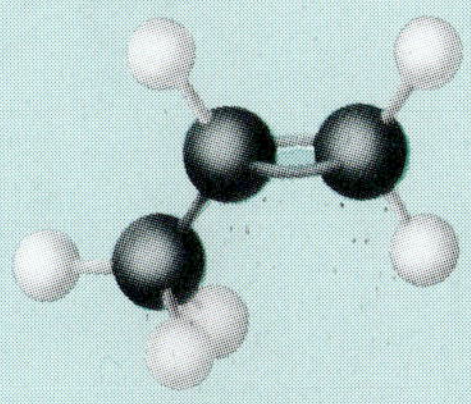

Space-filling model

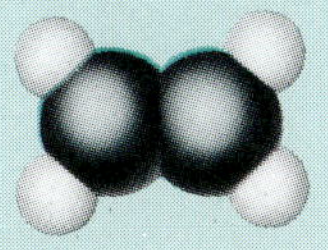

* These are the internationally accepted names based on the rules of the International Union of Pure and Applied Chemistry (IUPAC).

Hydrocarbons containing one or more double bonds are called **alkenes** (identified by the suffix *-ene*), and those with one or more triple bonds are called **alkynes** (identified by the suffix *-yne*). Table 8.5 shows the two simplest alkenes, and Table 8.6 shows two simple alkynes.

Alkenes and alkynes, which are very reactive because of their electron-rich multiple bonds, are used as high-powered fuels and as starting materials to make other organic compounds. For example, ethyne (commonly called *acetylene*) is used to produce an extremely hot flame for welding and a bright, luminous flame for miners' lamps (Figure 8.3). Manufacturers use ethene (commonly called *ethylene*) to make the plastic polyethylene (Section 14.2) and many other products. Food processors also use ethylene gas to make their produce ripen faster.

TABLE 8.6 TWO MEMBERS OF THE ALKYNE SERIES

	Ethyne	2-Butyne
IUPAC name	Ethyne	2-Butyne
Common name	Acetylene	
Molecular formula	C_2H_2	C_4H_6
Condensed structural formula	$CH \equiv CH$	$CH_3C \equiv CCH_3$
Expanded structural formula	$H - C \equiv C - H$	(expanded structure shown)
Ball-and-stick model	(model)	(model)
Space-filling model	(model)	(model)

Figure 8.3 A miner's acetylene lamp. The acetylene (ethyne) gas is lit by a spark from a sealed internal light mechanism. (Geological Museum, Colorado School of Mines)

AROMATIC HYDROCARBONS Some cyclic compounds also have multiple carbon-to-carbon bonds. The best-known example is benzene, a six-carbon compound. August Kekulé, a German chemist, first proposed a ring structure for benzene (see "Chemistry Spotlight").

Although the proposed structure (Figure 8.4) shows benzene with three double bonds in the ring, benzene doesn't react like alkenes. Thus, chemists eventually concluded that the ring atoms don't have double bonds. Instead, the three extra pairs of electrons making up the three double bonds move freely about in a doughnut-shaped, three-dimensional space surrounding the flat carbon ring, shared equally by all six ring atoms; it's as if each carbon were joined to its neighbor by an average of one and one-half bonds. This is why the formula for benzene now is often written with a circle in the center to represent those six electrons (Figure 8.5).

Condensed structural formulas for cyclic compounds have an appropriate geometric figure (usually a triangle, square, pentagon, or hexagon) to represent the carbon atoms in the ring. Hydrogen atoms bonded to the ring carbons are not shown; they are "understood" to be there. If some atom other than hydrogen is bonded to a ring carbon atom, however, it is shown in the formula.

CHEMISTRY SPOTLIGHT   KEKULÉ AND THE WHIRLING SNAKES

The structure of benzene was a mystery for a long time. Discovered in 1825, benzene has the formula C_6H_6. Scientists knew that the only way a six-carbon substance could have so few hydrogen atoms was for it to be highly unsaturated. But benzene didn't react like alkenes and alkynes; somehow, it was different.

According to legend, one evening in 1865 August Kekulé was sitting half-asleep in front of his fireplace, idly staring into the fire. As his mind wandered, he seemed to see atoms in the fire dancing around in snakelike movements. Suddenly one of the snakes seized its own tail and whirled around in a circular motion. Kekulé awoke and spent the rest of the night working out what he had seen. He proposed that benzene is a ring of six carbon atoms, having alternating single and double bonds, with each carbon bonded to one hydrogen atom.

Of his experience he wrote, "Let us learn to dream, gentlemen, then perhaps we shall find the truth. But let us beware of publishing our dreams till they have been tested by the waking understanding." Again, we see that scientific ideas can last only as long as they fit with the experimental data.

space-filling model

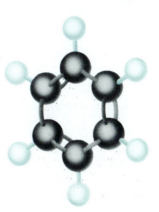

ball-and-stick model

Figure 8.4 Benzene (C_6H_6) is one of the most important aromatic hydrocarbons.

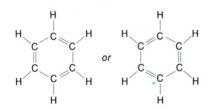

structural formula

condensed structural formula

Figure 8.5 Condensed structural formula for benzene.

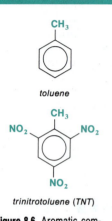

toluene

trinitrotoluene (TNT)

Figure 8.6 Aromatic compounds in which one or more groups replace hydrogen atoms in benzene.

Organic compounds containing a benzene or similar ring are called **aromatic compounds** (originally named because of their odors). Most aromatic compounds come from coal tar, a sticky black mixture produced when coal is heated in the absence of air. In many aromatic compounds, side chains of carbon or other atoms replace one or more hydrogen atoms

TABLE 8.7 SOME POLYCYCLIC AROMATIC COMPOUNDS

Condensed Structural Formula	Name	Some Uses or Effects
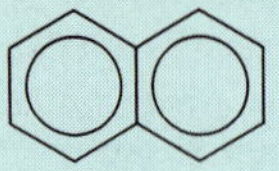	Naphthalene	Used in some mothballs
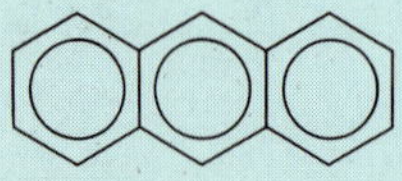	Anthracene	Used in dyes
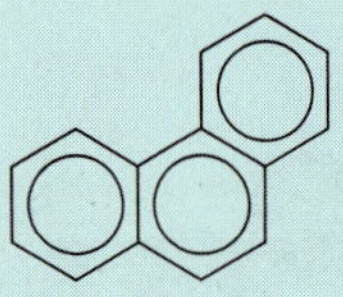	Phenanthrene	Used in dyes, explosives, and synthesis of drugs
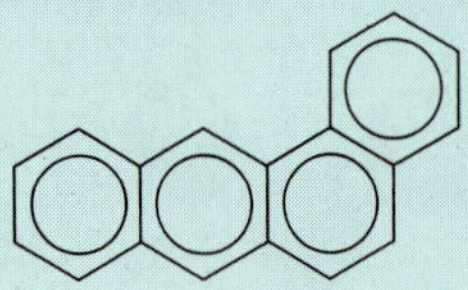	Benzanthracene	Basic framework for major carcinogenic polycyclic aromatic compounds
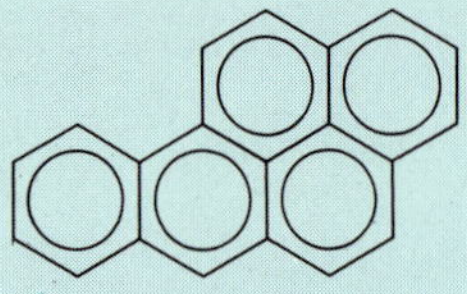	3,4-Benzopyrene	Active carcinogen found in cigarette smoke, wood smoke, and photochemical smog
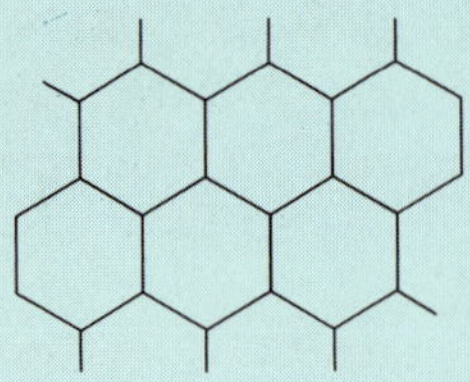	Chicken wire	Keeping chickens

in benzene. One example is toluene (Figure 8.6), which is used to make many other aromatic compounds. Replacing three hydrogens in toluene with nitro ($-NO_2$) groups produces trinitrotoluene, or TNT (Figure 8.6).

Benzene rings also can fuse together to form **polycyclic aromatic compounds**. The simplest is naphthalene ($C_{10}H_8$), which consists of two fused benzene rings (Table 8.7); it is used in one kind of mothballs. Additional aromatic rings can fuse together to produce an array of compounds. Table 8.7 shows some of these, including the carcinogenic (cancer-producing) 3,4-benzopyrene found in cigarette smoke.

8.3 Fossil Fuels and Petrochemicals: From Oil to Plastics

IMPORTANCE OF FOSSIL FUELS   We live in a unique period in history in which **fossil fuels**—natural gas, petroleum, and coal—are our major energy source. These fuels formed over millions of years as the remains of plants and animals were buried deep in the earth and slowly decomposed under the influence of heat and pressure.

Natural gas, petroleum, and coal also are important as raw materials for making products such as paints, plastics, pesticides, and drugs. These products, known as **petrochemicals**, account for the vast majority of the organic chemicals produced by industry. We now use more than 90 percent of petroleum and other fossil fuels for energy, but we will likely use an increasing percentage for petrochemicals as the supply—especially of petroleum and natural gas—dwindles in the coming decades.

COAL   From 55 to 95 percent of coal is carbon. Over many millions of years, coal developed in stages from peat to anthracite (hard coal), with a greater percentage of carbon occurring in each successive stage (Figure 8.7). The other material (besides carbon) is mostly hydrogen, oxygen, nitrogen, and sulfur compounds. Anthracite coal typically contains 1 to 3 percent sulfur, which can cause air pollution from sulfur oxides formed when it burns (Chapter 12).

Most of the coal mined each year is burned to provide heat and to supply steam for industry, transportation, and electricity. Much of the remainder is heated at 1000 to 1300°C in the absence of air:

$$\text{coal} \xrightarrow[\text{heat}]{\text{no air}} \text{coke} + \text{coal tar} + \text{coal gas}$$

Coke, the solid residue remaining after heating coal, is used to make iron and steel (Section 10.4). The liquid material produced, called coal tar, is

Figure 8.7 Stages in the formation of different types of coal over millions of years.

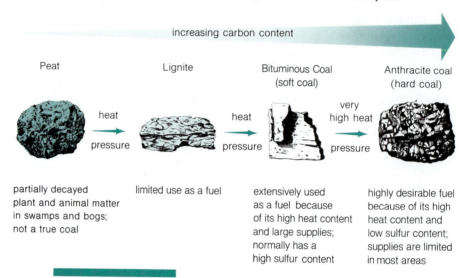

increasing carbon content

Peat	Lignite	Bituminous Coal (soft coal)	Anthracite coal (hard coal)

heat / pressure

heat / pressure

very high heat / pressure

partially decayed plant and animal matter in swamps and bogs; not a true coal

limited use as a fuel

extensively used as a fuel because of its high heat content and large supplies; normally has a high sulfur content

highly desirable fuel because of its high heat content and low sulfur content; supplies are limited in most areas

rich in aromatic compounds and is used to make dyes, plastics, pesticides, medicines, and other products. Coal gas, the material that vaporizes when coal is heated, is a useful fuel.

NATURAL GAS   The term *natural gas* refers to any gas obtained from the ground, but the term most commonly applies to natural deposits containing 50 to 99 percent methane (CH_4), with small amounts of ethane (C_2H_6), propane (C_3H_8), and longer-chain hydrocarbons (see Table 8.2). Propane and butane are removed, converted into liquids (called liquefied petroleum gas, or LPG), and stored in high-pressure cylinders for use as fuel. The remaining gas, which is mostly methane, is pumped into pipelines for distribution as a fuel.

Natural gas is the cleanest-burning of the fossil fuels. The reaction for its combustion is

$$CH_4 + 2O_2 \longrightarrow CO_2 + 2H_2O$$

PETROLEUM   The other major source of hydrocarbons is *petroleum*, or *crude oil*. Petroleum from oil wells often is a dark, greenish brown, foul-smelling liquid. It contains a mixture of saturated and aromatic hydrocarbons and small amounts of oxygen, sulfur, and nitrogen. The sulfur content of crude oil varies from less than 1 percent to as high as 5 percent, and it is a potential source of air pollution when petroleum products are burned (Chapter 12).

PETROLEUM REFINING   Crude oil is a mixture of such diverse materials that it isn't very useful by itself. Thus, oil refineries separate the mixture into fractions that are more homogeneous by the process of *fractional distillation*. In a furnace at the bottom of a huge fractionating column (Figure 8.8), which often rises as high as 30 m (100 ft), the crude petroleum is heated to about 315°C in the absence of oxygen.

Figure 8.8 Fractional distillation separates crude petroleum into its components on the basis of their different boiling points. (Modified from "Chemistry and Petroleum," American Petroleum Institute, New York)

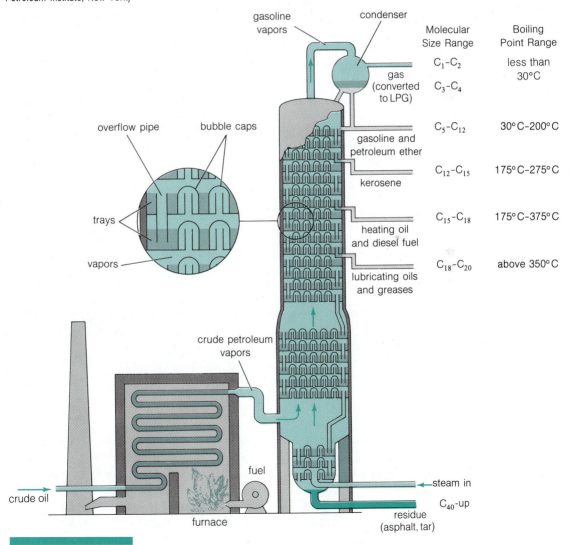

Hydrocarbon oil molecules are nonpolar, with only weak London forces (Section 5.1) between them, and their boiling points vary with size. During fractional distillation, the smaller, lower-boiling substances rise as vapor to the top of the column where they are collected. Larger molecules cool enough to liquefy as they rise in the column and are collected at a lower level. Thus, various hydrocarbons are separated on the basis of size and collected as liquid fractions at particular heights in the fractionating column (see Figure 8.8).

GASOLINE PRODUCTION   All the separated fractions are used commercially, but the greatest demand is for gasoline. Although the gasoline fraction makes up 25 to 45 percent of the petroleum distilled, this isn't enough to meet consumer demand. Distilling enough crude oil to satisfy this demand would produce large surpluses of other petroleum fractions. Furthermore, the gasoline fraction produced by fractional distillation is not of sufficient quality to use directly in our automobiles.

To overcome these problems, refineries carry out several other processes, the most common of which are *cracking*, *polymerization*, and *reforming* (Table 8.8). Cracking and polymerization change the size of molecules and enable refineries to adjust their supply of certain fractions. During late winter, for example, refineries start cracking heating oil and polymerizing hydrocarbons to produce extra gasoline for the spring and summer months.

Reforming causes hydrocarbons to become branched or cyclic. This is done especially with gasoline-size molecules (C_5 to C_9) because branched and cyclic products have less tendency to ignite prematurely in the engine

TABLE 8.8 MAJOR REFINING PROCESSES

Name	Petroleum Fraction Used	Description
Cracking	Heating oil (C_{15}–C_{18})	Breaks down larger hydrocarbons into C_6–C_8 gasoline molecules
Polymerization	Gases and petroleum ether (C_2–C_4)	Forms larger C_6–C_8 gasoline molecules from C_2–C_4 molecules
Reforming	Gasoline (C_5–C_{12})	Converts hydrocarbons into highly branched or cyclic compounds

and cause the engine to "knock." An example of the reforming process is

$$CH_3(CH_2)_6CH_3 \xrightarrow{\text{reforming}} \begin{array}{c} CH_3 \\ | \\ CH_3CCH_2CHCH_3 \\ | \quad | \\ CH_3 \ CH_3 \end{array}$$

n-octane   a structural isomer of
                        octane called "isooctane"

On the basis of its antiknock properties, a gasoline mixture is assigned an *octane number*; the higher the number, the lower the tendency to cause engines to knock. One isomer of octane, called isooctane (see the reaction above), has excellent antiknock properties and was given an octane value of 100; *n*-heptane, which performs very poorly, was assigned a value of 0. On this scale, a gasoline mixture that performs as well as a mixture of 90 percent isooctane and 10 percent heptane has an octane rating of 90. Most high-compression engines today require fuel with octane values of about 90.

The octane number of the gasoline fraction is too low to work in modern automobile engines. One solution was to use the additive *tetraethyl lead*, $Pb(C_2H_5)_4$ (Figure 8.9). Although it was an inexpensive way to improve the quality of low-grade gasoline, the lead reacts during combustion to form products that pass into the air. Because of the threat lead poses to human health, the Environmental Protection Agency is requiring that lead additives in gasoline be phased out. Unleaded gasoline contains larger amounts of aromatic hydrocarbons such as toluene (see Figure 8.6) and certain alcohols and ethers to boost its octane number.

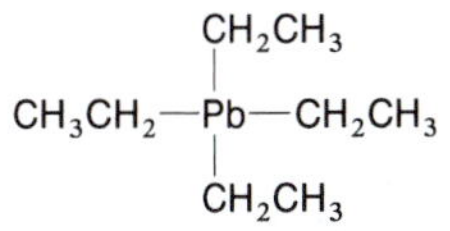

$$\begin{array}{c} CH_2CH_3 \\ | \\ CH_3CH_2-Pb-CH_2CH_3 \\ | \\ CH_2CH_3 \end{array}$$

Figure 8.9 Tetraethyl lead.

8.4 Functional Groups: Where the Action Is

Saturated hydrocarbons don't react readily. Their most familiar reaction—combustion—requires heat and a very reactive material (oxygen) before anything happens. But if other atoms or groups of atoms replace hydrogen atoms on the carbon chain or ring, the action picks up. These other groups are called **functional groups** because they determine most of the chemical behavior of the molecule.

Functional groups are the main reaction sites in a molecule. This greatly simplifies matters; for even if a molecule has tens, hundreds, or

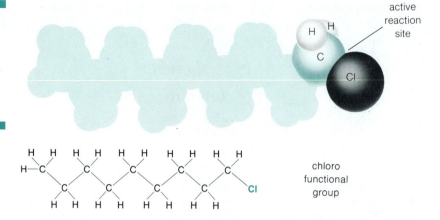

Figure 8.10 The functional group is the active reaction site on a carbon skeleton. Virtually ignoring the rest of the molecule, we can represent it as *R*—Cl.

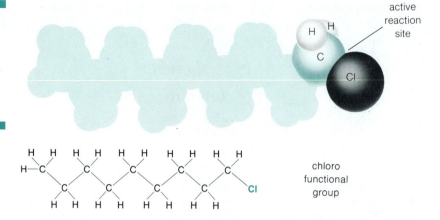

even thousands of carbon atoms, we only need to find its key functional groups to know how it will react (Figure 8.10). To highlight the functional groups, chemists use the symbol *R* to represent the *non*functional, hydrocarbon part of the molecule; thus, we can think of *R* as representing the "rest of the molecule we can ignore."

The art and science of organic chemistry is based on understanding the chemistry of the functional groups. In the rest of this chapter, we examine some of the most important functional groups. Table 8.9 presents their structures, names, and an example of each type.

8.5 Organic Halides: Refrigerants to Pesticides

The common halogens—fluorine, chlorine, bromine, and iodine—are in Group 17 (VIIA) in the periodic table and have seven valence electrons; thus, they form one covalent bond to become stable. When one or more halogen atoms bond to carbon, the compound is called an **organic halide**.

Chloroform, $CHCl_3$, is a liquid that readily evaporates and was once widely used as a general anesthetic. It has been replaced by safer anesthetics, such as halothane, some of which are also organic halides (Figure 8.11). Carbon tetrachloride, CCl_4, is a nonpolar liquid that was once used in dry cleaning to remove nonpolar stains such as grease and oil. Because it is toxic, it has been replaced by compounds such as tetrafluoroethylene, $CF_2{=}CF_2$.

Freons are one- or two-carbon compounds containing fluorine and chlorine. One example, called Freon-12, is CF_2Cl_2. These compounds are used as coolant liquids in refrigerators and air conditioners. Until the late 1970s, they were used in aerosol cans to expel hair spray, whipping cream, deodorant, and other products. Their use is declining because of evidence

TABLE 8.9 MAJOR FUNCTIONAL GROUPS (COLORED) IN ORGANIC COMPOUNDS

Type of Compound	Structural Formula*	Condensed Formula	Prefix or Suffix for Group Name	Example Structural Formula	IUPAC Name	Common Name
Halide	—Cl (or —Br, —F, —I)	RCl	Chloro- (or bromo-, fluoro-, or iodo-)	CH_3CH_2Cl	Chloroethane	Ethyl chloride
Alcohol	R—O—H	ROH	-ol	CH_3CH_2—O—H	Ethanol	Ethyl alcohol
Ether	R—O—R'	ROR'	ether	CH_3OCH_3	Methoxymethane	Dimethyl ether
Aldehyde	R—$\overset{\overset{\text{O}}{\|\|}}{C}$—H	RCHO	-al	$CH_3\overset{\overset{\text{O}}{\|\|}}{C}$—H	Ethanal	Acetaldehyde
Ketone	R—$\overset{\overset{\text{O}}{\|\|}}{C}$—$R'$	RCOR'	-one	$CH_3\overset{\overset{\text{O}}{\|\|}}{C}CH_3$	Propanone	Acetone
Carboxylic acid	R—$\overset{\overset{\text{O}}{\|\|}}{C}$—O—H	RCOOH	-oic acid	$CH_3\overset{\overset{\text{O}}{\|\|}}{C}$—O—H	Ethanoic acid	Acetic acid
Ester	R—$\overset{\overset{\text{O}}{\|\|}}{C}$—O—$R'$	RCOOR'	-oate	$CH_3\overset{\overset{\text{O}}{\|\|}}{C}$—O$CH_3$	Methyl ethanoate	Methyl acetate
Amine	R—$\overset{\overset{\text{H}}{\|}}{N}$—H	RNH$_2$	amine	$CH_3CH_2\overset{\overset{\text{H}}{\|}}{N}$—H	Aminoethane	Ethyl amine
Amide	R—$\overset{\overset{\text{O}}{\|\|}}{C}$—$\overset{\overset{\text{H}}{\|}}{N}$—H	RCONH$_2$	-amide	$CH_3\overset{\overset{\text{O}}{\|\|}}{C}$—$\overset{\overset{\text{H}}{\|}}{N}$—H	Ethanamide	Acetamide

* R and R' represent carbon groups, either chains or rings

Br F
| |
H—C—C—F
| |
Cl   F

Figure 8.11 Halothane, a general inhalation anesthetic.

that they are destroying the ozone (O_3) in the upper atmosphere that protects us from the sun's harmful ultraviolet rays (see Chapter 12).

Athletic trainers often use ethyl chloride (see Table 8.9) to relieve minor injuries. Sprayed on a bruised shin or forearm, ethyl chloride vaporizes quickly and takes with it heat from the skin. This reduces the soreness and leaves the bruised area feeling a bit numb. Another use of organic halides, particularly chlorine compounds, is in pesticides. The structures of several organochlorine pesticides, including DDT, are given in Section 17.2.

8.6 Oxygen-Containing Groups: Cocktails to Flavors

Figure 8.12 Structural formulas of ethanol (ethyl alcohol), C_2H_6O, and the simplest ether, dimethyl ether, C_2H_6O.

ALCOHOLS AND ETHERS   Oxygen is in Group 16 (VIA) in the periodic table and has six valence electrons, so it forms two covalent bonds to become stable. In organic compounds, oxygen can form either a double bond to carbon or two single bonds. When it forms single bonds, oxygen produces two functional groups—alcohols and ethers.

An **alcohol** contains oxygen with covalent bonds to carbon and hydrogen (Figure 8.12; see Table 8.9). We can represent an alcohol as R—O—H, where R stands for a carbon chain or ring. Alcohols are named by replacing the final e of the hydrocarbon name with the suffix -*ol*. Thus, the two-carbon alcohol is called ethanol (or ethyl alcohol).

When oxygen is bonded to two carbon atoms, the functional group is known as an **ether** (see Table 8.9 and Figure 8.12). Ethers can be represented as R—O—R', where R and R' may be the same or different carbon groups. To name ethers, we identify the carbon groups on each side of the oxygen atom, replace the -*ane* ending with the suffix -*yl*, and then add the word *ether*. Thus, the ether shown in Figure 8.12 is called dimethyl ether. [The prefix *di*- indicates that two methyl (C_1) groups are attached to the oxygen.]

ethanol (ethyl alcohol)

H H
| |
H—C—C—O—H
| |
H H

or

CH_3CH_2OH

or

dimethyl ether

H   H
| |
H—C—O—C—H
| |
H   H

or

CH_3—O—CH_3

Ethanol and dimethyl ether have the same molecular formula, C_2H_6O, so they are structural isomers. Their physical and chemical properties show how different structural isomers can be. Ethanol, for example, boils at 78°C, whereas dimethyl ether boils at −24°C—a difference of 102°. The main reason for ethanol's higher boiling point is that the —O—H group causes hydrogen bonding between ethanol molecules; this makes it harder (because more heat is required) to separate these molecules and form a gas. Ethers, on the other hand, cannot form hydrogen bonds because they have no —O—H group. Hydrogen bonding also enables ethanol, but not dimethyl ether, to dissolve readily in water.

Ethanol—commonly called *ethyl alcohol*, or *grain alcohol*—is the active ingredient in alcoholic drinks. Microorganisms such as yeasts convert sugar in fruits or vegetables into ethanol in the absence of air, a process called *fermentation*. Apples yield hard cider; grapes yield wine; beer and ale are made from malt (grain that has sprouted). The alcoholic content of these fermentation products varies from a few percent to about 15 percent. Fermentation stops at this point because yeasts don't survive concentrations of ethanol higher than 12 to 15 percent.

Drinks with a higher ethanol content, such as brandy, whiskey, and vodka, are made by distilling the fermentation mixture to obtain a high-ethanol fraction. The color and flavor of these liquors come from minor contaminants distilled with the ethanol or introduced while the liquor ages in wooden casks.

The ethanol concentration in an alcoholic beverage is often expressed as a "proof," which is twice the percentage (by volume) of ethanol. Thus, 80-proof whiskey contains 40 percent ethanol. In the seventeenth century, dealers tested whiskey by pouring it on gunpowder and igniting it. If the gunpowder exploded after the whiskey burned off, this was the "proof" of its high ethanol (and low water) content. Ethanol intended for purposes other than drinking, however, is mixed with small amounts of toxic or foul-smelling contaminants. This enables the product, called *denatured alcohol*, to be used for industrial purposes without being subject to the high taxes levied on drinkable ethanol.

The simplest alcohol, methanol (CH_3—OH), is often called *methyl alcohol* or *wood alcohol*. It is burned as a fuel in camp stoves and may someday become an important fuel for automobiles. It is toxic to drink and occasionally occurs in improperly made moonshine whiskey or is drunk by people who see the word *alcohol* and think it is ethanol. Some make that mistake only once, for drinking methanol can cause blindness and even death.

Several other alcohols are of practical interest (Figure 8.13). Isopropyl alcohol is widely used as rubbing alcohol and a disinfectant. Ethylene glycol is the main ingredient in permanent antifreeze; its two polar —O—H groups make it very soluble in water, so we can mix antifreeze

Figure 8.13 Some useful alcohols.

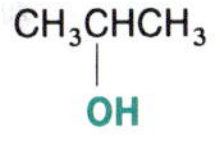

$$CH_3CHCH_3$$
$$|$$
$$OH$$

2-propanol
(isopropyl alcohol)

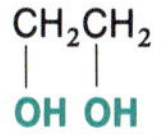

$$CH_2CH_2$$
$$|\quad|$$
$$OH\ OH$$

ethylene glycol

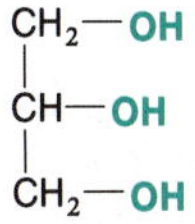

$$CH_2-OH$$
$$|$$
$$CH-OH$$
$$|$$
$$CH_2-OH$$

glycerine
(glycerol)

Figure 8.14 Two simple alcohols (phenols).

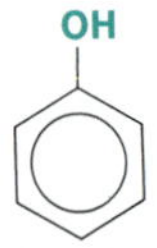

OH

phenol

OH

CH₃

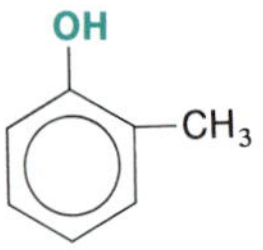

o-cresol

Figure 8.15 Two useful ethers.

CH₃CH₂—O—CH₂CH₃

diethyl ether

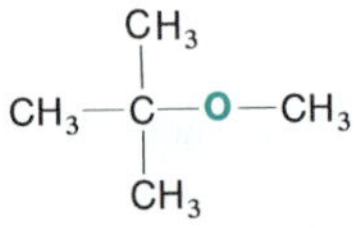

CH₃
|
CH₃—C—O—CH₃
|
CH₃

methyl t-butyl ether

and water in any desired proportion. Glycerine (glycerol), which has three alcohol groups, also forms strong hydrogen bonds with water. Glycerine is in many cosmetic preparations because its water-binding qualities help keep the skin soft and moist. Cholesterol (Figure 9.11), which you can recognize as an alcohol by the suffix -ol, is an important substance in blood and various organs in the body.

Phenol is the simplest aromatic compound that includes an alcohol group (Figure 8.14). In the nineteenth century, Joseph Lister, an English surgeon, discovered that using phenol on the skin or to disinfect surgical equipment was an effective way to prevent infections due to surgery. We use the antiseptic action of phenol derivatives in our mouthwashes, one of which bears Lister's name. Another derivative, hexachlorophene, is used widely in hospitals to combat infections. And o-cresol, a methylated version of phenol, is a key component in creosote, a wood preservative used for such things as fence posts, telephone poles, and railroad ties.

The best-known ether is diethyl ether (Figure 8.15), commonly called ether. In 1846 William Morton, a Boston dentist, publicly demonstrated that ether could be used as an anesthetic for surgery (Figure 8.16). This was the beginning of a new era in medicine, making possible medical and dental surgery that previously could not be done because it was too painful.

Diethyl ether is rarely used now as an anesthetic because it is so flammable and explosive. In fact, it is one of the most dangerous chemicals stored in laboratories. It vaporizes easily and spreads throughout a room. A lit match, a spark from an electrical appliance, or even a hot piece of metal near an open bottle of diethyl ether can cause an explosive fire.

Another important ether is methyl *t*-butyl ether (see Figure 8.15), which is used as an additive to increase the octane number of unleaded gasoline.

Figure 8.16 William Morton making the first public demonstration of ethyl ether as a general anesthetic at the Massachusetts General Hospital in 1846. Note the lack of gowns, masks, and gloves. The germ theory of disease had not yet been established. (National Library of Medicine)

ALDEHYDES AND KETONES If an oxygen atom forms a double covalent bond with a carbon atom, the group is known as a **carbonyl group** ($-\overset{\overset{\textstyle O}{\|}}{C}-$). If the carbonyl group is at the end of a carbon chain, the functional group $R-\overset{\overset{\textstyle O}{\|}}{C}-H$ is an **aldehyde**; if it is within the carbon chain, the functional group is a **ketone**, $R-\overset{\overset{\textstyle O}{\|}}{C}-R'$. The suffix *-al* is used in naming aldehydes, and the suffix *-one* is used for ketones (Figure 8.17).

Aldehydes and ketones are made by oxidizing alcohols. The carbon atom bonded to an alcohol group is converted into a carbonyl group; thus, compounds with an alcohol group on an end carbon become aldehydes, whereas compounds with the alcohol group bonded to an interior carbon atom become ketones. For example,

$$CH_3CH_2CH_2-OH \xrightarrow{\text{oxidation}} CH_3CH_2\overset{\overset{\textstyle O}{\|}}{C}-H$$

$$\textit{(an aldehyde)}$$

$$CH_3\overset{\overset{\textstyle OH}{|}}{C}HCH_3 \xrightarrow{\text{oxidation}} CH_3\overset{\overset{\textstyle O}{\|}}{C}CH_3$$

$$\textit{(a ketone)}$$

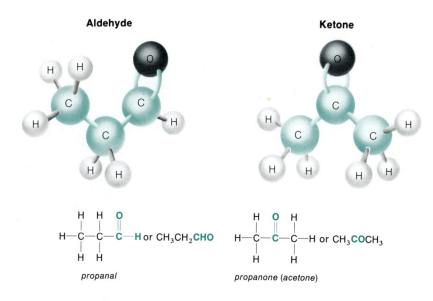

Aldehyde

Ketone

H—C—C—C—H or CH₃CH₂**CHO**

propanal

H—C—C—C—H or CH₃**CO**CH₃

propanone (acetone)

Figure 8.17 Comparison of the aldehyde and ketone functional groups.

The simplest aldehyde is methanal, $H-\overset{\overset{\displaystyle O}{\|}}{C}-H$, which has the common name *formaldehyde*. Methanol is oxidized into formaldehyde in the body, and it is the formaldehyde that makes methanol so toxic to drink. Formalin, a 40 percent solution of formaldehyde in water, is used to preserve biological specimens. If you have ever been in a biology laboratory, you have probably encountered its strong and invigorating odor.

Formaldehyde is used to help make clothes crease-resistant. Because of this, some people who wear "permanent-press" clothes suffer from skin irritation. Sweat frees formaldehyde gas from the clothes; this can cause rashes and irritation. Formaldehyde can also vaporize from formaldehyde–urea insulating material in buildings and cause respiratory discomfort in some people.

Ethanal (or acetaldehyde), the two-carbon aldehyde (see Table 8.9), is used for making such things as drugs and perfumes. Many longer-chain or aromatic aldehydes are used as perfumes or flavoring agents such as almond, vanilla, and cinnamon (Figure 8.18).

The simplest ketone is propanone (see Figure 8.17), commonly called *acetone*. Acetone and other ketones are used as solvents in removing varnish, paint, and fingernail polish. Small amounts are used to make plastics and perfumes (Figure 8.18), but some ketones have very unpleasant odors.

Ketones also are important in the body. For example, the breath of a person in a diabetic coma has the sweetish odor of acetone. By recognizing the odor and seeking medical attention for the unconscious person, you could save the person's life. Other important examples, which you can recognize as ketones by the suffix *-one*, are cortisone (Figure 19.17) and the sex hormones, progesterone and testosterone (Figure 19.18).

ORGANIC ACIDS AND ESTERS   If a second oxygen atom is bonded between a carbonyl carbon and a hydrogen atom, the functional group is a **carboxylic** (or **organic**) **acid**: $R-\overset{\overset{\displaystyle O}{\|}}{C}-O-H$. In naming carboxylic acids, we replace the last *e* in the hydrocarbon name with the suffix *-oic* and add the word *acid*. Figure 8.19 shows the structures of the two simplest organic acids, methanoic (or formic) and ethanoic (or acetic) acid.

Formic acid is one of the stinging substances produced by red ants, bees, and nettles. As Ogden Nash put it, "Would you be calm and placid/ If you were full of formic acid?" Formic acid also is used in the textile, rubber, and leather industries.

Vinegar is a weak solution of acetic acid in water. When wine or hard cider is exposed to air, its ethanol reacts with oxygen (is oxidized) to form acetic acid, which gives the beverage a sour or vinegary taste. We saw

Figure 8.18 Two aldehydes and a ketone used as flavoring agents.

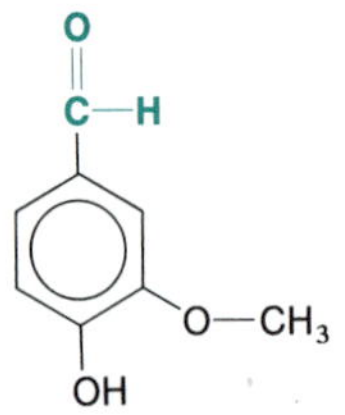

vanillin
(vanilla flavor)

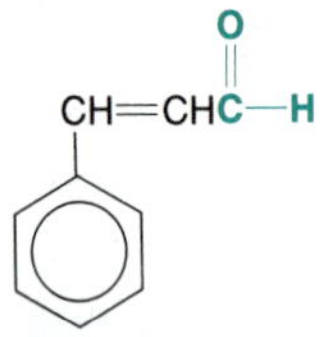

cinnamaldehyde
(cinnamon flavor)

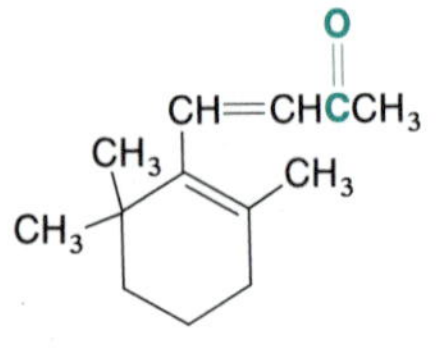

β-ionone
(aroma of violets)

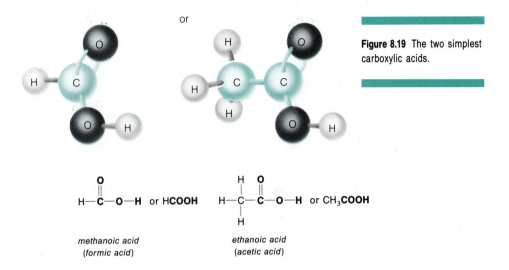

Figure 8.19 The two simplest carboxylic acids.

methanoic acid
(formic acid)

ethanoic acid
(acetic acid)

earlier that alcohols can be oxidized to aldehydes. The full reaction is

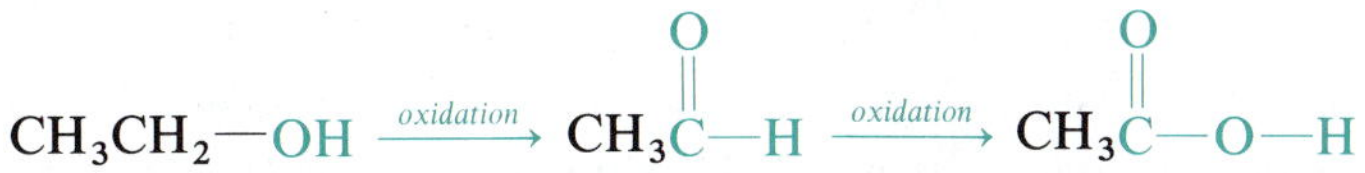

$$CH_3CH_2—OH \xrightarrow{oxidation} CH_3\overset{O}{\underset{}{C}}—H \xrightarrow{oxidation} CH_3\overset{O}{\underset{}{C}}—O—H$$

ethanol   *ethanal*   *ethanoic acid*
(ethyl alcohol)   *(acetaldehyde)*   *(acetic acid)*

Acids with four to ten carbon atoms have unpleasant odors. The four-carbon (butanoic or butyric) acid helps make rancid butter stink, and the eight-carbon (octanoic) and ten-carbon (decanoic) acids help provide the memorable fragrances of limburger cheese and of goats, respectively. Acids with twelve or more carbons (Figure 9.10) are called fatty acids; they don't vaporize readily, so we don't detect much odor.

Some substances have more than one carboxylic acid group (Figure 8.20). Citric acid, which gives citrus fruits a tangy, sour taste, has three carboxylic acid groups. The simplest example, however, is oxalic acid. Oxalic acid occurs naturally in rhubarb leaves and is toxic; people who eat too much of it can develop kidney stones made of insoluble calcium oxalate. This acid is also a reducing agent used to remove rust stains (Section 6.4).

When a carbon group replaces the hydrogen atom in a carboxylic acid group, the new functional group is known as an **ester**, $R—\overset{O}{\underset{}{C}}—O—R'$ (see Table 8.9). Esters form when organic acids and alcohols join together

Figure 8.20 Organic acids with more than one carboxylic acid group.

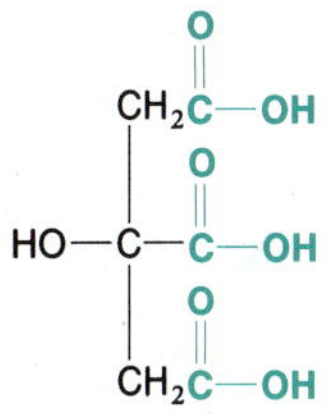

citric acid

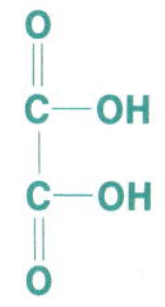

oxalic acid

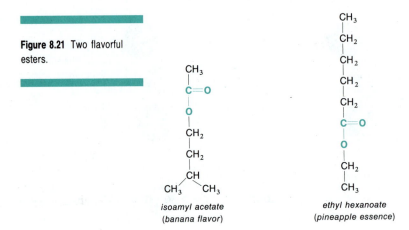

Figure 8.21 Two flavorful esters.

isoamyl acetate
(banana flavor)

ethyl hexanoate
(pineapple essence)

with the elimination of water:

$$CH_3—\overset{\overset{\displaystyle O}{\|}}{C}—\boxed{O—H + H}—O—CH_2—CH_3 \longrightarrow CH_3—\overset{\overset{\displaystyle O}{\|}}{C}—O—CH_2—CH_3 + H_2O$$

ethanoic acid   *ethanol*   *ethyl ethanoate*
(acetic acid)   *(ethyl acetate)*

Esters are solvents (as in fingernail polish remover and lacquer thinner) and flavoring agents. For example, they provide the pleasing aroma and part of the flavor of bananas, raspberries, pears, apricots, pineapples, and oranges (Figure 8.21). Esters also appear in perfumes, shampoos, and other personal products. Very large esters, called *polyesters* (Section 14.3), are used in the clothing industry. Esters also provide our most common form of fat, called triglycerides (Section 9.3).

8.7 Nitrogen-Containing Groups: From Drugs to Dynamite

Figure 8.22 Two simple amines.

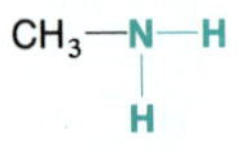

$$CH_3—\overset{\overset{\displaystyle }{|}}{\underset{\underset{\displaystyle H}{|}}{N}}—H$$

methylamine

$$CH_3CH_2—\overset{\overset{\displaystyle }{|}}{\underset{\underset{\displaystyle H}{|}}{N}}—H$$

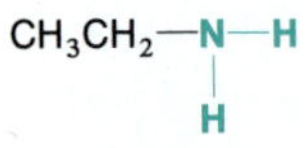

ethylamine

Because nitrogen is in Group 15 (VA) in the periodic table, its atoms have five valence electrons and form three covalent bonds to become stable. The simplest arrangement of nitrogen in an organic compound is the

amine functional group, $R—\overset{\overset{\displaystyle H \text{ (or } R)}{|}}{N}—H$ (or R) (see Table 8.9). Figure 8.22 shows the two simplest amines. Amines have the dubious distinction of providing the smells of fish and decaying flesh. Two aptly named amines, for example, have the common names of *cadaverine* and *putrescine*.

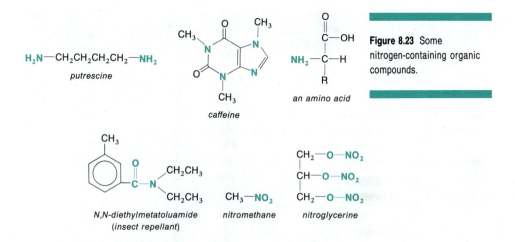

Figure 8.23 Some nitrogen-containing organic compounds.

Figure 8.23 shows several nitrogen-containing compounds. The suffix *-ine* often indicates a substance is an amine. Many drugs that affect the nervous system—such as nicotine, amphetamine, and mescaline—are amines (Chapter 20). The name *vitamin* came from an earlier (and mistaken) belief that all vitamins were amines. Some amines, such as nicotine, caffeine, and cocaine, include the nitrogen atom in a ring.

Amino acids, another important group of substances, contain both an amine group and a carboxylic acid group. Amino acids bond together to make the proteins in our bodies (Section 9.4); when this happens, the carboxylic acid group in one amino acid reacts with the amine group of another amino acid and eliminates a molecule of water:

$$H_2N-CH-\overset{\displaystyle O}{\overset{\displaystyle \|}{C}}-OH + H_2N-CH-\overset{\displaystyle O}{\overset{\displaystyle \|}{C}}-OH \longrightarrow$$
$$\underset{R}{\big|} \qquad\qquad\qquad \underset{R'}{\big|}$$

amino acid　　　　　*amino acid*

$$H_2N-CH-\overset{\displaystyle O}{\overset{\displaystyle \|}{C}}-N-CH-\overset{\displaystyle O}{\overset{\displaystyle \|}{C}}-OH + H_2O$$
$$\underset{R}{\big|} \qquad \underset{H}{\big|}\ \underset{R'}{\big|}$$

2 amino acids joined together
(a dipeptide)

The resulting functional group, $-\overset{\displaystyle O}{\overset{\displaystyle \|}{C}}-\overset{\displaystyle H\ (\text{or } R)}{\overset{\displaystyle |}{N}}-H$ (or **R**) is called an **amide** (see Table 8.9); in proteins, it is called a *peptide bond*.

Nitrogen also occurs in organic compounds as the **nitro group**, —NO_2. Many nitro compounds react with explosive force. Drag car racers, for example, use nitromethane as a motor fuel to get instant acceleration. Other explosives are trinitrotoluene (TNT) and nitroglycerine (see Figure 8.23).

CHARLES KETTERING   We have to date chopped away only a few fragments from the mountain of knowledge—fragments that have changed our way of life. But looming ahead of us, practically intact, lies a huge mass of fundamental facts, any one of which if uncovered could change our civilization.

Summary

Organic chemistry is the study of compounds that contain carbon. Many organic compounds are hydrocarbons, containing only carbon and hydrogen. Alkanes, a group of saturated hydrocarbons, contain only single carbon–carbon bonds. Unsaturated hydrocarbons contain at least one carbon–carbon double or triple bond. These compounds include alkenes (containing at least one double bond) and alkynes (containing at least one triple bond). Aromatic compounds contain a benzene or similar ring.

Our main source of hydrocarbons and other organic compounds are the fossil fuels—petroleum, natural gas, and coal.

Organic compounds containing elements besides carbon and hydrogen have a variety of structural arrangements, called functional groups, that largely determine the physical and chemical characteristics of the compound; Table 8.9 summarizes the functional groups. Organic halides contain halogens (elements in Group 17/VIIA); oxygen-containing compounds include alcohols, ethers, aldehydes, ketones, carboxylic acids, and esters; amines, amides, and nitro compounds contain nitrogen. Various properties and uses of these compounds are discussed.

Terms for Review

After completing this chapter, you should know and understand the meaning of the following terms:

alcohol (p. 226)	hydrocarbon (p. 209)
aldehyde (p. 229)	ketone (p. 229)
alkane (p. 209)	nitro group, $-NO_2$ (p. 234)
alkene (p. 215)	organic acid (p. 230)
alkyne (p. 215)	organic chemistry (p. 205)
amide (p. 233)	organic halide (p. 224)
amine (p. 232)	petrochemical (p. 219)
amino acid (p. 233)	polycyclic aromatic compound (p. 219)
aromatic compound (p. 218)	
carbonyl group (p. 229)	polyunsaturated compound (p. 214)
carboxylic acid (p. 230)	
ester (p. 231)	saturated hydrocarbon (p. 209)
ether (p. 226)	structural isomer (p. 213)
fossil fuel (p. 219)	unsaturated hydrocarbon (p. 214)
functional group (p. 223)	

Questions

Odd-numbered questions are answered at the back of this book.

1. Why is carbon able to form so many compounds? Why are the atoms with which it combines usually nonmetals?

2. Distinguish among alkanes, alkenes, and alkynes and give an example of each.

3. Distinguish between an alkene and an aromatic compound.

4. Define structural isomers. Write structural formulas for the two isomers that have the molecular formula C_2H_6O (one is an ether, and one is an alcohol).

5. Draw the structural formula for propene, which has the molecular formula C_3H_6. Draw structural formulas for any isomers of propene.

6. Explain why the following are all the same compound, *n*-hexane:

$$CH_3$$
$$|$$
$$CH_2CH_2CH_2CH_2CH_3$$

$$CH_3(CH_2)_4CH_3$$

$$CH_3 \qquad CH_3$$
$$| \qquad\quad |$$
$$CH_2CH_2CH_2CH_2$$

7. Write condensed structural formulas for the five structural isomers having the formula C_4H_8.

8. Examine the structure of toluene in Figure 8.6. Is toluene an ionic, polar, or nonpolar substance? Is it a solid, liquid, or gas at room temperature? Is toluene very soluble in water?

9. If a drop of gasoline and a drop of water were placed on the same warm surface, which one would evaporate first? Why?

10. Distinguish between or among the following: (a) fossil fuels and petrochemicals and (b) peat, bituminous coal, and anthracite coal.

11. Explain why the gasoline obtained directly from fractional distillation of crude oil must be upgraded. Describe the three main refining processes for doing this.

12. Heptane can be made into isoheptane by (a) fractional distillation, (b) reforming, (c) cracking, or (d) polymerization.

13. What is the basis for the octane number in gasoline?

14. Draw structural formulas for (a) isobutane, (b) *n*-octane, (c) benzene, (d) ethyne, (e) methanol, (f) methyl amine, (g) methanoic (formic) acid, (h) methyl chloride, (i) diethyl ether, and (j) methanal (formaldehyde).

15. Name the compounds with the following condensed formulas:

   a. $CH_3CH_2CH_2CH_3$
   b. $CH_2{=}CHCH_3$
   c. CH_3CH_2OH
   d.   ⬡

   e. CH_3CH_2Cl
   f. $CH_3CH_2NH_2$

16. Write electron dot structures (like the one on the top of Figure 8.1) for each of the molecules in Question 15.

17. What is a functional group? Why are they so important?

18. Locate and name the functional group or groups in the following molecules:

   a.

$$\text{⬡} \begin{array}{l} -O-\overset{\overset{\displaystyle O}{\|}}{C}-CH_3 \\ -\underset{\underset{\displaystyle O}{\|}}{C}-OH \end{array}$$

   aspirin

   b. $CH_2{-}OH$
         $|$
         $CH{-}OH$
         $|$
         $CH_2{-}OH$

   glycerine (glycerol)

c.

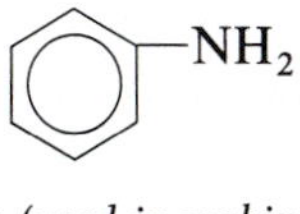

aniline (*used in making dyes*)

d. $CCl_3CH—OH$
              |
             OH

chloral hydrate (*a sedative*)

e.

$$\underset{CH_3}{\overset{CH_3}{\diagdown}}C{=}CHCH_2CH_2\underset{\underset{CH_3}{|}}{C}{=}\overset{\overset{O}{||}}{C}HCH$$

citral (*aroma of lemons*)

f.

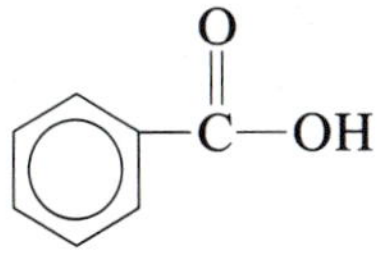

benzoic acid (*a food preservative*)

g.

$$CH_3\underset{\underset{NH_2}{|}}{C}H\overset{\overset{O}{||}}{C}{—}OH$$

alanine (*amino acid*)

h.

$$H\overset{\overset{O}{||}}{C}{—}O{—}CH_2CH_3$$

ethyl formate (*aroma of rum*)

19. In addition to carbon and hydrogen, carboxylic acids contain the element _____, and amines contain the element _____.

20. Explain why alcohols, organic acids, and amines have hydrogen bonding between molecules. What effect would this have on their boiling point?

21. Would ethanol or octanol dissolve better in water? Why?

22. Many esters have pleasant aromas. But polyesters, very large molecules containing many ester functional groups, typically don't have aromas. Why does the size of the molecule make a difference?

Topics for Discussion

1. Does *organic* have the same meaning in chemistry as in consumer products such as health foods? Are "organic" foods chemically organic?

2. Survey the items in your room and attempt to determine how many of them are based on petrochemicals.

3. Should we set aside a fraction, say 10 percent, of our oil reserves for use in the petrochemicals industry in future decades? What might be the consequences of (a) doing this and (b) not doing this?

4. What are some organic ingredients of commercial products? What functional groups do they contain?

CHAPTER 9

Biochemistry: Some Important Molecules of Life

General Objectives

1. What are the major types and functions of carbohydrates?

2. What are the major types and functions of lipids?

3. What are the structures and functions of proteins?

4. What are nucleic acids, and how are they related to the genetic code?

What do you have in common, chemically, with snakes and snails, but not with things like bricks and bowling balls? Do the chemicals in you have some unique quality that enables you to be alive? The branch of chemistry that deals with such questions is **biochemistry**—the study of the composition, structure, and reactions of living things. Thus, biochemists try to account for a rabbit's long ears, a caterpillar's cocoon, a rose's waxy petals, and a skunk's special perfume, in terms of the organism's chemistry.

Biochemists have learned that all organisms—no matter how different they look or act—are made from the same basic ingredients. The recipe calls for portions of carbohydrates, lipids, proteins, nucleic acids, water, and minerals, with just a pinch of vitamins. Of these ingredients, only minerals and water are common in nonliving things.

This chapter introduces four important substances of life: carbohydrates, lipids, proteins, and nucleic acids. (Vitamins and minerals are discussed in Chapter 18.)

9.1 Biochemistry: Where There's Life, There's Carbon

VITAL MOLECULES   What is so special about carbohydrates, lipids, proteins, and nucleic acids? Why do living things—and virtually nothing else in nature—make them? Do they have some quality that is the essence of life itself?

German chemist Friedrich Wöhler (Figure 9.1) discovered in 1828 that heat could change ammonium cyanate, a mineral material, into urea, which occurs in the urine of people and many other living things. He

Figure 9.1 Friedrich Wöhler (1800–1882). (E. F. Smith Memorial Collection, Center for History of Chemistry, University of Pennsylvania)

wrote to a friend, "I have to let out that I can make urea without needing a kidney, whether of man or dog."

His discovery, and many similar experiments that followed, convinced chemists that the chemicals of life do not possess some unique, mysterious quality that accounts for life itself. Indeed, scientists have chemically synthesized all sorts of biologically important substances—hormones, vitamins, and even genes—that work just as well in the body as the same chemicals made by cells. Vitamin C, for example, has the same structure whether it is isolated from rose hips or made in a test tube. Thus, the pedigree of a molecule makes no difference in how it functions.

What chemical qualities do carbohydrates, lipids, proteins, and nucleic acids have in common then? For one thing, they all are *organic* molecules. But the presence of carbon cannot be the only requirement of life. In fact, many of our ancestors who are not alive have that fact recorded on carbon-containing marble gravestones.

Another common feature is that the substances of life can all exist as large molecules, called **macromolecules**. Many macromolecules are **polymers**, which consist of hundreds or even thousands of small repeating molecular units, called **monomers**. Just as freight trains form when boxcars are coupled together, polymers form when monomers bond together. Proteins are polymers of amino acid monomers, nucleic acids are polymers of nucleotide monomers, and certain carbohydrates are polymers of simple sugar monomers (Figure 9.2). Lipids made from glycerol and fatty acids are not polymers, but they can be classified as macromolecules.

Being a macromolecule or polymer, however, doesn't ensure that a molecule is part of living things. For example, synthetic materials such as polyethylene, Styrofoam, and nylon are polymers (Chapter 14), but none is essential to life. Indeed, scientists have concluded that carbohydrates, lipids, proteins, and nucleic acids do not possess unique chemical proper-

Figure 9.2 The four major substances in cells are macromolecules composed of simpler building block molecules.

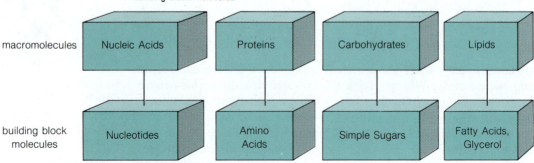

ties, nor do they follow different laws of nature. These remarkable organic materials are special in just one way: Life as we know it does not exist without them. Let's examine each of these substances individually.

9.2 Carbohydrates: Sugar and Rice and Everything Nice

FUEL FOR YOUR BODY Carbohydrates are compounds of carbon, hydrogen, and oxygen. The name comes from the early belief that carbohydrates consisted of hydrated carbon—that is, carbon bound to water molecules. For example, the formula for glucose, the most abundant simple carbohydrate in your body, can be written as $C_6H_{12}O_6$ or $C_6(H_2O)_6$. Table sugar, another carbohydrate, has the formula $C_{12}H_{22}O_{11}$ or $C_{12}(H_2O)_{11}$. Today we know that the latter way of writing these formulas is deceiving. Carbohydrates are not simple hydrates of carbon, although the name remains.

Carbohydrates are cyclic organic compounds that contain several alcohol groups (—O—H). They may be *simple sugars* (carbohydrates with a single ring) or *complex sugars* (polymers of simple sugar units). **Monosaccharides** are simple sugars. When two sugar units bond together, the carbohydrate is classified as a **disaccharide. Polysaccharides** contain hundreds or even thousands of monomer units of simple sugars joined together in varying chain lengths.

Bread, potatoes, noodles, corn, candy, and fruit are all rich in carbohydrates. We can get energy from carbohydrates, lipids, and proteins, but for most people carbohydrates are the major energy source. Glucose is the simple sugar your body uses directly for energy. Through a complex series of steps, glucose reacts with oxygen (is oxidized) to produce carbon dioxide, water, and energy:

$$C_6H_{12}O_6 + 6O_2 \longrightarrow 6CO_2 + 6H_2O + \textbf{energy}$$

Because the body converts other carbohydrates into glucose before oxidizing them, glucose is the carbohydrate given to people who need a direct source of energy in life-threatening situations.

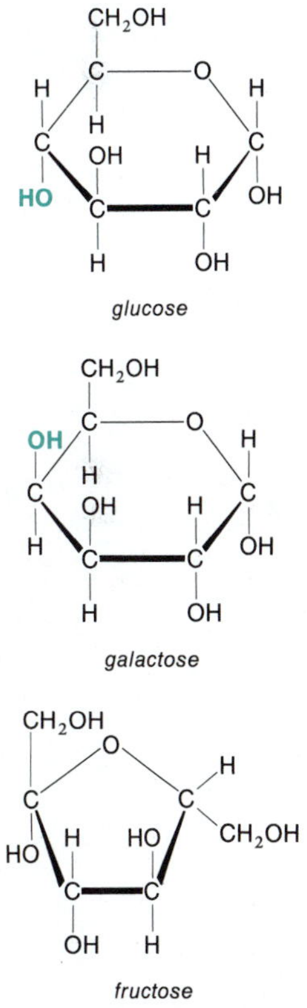

MONOSACCHARIDES Monosaccharides have three to seven carbon atoms. The most common are five- and six-carbon compounds called *pentoses* and *hexoses*, respectively. The suffix *-ose* designates sugars.

Two important pentoses, ribose and deoxyribose, are components of the nucleic acids, RNA and DNA. The three main hexoses in your body are glucose (blood sugar, or dextrose), fructose (fruit sugar, or levulose), and galactose. As Figure 9.3 shows, these sugars have different structures but the same molecular formula, $C_6H_{12}O_6$.

glucose

galactose

fructose

Figure 9.3 The three main hexoses in your body. Each has the same molecular formula, $C_6H_{12}O_6$.

Glucose is sometimes called *blood sugar* because it is the main sugar circulating in the bloodstream to provide energy throughout the body. After several hours without food, a normal adult has a blood sugar level in the range of 70 to 100 mg of glucose per 100 mL of blood (Figure 9.4).

People with *hypoglycemia* (or low blood sugar) have below-normal glucose levels and must carefully control their carbohydrate intake. If they fail to do so, they may feel sluggish or dizzy and may even faint because their brain cells don't get enough energy from the glucose in their blood. Skipping meals can do this, as can eating meals too rich in carbohydrates. In the latter case, the body overcorrects for the high levels of glucose entering the blood by secreting too much *insulin*, a hormone that lowers blood sugar levels.

People with *hyperglycemia* have above-normal levels of glucose in their blood. When the glucose level exceeds about 160 mg per 100 mL of blood, the kidneys begin to excrete glucose in the urine. This is one symptom (along with frequent urination and thirst) of *diabetes mellitus*.

The most common form of diabetes appears after the age of thirty or forty years, usually in people who are overweight. These diabetics don't produce enough insulin in their pancreas and may be able to control their hyperglycemia by careful diet and exercise or by oral drugs that stimulate the production of insulin.

In other cases, however, a diabetic must take regular injections of insulin. A diabetic who doesn't receive enough insulin can go into a coma, which is usually accompanied by a sweetish odor (from ketones) on the breath. The treatment is a shot of insulin. But a diabetic who injects too much insulin can go into shock because of acute hypoglycemia. Here the remedy is to give the person some quickly digestible carbohydrate such as candy, fruit, juice, or sugar. You might save someone's life by knowing how to detect the difference between a diabetic coma (sweetish odor on the breath), which requires insulin, and insulin shock, which requires orange juice or a lump of sugar.

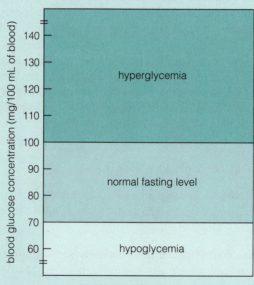

Figure 9.4 Conditions related to blood glucose levels.

PRACTICE EXERCISE

How do the structures of glucose and galactose differ?

SOLUTION

Notice the different orientation of the alcohol group (in color) in Figure 9.3. Other than this, glucose and galactose have the same structures.

Glucose is a moderately sweet sugar found naturally in a few foods such as honey and fruits, but it mostly forms in the body through the breakdown of starch polysaccharides. Your health and well-being, and indeed your life, depend on your body maintaining a certain level of glucose in the bloodstream (see "Chemistry Spotlight"). *Fructose*, the sweetest monosaccharide, is found in fruits and honey and is converted into glucose in your liver. *Galactose* does not occur as such in foods, but it forms in the body from the digestion of lactose, a disaccharide found in milk.

DISACCHARIDES Disaccharides consist of two monosaccharides joined together with the elimination of water (Figure 9.5). When both sugar units are hexoses such as glucose and fructose, the disaccharides have the

glucose + galactose ⟶ lactose + H_2O
glucose + fructose ⟶ sucrose + H_2O
glucose + glucose ⟶ maltose + H_2O

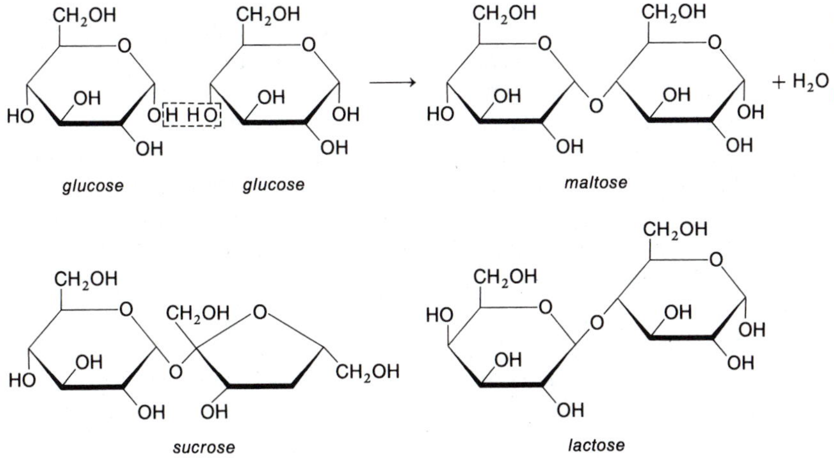

Figure 9.5 A disaccharide forms when two monosaccharides join and eliminate a molecule of H_2O.

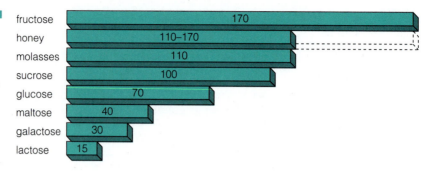

Figure 9.6 Approximate sweetness in comparison to sucrose, which is assigned a value of 100. On this scale, aspartame—an artificial sweetener with the trade name Nutrasweet—has a value of about 18,000, and saccharin has a value of about 40,000.

general formula $C_{12}H_{22}O_{11}$. The three most common disaccharides are sucrose, lactose, and maltose.

Sucrose, or table sugar, consists of glucose bonded to fructose. Sucrose occurs naturally in sugar cane and sugar beets, and it is more widely produced than any other pure organic compound in the world. Its sweet taste is the standard for all sweeteners, as shown in Figure 9.6. *Lactose* is milk sugar and makes up 4 to 6 percent of cow's milk and 5 to 8 percent of human milk. Its structure contains galactose and glucose. *Maltose* is the disaccharide made from two glucose units. It is produced in plants when seeds germinate, and it finds its way into malted milk and beer.

Because their polar alcohol groups form hydrogen bonds with water molecules, monosaccharides and disaccharides dissolve well in water (Section 5.3). This is why table sugar (sucrose) dissolves readily in a cup of coffee. And because glucose dissolves in blood, it travels throughout the body and provides a constant supply of energy to organs, such as the brain, that do not store their own fuel.

POLYSACCHARIDES   Polysaccharides are polymers composed of long chains of monosaccharide units, especially glucose. The most common polysaccharides are starch, glycogen, and cellulose.

Starch occurs in plants such as rice, potatoes, corn, and wheat, whereas *glycogen* occurs in animals, especially in liver and muscle tissue. Both starch and glycogen are made by joining hundreds or thousands of glucose molecules. They differ in the average number of glucose units and in the amount of branching present in their polymer chains. Starch is a mixture of two types of polysaccharides—amylose, a straight-chain polymer of several dozen to several hundred glucose molecules, and amylopectin, a branched-chain polymer consisting of about 1000 glucose molecules (Figure 9.7). Glycogen is like amylopectin but is more branched.

Most of your carbohydrate energy comes from eating starch and digesting it into glucose. Whenever this gives too much glucose for imme-

amylose (20–30%)

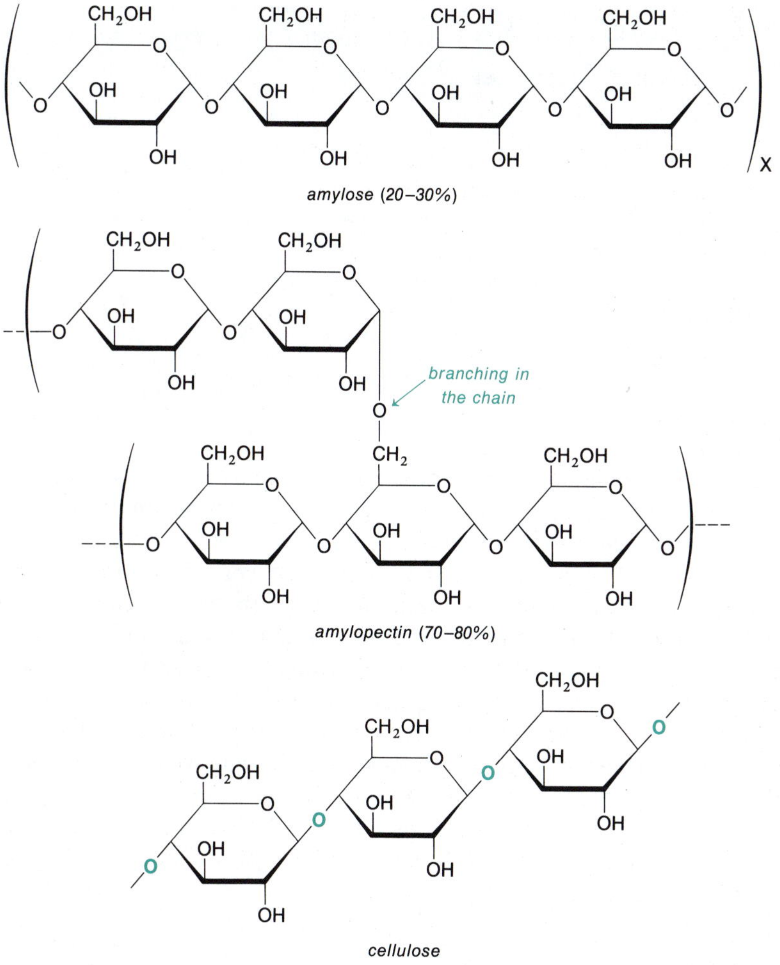

Figure 9.7 The two major polysaccharides in starch are made from glucose units. Notice the different way glucose units are connected in cellulose (in color).

branching in the chain

amylopectin (70–80%)

cellulose

diate use, your body converts some of the extra glucose into glycogen, which it stores in liver and muscle cells. Glycogen then changes back into glucose as required.

Cellulose consists of 100 to 3000 glucose molecules linked together in a straight chain. This polysaccharide is the main structural material in plants. The extensive hydrogen bonding between alcohol (—O—H) groups on neighboring cellulose molecules makes the material tough and insoluble in water. Cellulose fibers pack into layers in cell walls of plants, and other polysaccharides help cement them into a strong, stiff substance.

Glucose units in cellulose join in a slightly different way than in amylose (see Figure 9.7). Because of this subtle difference in structure, we cannot digest cellulose, and all its glucose monomers are unavailable to us. But certain animals such as cows, sheep, horses, and termites can digest cellulose because they contain bacteria with enzymes that break the bonds between glucose units.

We write on cellulose (paper), wear it (linen, cotton, and rayon), and build with it (wood). Wood is about 50 percent (by mass) cellulose, and cotton is almost pure cellulose. Paper towels, bath towels, and wash rags soak up and hold water because of capillary action and because the alcohol groups (—O—H) in their glucose units hydrogen bond with water molecules. From this we can predict that a chemical used to waterproof cotton or other cellulose fabrics will have few, if any, sites for hydrogen bonding.

9.3 Fats and Oils: Sometimes Too Much of a Good Thing

FATS AND OILS Fats, oils, waxes, and steroids belong to a class of organic compounds known as **lipids**—naturally oily or waxy nonpolar materials that dissolve in organic solvents but not in water. Although some lipids contain ester or other functional groups, they contain mostly carbon and hydrogen, which have similar electronegativities (Section 4.4); this is why lipids are nonpolar and dissolve only in nonpolar solvents.

Fats and **oils** are esters of long-chain organic acids known as fatty acids. **Fatty acids** are carboxylic acids containing twelve or more carbon atoms. Animal fats contain mostly saturated fatty acids, whereas vegetable oils typically contain unsaturated fatty acids. Plants also contain polyunsaturated fatty acids, which have two or more double bonds. Figure 9.8 shows examples of each type of fatty acid.

Fats and oils are complex molecules containing one, two, or three ester groups. They form when glycerol, which has three —O—H groups, reacts with up to three molecules of the same or different fatty acids. When glycerol combines with three fatty acids, the product is a *triglyceride* (Figure 9.9), the major component of fats and oils.

Figure 9.8 A comparison of saturated, unsaturated, and polyunsaturated fatty acids.

$$H-O-\overset{\overset{\textstyle O}{\|}}{C}-CH_2-CH_2-CH_2-CH_2-CH_2-CH_2-CH_2-CH_2-CH_2-CH_2-CH_2-CH_2-CH_2-CH_2-CH_2-CH_2-CH_3$$

stearic acid. $C_{17}H_{35}$ COOH (*saturated fatty acid*)

$$H-O-\overset{\overset{\textstyle O}{\|}}{C}-CH_2-CH_2-CH_2-CH_2-CH_2-CH_2-CH_2-\overset{\overset{\textstyle H}{|}}{C}=\overset{\overset{\textstyle H}{|}}{C}-CH_2-CH_2-CH_2-CH_2-CH_2-CH_2-CH_2-CH_3$$

oleic acid. $C_{17}H_{33}$ COOH (*unsaturated fatty acid*)

$$H-O-\overset{\overset{\textstyle O}{\|}}{C}-CH_2-CH_2-CH_2-CH_2-CH_2-CH_2-CH_2-\overset{\overset{\textstyle H}{|}}{C}=\overset{\overset{\textstyle H}{|}}{C}-CH_2-\overset{\overset{\textstyle H}{|}}{C}=\overset{\overset{\textstyle H}{|}}{C}-CH_2-\overset{\overset{\textstyle H}{|}}{C}=\overset{\overset{\textstyle H}{|}}{C}-CH_2-CH_3$$

linolenic acid. $C_{17}H_{29}$ COOH (*polyunsaturated fatty acid*)

Figure 9.9 Formation of a triglyceride by the reaction of glycerol with three fatty acids.

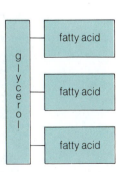

glycerol + three fatty acids ⟶ triglyceride + three molecules of water

TABLE 9.1 MAJOR DIFFERENCES BETWEEN FATS AND OILS

Fats	Oils
Mostly animal origin	Mostly vegetable origin
Solids at room temperature (melting point above 20°C)	Liquids at room temperature (melting point below 20°C)
More saturated (fewer C=C bonds)	More unsaturated (more C=C bonds)

TABLE 9.2 AVERAGE FATTY ACID COMPOSITION OF SOME FATS AND OILS

	Types of Fatty Acids (%)			
	Saturated	Monounsaturated	Polyunsaturated	Other
Animal Fats				
Butter	51	32	4	13
Beef tallow	47	50	2	1
Lard	41	50	6	3
Human	40	57	0	3
Vegetable Oils				
Coconut	65	8	0	17
Corn	14	52	34	0
Olive	9	84	5	2
Peanut	11	56	26	7
Safflower	7	19	73	1
Soybean	12	29	58	1
Sunflower	8	25	66	1

Table 9.1 summarizes the differences between fats and oils, and Table 9.2 compares the amount of unsaturation in their fatty acids. Animal fats (such as lard, tallow, suet, and bacon grease) tend to be solids at room temperature, whereas vegetable oils (such as olive oil, soybean oil, and corn oil) tend to be liquids. The amount of unsaturation in the molecules accounts for this because each double bond in a carbon chain causes the molecule to have a more rigid, bent shape. Just as rigid, bent coat hangers are more awkward than pencils to pack tightly together, the bent, unsaturated molecules do not pack as close together as saturated ones. Neither do they attract each other as much. Thus, the unsaturated oils are less dense than fats and are more likely to be liquids. Eliminating some of the unsaturation, however, turns plant oils (liquids) into soft solids like vegetable shortening (Figure 9.10).

Your body creates fats by breaking down excess glucose to two-carbon units and then reassembling these into fatty acids and eventually triglycerides, which are stored in fatty (adipose) tissue. This is why fatty acids typically contain only even numbers of carbons. Fats are used for insulating your body, for cushioning your internal organs (such as your kidneys) against mechanical damage, and for long-term energy storage.

When stored fats are oxidized in your body, they produce more than twice as much energy per gram as carbohydrates or proteins, and they

Figure 9.10 Partial hydrogenation of a vegetable oil into shortening. For simplicity, the effect on just one of the three fatty acids is shown.

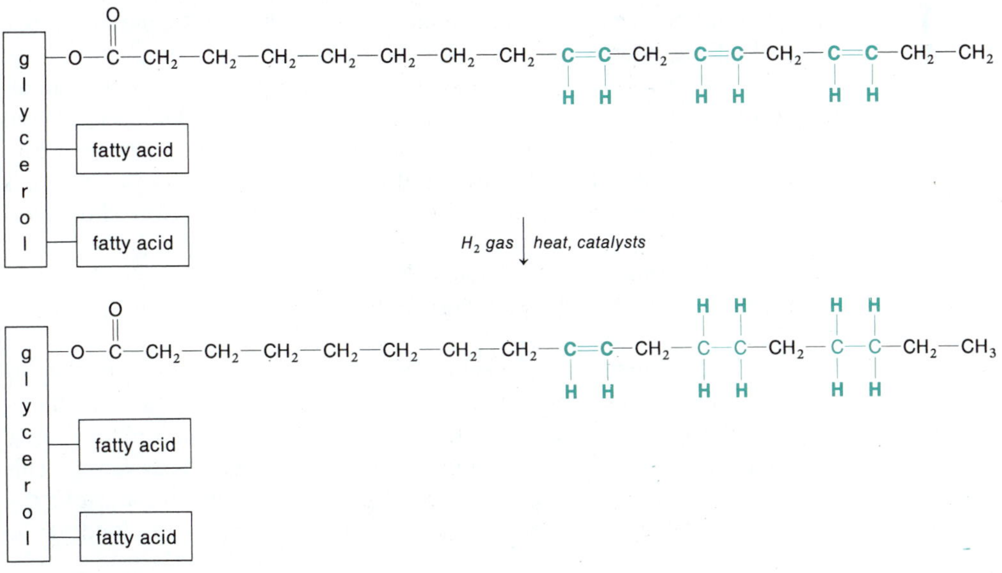

produce large amounts of water:

$$C_{55}H_{100}O_6 + 77O_2 \longrightarrow 55CO_2 + 50H_2O + \textbf{energy}$$
a triglyceride

A hibernating animal takes advantage of this by storing fat and gradually oxidizing it; this enables the animal to survive for long periods without a fresh supply of either energy or water. Camels use a similar strategy; they store fat in their humps to provide an internal water (and energy) reservoir for dry desert days and nights.

Lipids have other practical uses, too. Fats have long been used for making soap (Section 15.2). Waxes, which are esters of fatty acids and alcohols with long carbon chains, furnish living things with a slick, protective coating on such things as feathers, fur, skin, leaves, and fruit. Waxes also give a protective coating to manufactured products such as cars, skis, and furniture because their nonpolar structures make them impervious to water. Other lipids are critical materials in membranes, which provide boundaries to the watery interiors of cells.

Being nonpolar, lipids dissolve other nonpolar compounds. For example, butter is naturally yellow because it dissolves nonpolar pigments from grass. And nonpolar materials such as the pesticide DDT (Figure 17.5) and vitamin D (Figure 18.8) dissolve and can accumulate in fatty tissues of plants and animals.

STEROIDS   An important group of nonester lipids, the **steroids**, includes cholesterol, vitamin D, and several hormones. Figure 9.11 shows the characteristic four-ring carbon frame for steroids. Different steroids have different carbon chains and functional groups (especially alcohols, ketones, and alkenes) bonded to this basic carbon frame.

Cholesterol, the most abundant steroid in your body, contains an alcohol (—O—H) group, two methyl (—CH$_3$) groups, a double bond, and a branched eight-carbon chain bonded to the steroid frame (see Figure 9.11). Even with its alcohol group, cholesterol, like all other lipids, is essentially nonpolar and does not dissolve in water. So before cholesterol or other lipids can be suspended and pass through your bloodstream, they must combine with detergentlike materials (Section 15.1) such as proteins. Thus, much of the cholesterol in your blood is present as lipoprotein (lipid–protein) complexes.

Despite these mixing mechanisms, some cholesterol and other lipids can come out of suspension and gradually accumulate on the inner lining of blood vessesls such as arteries. As these deposits build up, they make the heart work harder to pump blood through the narrower vessels. This condition, called *atherosclerosis*, can lead to a heart attack or stroke if a blood clot lodges in a blood vessel and blocks the flow of blood to the heart or brain.

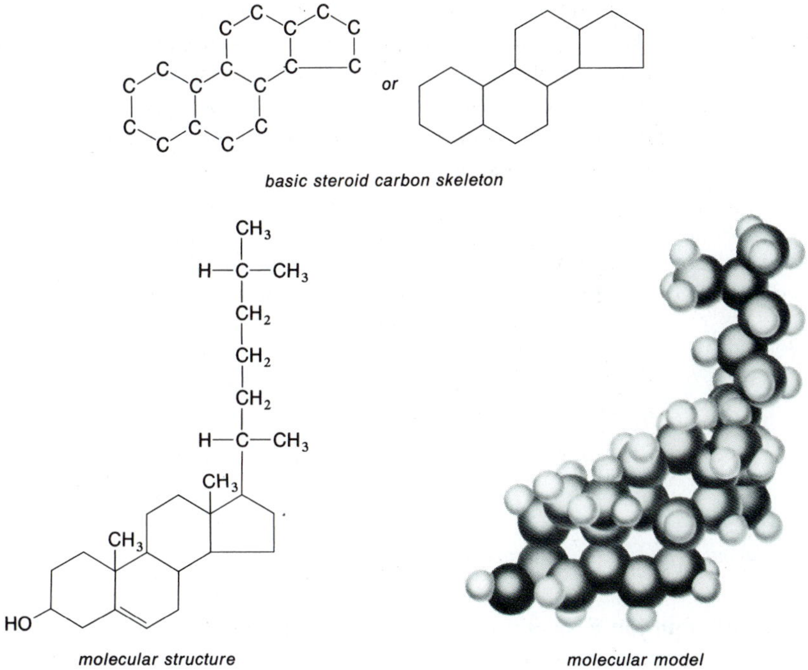

basic steroid carbon skeleton

Figure 9.11 The steroid carbon frame and the structure of cholesterol.

molecular structure

molecular model

Cholesterol deposits also can cause trouble in another way. The gallbladder normally stores cholesterol and natural detergent materials that help you digest lipids. But if there is too much cholesterol or too little detergent, cholesterol solidifies and forms gallstones, which can cause abdominal pain, severe nausea, and lipid indigestion.

But despite these problems, you could not live long without cholesterol. It is a critical component in cell membranes and is the starting material for making all your body's other steroids, such as vitamin D (for strong bones; Figure 18.8), cortisone (for carbohydrate and protein metabolism; Figure 19.17), and male and female sex hormones (Figure 19.18).

9.4 Proteins: Hair, Nails, Muscles, Skin, and Enzymes

NATURE AND FORMATION OF PROTEINS If we ranked chemicals according to their biological importance, we could find many reasons to put proteins at the top of the list. In fact, *protein* comes from a Greek word that means "first" or "foremost."

Every part of our bodies contains proteins. Hair, skin, nails, muscles, tendons, and blood vessels are made of protein; even our bones have a matrix of protein. And it's difficult to find any bodily process that doesn't involve proteins. Proteins function as enzymes to catalyze virtually all metabolic reactions in our bodies and as antibodies to protect us against diseases. Proteins also transport lipids in the blood, control the maintenance of cells, and enable us to move through muscle contraction.

Proteins are polymers formed by joining together amino acid monomer units. **Amino acids** consist of a carboxylic acid group, an amine group, and a side chain that can contain just about any functional group:

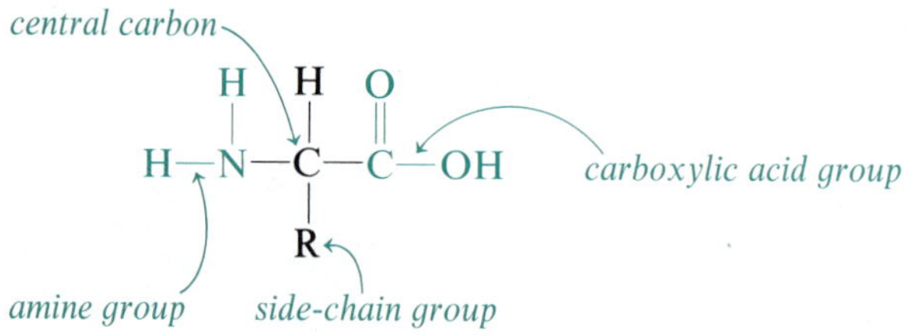

Although the R side-chain group could conceivably take almost any chemical form, only 20 different amino acids (some scientists count one or two more) occur in the proteins of living things. Table 9.3 shows the names, abbreviations, and structures of these 20 amino acids.

Proteins form in the body when enzymes (which are themselves proteins) catalyze reactions in which the carboxylic acid group of one amino acid reacts with the amine group of another amino acid with the elimination of water:

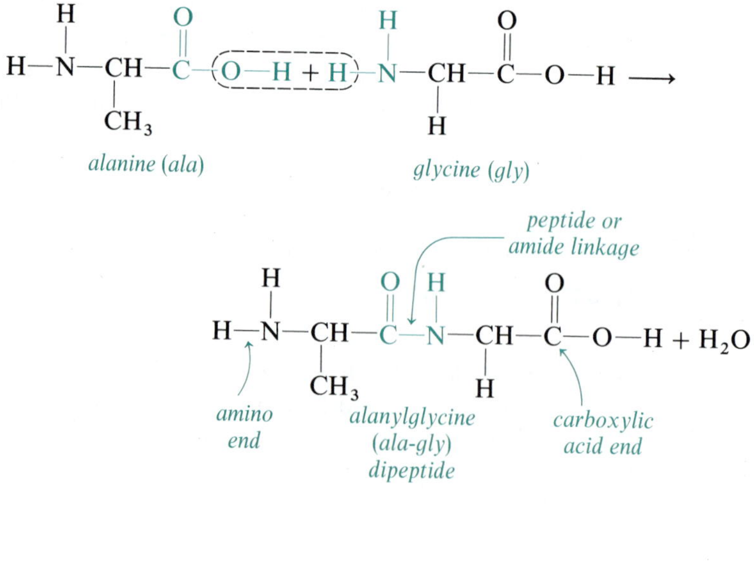

TABLE 9.3 COMMON AMINO ACIDS

Class	R Group	Name	Symbol
Nonpolar, linear or branched	—H	glycine	gly
	—CH_3	alanine	ala
	—$CH(CH_3)_2$	valine	val
	—$CH_2CH(CH_3)_2$	leucine	leu

	—$CHCH_2CH_3$
	     |
	     CH_3
		isoleucine	ile

Aromatic	—CH_2—⬡	phenylalanine	phe
	—CH_2—⬡—OH	tyrosine	tyr
	—CH_2—(indole)	tryptophan	trp
Sulfur-containing	—CH_2SH	cysteine	cys
	—$(CH_2)_2SCH_3$	methionine	met
Hydroxyl-containing (besides tyrosine)	—CH_2OH	serine	ser

	—$CHOH$
	     |
	     CH_3
		threonine	thr

Acidic (or amides)	—CH_2COOH	aspartic acid	asp
	—$(CH_2)_2COOH$	glutamic caid	glu
	$\overset{\displaystyle O}{\overset{\displaystyle \|}{—CH_2C}}NH_2$	asparagine	asn
	$\overset{\displaystyle O}{\overset{\displaystyle \|}{—(CH_2)_2C}}NH_2$	glutamine	gln
Basic	—CH_2—(imidazole, N)	histidine	his
	—$(CH_2)_4NH_2$	lysine	lys

	—$(CH_2)_3NHCNH_2$
	          ||
	          NH
		arginine	arg

Imino acid (complete structure)	(pyrrolidine ring)—COOH N | H	proline	pro

The group that holds together the two amino acids (shown in color above) is known as an amide, or *peptide bond*. The protein chain can grow indefinitely long because amino acids on each end of the chain always have either an amine or a carboxylic acid group available for forming a peptide bond with another amino acid.

Because each link in the chain can be any one of 20 different amino acids, an incredibly large number of proteins with different sequences of amino acids are possible. For example, the possible sequences for a chain only 15 amino acids long is more than 30,000,000,000,000,000,000. Typical proteins contain 50 to 1000 or more amino acid units. Imagine how many different combinations are possible in a chain 200 amino acids long.

THE ARCHITECTURE OF PROTEINS   The sequence of amino acids is known as the *primary protein structure*. The 20 amino acids serve as the alphabet to build words (proteins) of varying length and composition. Changing the sequence alters the properties of a protein, just as changing a letter in a word can make the word useless or nonsensical or give it another meaning.

Similarly, changing the amino acid sequence in a protein may produce a new protein that cannot function like the original one. For example, people with sickle-cell anemia have hemoglobin protein containing a different amino acid in just one position in the 146-amino acid sequence (Figure 9.12). That doesn't sound like much, but this single change alters the shape of the hemoglobin and red blood cells. The unusual, sickle-shaped cells that result (Figure 9.13) are more fragile, less able to bind oxygen, and occasionally clog small blood vessels.

This and many other examples demonstrate that the primary protein structure determines the overall shape and function of a protein. Proteins twist and turn and fold into wondrous shapes as their *R* groups interact, as hydrogen bonds form between polar groups, and as nonpolar amino acids tuck inside to avoid water molecules lurking near the protein's surface. And having the right shape is necessary for proteins to function normally.

The protein's shape also determines the shape of larger structures. Sickle-cell hemoglobin, as we have seen, changes the shape of red blood cells. Another example is hair. The protein molecules of hair, called *keratin*, can take on different shapes depending on the number of disulfide (—S—S—) bridges (Figure 9.14) within and between neighboring keratin molecules. This determines whether the hair is curly. Permanent waves can be put in hair by using reducing agents (Section 6.3) to disrupt the existing disulfide bridges in the hair. After the hair is styled, oxidizing agents are applied to form new disulfide bridges in different locations to make the curls permanent.

Figure 9.12 The sequence of 146 amino acids in human hemoglobin. People with sickle-cell anemia have the amino acid valine (val) in place of one glutamic acid (glu) unit (highlighted at the top of the diagram).

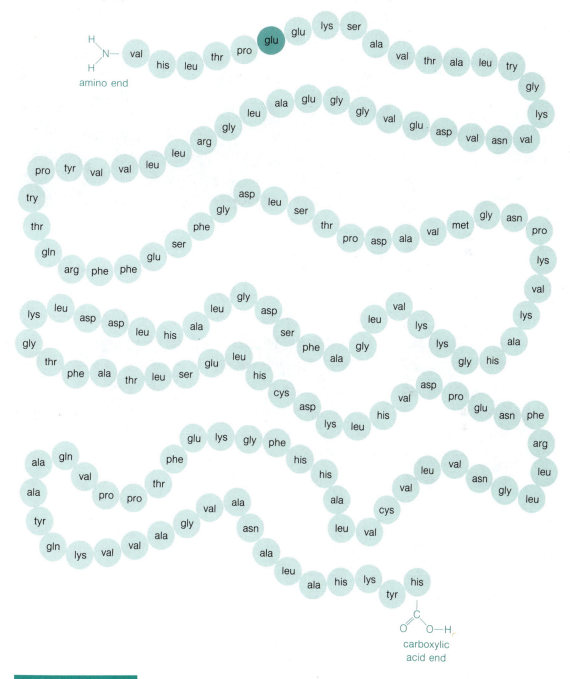

Figure 9.13 Comparison of normal red blood cells (left) and red blood cells from a person having sickle-cell anemia (right). (Science Source/Photo Researchers)

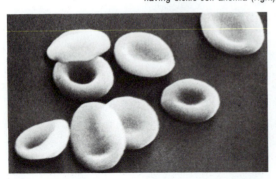 

Figure 9.14 Permanent waves can be created in straight hair by breaking the disulfide bonds between the cysteine amino acids in hair protein molecules, then reforming these linkages in different locations.

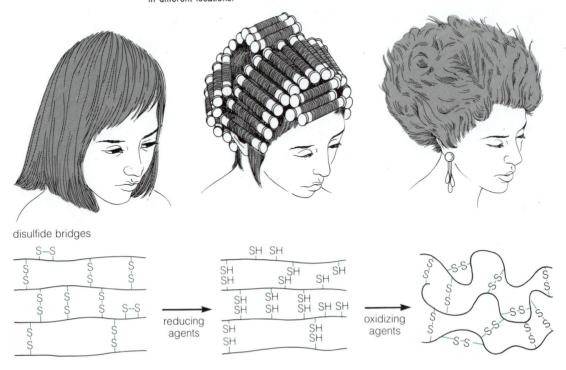

Heat and certain chemicals can disrupt a protein's shape and ability to function without breaking its peptide bonds; this is called *denaturation*. For example, milk curdles when an acid, such as lemon juice or vinegar, causes the molecules of casein, a milk protein, to change shape. Solvents such as alcohols and acetone can be used as disinfectants because they denature proteins in bacteria and viruses. Heating denatures proteins by breaking their weak hydrogen bonds. Boiling an egg, for example, changes its albumin proteins so dramatically that the egg white turns from liquid to solid.

ENZYMES   One of the most critical functions of proteins is to catalyze reactions in the body; such proteins are called **enzymes**. In the chemistry laboratory, we can accelerate reactions by increasing the temperature, increasing the pressure, or adding an acid or base catalyst. But these methods aren't effective in the body, which maintains a fairly constant temperature, pressure, and pH level. Under these mild conditions, our reactions wouldn't occur fast enough to keep us alive if we didn't use enzymes.

Like other proteins, enzymes must have just the right shape to function. The enzyme shape fits with the shape of the reactants in a very specific way, like a lock fits with a key (Figure 9.15). Because acids and bases

Figure 9.15 Crude lock and key model of enzyme action.

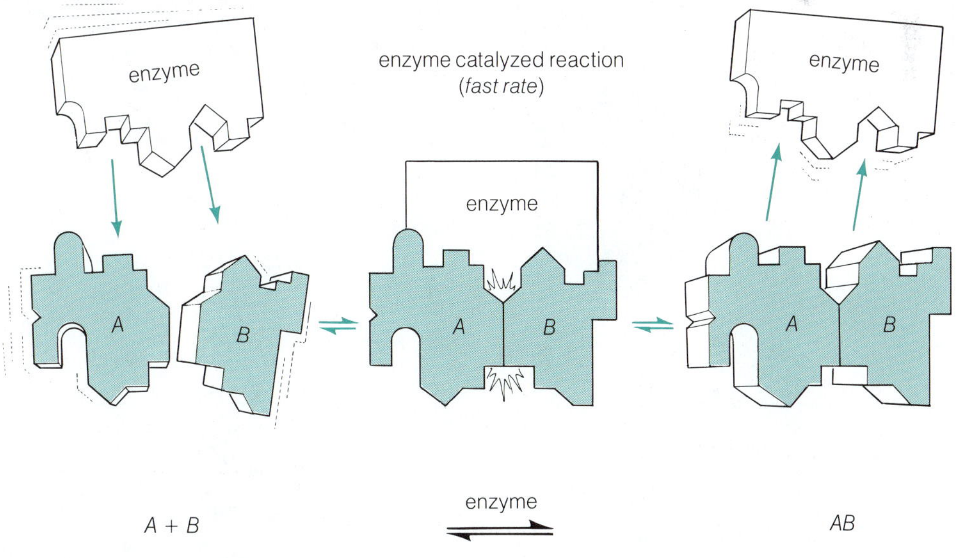

$A + B$ → enzyme → AB

denature proteins, we can understand why cells need to regulate their internal pH closely so that their enzymes can continue to work as catalysts.

9.5 Nucleic Acids: The Secret of Life?

DNA AND THE GENETIC CODE   Why are your eyes, hair, and skin the color that they are? How did you get these and all your other genetic traits from your parents? The answer lies in the class of molecules called **nucleic acids**, acidic substances in the cell nucleus that we now know as DNA and RNA.

In his best-seller *The Double Helix*, James Watson recalls, "I felt slightly queasy when at lunch Francis [his co-worker, Francis Crick] winged in . . . to tell everyone within hearing distance that we had found the secret of life." The "secret" Watson and Crick (Figure 9.16) discovered was the structure of deoxyribonucleic acid, or **DNA**, the hereditary stuff in the chromosomes of your cells that specifies your genetic characteristics.

DNA has two important functions. First, when one of your cells divides, its DNA must first *replicate*, or copy, itself so that both resulting cells will have the same genetic information (DNA) as the original cell. Second, DNA serves as the master blueprint for making proteins. Within each chromosome are **genes**—shorter sections of DNA that carry the in-

Figure 9.16 In 1953 James Watson (left) and Francis Crick (right) built models of the double-helix structure of DNA based on X-ray data from Maurice Wilkins and Rosalind Franklin. (From J. D. Watson, *The Double Helix*, Atheneum, New York, 1968)

Figure 9.17 A nucleotide unit.

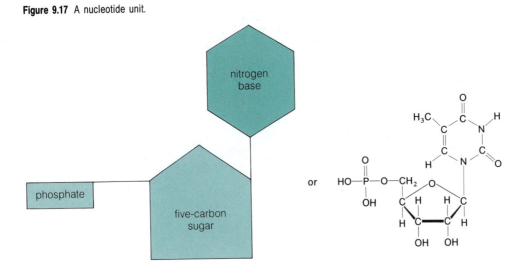

formation for making specific proteins in the cell. This process of protein synthesis also depends on the other type of nucleic acid called ribonucleic acid, or **RNA**. The proteins your cells make, in turn, cause your genetic features to appear. For example, a protein enzyme that helps synthesize a skin pigment can determine the color of your skin.

DNA and RNA are polymers. Their monomer units, called **nucleotides** (see Figure 9.2), consist of three parts—phosphate, a pentose (five-carbon sugar), and a nitrogen-containing base. Figure 9.17 shows symbolically how these parts bond together to make a nucleotide unit. Two kinds of pentose sugars are found in nucleotides: ribose (in RNA) and deoxyribose (in DNA) (Figure 9.18). The two types of nitrogen-containing bases are the single-ring pyrimidines and the double-ring purines (Figure 9.19). These act as bases because their nitrogen atoms tend to accept protons (Section 6.1). The purines (adenine and guanine) and one of the pyrimidines (cytosine) occur in both DNA and RNA. However, the other pyrimidines differ: Thymine occurs in DNA, whereas uracil appears only in RNA. So the structural differences between the nucleotide units in DNA and RNA lie in their pentose units and their base units.

Nucleotide units link to form long-chain molecules of DNA and RNA. Alternating phosphate and sugar units form a "backbone," with the various nitrogen bases extending outward from the chain (Figure 9.20). RNA usually is a single polynucleotide chain. But in 1953 Watson and Crick (using X-ray data provided by Maurice Wilkins and Rosalind Franklin) proposed that DNA normally occurs in cells in the form of a double helix consisting of two long strands of DNA wound around each other in a spiral, or helix (Figure 9.21). This structure looks much like a spiral staircase in which the sugar–phosphate backbone forms the outside handrail and the pairs of bases inside the helix form the steps. DNA molecules vary in the number and sequence of the base pairs in the steps, and they may have as many as 30,000 steps. This makes DNA the largest of all the natural organic molecules.

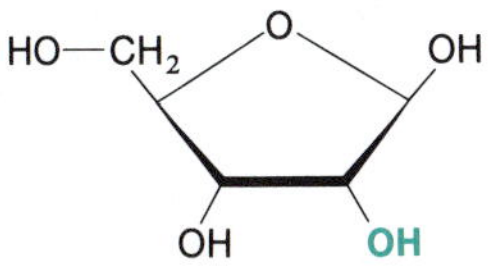

ribose
(sugar unit in RNA)

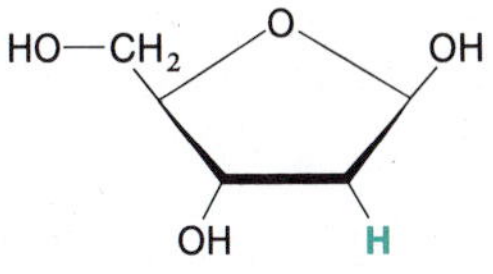

deoxyribose
(sugar unit in DNA)

Figure 9.18 Structure of the sugar units ribose and deoxyribose found, respectively, in RNA and DNA. The only difference is the colored group.

Figure 9.19 The five common nitrogen bases in DNA and RNA. Thymine (T) occurs only in DNA, whereas uracil (U) is only in RNA. The same carbon–nitrogen ring (in color) is in all these bases.

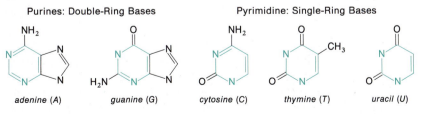

Purines: Double-Ring Bases

Pyrimidine: Single-Ring Bases

adenine (A)   guanine (G)   cytosine (C)   thymine (T)   uracil (U)

Figure 9.20 The detailed structure of a segment of a DNA strand.

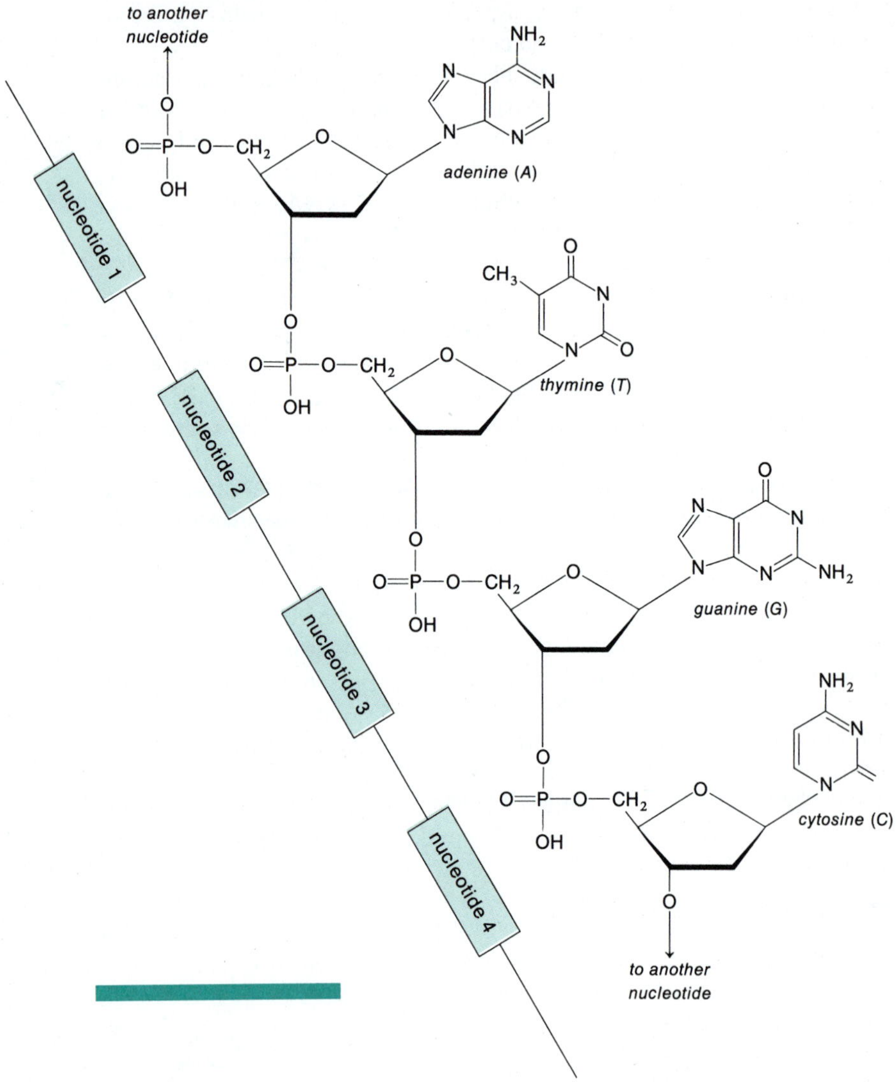

The "glue" holding the two strands together is the extensive hydrogen bonding between polar groups in each pair of bases. Watson and Crick deduced that only certain bases pair with one another. Adenine (A) forms two hydrogen bonds with only thymine (T), whereas guanine (G) forms three hydrogen bonds with only cytosine (C) (Figure 9.22). Although these

Figure 9.21 Representation of a DNA double helix.

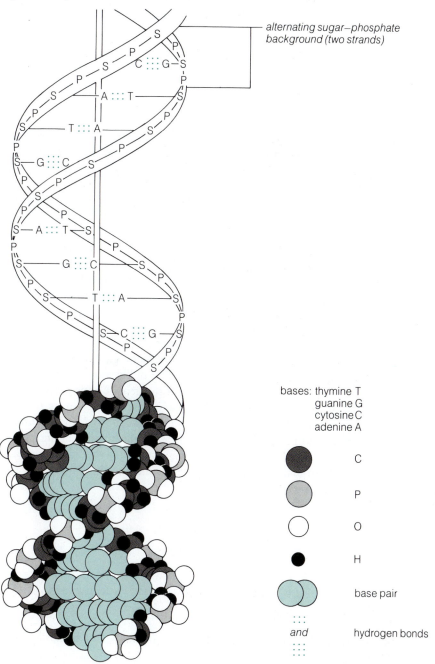

alternating sugar–phosphate
background (two strands)

bases: thymine T
guanine G
cytosine C
adenine A

C

P

O

H

base pair

and hydrogen bonds

Figure 9.22 Hydrogen bonds (dotted lines in color) form between A and T and between C and G in the DNA double helix.

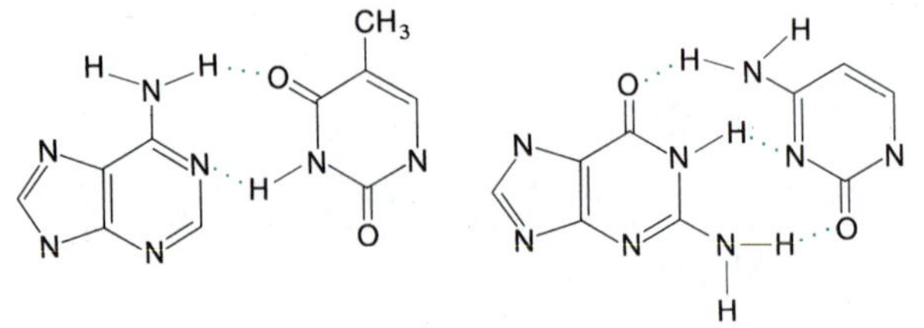

adenine (A)   thymine (T)   guanine (G)   cytosine (C)

hydrogen bonds are weaker than the covalent bonds holding together nucleotides in each strand, there are so many of them that they provide enough energy to hold the double helix intact at body temperature. This model of DNA as a double helix with hydrogen-bonded base pairs brought a Nobel Prize to Watson, Crick, and Wilkins in 1962. The model helps us understand not only how DNA looks but also how it works.

Today we know the entire genetic code—how to translate a specific sequence of nucleotide units in DNA into a specific amino acid sequence in a protein. For example, the sequence CTC (cytosine–thymine–cytosine) in DNA specifies the amino acid *glutamic acid*, whereas the slightly different sequence CAC (cytosine–adenine–cytosine) specifies the amino acid *valine*.

In a gene carrying hundreds or even thousands of nucleotide units, changing just one unit—such as CTC into CAC—might not seem significant, but sometimes it is. That very change, in a strategic place, spells the difference between normal and sickle-cell hemoglobin.

Although the search for an understanding of life processes has merely begun, it is easy to see why this cracking of the genetic code is considered one of the great scientific achievements of recent years. But as we gain more control over genetic processes, some important and highly controversial ethical and political questions arise. Should we alter human genes (Section 22.1)? If so, how far should we go, and who should decide such questions? There is little doubt that in your lifetime these questions will become major issues.

RENÉ DUBOS   Each human being is unique, unprecedented, unrepeatable. The species *Homo sapiens* can be described in the lifeless words of physics and chemistry but not the man of flesh and bone.

Summary

All living things contain carbohydrates, lipids, proteins, nucleic acids, vitamins, minerals, and water. The first four are macromolecules, and carbohydrates, proteins, and nucleic acids exist in the body as polymers.

For most people, carbohydrates are the major source of energy in the diet. Carbohydrates are classified as monosaccharides (such as glucose, fructose, and galactose), disaccharides (sucrose, lactose, and maltose), and polysaccharides (starch, glycogen, and cellulose).

Lipids are nonpolar substances and include fats, oils, waxes, and steroids. Fats and oils are triglycerides. Oils are more unsaturated than fats and thus are liquids at room temperature. Steroids have a four-ring structure. Cholesterol is the steroid from which other steroids are made in the body.

Proteins are polymers made from twenty different amino acids. Proteins provide much of the body's structure, and protein enzymes catalyze virtually all chemical reactions in the body. The sequence of amino acids in a protein determines its shape and ability to function.

The nucleic acids, DNA and RNA, are polymers of nucleotides. DNA carries the genetic (hereditary) information and specifies which proteins the cell can make. DNA exists as a double helix.

Terms for Review

After completing this chapter, you should know and understand the meaning of the following terms:

amino acid (p. 252)

biochemistry (p. 239)

carbohydrate (p. 241)

disaccharide (p. 241)

DNA (p. 258)

enzyme (p. 257)

fat (p. 246)

fatty acid (p. 246)

gene (p. 258)

lipid (p. 246)

macromolecule (p. 240)

monomer (p. 240)

monosaccharide (p. 241)

nucleic acid (p. 258)

nucleotide (p. 259)

oil (p. 246)

polymer (p. 240)

polysaccharide (p. 241)

protein (p. 252)

RNA (p. 259)

steroid (p. 250)

Questions

Odd-numbered questions are answered at the back of this book.

1. In terms of chemical makeup, how do you differ from a dog? A rock?

2. What are the building block molecules of (a) carbohydrates, (b) lipids, (c) proteins, and (d) nucleic acids? Which of the four does not form polymers?

3. Which of the following are structural isomers? (a) fructose, (b) ribose, and (c) glucose.

4. What do starch and cellulose have in common? How are they different?

5. Celery and potato chips are both made from polymers of glucose. For a person on a diet, why is celery a good snack and potato chips a poor snack?

6. How does your body maintain a fairly constant level of blood glucose even though you only eat foods that supply glucose a few times a day?

7. The lipids include fatty acids, triglycerides, waxes, and steroids. What chemical feature do all these substances have in common?

8. Are all fatty acids organic acids? Are all organic acids fatty acids? Explain each answer.

9. Write the structural formula for the triglyceride made from linolenic, stearic, and oleic fatty acids (see Figure 9.8). Would this be more likely to occur in a fat or an oil?

10. What are steroids? Give two examples of steroids in your body.

11. When we use our food for fuel, we oxidize it to carbon dioxide and water. Which fuel—carbohydrates or fats—is already partially oxidized and thus supplies *less* energy when it is oxidized completely? [Compare the structures of some carbohydrates (Figure 9.3) and fats (Figure 9.8) to see which one already has many of its carbon atoms bonded to oxygen atoms.]

12. What are some major functions of proteins in the body?

13. What functional groups are always present in any amino acid? How do amino acids differ from one another?

14. Table 9.3 shows the structure of the *R* group for the amino acid phenylalanine. Draw the complete structural formula for phenylalanine.

15. Draw the structural formula for two phenylalanine molecules joined together by a peptide bond.

16. What is primary protein structure?

17. Describe the cause of sickle-cell anemia and discuss its effects.

18. What are enzymes? What would be the effect of heat on enzyme action?

19. What are the three components of a nucleotide?

20. What are the structural differences between RNA and DNA?

21. How is the double helix of DNA held together? Are the bonds strong or weak compared to the bonds between nucleotides in each strand?

22. What is the function of DNA in a cell?

Topics for Discussion

1. Section 9.4 gives reasons for ranking proteins first among the most important biological molecules. What reasons might be given for putting nucleic acids at the top of the list? Carbohydrates? Lipids?

2. Many body substances can be chemically synthesized. Are these synthetic proteins, carbohydrates, lipids, and vitamins as wholesome and beneficial as their natural counterparts? Why or why not?

3. Do you think someday it will be possible to chemically synthesize a simple form of life? Why or why not?

PART III

Chemistry, Resources, and the Environment

JOHN F. KENNEDY

Our entire society rests upon—and is dependent upon—our water, our land, our forests, and our minerals. How we use these resources influences our health, security, economy, and well-being.

CHAPTER 10

Mineral Resources: Using the Earth's Raw Materials

General Objectives

1. How do the carbon, oxygen, nitrogen, and phosphorus chemical cycles help sustain life on earth?

2. What is a resource, and what are the two major types of resources?

3. What useful products are made from silicon minerals?

4. How do we get metals such as copper, iron, and aluminum from minerals in the earth's crust?

5. How can we deal with the vast quantities of solid and hazardous wastes we produce?

What keeps all living things alive on this tiny planet as it hurtles through space? Life as we know it exists only near the earth's surface in a thin film about 14 km (9 miles) thick. In this film is all the water, minerals, oxygen, nitrogen, phosphorus, and other chemicals that we need in order to live.

If we think of the earth as an apple, then the apple's skin holds all the supplies to maintain life. Everything in this skin is interdependent. Air helps purify the water; water helps keep plants and animals alive; and plants help keep animals alive and help renew the air.

Although human and natural processes constantly change the earth's chemicals from one form to another, our total supply of elements is essentially fixed. In the next four chapters, we look at the use and abuse of the earth's mineral, energy, air, and water resources. We begin by looking at some of the earth's key chemical cycles and examine how we use mineral raw materials from the earth's crust.

10.1 Chemistry of the Earth: Chemical Cycling and Energy Flow

OUR LIFE-SUPPORT SYSTEM Life on earth depends on chemical cycling and on a one-way flow of energy from the sun through the materials and living things on earth (Figure 10.1). In accordance with the second law of thermodynamics (Section 7.2), this process degrades high-quality solar energy into low-quality heat in the environment, and this heat then flows back into space.

Figure 10.1 Life on earth depends on the cycling of critical chemicals (solid lines) and the one-way flow of energy (dashed lines).

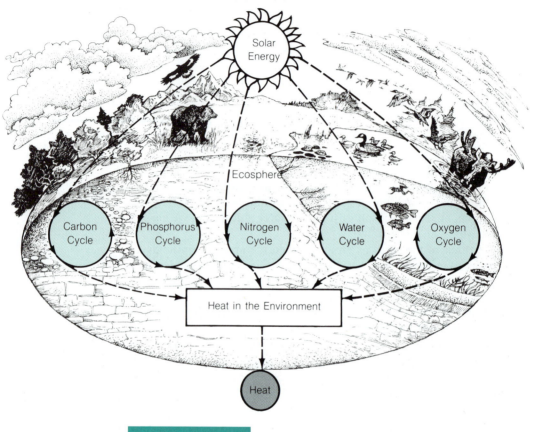

We can view this life-support system as having four main parts (Figure 10.2): (1) the **atmosphere** (air) above the earth's surface; (2) the **hydrosphere**, containing all of earth's liquid, solid, and gaseous water; (3) the **lithosphere**, consisting of the earth's outer *crust* of soil and rock, its *mantle* of partially molten rock beneath this crust, and its inner *core* of molten rock; and (4) the **biosphere**, consisting of all life.

The chemical ingredients of life can almost be summed up in five words: *oxygen, carbon, hydrogen, nitrogen,* and *phosphorus.* These five elements make up 97 percent of the mass of your body (Figure 10.3). Your body needs many other elements for good health and survival, but mostly in only minute amounts (Section 18.2). Now let's look briefly at the carbon,

Figure 10.2 Our life-support system: general structure of the earth.

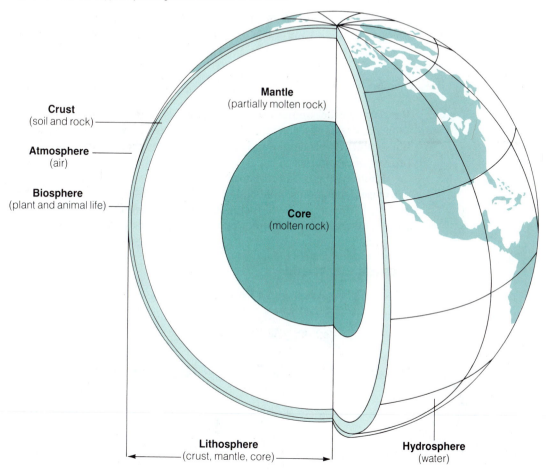

Figure 10.3 Percentage by mass of elements in the human body.

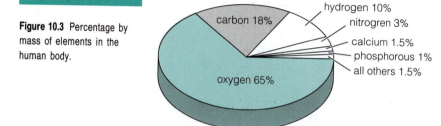

oxygen, nitrogen, and phosphorus cycles. (The water cycle is discussed in Section 13.2.)

CARBON AND OXYGEN CYCLES   In *photosynthesis*, green plants absorb solar energy and use it to produce oxygen (O_2) gas and carbohydrates such as glucose ($C_6H_{12}O_6$):

$$6CO_2 + 6H_2O + \textbf{energy} \underset{respiration}{\overset{photosynthesis}{\rightleftharpoons}} C_6H_{12}O_6 + 6O_2$$

People and other animals then get organic nutrients by eating the plants. Plants and animals generate the energy they need by metabolizing a portion of the organic compounds back into carbon dioxide and water in the process of *respiration*.

Photosynthesis and respiration form the basis of the **carbon** and **oxygen cycles**, shown in simplified form in Figure 10.4. Notice also that some carbon is tied up in fossil fuels such as petroleum (Section 11.3) and in carbonate rocks such as limestone ($CaCO_3$). This carbon returns to the

Figure 10.4 Simplified version of the carbon and oxygen cycles, showing chemical cycling (solid arrows) and one-way energy flow (open arrows).

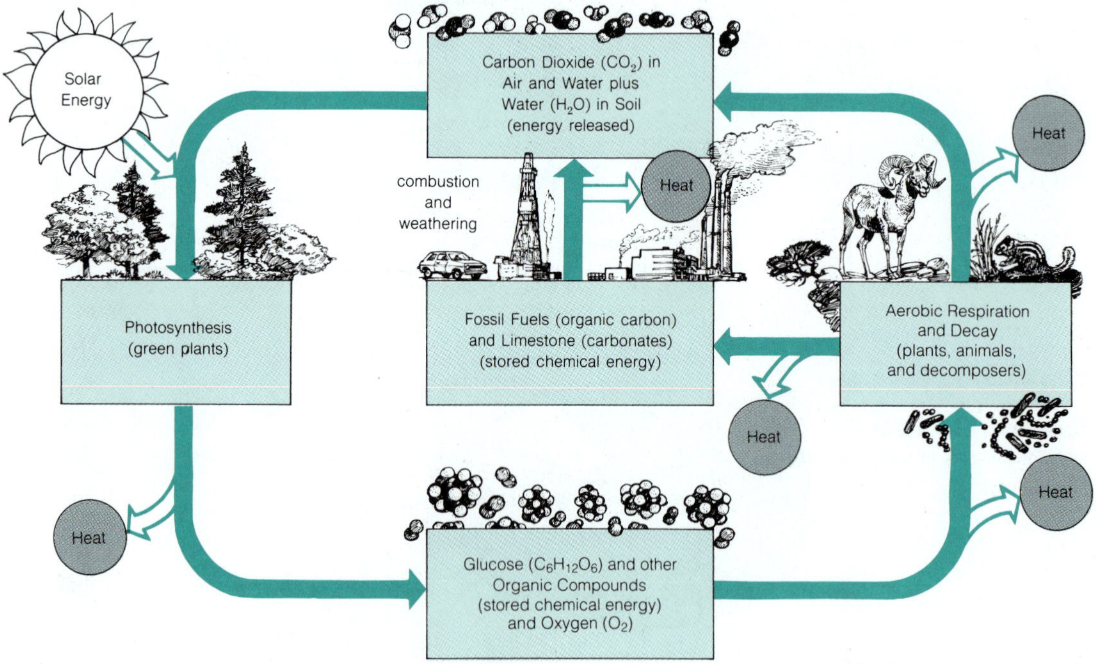

cycle as carbon dioxide when fossil fuels are burned or heat and acids act on carbonate rock formations.

NITROGEN CYCLE   The amount of usable nitrogen in the soil often limits the growth of plants (Section 17.1). Although nitrogen gas (N_2) makes up about 78 percent of the volume of the earth's atmosphere, this form of nitrogen is useless to most plants and animals. But certain bacteria, blue-green algae, and lightning can convert gaseous N_2 into usable nitrate salts (containing NO_3^- ions).

Figure 10.5 shows a simplified form of the **nitrogen cycle**. Nitrate salts dissolve easily in soil water and are taken up by plant roots. The plants convert nitrates into organic nitrogen-containing compounds necessary for life, such as proteins, nucleic acids (DNA and RNA) (Chapter 9), and vitamins. Animals then get their nitrogen-containing materials by eating plants or other animals that have eaten plants. Animals, soil bacteria, and fungi convert some nitrogen compounds to ammonia (NH_3) and salts containing ammonium (NH_4^+) ions. These forms of nitrogen are eventually oxidized into N_2 gas to renew the cycle.

Figure 10.5 Simplified diagram of the nitrogen cycle.

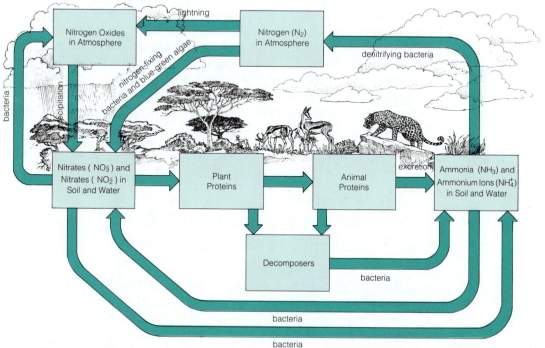

Figure 10.6 Mining of phosphate rock for use in producing commercial inorganic fertilizers. (U.S. Geological Survey)

PHOSPHORUS CYCLE   Phosphorus, mainly in the form of phosphate (PO_4^{3-}) ions, is an essential nutrient for plants and animals. Your body needs phosphorus for your DNA, RNA (Section 9.5), cell membranes, bones, and teeth (Section 16.1). Forms of phosphorus cycle through the air, water, soil, and living organisms through the **phosphorus cycle**.

We dig up large amounts of phosphate rock each year (Figure 10.6) and use most of it to make fertilizers (Section 17.1) and phosphate compounds for detergents (Section 15.4). Some phosphates are released by the slow breakdown of phosphate rocks or from fertilizers dissolved in soil water and are taken up by plant roots. Animals then get their phosphorus by eating plants or animals that have eaten plants. Much of the phosphorus eventually returns to the soil, rivers, and oceans in animal wastes and the decay of dead plants and animals.

10.2 Resources: Transforming the Earth's Raw Materials into Useful Products

WHAT IS A RESOURCE?   A **resource** is any form of matter or energy that is obtained from the physical environment to meet human needs. Although some inorganic materials such as sand, gravel, and building stone can be used pretty much as they are, most raw materials must be processed physically and chemically to make useful products (Figure 10.7).

Figure 10.7 General view of extracting and processing raw materials to make the products we use and then discard or recycle.

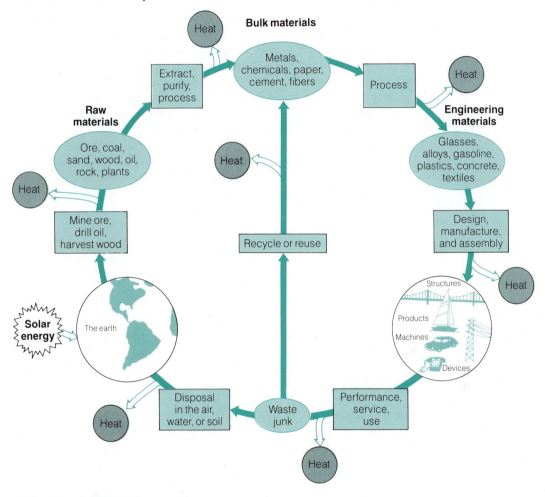

Human ingenuity then is a key factor that enables raw materials to be resources and to be made into useful products. For example, oil was a useless liquid until we learned how to find it, extract it, and make it into products such as gasoline, home heating oil, road tar, and other materials (Section 11.3). Similarly, coal and uranium fuels were once useless rocks.

Another important factor is cost. Something will not be widely used as a resource unless it can be found, removed from the earth, purified as needed, and then transformed into products we can use at an affordable price. For example, the ideal high-gloss, rust-free coating for cars is gold plate. But no matter how useful, it is too expensive for that purpose. Thus, a primary driving force in chemical technology is to find ways to make raw materials into useful products as cheaply as possible.

Abundance is also important. For centuries, and especially since the Industrial Revolution in the 1800s, we have consumed resources as if the earth's raw materials were inexhaustible. To produce products as cheaply as possible, industrialized nations have used the earth's matter and energy resources at a faster and faster rate. As long as raw materials seem plentiful, any useful substance will continue to be a resource. But once the supply has been depleted, the material—like a gold-plated car—becomes too costly to remain a resource.

Finally, safety is a factor in determining what constitutes a resource. For example, the technology and raw materials to produce nuclear energy are readily available. But widespread concern about its safety, and the resulting increase in cost, have kept it from becoming a resource (Section 11.4).

TYPES OF RESOURCES   Resources can be classified as nonrenewable or renewable. A *nonrenewable resource* either is not replaced by natural chemical cycles or is renewed more slowly than it is being used. Minerals and fossil fuels are two examples.

A *renewable resource* either comes from an essentially inexhaustible source (such as solar energy) or can be renewed fairly rapidly by natural chemical cycles or by human recycling activities. Examples include fresh air, fresh water, and fertile soil. But if a normally renewable resource such as fertile soil is depleted faster than it is replaced, it can become a nonrenewable resource.

SOIL: THE BASE OF LIFE   *Soil* is a complex mixture of decaying organic matter, water, air, living organisms, and tiny particles of inorganic minerals and rocks. The thin layer of topsoil—at most only a meter or two thick— provides nutrients for plants, which directly or indirectly provide the food that we and other animals use to stay alive and healthy.

Soils consist of a series of layers, as shown in Figure 10.8. Most soils contain from 1 to 7 percent organic matter in the two top layers, consisting of dead plants, animal droppings, and the like. Bacteria and other microorganisms break down some of these organic compounds into simpler chemical forms such as nitrate (NO_3^-), phosphate (PO_4^{3-}), and sulfate (SO_4^{2-}) ions that dissolve in soil water and are taken up as nutrients by roots of plants. Inorganic compounds in soil—such as quartz (silicon dioxide, SiO_2), alumina (aluminum oxide, Al_2O_3), iron(III) oxide (Fe_2O_3), and limestone ($CaCO_3$)—make up about 45 percent of a typical sample of soil. These compounds are mostly in the lower layers.

Figure 10.8 Generalized profile of soil. These horizontal layers vary in number, composition, and thickness, depending on the type of soil.

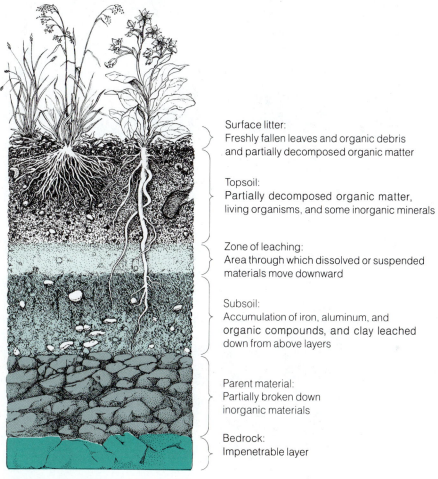

Surface litter:
Freshly fallen leaves and organic debris
and partially decomposed organic matter

Topsoil:
Partially decomposed organic matter,
living organisms, and some inorganic minerals

Zone of leaching:
Area through which dissolved or suspended
materials move downward

Subsoil:
Accumulation of iron, aluminum, and
organic compounds, and clay leached
down from above layers

Parent material:
Partially broken down
inorganic materials

Bedrock:
Impenetrable layer

Despite its importance, soil has been one of the most abused resources. The two main causes of soil erosion are blowing wind (Figure 10.9) and flowing water (Figure 10.10). But farming, construction, logging, mining, overgrazing of livestock, and other human activities remove the protective ground cover and accelerate the rate of erosion. According to the Soil Conservation Service, about one-third of the original topsoil on U.S. croplands has been washed or blown into rivers, lakes, and oceans. And the topsoil is eroding about seven times faster than it is being replaced.

Figure 10.9 Wind blowing soil off Iowa farmland (Don Ultang, U.S. Department of Agriculture)

Figure 10.10 Extensive soil erosion and gully formation caused by flowing water. (U.S. Department of Agriculture, Soil Conservation Service)

MINERALS AND ORES   **Minerals** are naturally occurring inorganic substances found as various types of rock in the earth's crust. Some are *nonmetallic minerals* such as sand (mostly silicon dioxide, SiO_2), diamond (one form of carbon), and graphite (another form of carbon). *Metallic minerals* occur mostly as compounds of metals such as copper, iron, and aluminum.

Natural deposits of minerals from which one or more elements can be obtained profitably are called **ores**. High-grade ores contain high percentages of the desired substance, whereas those with low percentages are called low-grade ores. In Sections 10.3 and 10.4, we see how to obtain elements such as silicon, copper, iron, and aluminum from their ores.

10.3 Nonmetallic Mineral Resources: Silicon

AN ABUNDANT AND USEFUL METALLOID ELEMENT   Silicon (Si) is the earth's second most abundant element and makes up about 26 percent of the earth's crust (Figure 10.11). We use silicon in glass, concrete, asbestos, and in silicone plastics (Section 14.4). In addition, the tiny microchips that make up the heart of computers and other modern electronic marvels wouldn't exist if it weren't for the remarkable properties of the metalloid element silicon.

The most common compound of silicon in nature is silicon dioxide (SiO_2), or silica. Sand is an impure form of silica. The silicon dioxide in sand and other silica rocks can be reduced to silicon by an oxidation—reduction reaction (Section 6.3) with carbon (a reducing agent):

$$SiO_2 + 2C \xrightarrow[\text{temperature}]{\text{high}} Si + 2CO$$

silica   reducing   free
          agent   element

Current passes through the mixture of silica and carbon in an electric furnace to produce the high temperature.

The resulting silicon is a shiny, dark, silvery solid with a melting point of 1410°C. Silicon's ability to conduct electricity falls somewhere

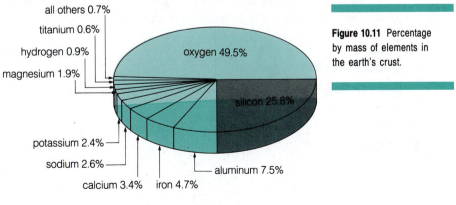

all others 0.7%
titanium 0.6%
hydrogen 0.9%
magnesium 1.9%
oxygen 49.5%
silicon 25.8%
potassium 2.4%
sodium 2.6%
calcium 3.4%   iron 4.7%
aluminum 7.5%

earth's crust

Figure 10.11 Percentage by mass of elements in the earth's crust.

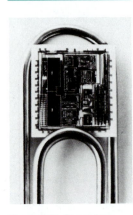

between the high conductivity of metals and the low to nonexistent conductivity of nonmetals. A material such as silicon that conducts only a limited amount of electricity is called a **semiconductor**.

Silicon is the primary material in electronic devices such as the tiny chips in computers and electronic calculators (Figure 10.12). The silicon produced in an electric furnace must be purified further before it can be used in such devices. The first step is to react the silicon with chlorine gas to convert it into silicon tetrachloride:

$$Si + 2Cl_2 \longrightarrow SiCl_4$$

$SiCl_4$ has a low boiling point (59°C), so it is readily vaporized by heat and leaves behind the unreacted impurities that were in the silicon. $SiCl_4$ then reacts with hydrogen gas (a reducing agent) to yield much purer silicon:

$$SiCl_4 + 2H_2 \longrightarrow Si + 4HCl$$

Figure 10.12 Comparison of a microcircuit printed on a tiny silicon chip with a standard-size paper clip. (Courtesy AT&T Bell Laboratory)

Although quite pure, this silicon must be purified even more before it can be in computer chips. Slowly passing a heating coil over a bar of silicon melts successive wafer-thin sections of the bar and then allows them to recrystallize. In this process, known as *zone refining* (Figure 10.13), the impurities collect in the moving, thin molten zone while the silicon that solidifies behind this zone is extremely pure. The end of the bar, where the impurities become concentrated, is then cut off. This process can be repeated several times to get the desired purity.

GLASS If you tried to melt glass, you would discover that it doesn't have a distinct melting point; it just gradually gets softer. At 600 to 800°C, window glass gets soft enough to mold into almost any shape.

Glass is mostly silicon dioxide, SiO_2. Ordinary window glass is made by melting sand (SiO_2) with other substances such as sodium carbonate

Figure 10.13 Zone-refining apparatus used to obtain extremely pure materials.

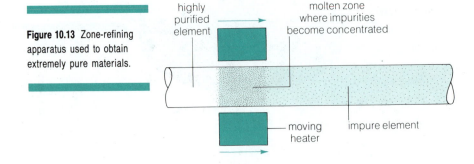

highly purified element

molten zone where impurities become concentrated

moving heater

impure element

(Na_2CO_3) and limestone ($CaCO_3$). When the resulting mixture cools, it hardens into an arrangement that lacks the regular pattern of crystals. This noncrystalline structure is why glass doesn't melt at a distinct temperature.

Many thousands of recipes exist for glass. A variety of additives give different types of glass their distinctive properties (Table 10.1).

Glasses that darken in bright light contain a silver compound such as AgCl or AgBr. Bright sunlight decomposes some of the silver compound to produce tiny bits of dark metallic silver:

$$AgCl \xrightleftharpoons{light} Ag + Cl$$

The silver and chlorine atoms remain trapped in the glass. When the light intensity decreases, the glass lightens as the reverse reaction occurs and the atoms recombine to form AgCl. Recall (Section 6.4) that the reaction of light with silver compounds is also used in photography.

CEMENT   Cement is a mixture mostly of silicon dioxide (SiO_2), limestone ($CaCO_3$), and alumina (Al_2O_3). The mixture is heated to about 1500°C and cooled. Then the powder is mixed with gypsum ($CaSO_4 \cdot 2H_2O$) to make cement. Adding water to this mixture triggers a complicated hardening process.

Cement will set into a stronger, harder mass if sand and gravel are added; the resulting material is concrete. Cement and concrete set up and harden even under water, so they can be used to make bridge piers and other underwater structures.

TABLE 10.1 SOME INGREDIENTS IN DIFFERENT TYPES OF GLASS	
Ingredient	Effect
Barium oxide (BaO)	Adds sparkle; used in food bottles.
Boron oxide (B_2O_3)	Glass expands very little when heated; used in Pyrex glassware and other cookware.
Chromium oxide (Cr_2O_3)	Makes glass green.
Cobalt(II) oxide (CoO)	Makes glass blue.
Lead(II) oxide (PbO)	Increases density and reflections; used in crystal, cut glass, and lenses.
Tin(IV) oxide (SnO_2)	Makes glass opaque.

10.4 Metallic Mineral Resources: Getting Metals from Ores

METALLURGY   Only a few, fairly unreactive metals such as silver and gold occur in nature as a pure, free element. Most metals are in mineral deposits combined with nonmetals in the form of oxides, sulfides, chlorides, carbonates, phosphates, and other compounds. In these ores, the metals exist as positively charged ions.

Metallurgy is the series of processes that produces a free metal from its ore. These processes include

- Mining (Figure 10.14).
- Preliminary treatment to concentrate metallic ore by removing sand, silicate (SiO_4^{4-}) minerals, and other undesirable impurities.
- Reducing the positively charged metal ions to the free metal.
- Refining (purifying) the free metal.
- Mixing certain amounts of other substances with the purified metal to form a material—an *alloy*—with the desired properties such as structural strength or resistance to corrosion.

Figure 10.14 Removing iron ore from an open-pit mine. (American Iron and Steel Institute)

REMOVING OTHER IMPURITIES BEFORE OR DURING REDUCTION   An ore is concentrated by crushing, grinding, washing, and various separation procedures. Sometimes other impurities must be removed before or during reduction of the concentrated ore to the free metal. This is often done by adding a *flux*, a chemical that reacts with nonmetallic impurities to produce a low-melting substance called *slag*. For example, many concentrated ores still contain silicon dioxide as an impurity. Reaction with $CaCO_3$ (limestone) at high temperatures in a two-step process removes this impurity:

$$CaCO_3 \xrightarrow[\text{temperature}]{\text{high}} CaO + CO_2$$
limestone
flux

$$SiO_2 + CaO \xrightarrow[\text{temperature}]{\text{high}} CaSiO_3$$
impurity   *molten slag*

The molten slag is easily removed because it floats on top of the more dense molten ore or metal.

REDUCING THE MOLTEN ORE   The method for reducing a molten metal ore depends on how easy it is to reduce the metal ions to the free metal, as summarized in Table 10.2. *Roasting*, a process in which the ore is heated in air, converts most sulfide ores to an oxide. For example,

Roasting: $\quad 2ZnS \;\; + 3O_2 \;\; \xrightarrow[\text{temperature}]{\text{high}} \;\; 2ZnO + 2SO_2$

$\qquad\qquad\quad$ *sulfide ore* $\qquad\qquad\qquad\qquad$ *oxide*

TABLE 10.2 REDUCTION PROCESSES FOR SELECTED METAL IONS IN ORES

Name of Metal	Oxidized Metal Ion in Ore	Reduced Free Metal
Hard to Reduce: Electrolytic Reduction of Molten Ore Compound		
Lithium	$Li^+ \longrightarrow$	Li
Potassium	$K^+ \longrightarrow$	K
Calcium	$Ca^{2+} \longrightarrow$	Ca
Sodium	$Na^+ \longrightarrow$	Na
Magnesium	$Mg^{2+} \longrightarrow$	Mg
Aluminum	$Al^{3+} \longrightarrow$	Al
Moderately Hard to Reduce: Chemical Reduction with Coke (C) or Carbon Monoxide (CO)		
Manganese	$Mn^{2+} \longrightarrow$	Mn
Zinc	$Zn^{2+} \longrightarrow$	Zn
Chromium	$Cr^{3+} \longrightarrow$	Cr
Iron	$Fe^{3+} \longrightarrow$	Fe
Cobalt	$Co^{2+} \longrightarrow$	Co
Nickel	$Ni^{2+} \longrightarrow$	Ni
Tin	$Sn^{2+} \longrightarrow$	Sn
Lead	$Pb^{2+} \longrightarrow$	Pb
Easy to Reduce: Easily Reduced by Roasting in Oxygen Gas or Occurs as Free Metal		
Copper	$Cu^+, Cu^{2+} \longrightarrow$	Cu
Mercury	$Hg^{2+} \longrightarrow$	Hg
Silver		Ag (occurs as free metal)
Platinum		Pt (occurs as free metal)
Gold		Au (occurs as free metal)

Figure 10.15 Blast furnace for reducing iron ore to pig iron.

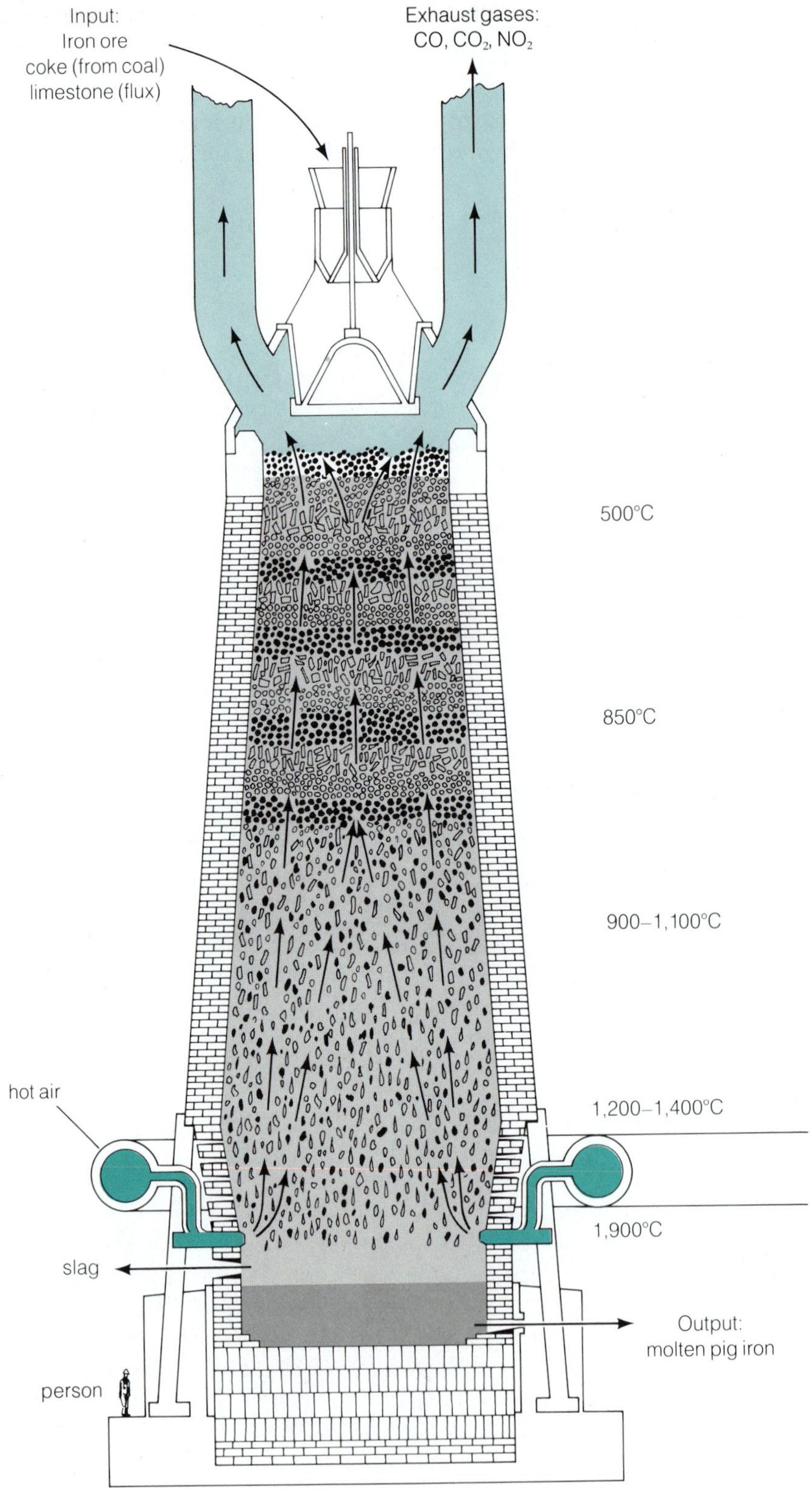

Input:
Iron ore
coke (from coal)
limestone (flux)

Exhaust gases:
CO, CO$_2$, NO$_2$

500°C

850°C

900–1,100°C

1,200–1,400°C

1,900°C

hot air

slag

person

Output:
molten pig iron

Roasting produces the free metal directly if the sulfide ore contains metallic ions (such as Cu^+) that are easily reduced:

Roasting: $\quad Cu_2S \; + O_2 \; \xrightarrow[\text{temperature}]{\text{high}} \; 2Cu \; + SO_2$

$\qquad\qquad$ *sulfide ore* $\qquad\qquad\qquad$ *free metal*

The sulfur dioxide produced by roasting can be a serious air pollutant (Section 12.1).

Some oxides can be reduced to the free metal by a reducing agent such as coke (impure carbon) or carbon monoxide (CO) at a high temperature (see Table 10.2). For example,

$Fe_2O_3 + \qquad 3C \qquad \xrightarrow[\text{temperature}]{\text{high}} \qquad 2Fe \; + 3CO$

oxide $\quad$ *reducing agent* $\qquad\qquad$ *free metal*
$\qquad\qquad$ *(coke)*

Iron ore (Fe_2O_3) is reduced in gigantic *blast furnaces* 33 m (110 ft) high (Figure 10.15). The reaction above is the net reaction for a complex sequence of reactions that take place in this furnace. Periodically, the resulting metal, called *pig iron*, is removed from the furnace. Pouring pig iron into a mold and letting it cool produces *cast iron*.

Pig iron contains too much carbon and other impurities for most uses, so further steps are necessary to make it into steel. Reacting pig iron with pure oxygen (*basic oxygen furnace*; Figure 10.16) or with an oxidizing agent (*electric furnace*) converts carbon impurities to carbon monoxide (and some carbon dioxide):

$2C + O_2 \; \xrightarrow[\text{temperature}]{\text{high}} \; 2CO$

Thousands of different types of steel with desired properties such as toughness, hardness, and resistance to wear, shock, and corrosion can be made by (1) varying the amount of carbon removed, (2) subjecting the steel to various heat and mechanical treatments, and (3) adding small amounts of other metals to form steel alloys. Corrosion-resistant stainless steels, for example, are low in carbon and contain more than 12 percent chromium.

ELECTROLYTIC REDUCTION OF THE MOLTEN ORE $\quad$ Chemical reduction is not possible for some ores (see Table 10.2), so they are reduced by passing an electrical current through the molten ore. This is called *electrolytic reduction*.

One example is the reduction of aluminum oxide (Al_2O_3) ore (bauxite) to aluminum. The method is known as the *Hall process*, after its inventor, Charles Martin Hall (Figure 10.17 and the "Chemistry Spotlight"). In this process, large amounts of electrical current pass through a molten mixture of concentrated bauxite ore. A flux is used to dissolve

Figure 10.16 Molten pig iron from a blast furnace being poured into a basic oxygen furnace for conversion into steel. (American Iron and Steel Institute)

Figure 10.17 Charles M. Hall (1863–1914), who invented an electrolytic reduction process to convert aluminum oxide ore (bauxite) into aluminum metal. (Courtesy ALCOA)

Figure 10.18 Hall cell for the electrolytic production of aluminum from molten aluminum oxide.

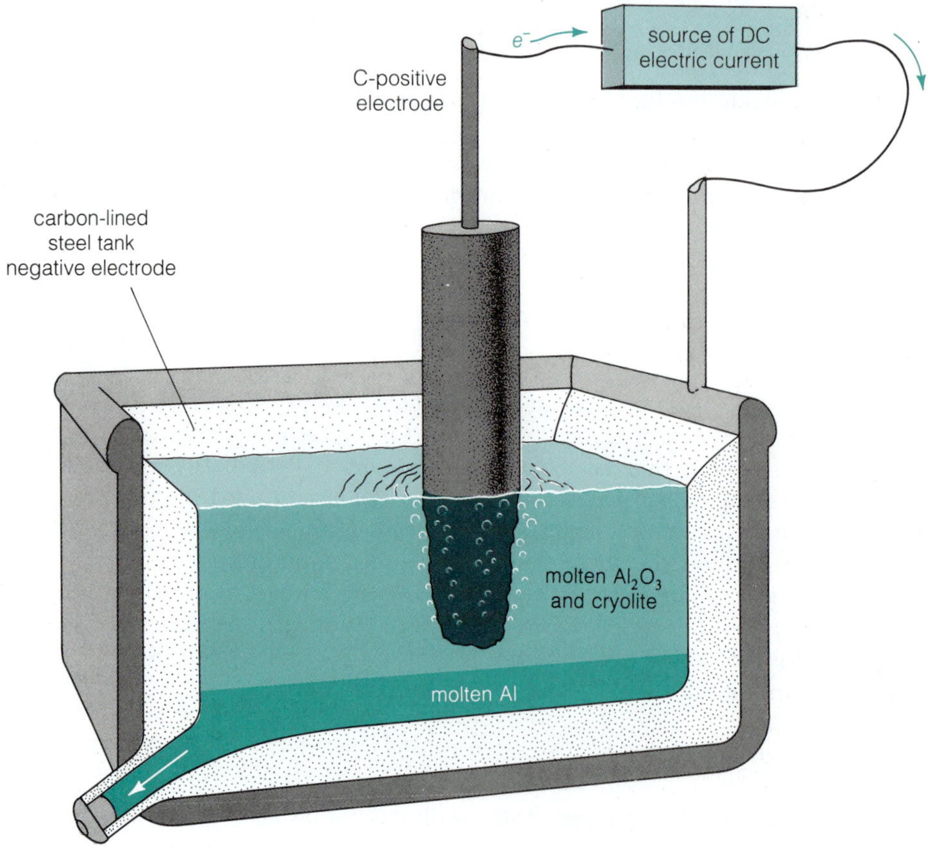

e^-

source of DC
electric current

C-positive
electrode

carbon-lined
steel tank
negative electrode

molten Al_2O_3
and cryolite

molten Al

Figure 10.19 Aluminum is so light and such an excellent reflector of light and radiant energy that aluminized clothing can be used in many industries to protect workers against heat and flames. (Courtesy ALCOA)

the aluminum oxide, lower its melting point, and remove impurities (Figure 10.18):

$$2Al_2O_3 \xrightarrow[\text{current}]{\text{electrical}} 4Al + 3O_2$$

molten bauxite   *free metal*

The aluminum produced by this process is an important industrial metal because of its lightness, strength, high electrical conductivity, ability to reflect light and heat (Figure 10.19), and resistance to corrosion.

CHEMISTRY SPOTLIGHT   **THE HALL PROCESS—WHY IT CAN PAY TO LISTEN TO YOUR CHEMISTRY TEACHER**

Figure 10.20 Replica of the woodshed laboratory where Charles M. Hall discovered the electrolytic reduction process for making aluminum. (Courtesy ALCOA)

In 1885 aluminum could be produced from its ore only on a small scale, and it cost more than silver. Today it costs less than 50¢ per pound. The credit goes to Charles Martin Hall (see Figure 10.17), who studied chemistry at Oberlin College in Ohio. In his freshman year, Hall began going to his chemistry professor, Frank F. Jewett, to buy small items of equipment and chemicals for his home laboratory. Inspired by Hall's industry and potential as a scientist, Jewett took the young man under his wing and gave him work space in his college laboratory.

One day while speaking to his students in the laboratory, Jewett remarked that if anyone could invent a way to produce aluminum economically on a commercial scale, that person would benefit humanity and at the same time grow rich. Turning to a classmate, Charles Hall said, "I'm going for that metal."

With help and equipment from Professor Jewett, he began working on this problem in his senior year. After he graduated, he borrowed the equipment and continued to work in a laboratory in a woodshed behind his father's parsonage (Figure 10.20). About six months later, in 1886—at age twenty-three—he walked into his former teacher's office with a dozen little globules of aluminum made by the electrolytic reduction process that now bears his name. In France about the same time, another chemistry student, Paul L. T. Heroult (also twenty-three years old), discovered the same process quite independently of Hall.

Hall, whose careful research notes enabled him to receive the patent for this process, went on to cofound Pittsburgh Reduction Company, which later became Aluminum Company of America (ALCOA). He died a multimillionaire in 1914, bequeathing large gifts to his community, to Oberlin College, and to higher education in general.

Aluminum metal resists corrosion because, upon its exposure to air, a continuous, tightly adhering, thin coating of Al_2O_3 forms on its surface and protects it against further oxidation (corrosion). Even some highly corrosive acids can be shipped in containers made of aluminum, thanks to its automatically formed oxide coating.

10.5 Disposal of Used Resources: Solid and Hazardous Wastes

ENVIRONMENTAL IMPACT OF RESOURCE USE   The mining, processing, and use of mineral resources disturbs the land, pollutes the air and water, and leaves us with solid-waste materials, some of which are hazardous. According to the second law of thermodynamics, this sort of degradation is inevitable. But most land disturbed by mining can be reclaimed to some degree, and existing environmental protection laws require it. In the remainder of this chapter, we look briefly at how the solid and hazardous wastes produced by resource use can be handled.

SOLID WASTES: WHAT TO DO WITH IT   *Solid waste* is any unwanted or discarded material that is solid. It includes yesterday's newspaper, today's junk mail, dinner scraps, leaves and grass clippings, nonreturnable bottles and cans, abandoned cars, animal manure, mining and industrial wastes, and an array of other cast-off materials. Figure 10.21 shows the average composition of the stuff we throw away.

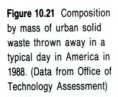

Figure 10.21 Composition by mass of urban solid waste thrown away in a typical day in America in 1988. (Data from Office of Technology Assessment)

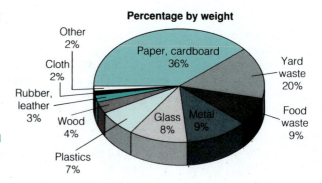

Percentage by weight

Other 2%
Cloth 2%
Rubber, leather 3%
Wood 4%
Plastics 7%
Glass 8%
Metal 9%
Paper, cardboard 36%
Yard waste 20%
Food waste 9%

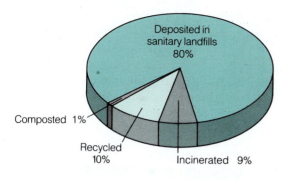

Figure 10.22 Fate of solid waste in the United States. (Data from Environmental Protection Agency)

The United States annually produces an average of 40 metric tons (44 tons) of solid waste per person. Most (about 98 percent) of this comes from mining, oil and natural gas production, and industry. Those wastes are either left where they were produced, stored in pits or lagoons, or dumped at designated sites. As Figure 10.22 shows, most of the solid wastes produced in homes and businesses are carted away by garbage trucks and put in sanitary landfills (Figure 10.23).

Figure 10.23 A sanitary landfill. Wastes are spread in a thin layer and then compacted with a bulldozer. A scraper (foreground) covers the wastes with a fresh layer of soil at the end of each day. Portable fences catch and hold debris blown by the wind.

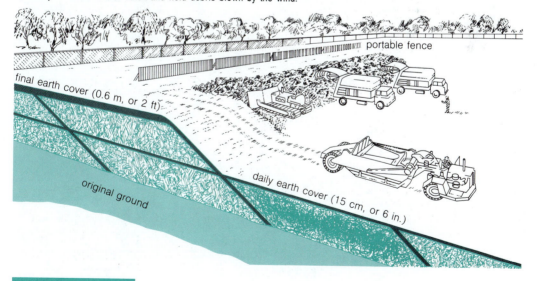

But we have to do better. Cities and industries are running out of space for sanitary landfills, and some landfills may pollute underground drinking-water supplies (Section 13.6). Many citizens also oppose building incinerators anywhere near them, for incinerators can pollute the air unless they are properly designed, maintained, and monitored.

An alternative is to recycle or reuse more of our solid wastes (Table 10.3). *Recycling* is collecting, reprocessing, and refabricating products

TABLE 10.3 THREE SYSTEMS FOR HANDLING DISCARDED MATERIALS

Item	Throwaway System	Resource Recovery and Recycling System	Low-Waste System
Glass bottles	Dump or bury	Grind and remelt; remanufacture; convert to building materials	Ban all nonreturnable bottles and reuse (rather than remelt and recycle) bottles
Bimetallic "tin" cans	Dump or bury	Sort, remelt	Limit or ban production; use returnable bottles
Aluminum cans	Dump or bury	Sort, remelt	Limit or ban production; use returnable bottles
Cars	Dump	Sort, remelt	Sort, remelt; tax cars lasting less than 15 years, weighing more than 818 kg (1800 lb), and getting less than 13 km/L (30 mpg)
Metal objects	Dump or bury	Sort, remelt	Sort, remelt; tax items lasting less than 10 years
Tires	Dump, burn, or bury	Grind and revulcanize or use in road construction; incinerate to generate heat and electricity	Recap usable tires; tax all tires not usable for at least 64,000 km (40,000 miles)
Paper	Dump, burn, or bury	Incinerate to generate heat	Compost or recycle; tax all throwaway items; eliminate overpackaging
Plastics	Dump, burn, or bury	Incinerate to generate heat or electricity	Limit production; use returnable glass bottles instead of plastic containers; tax throwaway items and packaging
Yard wastes	Dump, burn, or bury	Incinerate to generate heat or electricity	Compost; return to soil as fertilizer; use as animal feed

such as a nonreturnable aluminum can. *Reuse* is using a product (such as a returnable bottle) over and over again in its original form until it is broken or lost. Compared with making products from virgin resources, recycling and reuse generally harm the environment less and require less energy if the resources aren't too widely scattered and mixed with other materials.

Although surveys indicate that most people in the United States favor more recycling, the 10 percent of urban and industrial solid wastes they recycle (see Figure 10.22) compares to figures of 40 to 60 percent in Japan and many other industrialized nations. Three factors that hinder recycling in the United States are (1) older manufacturing processes that cannot use recycled materials, (2) economic incentives that encourage the mining and use of virgin resources, and (3) fluctuating prices for recycled materials because of variable demand. If such barriers were removed, U.S. municipalities could recycle at least 35 percent of their solid wastes.

HAZARDOUS WASTES: NOT IN MY BACKYARD   Each year, by conservative estimates, an average of 1000 kg (1.1 tons) of hazardous waste are produced for each resident in the United States—enough to fill the New Orleans Superdome from floor to ceiling twice a day. A *hazardous waste* is any discarded material that poses a substantial threat to human health or to the environment when handled improperly. These wastes include

— Toxic substances (cyanides; nerve gases; compounds of lead, mercury, and arsenic; and some pesticides).

— Flammable substances (waste oils, dry cleaning and industrial solvents).

— Corrosive substances (acids and bases).

— Radioactive substances.

— Infectious materials (wastes from hospitals and research laboratories).

— Explosive or dangerously reactive substances.

Most (about 95 percent) of the regulated hazardous wastes in the United States has been disposed in lagoons, deep underground wells (Figure 10.24), sanitary landfills (see Figure 10.23) not designed to handle such wastes, and (more recently) secured landfills designed for such wastes (Figure 10.25). But increased regulations, higher land costs, and opposition by citizens who do not want hazardous wastes stored near them have made land disposal less attractive.

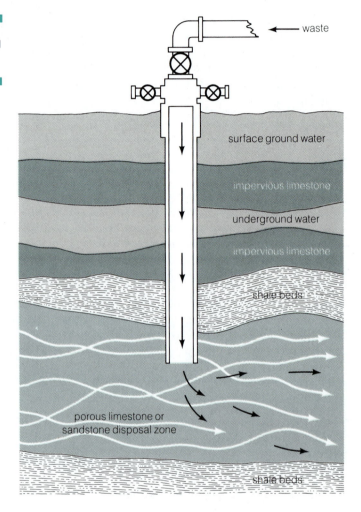

Figure 10.24 An underground waste-disposal well.

waste

surface ground water

impervious limestone

underground water

impervious limestone

shale beds

porous limestone or
sandstone disposal zone

shale beds

The Environmental Protection Agency (EPA) estimates that since 1950 about 70 percent of the hazardous wastes in the United States have been disposed of improperly. In 1977, for example, residents of a suburb of Niagara Falls, New York, discovered hazardous industrial wastes oozing into their neighborhood. The wastes came from the Love Canal chemical dump used for more than a decade by the Hooker Chemicals and Plastics Corporation. The EPA has identified 20,000 hazardous waste sites in the United States and estimated that at least 1200 of these sites could pose health hazards to nearby residents or threaten the environment. Only a handful have been cleaned up so far.

Figure 10.25 A secure landfill for the long-term storage of hazardous wastes.

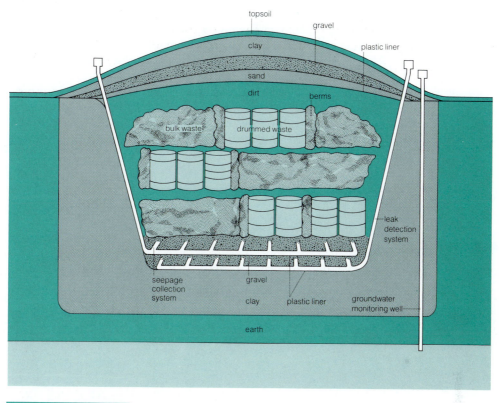

Many hazardous wastes are dumped illegally in municipal landfills, rivers, sewer drains, wells, empty lots and fields, warehouses, old quarries, and abandoned mines; others are spread along roadsides or mixed with heating oil and burned in the boilers of schools, hospitals, offices, and apartment buildings. According to the EPA, about 13,000 accidental spills of hazardous materials occurred in 1980–1990 as the wastes were transported (mostly by train or truck) to disposal sites.

The National Academy of Sciences has outlined three basic ways to deal with hazardous wastes: (1) prevention of wastes by reducing, recycling, and reusing wastes; (2) conversion to less hazardous wastes; and (3) permanent storage. The EPA estimates that 20 percent of the hazardous wastes in the United States could be recycled, reused, or exchanged with another industry that can use the material as a resource. Only 4 percent of the wastes are currently handled in this way.

A promising but as yet little-used approach in the United States is to convert hazardous waste to less hazardous or nonhazardous materials. Chemical treatments include (1) neutralizing acidic or basic wastes, (2) oxidizing or reducing wastes into different substances, and (3) removing toxic metals and other compounds by precipitation or absorption (using a chemical such as activated carbon). Denmark has twenty-one facilities that detoxify about 75 percent of the nation's hazardous wastes.

Another approach is incineration. The EPA estimates that about 60 percent of all U.S. hazardous wastes could be incinerated. With proper air-pollution controls, incineration is potentially the safest way to dispose of most hazardous wastes. But it is also the most expensive, and it isn't suitable for some wastes. Currently, the Netherlands incinerates about half of its hazardous wastes.

DENIS HAYES

Waste is a human concept. In nature nothing is wasted, for everything is part of a continuous cycle. Even the death of a creature provides nutrients that will eventually be reincorporated in the chain of life.

Summary

Our life-support system consists of the atmosphere, hydrosphere, lithosphere, and biosphere. Life is sustained by a flow of energy from the sun and from the cycling of elements, especially oxygen, carbon, hydrogen, nitrogen, and phosphorus.

Matter and energy are resources if they are useful, inexpensive, abundant, and safe. Resources are nonrenewable unless the supply is adequately replenished as it is used. Two important resources in the earth's crust are soil and minerals.

Silicon is a resource. Its compounds can be processed to make glass, cement, and chips for computers and calculators. Metallic ores are typically purified and reduced to the free metal elements, which are used in many ways. Reduction is done by roasting in oxygen

(O₂) gas, heating with coke (C) or carbon monoxide (CO) gas, or electrolytic reduction.

Most municipal solid wastes are stored in sanitary landfills. Hazardous wastes are stored in secured landfills, lagoons, or deep underground wells. Much more solid waste could be recycled and reused. More hazardous waste could be converted into less toxic materials by incineration or by other chemical treatments.

Terms for Review

After completing this chapter, you should know and understand the meaning of the following terms:

atmosphere (p. 270)	lithosphere (p. 270)	oxygen cycle (p. 272)
biosphere (p. 270)	mineral (p. 278)	phosphorus cycle (p. 274)
carbon cycle (p. 272)	nitrogen cycle (p. 273)	resource (p. 274)
hydrosphere (p. 270)	ore (p. 278)	semiconductor (p. 280)

Questions

Odd-numbered questions are answered at the back of this book.

1. Classify each of the following as part of the atmosphere, biosphere, hydrosphere, or lithosphere: (a) apple, (b) uranium ore, (c) caterpillar, (d) polar ice caps, (e) O₂ gas, (f) you, (g) bricks, and (h) clouds.

2. When you turn on a light or a television set, what effect are you having on the carbon and oxygen cycles?

3. List the (a) oxidized and (b) reduced forms of nitrogen gas (N₂) shown in Figure 10.5.

4. Which of the following is currently a resource? (a) Slag from a blast furnace, (b) redwood forests, (c) diamonds, and (d) broken glass.

5. For the item(s) in Question 4 that you classified as a resource, is the resource renewable or nonrenewable?

6. How can a renewable resource be converted to a nonrenewable resource?

7. Explain why the use of a resource is closely tied to economics and human ingenuity.

8. Give two examples of (a) materials that were once considered useless but are now considered resources, (b) materials that were once resources but are now considered useless, (c) renewable resources, (d) nonrenewable resources, and (e) re-

newable resources that we could convert to nonrenewable resources.

9. List the two main natural causes of soil erosion.

10. How are nitrates produced in soil?

11. What are the major natural sources of silicon?

12. What is the reducing agent for producing silicon from SiO_2?

13. A substance that conducts only a limited amount of electricity is called a(n) _____.

14. Why does glass lack a distinct melting point?

15. Outline the chemistry involved in (a) converting iron ore to pig iron and (b) converting pig iron to steel.

16. Write a balanced reaction for the roasting of Cu_2S ore to produce free copper (Cu) and sulfur dioxide (SO_2).

17. Write a balanced reaction for the reduction of zinc oxide (ZnO) with coke to produce free zinc.

18. Identify whether ores of the following metals can be reduced to the free metal by roasting, chemical reduction, or electrolytic reduction: (a) zinc, (b) mercury, (c) potassium, and (d) magnesium.

19. Do you think it takes more energy to produce a ton of aluminum or a ton of steel? Why?

20. Summarize the environmental effects involved in (a) mining an ore and (b) purifying and reducing an ore to a free element.

21. Distinguish between recycling and reuse of a resource and give an example (not given in the text) of each.

22. How does the second law of thermodynamics limit what can be accomplished by recycling?

23. Compare the throwaway, recycling, and low-waste approaches for handling (a) glass bottles and (b) aluminum cans. Which approach do you favor? Why?

24. What are the major advantages and disadvantages of (a) land disposal of hazardous wastes, (b) chemical conversion to less hazardous or nonhazardous substances, (c) incineration, and (d) recycling and reuse. Which approach(es) do you favor? Why?

Topics for Discussion

1. As we will discuss in Chapter 12, the concentration of carbon dioxide in the atmosphere is rising. Examine the carbon cycle (see Figure 10.4) to identify human activities that (a) increase the amount of CO_2 in the atmosphere and (b) could be taken to counteract the increase in CO_2.

2. Which of the following steps do you think your college and your local city government should take? (a) Have mandatory recycling programs; (b) require that a certain fraction of all paper purchases contain recycled fiber; (c) sell soft drinks only in returnable bottles.

3. What are some of the major problems confronting the steel industry in the United States? What are some possible solutions?

4. Would you oppose locating a secured landfill or incinerator for hazardous wastes near your community? What alternatives would you suggest?

CHAPTER 11

Energy Resources:
The Times Are Changing

General Objectives

1. What are the major sources of energy, and how do we use them?

2. What are the concepts of energy quality, energy efficiency, and net energy? How do we use them to evaluate energy alternatives?

3. What are the major nonrenewable energy sources? What are the advantages and disadvantages of using them?

4. What are the major renewable energy sources? What are the advantages and disadvantages of using them?

When you turn on an air conditioner, drive a car, eat breakfast, or listen to the radio, you use energy. The amounts and types of available energy shape not only your life-style but also national and world economic systems. This chapter examines our energy resources and how fast we and the rest of the world are using them. We also see how the two laws of thermodynamics (Chapter 7) can help us evaluate our energy options.

11.1 Energy Resources: Will There Be Enough?

THE SUN: SOURCE OF ENERGY FOR LIFE ON EARTH The sun is a medium-size star composed mostly of hydrogen. A gigantic nuclear fusion reactor, the sun converts each second about 4.1 billion kg (4.5 million tons) of its mass into energy. The sun has probably existed for at least 6 billion years, and it has enough hydrogen left to keep going for at least another 8 billion years. The sun then is a virtually inexhaustible source of energy that sustains life on earth. Green plants and certain bacteria use solar energy in photosynthesis to make sugars and other carbon compounds. The rest of us get our chemical energy by eating other plants and animals.

The sun also heats earth and our buildings, free of charge. Without this energy, the average temperature outside would be a frigid $-240°C$ $(-400°F)$. But who would be around to measure it?

Most people think of solar energy only as direct energy from the sun. But solar energy also is responsible for other forms such as wind, flowing water, and biomass (energy stored in plants). All these energy resources are renewable. The nonrenewable energy resources include fossil fuels, nuclear energy, and geothermal energy (heat from within the earth).

ENERGY USE Each of us needs about 2000 kilocalories (kcal) of energy a day to survive (Figure 11.1). But today a person who lives in the United

Figure 11.1 Average daily per-capita energy use of various cultures.

Society	Kilocalories per Person per Day
Modern Industrial (United States)	230,000
Modern Industrial (other developed nations)	125,000
Early Industrial	60,000
Advanced Agricultural	20,000
Early Agricultural	12,000
Hunter-Gatherer	5,000
Primitive	2,000

States uses an average of 230,000 kcal a day—115 times the survival level. The 4.8 percent of the world's population living in the United States uses about 25 percent of the world's commercial energy. At the other extreme, India, with about 16 percent of the world's people, uses only about 1.5 percent of the world's commercial energy. In 1989 the 249 million U.S. population used more energy for air conditioning than the 1.1 billion Chinese used for all purposes.

Most (nearly 90 percent) of the energy used in the United States comes from burning three nonrenewable fossil fuels—coal, oil, and natural gas (Figure 11.2). The pattern worldwide (Figure 11.3) is similar except that biomass is a more important factor, especially in the less industrialized countries.

11.2 Some Energy Concepts: Energy Quality, Energy Efficiency, and Net Energy

ENERGY QUALITY   According to the second law of thermodynamics (Section 7.2), whenever we use energy to do work, energy changes into a lower quality or less useful form—most commonly low-temperature heat.

Figure 11.2 Changes in consumption of nonrenewable (shaded) and renewable (unshaded) energy resources in the United States between 1850 and 1986. Relative circle size indicates the total amount of energy used. (Source: U.S. Department of Energy)

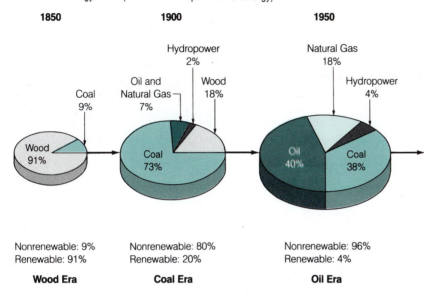

1850

Coal
9%

Wood
91%

Nonrenewable: 9%
Renewable: 91%

Wood Era

1900

Hydropower
2%

Oil and
Natural Gas
7%

Wood
18%

Coal
73%

Nonrenewable: 80%
Renewable: 20%

Coal Era

1950

Natural Gas
18%

Hydropower
4%

Oil
40%

Coal
38%

Nonrenewable: 96%
Renewable: 4%

Oil Era

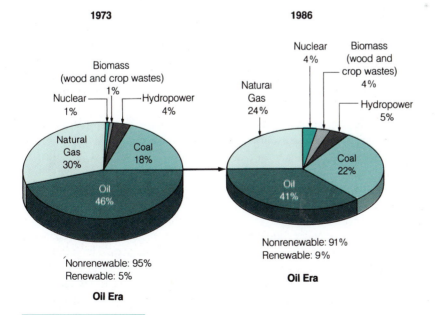

1973

Biomass
(wood and crop wastes)
1%

Nuclear
1%

Hydropower
4%

Natural
Gas
30%

Coal
18%

Oil
46%

Nonrenewable: 95%
Renewable: 5%

Oil Era

1986

Nuclear
4%

Biomass
(wood and
crop wastes)
4%

Natural
Gas
24%

Hydropower
5%

Coal
22%

Oil
41%

Nonrenewable: 91%
Renewable: 9%

Oil Era

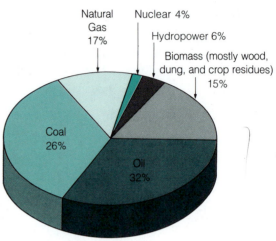

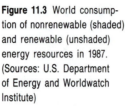

Figure 11.3 World consumption of nonrenewable (shaded) and renewable (unshaded) energy resources in 1987. (Sources: U.S. Department of Energy and Worldwatch Institute)

Nonrenewable: 79%
Renewable: 21%

TABLE 11.1 ENERGY QUALITY

Very High Quality
Electricity
Heat above 2500°C
Nuclear fission
Nuclear fusion
Concentrated sunlight
High-velocity wind

High Quality
Heat above 1000°C
Hydrogen gas
Natural gas
Oil
Gasoline
Coal
Food

Moderate Quality
Heat above 100°C
Normal sunlight
Moderate winds
Fast-flowing water
Wood and crop wastes
Intense geothermal

Low Quality
Heat below 100°C
Dispersed geothermal

Different forms of energy vary considerably in their quality (Table 11.1). High-quality energy—electricity, oil, gasoline, nuclear, high-temperature heat, intense sunlight, and wind—is concentrated, whereas low-quality energy is dispersed. For example, more energy is stored in the Atlantic Ocean than in all the oil in Saudi Arabia, but the ocean's energy is low-temperature heat, which is so widely dispersed that we can't do much with it.

ENERGY EFFICIENCY **Energy efficiency** is the percentage of total energy input that does useful work and is not converted into waste heat. Because energy conversions vary considerably in their efficiencies (see Figure 7.5), we can reduce waste by using the most efficient processes or by making processes more efficient.

The energy efficiency of a home-heating system, for example, depends on the efficiency of each step in the system, including extracting the fuel, purifying and upgrading it to a useful form, transporting it, and finally using it. Figure 11.4 shows how the overall energy efficiencies are determined for heating a home passively with solar energy through the use of south-facing windows and with electricity produced at a nuclear power plant.

You can save energy and money by buying energy-efficient refrigerators, washers, furnaces, air conditioners and other devices. The initial cost is usually higher, but in the long run such devices save money. For example, new fluorescent bulbs that resemble the screw-in incandescent bulbs cost more than incandescent bulbs, but they are much more efficient and

Figure 11.4 Comparison of the energy efficiency of two types of space heating. The overall efficiency is obtained by multiplying the percentage shown inside the circle for any step by the energy efficiency for that step (shown in parentheses).

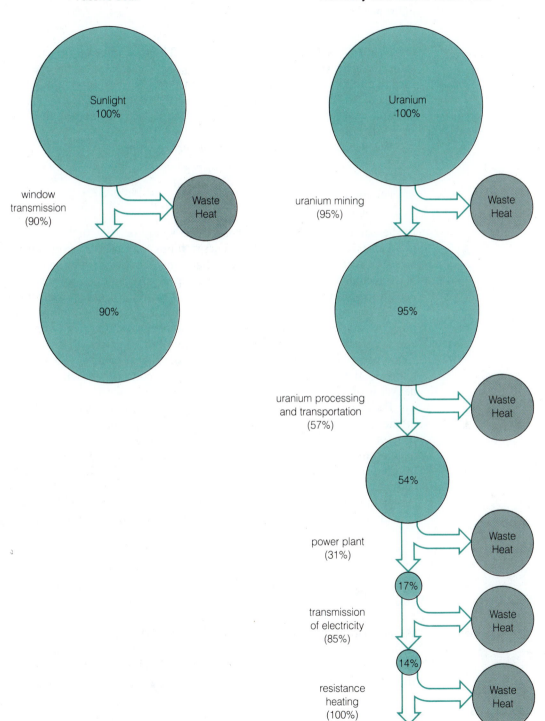

last five years or more. Over a five-to-ten-year period, the fluorescent bulbs actually cost less to buy and use.

NET ENERGY: IT TAKES ENERGY TO GET ENERGY   The true value of any energy source is its **net energy**—the total energy of the resource minus the energy used to find, process, concentrate, and transport it. For example, if it takes 9 units of fossil fuel (or other) energy to deliver 10 units of nuclear, solar, or additional fossil fuel energy (perhaps from a deep well at sea), the net energy is only 1 unit.

We also can express this as the *net energy ratio*:

$$\text{net energy ratio} = \frac{\text{energy produced}}{\text{energy used to produce it}}$$

In the example above, the net energy ratio would be 10/9, or 1.1.

Table 11.2 lists estimated net energy ratios for various energy alternatives. Currently, fossil fuels have high net energy ratios because so much of them come from cheap and accessible deposits. When those sources are depleted, however, the ratios will decline. The low net energy ratio for nuclear energy includes the large amounts of energy to build and operate power plants; to process, transport, and store nuclear fuels and wastes; and to meet complex safety and environmental requirements. Large-scale solar energy power plants also have low ratios; solar energy is so widely scattered that it takes considerable energy to collect, concentrate, and deliver it on a large scale.

Perhaps the most striking example of net energy is food production in the United States. If we consider all the energy we spend to grow, process, package, transport, refrigerate, and cook our food, we find that it takes about 9 cal of fossil fuel energy to put only 1 cal of food energy on the table—a net energy *loss* of 8 cal per calorie of food. "Primitive" cultures, on the other hand, produce 5 to 50 food cal for each calorie of energy put into the food system. Feeding the entire world with U.S.-type agriculture would empty the world's known oil reserves in thirteen years.

11.3 Fossil Fuels: How Long Will They Last?

ARE WE RUNNING OUT?   How much longer can world and U.S. supplies of oil, natural gas, and coal last? No one knows for sure. The answer depends on how much more will be discovered, how expensive it will be to recover, and how fast we will use it up.

TABLE 11.2 NET ENERGY RATIOS OF VARIOUS SYSTEMS	
System	**Net Energy Ratio**
Space Heating	
Passive solar	5.8
Natural gas	4.9
Oil	4.5
Active solar	1.9
Coal gasification	1.5
Electric resistance heating (fossil fuel plant)	0.4
Electric resistance heating (nuclear plant)	0.3
High-Temperature Industrial Heat	
Surface-mined coal	28.2
Natural gas	4.9
Oil	4.7
Coal gasification	1.5
Direct solar (concentrated by mirrors, or other devices)	0.9
Transportation	
Natural gas	4.9
Gasoline	4.1
Biofuel (ethyl alcohol)	1.9
Coal liquefaction	1.4
Oil shale	1.2

SOURCES: Colorado Energy Research Institute, *Net Energy Analysis: An Energy Balance Study of Fossil Fuel Resources*, 1976; and Howard T. Odum and Elisabeth C. Odum, *Energy Basis for Man and Nature*, 3rd ed., New York: McGraw-Hill, 1981.

Table 11.3 shows estimates of how soon we will use up the world's supply of fossil fuels. If 80 percent of the supply is the maximum we can use up before it becomes too expensive to recover, then the world's known reserves of crude oil will cease to be a resource early in the next century. All estimated supplies (three times today's known reserves) will last a few decades more. Natural gas is projected to last a bit longer. Coal is the most abundant fossil fuel, so its use is likely to increase.

HEAVY OILS FROM OIL SHALE AND TAR SANDS   **Oil shale** is an underground rock that contains varying amounts of a solid, waxy mixture of heavy hydrocarbon compounds. Shale can be mined like coal, crushed, and then

TABLE 11.3 ESTIMATED DEPLETION OF THE WORLD'S FOSSIL FUELS		
	Year of 80% Depletion at	
Resource	1984 Rate	2% Annual Increase
Oil		
Known reserves	2013	2006
Total recoverable reserves	2073	2037
Natural gas		
Known reserves	2045	2022
Total recoverable reserves	2190	2070
Coal		
Known reserves	2210	2055
Total recoverable reserves	2880	2140

SOURCES: American Petroleum Institute and U.S. Department of Energy.

Figure 11.5 Sample of oil shale rock and the shale oil extracted from it. (Courtesy of Chevron Shale Oil Company)

heated to produce a thick, dark brown liquid called *oil shale* (Figure 11.5). After treatments to remove impurities and increase its flow rate, shale oil can travel through pipelines to a refinery.

The world's largest known deposits of oil shale are in the United States, with at least 80 percent of those deposits on federal lands in Colorado, Utah, and Wyoming. Enough is probably there to supply the United States with crude oil for forty years at the current rate of use. Significant amounts of oil shale are also in Canada, China, and the Soviet Union.

But oil shale is not yet a significant energy resource. Three major factors have limited its use: (1) environmental damage to the land, water, and air; (2) a low net energy yield (see Table 11.2); and (3) high cost. Major oil companies have abandoned most of their pilot shale oil plant projects in the United States because they were too expensive—without large government subsidies.

Tar sands (or oil sands) are mixtures of clay, sand, water, and a black, high-sulfur, gooey oil. Tar sands are removed by surface mining and heated with steam at high pressure to liquefy the oil so that it floats to the top. Then the oil is purified, upgraded, and refined.

The world's largest known deposits of oil sands lie in a cold, desolate area along the Athabasca River in Alberta, Canada. Since 1985 two plants

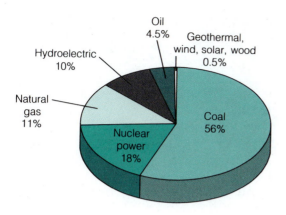

Oil
4.5%

Hydroelectric
10%

Geothermal,
wind, solar, wood
0.5%

Natural
gas
11%

Coal
56%

Nuclear
power
18%

Figure 11.6 Electricity generation in the United States in 1987. (Source: U.S. Department of Energy)

there have been supplying 12 percent of Canada's oil demand at a lower cost than for crude oil. The deposits could supply Canada's entire oil needs for about thirty years at current consumption rates.

But producing oil from tar sands has many problems. Deep deposits are too expensive to remove, and the net energy yield is low. In addition, mining tar sands produces an immense amount of solid waste, and processing the oil contaminates the nearby air and water.

INCREASED USE OF COAL   Coal is the most abundant fossil fuel in the world. In the United States, 70 percent of the coal is burned in power plants to generate electricity. In fact, coal is the largest source of electricity in the country (Figure 11.6). But three major problems limit our use of coal:

— Coal is expensive to move because it doesn't flow through pipelines.

— Burning coal pollutes the air (Chapter 12) and water (Chapter 13).

— Mining coal is dangerous and harms the land.

Underground mining is expensive and dangerous because of potential cave-ins and explosions. A single spark can ignite underground air laden with coal dust or methane gas. In addition, miners become disabled when particulate material accumulates in their lungs, a condition known as *black lung disease.*

Most coal now is mined by removing the soil and rocks above the deposits (Figure 11.7). Surface (or strip) mining removes at least 90 percent of the coal in a deposit, costs less per ton of coal removed (including the cost of restoring the land), and is less hazardous for miners. But without adequate reclamation efforts, surface mining can devastate the land and nearby water (Figure 11.8).

Figure 11.7 A giant shovel used for strip-mining coal. As long as a football field, it can dig out a 325-ton load of land every fifty-five seconds and drop it the equivalent of a city block away. (National Coal Association/Bueyres Erie Co.)

Figure 11.8 Effects of area strip mining of coal in Missouri. (National Archives)

BURNING COAL MORE CLEANLY AND EFFICIENTLY   A more efficient, cleaner, and cheaper way to burn coal is to use *fluidized-bed combustion* (FBC). A stream of hot air blown into a boiler from below suspends a mixture of sand, powdered coal, and crushed limestone (Figure 11.9). The upward flow of air churns and tumbles this powdered mixture to give it the consistency of thin oatmeal.

The powdered coal burns very efficiently while the limestone ($CaCO_3$) is converted into calcium oxide (CaO):

$$CaCO_3 \longrightarrow CaO + CO_2$$

CaO then reacts with sulfur dioxide released from sulfur impurities in the coal to produce dry, solid calcium sulfate ($CaSO_4$):

$$2SO_2 + 2CaO + O_2 \longrightarrow 2CaSO_4$$

This removes 90 to 98 percent of sulfur dioxide, a major pollutant in the burning of coal. Commercial FBC boilers are expected to begin replacing conventional coal boilers in the mid-1990s.

SYNFUELS FROM COAL GASIFICATION AND LIQUEFACTION   **Coal gasification** is the conversion of coal into natural gas, a more portable fuel. One method converts coal into coke and then heats the coke with steam to produce a gaseous mixture of carbon monoxide (CO) and hydrogen (H_2). The reaction is

$$H_2O \text{ (steam)} + C \text{ (coke)} \longrightarrow CO + H_2$$

Because of its low heat content, this gas isn't worth transporting by pipeline and is normally used only on site by the industries that make it.

Figure 11.9 Fluidized-bed combustion of coal.

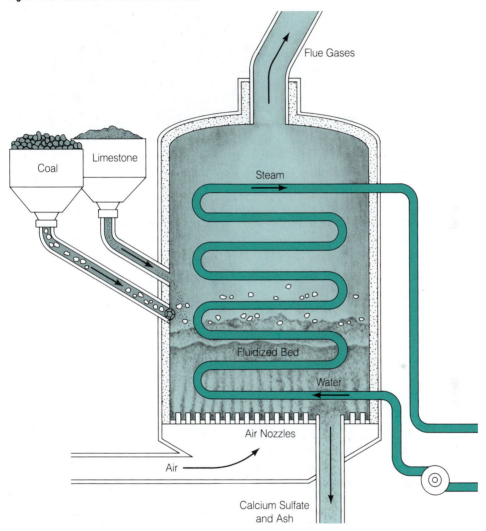

A more useful process is to react the products (CO and H_2) from the reaction above to make methane (CH_4):

$$C + 2H_2 \longrightarrow CH_4$$
$$CO + 3H_2 \xrightarrow{catalyst} CH_4 + H_2O$$

Figure 11.10 This government-supported coal gasification plant in Beulah, North Dakota, was closed in 1985 because synfuels cost two to three times as much as conventional oil. (U.S. Department of Energy)

This methane, called *synthetic natural gas* (SNG), is cleaner to burn, has a high heating value, and can be transported by pipeline.

Coal liquefaction is the conversion of coal into liquid fuels such as methanol or synthetic gasoline. For example, the products (CO and H_2) of the first reaction shown above for coal gasification can be combined in two steps to make methanol:

$$CO + H_2 \longrightarrow \underset{\text{formaldehyde}}{H-\overset{\displaystyle O}{C}-H}$$

$$\underset{}{H-\overset{\displaystyle O}{\underset{\|}{C}}-H} + H_2 \longrightarrow \underset{\text{methanol}}{CH_3OH}$$

A coal liquefaction plant supplies 10 percent of South Africa's liquid fuels at a cost similar to that for oil.

Making synthetic natural gas or oil (called *synfuels*) from coal is technically possible, but it is still too expensive for widespread use (Figure 11.10). In addition, useful energy is lost as heat during the conversion of coal to other fuels. Coal gasification, for example, reduces by 30 to 40 percent the usable energy available from our coal supplies.

11.4 Nuclear Energy: Here to Stay?

NUCLEAR FISSION REACTORS   Recall (Section 3.7) that large amounts of energy—mostly high-temperature heat—are released when the nuclei of certain atoms undergo *nuclear fission*. Inside *nuclear reactors*, a fissionable isotope (uranium-235 or plutonium-239) releases energy when a slow-moving neutron enters its nucleus and splits the nucleus into lighter fragments (see Figures 3.8 and 3.9), most of which are radioactive.

Figure 11.11 shows a light-water reactor, the most common type of nuclear reactor. Its core contains fuel and the devices that control the rate of fission. A heat-exchange system cools the reactor and transfers the heat to a system that produces steam. The steam then spins the blades of a

Figure 11.11 A nuclear power plant with a light-water reactor.

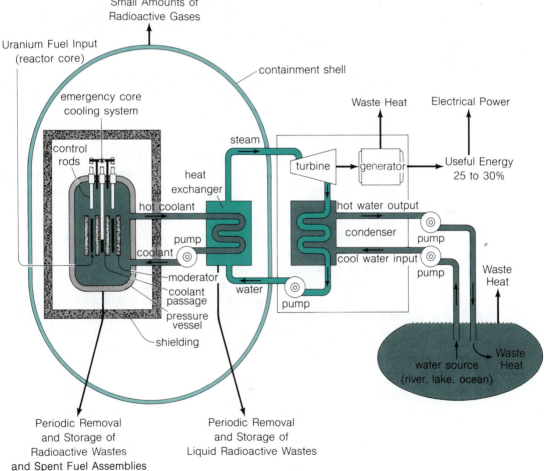

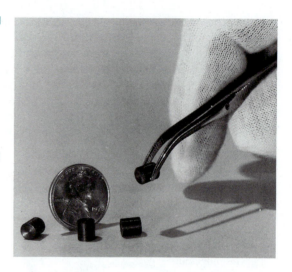

Figure 11.12 Fuel pellets of enriched uranium-235 oxide weigh less than 1 g (1/30 oz) but have the energy equivalent of 266 kg (585 lb) of coal. (Westinghouse Hanford Company)

turbine, which runs an electrical generator to produce electricity. The core and heat exchanger are surrounded by a containment system that protects workers from the reactor's intense radiation and prevents radiation from leaking into the environment.

The core typically consists of about 180 fuel assemblies, each containing about 200 long, thin **fuel rods** packed with eraser-sized pellets of uranium oxide (UO_2) (Figure 11.12). The fuel is about 3 percent fissionable uranium-235 and 97 percent nonfissionable uranium-238. Interspersed with the fuel assemblies are **control rods** made of materials that capture neutrons. Moving these rods in and out of the reactor regulates the fission rate and thus the amount of power produced. Inserted completely, control rods stop the fission altogether to remove spent fuel, to make repairs, or in case of an accident.

Water functions as a **moderator**, circulating between fuel rods and fuel assemblies to slow neutrons emitted by fission; this increases the chances of neutrons continuing the fission chain reaction. Water also cools fuel rods to prevent them and other core materials from melting, and it carries the heat out of the core and into the heat exchanger.

THE NUCLEAR FUEL CYCLE   The nuclear power plant is only one part of the **nuclear fuel cycle** (Figure 11.13). In evaluating the overall cost, net energy, and safety of nuclear power, we need to look at the entire cycle—not just the nuclear plant itself.

The fuel cycle begins with the mining and milling of uranium ore. The ore (U_3O_8) is crushed, ground, and chemically concentrated. Then the uranium oxide is converted into gaseous uranium hexafluoride (UF_6) and treated to increase the concentration of uranium-235 from 0.7 percent to

Figure 11.13 The nuclear fuel cycle.

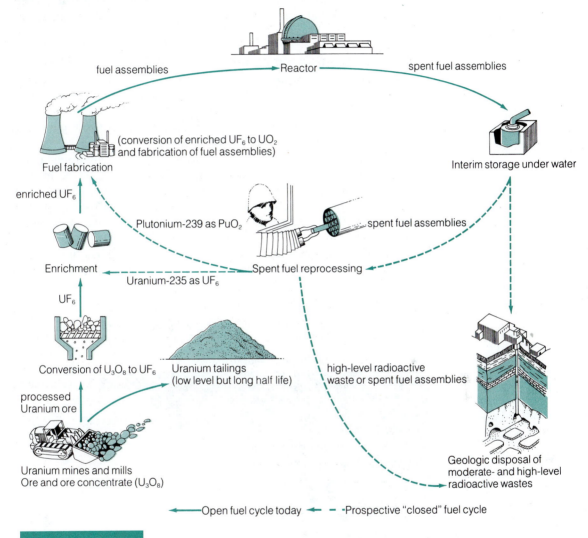

fuel assemblies → Reactor → spent fuel assemblies

(conversion of enriched UF_6 to UO_2 and fabrication of fuel assemblies)

Fuel fabrication

Interim storage under water

enriched UF_6

Plutonium-239 as PuO_2

spent fuel assemblies

Enrichment ← Spent fuel reprocessing

Uranium-235 as UF_6

UF_6

Conversion of U_3O_8 to UF_6

Uranium tailings (low level but long half life)

high-level radioactive waste or spent fuel assemblies

processed Uranium ore

Uranium mines and mills Ore and ore concentrate (U_3O_8)

Geologic disposal of moderate- and high-level radioactive wastes

◄──── Open fuel cycle today ◄─ ─ ─ Prospective "closed" fuel cycle

about 3 percent. Once converted into uranium dioxide (UO_2), the enriched (in uranium-235) material is made into pellets, loaded into fuel rods, and put into fuel assemblies, which are shipped to commercial power plants.

Each year about one-third of the fuel assemblies in the reactor core need to be replaced. Each spent fuel rod contains about 1 percent fissionable plutonium-239, a material that forms when neutrons bombard some of the uranium-238 in each rod. Spent fuel assemblies are stored in concrete-lined pools of water until they lose most of their radioactivity. Then they are removed and stored or sent to a fuel-reprocessing plant, where plutonium-239 and remaining uranium-235 are recovered and made into fuel for nuclear reactors.

NUCLEAR POWER: THE GREAT DEBATE   In the United States, nuclear reactors have generated nearly as much energy in the form of public debate as in the form of electricity. In 1987 417 reactors in thirty-three countries produced 17 percent of the world's electricity, 18 percent of U.S. electricity (Figure 11.14), and about 4 percent of the world's energy (see Figure 11.3). How much it will produce in the future is uncertain.

One issue is safety. A reactor cannot blow up like an atomic bomb because it doesn't have enough fissionable fuel in the right position to create a runaway chain reaction. But a reactor core could lose its cooling water through a break in one of the pipes; and if the backup cooling system also failed, the reactor core could overheat and melt through its containers.

In 1979 an improbable series of mechanical failures and human operator errors caused a partial meltdown and release of small amounts

Figure 11.14 Use of nuclear fission to produce electricity in various countries in 1987. (Source: Atomic Industrial Forum and International Atomic Energy Agency)

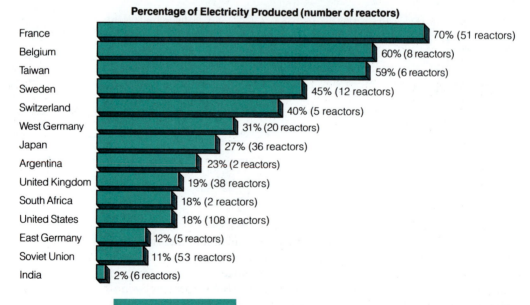

Percentage of Electricity Produced (number of reactors)

France	70% (51 reactors)
Belgium	60% (8 reactors)
Taiwan	59% (6 reactors)
Sweden	45% (12 reactors)
Switzerland	40% (5 reactors)
West Germany	31% (20 reactors)
Japan	27% (36 reactors)
Argentina	23% (2 reactors)
United Kingdom	19% (38 reactors)
South Africa	18% (2 reactors)
United States	18% (108 reactors)
East Germany	12% (5 reactors)
Soviet Union	11% (53 reactors)
India	2% (6 reactors)

of radiation from a reactor at the Three Mile Island nuclear plant in Pennsylvania. In 1986 a much larger nuclear accident occurred at Chernobyl in the Soviet Union, killing 31 people, hospitalizing 237, and causing the evacuation of 115,000 others. Graphite in the core caught fire, a partial meltdown occurred, and large amounts of radiation escaped into the air. As a result of such accidents, many people now have less confidence in the safety of nuclear power.

Another issue is storing radioactive wastes. Materials having short half-lives (Section 3.3) can safely be returned to the environment after a few days or weeks. But high-level wastes containing cesium-137 and strontium-90 must be stored for several hundred years (ten to twenty times their half-lives). And if reprocessing plants aren't developed in the United States, material containing plutonium-239 would need to be stored for thousands of years.

Where will it go? The most likely answer is to concentrate the waste, convert it into a dry solid, fuse it with glass or ceramic material, and seal it in a metal canister. Canisters would be buried permanently in underground salt, granite, or basalt formations that have been stable for millions of years and are expected to remain so. The United States hopes to have the first underground site (in Nevada) for high-level radioactive wastes operating before 2010. But the debate continues about whether disposal sites are safe enough.

Since 1958 the United States has been giving away and selling nuclear technology to other countries. But some of the nuclear information, components, and materials intended for power plants can be used instead to build nuclear weapons. Now it appears that several nations have done this. They join five other nations (United States, Soviet Union, Great Britain, France, and China) that have built and tested nuclear devices.

Cost is the primary reason, however, for the slow growth in nuclear energy production. Construction costs have soared. In addition, nuclear power plants in the United States have operated at only about 60 percent of their capacity because of breakdowns, lengthy maintenance operations, and the need to comply with more stringent federal safety standards since the Three Mile Island accident. In 1987 each kilowatt hour of electricity from a new coal plant with the latest air-pollution control equipment cost an average of 6¢; from a new nuclear plant, the cost was 13.5¢.

BREEDER REACTORS   Only 0.7 percent of the uranium in the earth is fissionable uranium-235; more than 99 percent is uranium-238. The world's supply of uranium-235 should last for at least a century at the present rate of use. But our supply of nuclear fuel could last 1000 years or longer if we could use nonfissionable uranium-238.

An experimental reactor, called a **breeder reactor**, does exactly that: It produces electricity by fission while making new fissionable fuel (plutonium-239) from nonfissionable and abundant uranium-238:

Step 1: $^{238}_{92}U$ $+ ^{1}_{0}n \longrightarrow$ $^{239}_{92}U$

nonfissionable   *unstable*

Step 2: $^{239}_{92}U \longrightarrow ^{\ \ 0}_{-1}e + ^{239}_{93}Np$

unstable

Step 3: $^{239}_{92}Np \longrightarrow ^{\ \ 0}_{-1}e + ^{239}_{94}Pu$

fissionable

A breeder reactor looks something like the reactor in Figure 11.11, except that its core contains a different fuel mixture and its two heat-exchange loops contain liquid sodium instead of water. A third loop, containing water, drives the turbine.

The future of breeder reactors is uncertain, however. A few breeder reactors have been in operation in Great Britain, the Soviet Union, West Germany, Japan, and France. But they have been plagued by technical problems and high costs.

NUCLEAR FUSION   Someday scientists hope to use *nuclear fusion* (Section 3.7) to produce electricity. At temperatures of 100,000,000°C or more, nuclei of lightweight hydrogen atoms can fuse into heavier helium nuclei, converting tiny amounts of their mass into huge amounts of energy.

The advantages of nuclear fusion are clear. Once initiated, nuclear fusion releases four times as much energy per gram of fuel as fission, and about 10 million times as much as burning fossil fuels. The fuel supply (hydrogen-2 and hydrogen-3 from water) is abundant. And fusion produces less radioactive waste than does fission.

Most research has focused on the deuterium–tritium fusion reaction because it has the lowest ignition temperature (about 100,000,000°C). In this reaction, a hydrogen-2 or deuterium (D) nucleus and hydrogen-3 or tritium (T) nucleus fuse to form a larger helium nucleus, a neutron, and energy (Figure 11.15):

$$^{3}_{1}H + ^{2}_{1}H \longrightarrow ^{4}_{2}He + ^{1}_{0}n + \textbf{energy}$$

Many technical problems have to be solved before fusion becomes an important source of net energy. The high temperature (ten times hotter than the sun's interior) creates a gaslike *plasma*, in which deuterium and tritium atoms are so energetic—so hot—that nuclei lose their electrons. Finding a way to squeeze the plasma together long enough and at a high enough density for the positively charged fuel nuclei to fuse is a difficult problem. No physical walls can confine the plasma, for any known material would vaporize at that temperature.

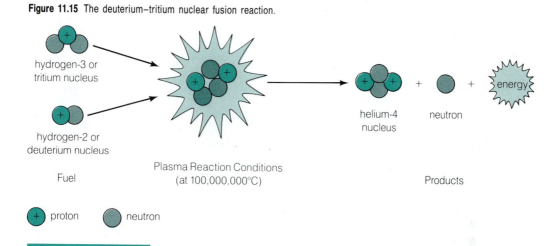

Figure 11.15 The deuterium–tritium nuclear fusion reaction.

So far, the most promising approach is magnetic confinement, in which powerful electromagnetic fields confine and force together the plasma nuclei within a vacuum—a sort of invisible "bottle." One idea is to squeeze the plasma into the shape of a large *toroid*, or doughnut (Figure 11.16). Such a reactor is known as *tokamak* (a shortened form of the Russian words for "toroidal magnetic chamber"); its design was pioneered by Soviet physicists.

In 1989 two scientists at the University of Utah announced another idea, called "cold fusion." They reported that fusion occurs at room temperature in an electrochemical process using deuterium-enriched water (D_2O) and a palladium (Pd) electrode. Efforts to duplicate their findings have had mixed results, however, and the idea now looks less promising.

Controlled, high-temperature nuclear fusion is still at the laboratory stage after forty years of research. Most energy experts don't expect nuclear fusion to become a significant source of net energy until 2050 or later.

11.5 Alternative Energy Resources: Working with Nature

DIRECT SOLAR ENERGY: SOME ADVANTAGES AND DIFFICULTIES Huge amounts of solar energy come to us free, and we can use it without seriously polluting the air and water. Unlike fossil fuels and nuclear energy, solar energy

Figure 11.16 The magnetic confinement approach to nuclear fusion in a *tokamak* reactor.

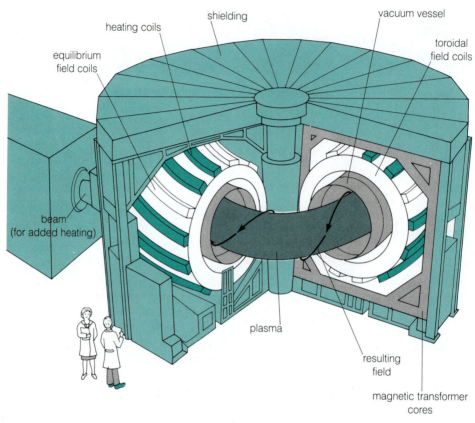

doesn't increase the amount of waste heat released to the environment; we simply use the energy more before the heat flows out to colder places in the universe.

But solar energy also has some disadvantages. It is more abundant in some locations than in others, and it isn't available at night. So we need a way to store the energy for nights and cloudy days. Furthermore, solar energy is widely dispersed (has high entropy), and concentrating it takes energy and money. This lowers the net energy gain.

One approach, called a *passive solar heating system*, relies on the building's design and composition to capture and store the sun's energy. Figure 11.17 shows some ways to do this. Today nearly 500,000 homes have passive solar designs.

Figure 11.17 Examples of passive solar design.

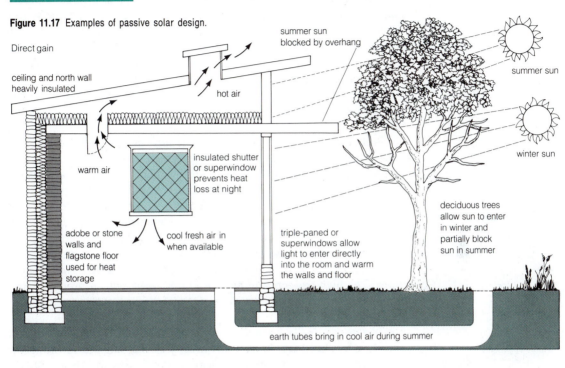

Direct gain

ceiling and north wall heavily insulated

hot air

summer sun blocked by overhang

summer sun

winter sun

insulated shutter or superwindow prevents heat loss at night

warm air

adobe or stone walls and flagstone floor used for heat storage

cool fresh air in when available

triple-paned or superwindows allow light to enter directly into the room and warm the walls and floor

deciduous trees allow sun to enter in winter and partially block sun in summer

earth tubes bring in cool air during summer

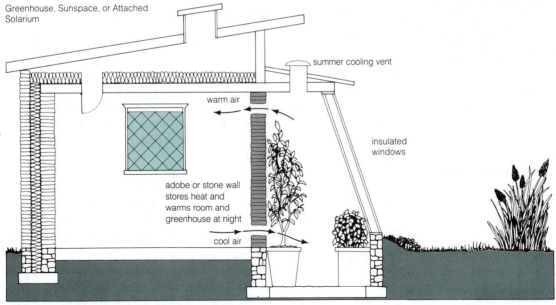

Greenhouse, Sunspace, or Attached Solarium

summer cooling vent

warm air

insulated windows

adobe or stone wall stores heat and warms room and greenhouse at night

cool air

Another approach, called an *active solar heating system*, is to mount solar collectors on the roof and angle them to capture the sun's rays. Collectors have a glass or other transparent lid and a dark bottom that absorbs solar radiation and converts it into heat. Then air or water (with antifreeze added in cold climates) is pumped through coils in the collector to transfer the trapped heat to the building's radiators and ducts. The heat also can be stored in a large bed of rocks or an insulated hot water tank for later use. High costs, however, have limited the use of active solar systems.

Still another idea is to concentrate the sun's energy with mirrors. This is like using a magnifying glass to focus enough sunlight on a small area to ignite paper or wood chips. A modern system for concentrating solar energy to produce heat or electricity is called a **power tower**. Several power towers have been built in the United States to produce electricity. Five are in the Mojave Desert near Barstow, California (Figure 11.18), and others are in the desert near Albuquerque, New Mexico, and near Los Angeles. At the Barstow plant, 2000 computer-controlled mirrors track the sun and focus its rays on a central boiler perched atop a twenty-story tower. The high-temperature heat produces steam, which produces electricity. But this electricity is expensive, and the net energy ratio is low (see Table 11.2).

Photovoltaic cells, commonly called **solar cells**, convert solar energy directly into electricity. Solar cells already power satellites orbiting the earth and provide electricity for at least 15,000 homes worldwide, mostly in isolated areas where running electrical lines to individual dwellings is too expensive.

A photovoltaic cell uses semiconductors (Section 10.3) that conduct electricity, but only slightly. Each cell is made of two layers of semiconductor material containing highly purified silicon (with added impurities), separated by junctions. Solar energy striking the silicon atoms knocks electrons free, producing a small, direct electrical current across the two layers. Because the electricity produced by a single cell is very small, many cells are wired together in a solar panel. Several panels are wired together and mounted on a roof facing the sun or on a rack that tracks the sun (Figure 11.19).

Solar cells are expensive, but the price is dropping as scientists develop cells that operate more efficiently and are cheaper to produce. If prices continue to fall as projected, solar cells may become cost competitive by the year 2000, and they could be generating 20 to 30 percent of the world's electricity by 2050.

Figure 11.18 Power tower used to generate electricity in the Mojave Desert near Barstow, California. (Sandia National Laboratories, Livermore, CA)

ENERGY FROM BIOMASS   Plants use solar energy to grow, and then we can use the plants for energy. **Biomass** is organic matter that can be burned

Figure 11.19 Use of photovoltaic (solar) cells to produce electricity.

Single Solar Cell

boron-doped silicon

junction

phosphorus-doped silicon

sunlight

cell

DC electricity

Panel of Solar Cells

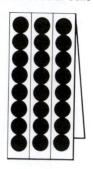

Array of Solar Cell Panels on a Roof

photovoltaic panels

power lines

panel wire

to breaker panel (inside house)

inverter (converts DC to AC)

battery bank (located in shed outside house, due to explosive nature of battery gases)

directly as fuel or converted into more convenient gaseous or liquid fuels, called *biofuels*.

Wood (or charcoal made from wood) is the primary energy source for cooking and heating for about 80 percent of the people in less developed countries. Indeed, biomass (mostly wood) provides about 15 percent of the world's energy (see Figure 11.3), but only about 4 percent in the United States (see Figure 11.2). Nevertheless, 10 percent of U.S. households is heated entirely with wood, and another 30 percent use wood for part of their heat.

Wood has a moderate to high net energy when it is collected and burned near its source. But small wood stoves pollute the air unless they are equipped with a catalytic combuster or other device that burns off most polluting gases.

Biomass also includes crop residues (the inedible, unharvested portions of food crops), animal manure, and organic urban wastes that can be collected and burned or converted into biofuels. Hawaii, for example, burns enough bagasse (the brownish, fibrous residue from sugarcane) to produce almost 10 percent of its electricity. But in most areas, plant residues require so much energy to collect, dry, and transport that they have a low net energy.

One solution is to establish large energy plantations where fast-growing, high-yield trees, grasses, or other crops could .be grown and harvested (Figure 11.20). This biomass could be burned directly, converted into biofuels, or made into other products. Another idea is to grow and harvest aquatic plants such as algae, water hyacinths, and kelp seaweed. But the net energy may still be a problem.

Plants, organic wastes, and other forms of solid biomass can be made into fuels such as methane, methanol, and ethanol. In the absence of air, bacteria convert the carbon in organic materials into methane (CH_4), which can be burned directly or made into methanol by the following reactions:

Figure 11.20 These fast-growing plants (*Euphorbia lathyris*) produce oil-like hydrocarbons. One acre of plants can produce about ten barrels of oil a year. (Gene Ekke Calvin)

$$CH_4 + H_2O \text{ (steam)} \longrightarrow CO + 3H_2$$

$$CO + 2H_2 \longrightarrow CH_3OH$$
$$\text{\textit{methanol}}$$

Yeasts metabolize carbohydrates to ethanol by a process called **fermentation**. A simplified reaction is

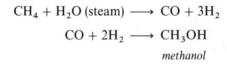

$$C_6H_{12}O_6 \longrightarrow 2CO_2 + 2CH_3CH_2OH$$
(*a carbohydrate*)   *ethanol*

Fermenting the sugars in sugarcane, sugar beets, sorghum, and corn produces a mixture of ethanol, water, and marketable CO_2 gas. This mixture is then distilled to produce ethanol.

Once we deplete our supplies of crude oil, we will probably need a liquid substitute for gasoline and diesel fuel. Some analysts see the biofuels methanol and ethanol as an answer. Two important advantages of these fuels are that they can be burned directly in automobile engines (1) without needing costly and toxic additives to boost octane ratings and (2) without causing as much air pollution as gasoline does.

Pure ethanol can be burned in today's cars with little engine modification. Gasoline can also be mixed with 10 to 23 percent (by volume) ethanol to make *gasohol*, a form of unleaded gasoline that burns in conventional engines. In 1988 ethanol accounted for 62 percent of the automotive fuel consumption in Brazil. In the United States, gasohol containing 10 percent ethanol accounts for 8 percent of all gasoline sales—25 to 35 percent in Illinois, Iowa, Kentucky, and Nebraska. Most of the ethanol in gasohol is made by fermenting corn. New, energy-efficient distilleries are lowering production costs, so gasohol may soon be able to compete, without tax breaks, with unleaded gasoline.

WIND **Wind energy** is produced by the unequal heating of the earth's surface and atmosphere by the sun. The earth's rotation helps give wind its characteristic flow pattern.

Today's wind machines range from simple water-pumping devices made of cloth and wood to large wind turbines with blade spans of up to 100 m (330 ft). With blades of wood, metal, or fiberglass, these modern wind turbines are stronger and lighter than older models (Figure 11.21).

Wind machines operate most efficiently in winds of 14 to 24 miles per hour. Thus, the most common sites are mountain passes and coastlines. One disadvantage is that large towers dotting the landscape in these areas are unsightly. Without proper design, they may be noisy and can interfere with local television reception. But wind machines have a favorable net energy yield, and they don't pollute the air or water.

Blessed with windy mountain passes and other favorable sites, California uses wind farms (Figure 11.22) to generate 70 percent of the world's wind-produced electricity. The California Energy Commission projects that the state will produce 8 percent of its electricity from wind power by 2000. Worldwide, however, wind energy will be a major energy factor only in those areas that have suitable winds.

GEOTHERMAL ENERGY Radioactive elements deep within the earth decay and produce heat that slowly flows into buried rock formations. Under intense pressure from the molten interior of the earth, some of this **geothermal energy** escapes through hot springs, geysers, and volcanoes. Over

Figure 11.21 An old windmill and a modern wind turbine in Clayton, New Mexico. (U.S. Department of Energy)

Figure 11.22 Array of modern wind turbines in California. (Courtesy of H. S. Windpower, Inc.)

millions of years, it also forms essentially nonrenewable deposits of dry steam, wet steam (a mixture of steam and water droplets), and hot water. Drilling geothermal wells brings this steam and hot water to the surface.

Although geothermal energy is not a major energy source worldwide, about twenty countries use it to produce electricity and heat. The best deposits in the United States are in California and the Rocky Mountain states. The Geysers steam field, located about 145 km (90 miles) north of San Francisco, has been producing electricity since 1960 more cheaply than fossil fuel and nuclear plants can. It supplies 6 percent of northern California's electricity. The Geysers field contains dry-steam deposits, which are the preferred type but also the rarest. Only dry-steam wells can be tapped easily and economically at present.

Underground wet-steam deposits contain liquid water under such high pressure that its temperature is higher (180 to 370°C) than its normal boiling point (100°C). When this superheated water comes to the surface, 10 to 20 percent of the flow flashes into steam because of the decrease in pressure. The steam is separated and goes to a turbine to produce electricity (Figure 11.23). The remaining hot water, which is often salty, and the condensed steam are usually injected back into the earth. The largest wet-steam power plant in the world is in New Zealand.

Hot water is the most common form of usable geothermal energy. Such deposits heat almost all the homes and buildings in Reykjavik, Iceland, which has a population of about 85,000 people. Hot water deposits also heat homes and farm buildings and dry crops at 180 locations in the United States, mostly in the West.

Environmental effects vary widely from site to site. The underground salty water can pollute surface waters. Most wells also pollute the air by releasing small amounts of hydrogen sulfide (which smells like rotten eggs),

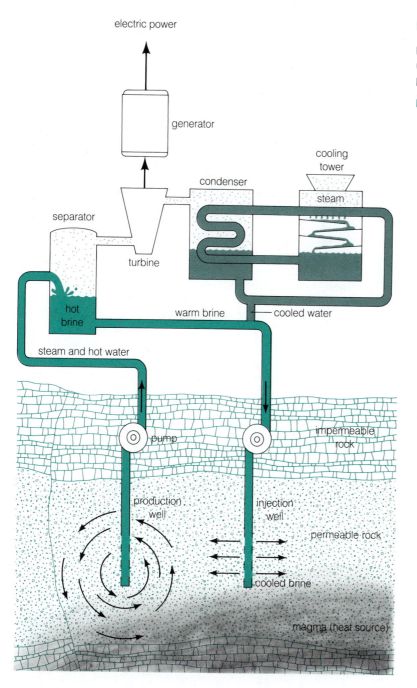

electric power

generator

separator

turbine

condenser

cooling tower

steam

hot brine

warm brine

cooled water

steam and hot water

pump

impermeable rock

production well

injection well

permeable rock

cooled brine

magma (heat source)

Figure 11.23 Extracting and using geothermal energy to produce electricity.

dirt particles, and radioactive radon-222 and radium-226 that escape in the steam. The ear-splitting hiss of escaping steam causes noise pollution, but the wells usually aren't near populated areas. Most experts consider the environmental effects to be less or no greater than those from fossil fuel plants.

If we developed our known geothermal reserves, they could supply about 5 percent of our electricity. Though it can make a significant contribution in local areas that have favorable deposits, geothermal energy will not be a major energy source worldwide in the next thirty years.

HYDROGEN GAS AS A FUEL   Some scientists have suggested that we use hydrogen gas (H_2) to fuel cars and heat homes and buildings when oil and natural gas run out. Hydrogen gas does not occur in significant quantities in nature. But a variety of chemical processes can be used to decompose water (Figure 11.24):

$$2H_2O + \textbf{energy} \longrightarrow 2H_2 + O_2$$

Then the H_2 fuel could be burned to propel automobiles or produce electricity and heat:

$$2H_2 + O_2 \longrightarrow 2H_2O + \textbf{energy}$$

Figure 11.24 The hydrogen energy cycle. Hydrogen gas must be produced by using electricity, heat, or perhaps solar energy to decompose water—thus reducing its net energy yield.

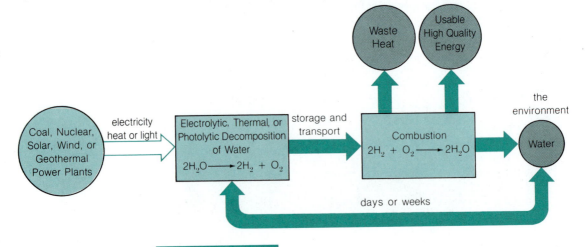

Although hydrogen gas is highly explosive, most analysts believe that we could learn how to handle it safely, as we have learned to manage gasoline and natural gas.

If you look at the two reactions above and remember the second law of thermodynamics, however, you will recognize a problem. Useful energy is lost in making water into hydrogen and then converting hydrogen back into water. In other words, the net yield of useful energy is *negative*. So using H_2 as a fuel only makes sense, in terms of energy, if it is tied to another source of energy that is abundant, cheap, and environmentally safe.

Chemists are trying to develop efficient cells that use ordinary light or sunlight to split water molecules into hydrogen and oxygen gases. If they succeed, affordable commercial cells may become available sometime after 2000.

CHEMISTRY SPOTLIGHT   WORKING WITH NATURE AND KEEPING IT SIMPLE

As one of the authors of this book (Miller), I am writing from an office that my wife, Peggy, and I designed to work with nature.

First, we purchased an old school bus and then sold the tires for the same price we paid for the bus. We built an insulated foundation, put heavy insulation around the entire bus, and added a wooden outside frame. A south-facing greenhouse or *passive solar collector* (Figure 11.17, bottom) with double-paned, conventional sliding glass windows (for ventilation) was then attached to the side of the bus. Direct solar energy provides at least 60 percent of our heating

needs during winter. A small wood stove provides backup heat as needed.

Four plastic pipes were buried about 5.5 m (18 ft) underground, extending down a gently sloping hillside away from the bus until their ends emerge some 31 m (100 ft) away. These earth tubes (Figure 11.17, top) were connected to a duct system containing a small fan. The fan slowly draws outside air through the buried tubes (which are surrounded by earth at 55°F), thus cooling the office to about 72°F. This natural air conditioning costs about $1 per summer to run the fan. Several large oak and other deciduous

trees in front of the windows provide shade during the summer and drop their leaves to let the sun in during winter.

Hot water is provided by solar collectors connected to a water stove. In winter, additional heating comes from a tankless, instant heater fueled by liquid-propane natural gas. We have reduced water use by installing water-saving faucets and by using a waterless, composting toilet (normally the biggest user of water in a typical household).

Working with nature is not only fun but also conserves resources and helps protect the environment.

TABLE 11.4 EVALUATION OF ENERGY ALTERNATIVES FOR THE UNITED STATES (Shading Indicates Favorable Conditions)

	Estimated Availability		Estimated Net Energy	Projected Cost of Entire System	Environmental Impact of Entire System
	Short Term (1990–2010)	Long Term (2010–2050)			
Nonrenewable Resources					
Fossil fuels					
Petroleum	High (with imports)	Low	High but decreasing	High for new domestic supplies	Moderate
Natural gas	High (with imports)	Moderate (with imports)	High but decreasing	High for new domestic supplies	Low
Coal	High	High	High but decreasing	Moderate but increasing	High
Oil shale	Low	Low to moderate	Low to moderate	High	High
Tar sands	Low	Poor to fair (imports only)	Low	High	Moderate to high
Biomass (urban wastes for incineration)	Low	Moderate	Low to fairly high	High	Moderate to high
Synthetic natural gas (SNG) from coal	Low	Low to moderate	Low to moderate	High	High (increases use of coal)
Synthetic oil and alcohols from coal and organic wastes	Low	High	Low to moderate	High	High (increases use of coal)
Nuclear energy					
Conventional fission	Low to moderate	Low to moderate	Low to moderate	High	High
Breeder fission	None	Moderate	Unknown, but probably moderate	High	High
Fusion	None	None to low (if developed)	Unknown	High	Unknown (probably moderate)
Geothermal energy (trapped pockets)	Poor	Poor	Low to moderate	Moderate to high	Moderate to high

Energy Resource	Estimated Availability		Estimated Net Energy	Projected Cost of Entire System	Environmental Impact of Entire System
	Short Term (1990–2010)	Long Term (2010–2050)			
Renewable Resources					
Improving energy efficiency	High	High	High	Low	Decreases impact of other sources
Water power (hydroelectricity)					
New large-scale dams and plants	Low	Very low	Moderate to high	Moderate to high	Low to moderate
Reopening abandoned small-scale plants	Moderate	Low	High	Moderate	Low
Tidal energy	None	Very low	Unknown (moderate?)	High	Low to moderate
Ocean thermal gradients	None	Low to moderate (if developed)	Unknown (probably low to moderate)	Probably high	Unknown (probably moderate)
Solar energy					
Low-temperature heat (for homes and water)	Moderate	High	Moderate to high	Moderate to high	Low
High-temperature heating	Low	Moderate to high	Moderate	Very high initially (but probably declining fairly rapidly)	Low to moderate
Photovoltaic production of electricity	Low to moderate	High	Moderate	High initially but declining fairly rapidly	Low
Wind energy					
Home and neighborhood turbines	Low	Moderate to high	Moderate	Moderate	Low
Large-scale power plants	None	Probably low	Low	High	Low to moderate?
Geothermal energy (low heat flow)	Very low	Low to moderate	Low	High	Moderate to high
Biomass (burning wood, crop, food, and animal wastes)	Moderate	Moderate to high	Moderate	Moderate	Moderate to high
Biofuels (alcohols and natural gas from plants and organic wastes)	Low to moderate	Moderate to high	Low to moderate	Moderate to high	Moderate to high
Hydrogen gas (from coal or water)	None	Moderate	Low	Variable	Variable

11.6 Our Energy Options: An Overview

In sorting out our energy options for the future, we need to ask

1. What are the realistic options?
2. What fraction of the demand can each potential energy source supply as net useful energy?
3. What financial and environmental costs are involved in each energy source?

Table 11.4 evaluates energy alternatives for the United States. Keep in mind that energy experts disagree about energy projections and estimates, and new data and innovations may change some information in this table. It does, however, provide a useful overview of our options.

From Table 11.4, we can see that our best option, in terms of net energy and environmental impact, is to increase energy efficiency. Phasing out low-gas-mileage cars, increasing mass transit, constructing energy-conserving buildings, and changing to less energy-intensive manufacturing and farming processes can save large amounts of energy. This in turn will lessen the environmental impact of other energy sources.

Natural gas, oil, hydroelectric, solar, and wind power offer good combinations of moderate to high net energy and low to moderate environmental impact. But we will probably run out of natural gas and oil in the next century (see Table 11.3), and coal has high environmental costs. Hydroelectric power provides only about 5 percent of our total energy and 12 percent of our electricity, and these percentages are likely to decline because we have already dammed most of the best sites. Solar energy will be expensive (at least initially) but will become increasingly important. Wind power will be used more in favorable locations, and geothermal energy will be useful in a few areas. The future of nuclear power is uncertain (Section 11.4).

One trend is reasonably clear. In the next century, we will rely less on oil, and perhaps less on natural gas, and shift to a more diverse combination of energy sources. Energy production will become more localized and variable, with each area taking better advantage of its own climatic conditions and energy resources. And as the cost of energy rises, using it efficiently will become even more important.

BRUCE HANNON   A country that runs on energy cannot afford to waste it.

Summary

The sun is the main source of energy that sustains life on earth. Renewable energy resources include solar, wind, flowing water, and biomass. Nonrenewable resources include fossil fuels, geothermal, and nuclear energy. Fossil fuels are our major energy source now, but natural gas and petroleum are expected to be depleted in the next century. Thus, we need to develop alternatives.

The first and second laws of thermodynamics limit what we can do in using energy and developing new sources. Energy quality is reduced when we do work. The percentage of energy that does work is the energy efficiency of a process. Net energy is the energy of a resource minus the energy needed to use the resource. Net energy is an important criterion in evaluating energy resources.

Table 11.4 summarizes the advantages and disadvantages of our current and potential future energy sources. Improving energy efficiency is an especially effective way to conserve our energy resources.

Terms for Review

After completing this chapter, you should know and understand the meaning of the following terms:

biomass (p. 320)

breeder reactor (p. 315)

coal gasification (p. 308)

coal liquefaction (p. 310)

control rod (p. 312)

energy efficiency (p. 302)

fermentation (p. 322)

fuel rod (p. 312)

geothermal energy (p. 323)

moderator (p. 312)

net energy (p. 304)

nuclear fuel cycle (p. 312)

oil shale (p. 305)

power tower (p. 320)

solar cell (p. 320)

tar sands (p. 306)

wind energy (p. 323)

Questions

Odd-numbered questions are answered at the back of this book.

1. Explain why most oil in the Mideast has a higher net energy yield than most oil drilled in the United States.

2. Explain net energy in terms of the first and second laws of thermodynamics.

3. Why does making hydrogen gas (H_2) from water, and then burning the H_2 as fuel, necessarily have a negative net energy?

4. What factors reduce the net energy for nuclear energy?

5. List some factors that reduce the net energy for burning wood in a wood stove for heat.

6. Compare the energy efficiency of a light bulb and that of a fluorescent light (see Figure 7.5).

7. Criticize the following statements:
   a. Hydroelectric power will become a more major energy source in the future.
   b. Switching from the internal combustion engine to electric cars can help solve our energy problem.
   c. Electricity is a more efficient way to heat homes than burning natural gas or oil.

8. Criticize the following statements:
   a. Electricity is clean heat.
   b. The world's supply of fossil fuels will be used up during the next century.
   c. The amount of nuclear energy we can produce is limited by the amount of uranium-235 in the earth.

9. What disadvantage do each of the following have because of the second law of thermodynamics? (a) Coal gasification and (b) coal liquefaction.

10. Why is the net energy lower for oil shale than for petroleum?

11. What is the function of each of the following in a nuclear reactor? (a) Fuel rods, (b) moderator, and (c) control rods.

12. List the major advantages and disadvantages of using nuclear fusion as a major energy source as compared with using nuclear fission.

13. The radioactive waste from nuclear fusion is expected to be less than for nuclear fission. Nevertheless, there will be some waste. Look at the nuclear reaction for fusion. Which product of the reaction do you think could produce radioactive wastes? How could it do this?

14. What nonfissionable substance can be used as fuel in a breeder reactor?

15. Distinguish between an active and a passive solar heating system.

16. Which has the most favorable net energy ratio? (a) Active solar, (b) power tower (direct solar), or (c) passive solar (see Table 11.2).

17. To be a suitable, sole source of heat for a home on cold days, which of the following would *not* also need an energy storage system? (a) Windmill, (b) electricity, (c) passive solar, (d) active solar, (e) geothermal, and (f) natural gas.

18. What is biomass? Compare the importance of biomass as an energy source in industrial and in less developed countries.

19. Gasohol is widely used in the midwestern United States. What is the major source of ethanol there?

20. What is the main limitation in the development of wind energy as a major energy source?

21. What is the main limitation in the development of geothermal energy as a major energy source?

22. What energy sources are likely to provide a larger percentage of energy in the United States in the future? Which sources are likely to provide a smaller percentage? Why?

Topics for Discussion

1. Suppose a nuclear power plant is proposed for your town or city. What are the major questions you want to ask?

2. What is the price for electricity where you live? How does this compare with the price in other areas of the United States? Why is your rate high (or low)? What are the main sources of your electricity?

3. What tax credits or other incentives would you favor to stimulate energy conservation or the development of new sources of energy? Explain. Which new energy sources do you favor the most? Why?

4. What is the current U.S. energy plan?

5. What would you be willing to give up to avoid using more hazardous or more polluting forms of energy production?

CHAPTER 12

Air Resources and Air Pollution: Can We Breathe Easy?

General Objectives

1. What are the major types and sources of air-polluting chemicals?

2. What is industrial smog? How can it be controlled?

3. What is photochemical smog? How can it be controlled?

4. What are the major effects of air pollution? What causes acid rain, the greenhouse effect, and depletion of the ozone layer? What can we do about it?

Take a deep breath. If the air you just took in was not polluted at least a little, you are among a small minority breathing clean air. You can breathe polluted air almost anywhere—in Denver, Dodge City, Dublin, Damascus, Dortmund, Dunedin, and Djakarta.

Air is one of our most important renewable resources. But it can be overloaded with chemicals to the point of causing harm. The air you just breathed contains gases, minute droplets of liquids, and tiny particles. Studies have detected more than 2000 different substances in urban air. **Air pollution** is air that has chemicals or heat in high enough concentrations to harm humans, other living things, or materials.

We can do much to make our air cleaner. Steps taken since 1970 have improved the air quality in most parts of the United States. This chapter examines how air is polluted and how we can make our air cleaner.

12.1 Major Air Pollutants: Types and Sources

OUR AIR RESOURCES: THE ATMOSPHERE   The air we live in is finite. About 95 percent of its mass is in a layer called the *troposphere*, which extends only 8 to 12 km (5 to 7 miles) above the earth's surface. If the earth were an apple, our air supply would be no thicker than the apple's skin.

Table 12.1 lists the normal composition of our atmosphere near sea level. Air contains mostly nitrogen (N_2) and oxygen (O_2), with small amounts of argon (Ar), carbon dioxide (CO_2), and trace amounts of other elements and compounds. When harmful amounts of any substance, including natural components, enter the atmosphere, air becomes polluted.

OUR POLLUTED AIR   Several tragic episodes of air pollution have occurred in this century. In 1911 at least 1150 Londoners died from the effects of   335

TABLE 12.1 COMPOSITION OF CLEAN, DRY AIR AT SEA LEVEL*	
Substance	Quantity (in ppm)
Nitrogen (N_2)	781,000
Oxygen (O_2)	209,000
Argon (Ar)	9,300
Carbon dioxide (CO_2)	350
Neon (Ne)	18
Helium (He)	5
Methane (CH_4)	2
Krypton (Kr)	1
Hydrogen (H_2)	0.5
Nitric oxide (NO)	0.4
Carbon monoxide (CO)	0.1
Ozone (O_3)	0.02

* Moist air can have as much as 4000 ppm of water (H_2O).

coal smoke. The author of a report on this disaster invented the word **smog** to describe the mixture of smoke and fog that hung over London. A similar incident in London killed 4000 people in 1952, and further disasters in 1956, 1957, and 1962 killed a total of about 2500 people. As a result, London has taken strong measures against air pollution and has much cleaner air today.

The first known U.S. air-pollution disaster was in 1948, when fog and fumes from steel mills and zinc smelters hung over Donora, Pennsylvania, for five days. Twenty people died and nearly half of the 14,000 residents became ill. In 1963 high concentrations of air pollutants in New York City killed about 300 people and injured thousands. Other episodes during the 1960s in New York, Los Angeles, and other large cities have led to much stronger pollution-control programs.

MAJOR POLLUTANTS AND SOURCES   Pollution can come from nature. Natural decay processes, winds, and volcanic eruptions put into the air huge amounts of particulate matter, carbon monoxide, carbon dioxide, methane (CH_4), and hydrogen sulfide (H_2S). In 1986, for example, the sudden release of perhaps several thousand tons of asphyxiating carbon dioxide

TABLE 12.2 MAJOR TYPES OF AIR POLLUTANTS

Class of Pollutants	Typical Members of the Class
Particulates (solid particles or liquid droplets suspended in air)	Smoke, soot, dust, asbestos, metallic particles (such as lead, beryllium, cadmium), oil, salt spray, sulfate salts
Carbon oxides	Carbon monoxide (CO), carbon dioxide (CO_2)
Sulfur oxides	Sulfur dioxide (SO_2), sulfur trioxide (SO_3)
Nitrogen oxides	Nitric oxide (NO), nitrogen dioxide (NO_2)
Hydrocarbons (compounds containing carbon and hydrogen)	Methane (CH_4), butane (C_4H_{10}), benzene (C_6H_6)
Photochemical oxidants	Ozone (O_3), PANs (a group of peroxyacyl nitrates), various aldehydes
Other inorganic gases and compounds	Hydrogen fluoride (HF), hydrogen sulfide (H_2S), ammonia (NH_3), sulfuric acid (H_2SO_4), nitric acid (HNO_3)
Other organic compounds	Chlorinated hydrocarbons (DDT), 3,4-benzopyrene, alcohols, organic acids
Heat	—

(CO_2) gas from Lake Nyos in Cameroon killed 1746 people. But natural sources are dispersed throughout the world, and except for occasional disasters, they rarely cause serious damage. In contrast, pollution from human activities tends to be concentrated in urban areas where it can harm the most people.

An *air pollutant* is any form of matter or energy in the air that does harm. Air pollutants include toxic substances. However, they also include beneficial substances (such as carbon dioxide and ozone, O_3) that may be present in the wrong concentration or in the wrong location.

Table 12.2 lists the major air pollutants from human activities. The first six classes account for nearly all air pollution in the United States in terms of mass. Carbon monoxide accounts for about half of the emissions (Figure 12.1), and its major source is transportation. Although sulfur oxides and particulate matter rank lower in terms of total emissions, these pollutants are more harmful than carbon monoxide. Their sources

Figure 12.1 Emissions of major outdoor air pollutants in the United States (left) and their major sources (right). (Data from Environmental Protection Agency)

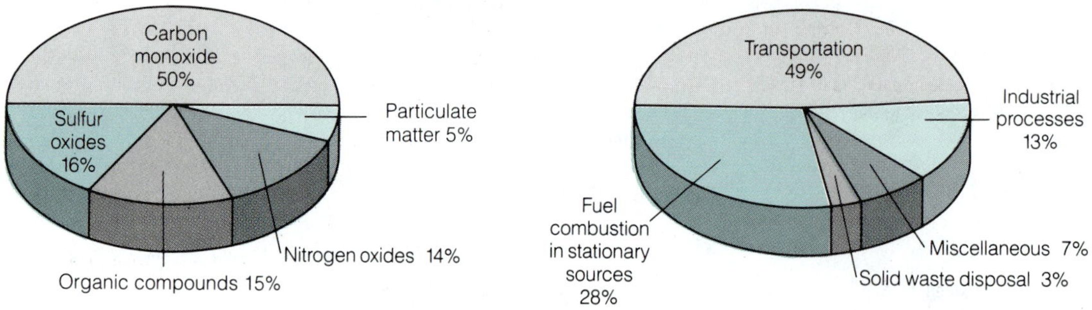

are stationary fuel combustion (especially coal-burning power plants) and industries such as pulp and paper mills, iron and steel mills, smelters, petroleum refineries, and chemical plants.

INDOOR AIR POLLUTION   To escape air pollution you might go home, close the doors and windows, and breathe what you believe is clean air. In fact, in a typical day you breathe mostly indoor air. But scientists have found that some indoor air is more dangerous than outdoor air.

Table 12.3 lists six of the most important indoor air pollutants, their sources, their effects on health, and ways you can reduce your exposure to them. These pollutants accumulate especially in mobile homes and in energy-efficient, airtight houses that don't bring in enough fresh air. Mobile homes also have more plywood and other materials that contain formaldehyde and other organic compounds that vaporize.

An average of 10 to 30 percent of U.S. homes may have unsafe levels of radon-222 gas. Radon accounts for 55 percent of the average total radiation exposure of people in the United States. In 1988 the U.S. Surgeon General's Office recommended that everyone living in a detached house, townhouse, mobile home, or first three floors of an apartment building test for radon-222. Testing kits cost $10 to $50.

Also notice in Table 12.3 that combustion gases are in indoor air. Kerosene heaters, gas stoves, and wood stoves that aren't properly vented release carbon monoxide, nitrogen oxides, and particulate material indoors. A study in China correlated lung cancer rates with high levels of cancer-causing organic compounds in the air of poorly ventilated buildings heated by burning coal.

A 1988 study by the Environmental Protection Agency (EPA) found that people in the United States are exposed to higher concentrations of

TABLE 12.3 MAJOR INDOOR AIR POLLUTANTS

Pollutant	Description	Health Effects	Sources in Homes	To Reduce Exposure
Radioactive radon-222 and its decay products polonium-218, lead-214, and bismuth-214	Odorless, colorless, radioactive gases, which occur naturally in the earth's crust	May cause lung and nasal cancers	Earth and rock beneath home; stone, brick, sand, and concrete block used for construction	Increase ventilation: open windows and crawlspace vents; add crawlspace vents; install air-to-air heat exchanger; seal floors and their openings
Formaldehyde	Strong-smelling, colorless, water-soluble gas	Nose, throat, and eye irritation; possibly nasal cancer	Various materials, including urea-formaldehyde foam insulation, particle board, plywood, furniture, drapes, and carpet	Don't use formal-dehyde materials, or use materials that are relatively low in formaldehyde
Asbestos	Fireproof, strong but crumbly mineral fiber	Skin irritation; lung and abdominal cancer and asbestosis (lung disease)	Some wall, ceiling, pipe, and boiler insulation; heat shields; vinyl floor material; patching material; and texture paint	Don't use asbestos materials; don't breathe asbestos fibers when crumbly asbestos materials are disturbed (such as during remodeling)
Combustion Gases				
Carbon monoxide	Colorless, odorless, tasteless gas	Lung ailments; impaired vision and brain functioning; fatal in very high concentrations	Unvented kerosene heaters, wood stoves; unvented gas stoves; attached garages	Install air-to-air heat exchanger; increase ventilation (see radon-222); be sure stoves are properly vented; don't leave car idling in garage; add catalytic oxidizers to wood stoves
Nitric oxide	Colorless, tasteless gas	Lung damage; lung disease after long exposure	Kerosene heaters; unvented gas stoves	Keep gas appliances properly adjusted; increase ventilation (see radon-222); install air-to-air heat exchanger
Combustion particles	Tiny smoke particles; polycyclic hydrocarbons such as benzopyrene	Lung cancer, emphysema, heart disease, irritation, respiratory infections	Tobacco smoke; wood smoke; unvented gas appliances; kerosene heaters	Avoid smoking tobacco indoors; vent combustion appliances outdoors; change air filter regularly

toxic organic chemicals in indoor air than in outdoor air, even in heavily industrialized areas. These substances come from consumer products and building materials such as paint, cleansers, cosmetics, adhesives, resins, and insulation. The most significant hazard was benzene (Section 8.2), a cancer-causing agent that enters indoor air mostly when people smoke cigarettes.

HEAT: THE ULTIMATE POLLUTANT   According to the second law of thermodynamics (Section 7.2), whenever energy does work, some of the energy changes into heat; the heat then flows into the atmosphere on its way back into space. So when you take a breath, move your arm, turn on a light, drive a car, or use an air conditioner, you add heat to the atmosphere.

Each person heats the atmosphere at least as much as a 100-watt light bulb would. The average rate per person is about the same as five 100-watt bulbs. Because they use so much energy, each person in the United States injects an average heat load equivalent to a hundred 100-watt bulbs.

All this heat makes a difference. Anyone who lives or works in a city knows that it is typically warmer there than in nearby suburbs or rural areas. Concrete, brick, and asphalt structures absorb heat during the day and release it slowly at night. Tall, closely spaced buildings slow the wind and reduce the rate of heat loss. As a result, a dome of heat hovers over a city, creating what is called an *urban heat island* (Figure 12.2).

Figure 12.2 The urban heat island.

Figure 12.3 In a thermal inversion (right), a warm layer of air traps pollutants in cool air that cannot rise to carry away the pollutants.

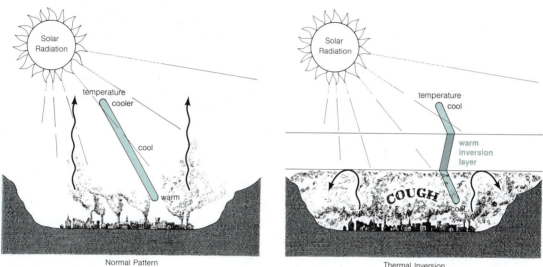

Normal Pattern

Thermal Inversion

HEAT AND AIR POLLUTION   The intensity of air pollution depends in part on the local climate and topography. In rainy areas, for example, water helps rinse pollutants from the air. Wind also helps disperse pollutants, but people living in valleys and basins (such as Los Angeles) don't enjoy this benefit.

Air temperature affects how the air moves up and down. As the sun warms the earth, the escaping heat also warms the air nearest the earth's surface. Normally, this warm air expands and rises, carrying low-lying pollutants up and away from the ground. Cooler air from above then sifts down into the low-pressure area created when the hot air rises (Figure 12.3, left). This continual up-and-down mixing helps keep pollutants from reaching dangerous levels.

But sometimes (often at night) a layer of dense, cool air moves into an urban area or valley and becomes trapped under the light, warm air it displaces. This condition is called a **thermal inversion** (Figure 12.3, right, and Figure 12.4). In effect, a warm lid covers the region. When pollutants cannot escape, they slowly accumulate to harmful levels.

Inversions usually last for only a few hours. But they can last for days, especially in the fall and winter when it takes longer for air near the earth to get warm enough to rise through the inversion layer. At the same time, more pollutants enter the air as people burn fuels for heat. It is not

Figure 12.4 Two faces of New York City. The almost clear view (left) was photographed on a Saturday afternoon. The effect of more cars in the city and a thermal inversion is shown in the photograph (right) taken the previous day. (*New York Daily News*/ Environmental Protection Agency)

surprising then that most air-pollution disasters—such as those in London and in Donora, Pennsylvania—occurred during lengthy thermal inversions during the fall or winter.

12.2 Industrial Smog: Gray Cities

TYPES OF SMOG   Table 12.4 shows how the most serious types of air pollutants can be classified as either industrial smog or photochemical smog. Most urban areas suffer from both types, but often one type predominates.

Sulfur dioxide and particulate matter are the main ingredients of **industrial smog**. Particulate matter makes the air gray over cities that depend heavily on coal and oil for heating, manufacturing, and producing electricity. Cities such as London, Chicago, and Pittsburgh used to be gray, but they have taken effective steps to reduce pollution. Industrial smog remains a way of life, however, in China and eastern European countries such as Poland and Czechoslovakia.

FORMATION OF INDUSTRIAL SMOG   Human activities provide about one-third of the air's sulfur compounds and nearly all of its sulfur dioxide. About two-thirds of the sulfur dioxide comes from burning coal and oil to produce electricity. The rest comes mostly from industries such as metal smelters (Section 10.4) and petroleum refineries (Section 8.3).

Coal and oil contain small amounts (0.5 to 5 percent) of sulfur as impurities. When these fuels burn, their sulfur reacts with oxygen to pro-

TABLE 12.4 BASIC TYPES OF SMOG

Characteristic	Industrial Smog (Section 12.2)	Photochemical Smog (Section 12.3)
Climate	Cool, humid air	Warm, dry air and sunny climate
Chief pollutants	Sulfur oxides, particulate matter	Carbon monoxide, nitrogen oxides, aldehydes, PANs, ozone
Main sources	Burning oil and coal	Burning gasoline
Time of worst episodes	Winter months (especially in the early morning)	Summer months (especially around noon)

duce sulfur dioxide:

$$S + O_2 \longrightarrow SO_2$$

A pollutant such as SO_2 that enters the air directly in that form (as SO_2) is called a **primary pollutant**.

Within a few days, 5 percent or more of the sulfur dioxide reacts to form sulfur trioxide (SO_3):

$$2SO_2 + O_2 \longrightarrow 2SO_3 \text{ (moderate rate)}$$

The SO_3 reacts almost immediately with water vapor to produce tiny droplets of sulfuric acid (H_2SO_4) in the air:

$$SO_3 + H_2O \longrightarrow H_2SO_4 \text{ (rapid rate)}$$

Pollutants such as SO_3 and H_2SO_4, which are produced by a reaction of a primary pollutant (SO_2) with other air components, are called **secondary pollutants**.

The mist of sulfuric acid can irritate lungs, corrode metals, and eat away building materials (Section 12.4). Sulfuric acid droplets also react with ammonia (NH_3) in the air to form solid particles of another secondary pollutant, ammonium sulfate ($(NH_4)_2SO_4$):

$$H_2SO_4 + 2NH_3 \longrightarrow (NH_4)_2SO_4$$

These particles fall to earth directly or dissolve in rainwater. They provide plants with essential nutrients such as nitrogen. But they also bind sulfuric acid droplets in the air, and the combination particles, breathed into the lungs, may do more harm than either pollutant would acting alone.

Industrial smog contains many other particles of various sizes. Large particles, as in dust and volcanic emissions, tend to fall out of the atmosphere quickly and aren't very harmful to humans. Coal dust and fly ash from burning are medium-sized particles (Figure 12.5); they stay in the air longer but can be removed by methods discussed in the next section.

Fine particles, which come from sea salt, oil, tobacco smoke, and other sources, are more hazardous. They stay in the air long enough to travel all over the world, and they are small enough to penetrate into your lungs. They also bring some unwanted visitors: toxic pollutants that have adsorbed onto their surfaces.

Figure 12.5 Medium-sized visible particulates from smokestacks such as this one in the Saar industrial region in West Germany. The particles can be removed effectively, as shown in Figure 12.7. (United Nations)

CONTROLLING INDUSTRIAL SMOG   The main ways to control sulfur oxide emissions are to burn less sulfur-containing fuels and to intercept sulfur oxides before they leave smokestacks. These options include the following:

— Use energy more efficiently and thus lessen the demand for fuel combustion. (For example, replacing all refrigerator/freezers in the United States with the most efficient units on the market would save the energy output of eighteen large coal-fired power plants.)

— Whenever possible, shift from coal to a mix of other energy sources such as nuclear, solar, wind, hydropower, or geothermal energy (Chapter 11).

— Burn low-sulfur coal, remove sulfur from fuels before burning them, or convert coal onto gaseous or liquid fuels that contain much less sulfur (Section 11.3).

— Use fluidized-bed combustion (see Figure 11.9) to burn coal more efficiently and cleanly.

— Remove sulfur oxides before they leave smokestacks.

Devices called **scrubbers** can remove sulfur oxides during burning. In one type of scrubber, stack gases pass through a chamber containing a fluidized bed of limestone ($CaCO_3$) that is heated to produce lime (CaO):

$$CaCo_3 \xrightarrow{heat} CaO + CO_2$$

The lime then reacts with sulfur dioxide, and with any sulfur trioxide formed in the stack from SO_2, to produce solid calcium sulfite ($CaSO_3$) or calcium sulfate ($CaSO_4$), which can be removed:

$$CaO + SO_2 \longrightarrow CaSO_3$$

$$CaO + SO_3 \longrightarrow CaSO_4$$

In the United States, all new coal-burning power plants are required to use scrubbers that remove at least 70 percent of the SO_2 from smokestacks. But this standard doesn't apply to plants built before 1972.

The first four approaches to reduce sulfur oxide pollutants also reduce emissions of particulate material. Figure 12.6 shows four additional ways to remove large and medium-sized particles: electrostatic precipitators, baghouse filters, cyclone separators, and wet scrubbers. Electrostatic

Figure 12.6 Four commonly used methods to remove particulates from exhaust gases.

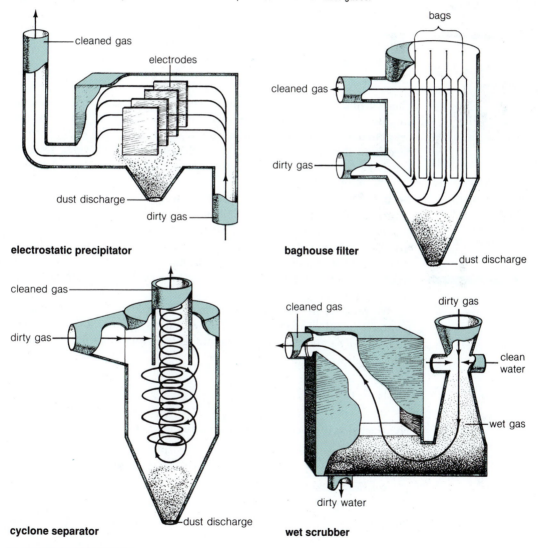

Figure 12.7 A stack with the electrostatic precipitator turned off (top) and with the precipitator operating (bottom). (Eastman Kodak Company)

precipitators, for example, create an electric field that charges the particles so they are attracted to electrodes and removed from the exhaust. If you wonder how well they work, look at Figure 12.7.

All four methods collect solid wastes that must be disposed. Except for cyclone separators, these methods are expensive and require energy. All remove large particles effectively, but only baghouse filters remove many of the hazardous fine particles. As a result, the total particulate emissions are decreasing more rapidly than the amount of fine particles.

12.3 Photochemical Smog: Cars + Sunlight = Tears

Photochemical smog contains a mixture of primary and secondary pollutants (see Table 12.4). It predominates in areas that have sunny, warm, dry climates and heavy automobile traffic. You can see the telltale brown haze from this smog hovering over such cities as Los Angeles, Denver, Salt Lake City (Figure 12.8), Sydney, Mexico City, and Buenos Aires.

THE CHEMISTRY OF PHOTOCHEMICAL SMOG　Nitrogen oxides come from both natural and human activities. Soil bacteria and lightning, for example, produce 10 times more nitrogen oxides than do humans. We cannot control these natural sources, which are widely dispersed and do little harm. But our activities add further emissions, mostly in populated areas. As a result, nitrogen oxide levels in urban air can be 10 to 100 times higher than in rural air.

We produce nitrogen oxides when we burn fossil fuels at high temperatures in automobile engines or power plants. Under these conditions, the normally unreactive N_2 in the air reacts with O_2 to produce nitric oxide (NO):

$$N_2 + O_2 \longrightarrow 2NO$$

NO then slowly reacts with oxygen in the air to form NO_2:

$$2NO \ + \ O_2 \ \longrightarrow \ 2NO_2 \quad \text{(\textit{slow to moderate rate})}$$
$$\text{\textit{colorless} \quad \textit{colorless} \quad \textit{yellow-brown}}$$

Nitrogen dioxide (NO_2) has a pungent, choking odor. You can see its presence in the characteristic brownish haze that hangs over cities such as Salt Lake City (see Figure 12.8).

Just as SO_3 can react with water to form sulfuric acid, small amounts of nitrogen dioxide react with water vapor to form nitric acid (HNO_3):

$$3NO_2 + H_2O \longrightarrow 2HNO_3 + NO$$

Some nitric acid washes out of the atmosphere as acid deposition (Figure 6.3 and Section 12.4). Nitric acid also reacts with ammonia in the air to form particles of ammonium nitrate, NH_4NO_3, that eventually fall directly to earth or wash out in the rain:

$$HNO_3 + NH_3 \longrightarrow NH_4NO_3$$

Most problems with photochemical smog come from NO_2 and other secondary pollutants that form in the presence of sunlight (Figure 12.9).

Figure 12.8 A heavy band of photochemical smog over Salt Lake City, Utah. (U.S. Department of the Interior, Bureau of Land Reclamation)

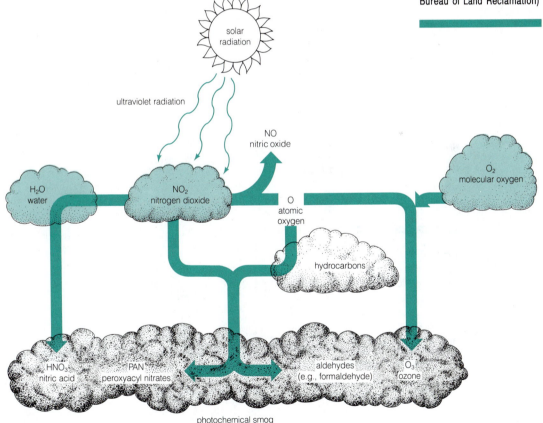

Figure 12.9 Simplified scheme of the formation of photochemical smog.

During a morning rush hour, for example, NO from automobiles builds up to high levels and reacts with oxygen gas (O_2) to form NO_2. Then as the sun rises, its ultraviolet (UV) rays convert the NO_2 back into NO and an atom of oxygen:

$$NO_2 \xrightarrow{\text{UV radiation}} NO + \quad O$$

$$\textit{oxygen atom}$$

Oxygen atoms are very reactive because they need two additional valence electrons to become stable. Some of these atoms react with O_2 in the air to produce ozone (O_3), with peak levels occurring around 10 AM (Figure 12.10):

$$O + O_2 \longrightarrow O_3$$

Ozone is very beneficial in air high above the earth (Section 12.5), but near ground level it does considerable harm. Ozone irritates eyes, harms lungs, and attacks materials such as rubber. In fact, the concentration of ozone at ground level is a common measure of the severity of photochemical smog.

Hydrocarbons, which enter the air mostly from spilled gasoline or from automobile exhaust, react with ozone and nitrogen oxides to produce pollutants called peroxyacyl nitrates (**PANs**):

$$
\begin{array}{c}
\quad\; O \\
\quad\; \| \\
R - C - O - O - NO_2
\end{array}
$$

$$(\textit{a PAN})$$

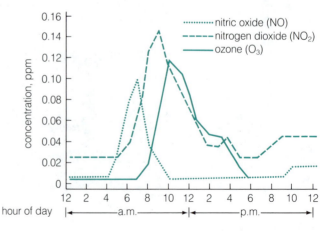

Figure 12.10 Atmospheric concentrations of nitric oxide, nitrogen dioxide, and ozone with time of day in Los Angeles. (Source: National Air Pollution Control Administration)

Hydrocarbons also react with oxygen atoms or with ozone to produce aldehydes such as formaldehyde (Section 8.6). Together, this gaseous soup of secondary pollutants constitutes photochemical smog (see Figure 12.9). Mere traces of ozone, PANs, and aldehydes in the air can seriously damage crops and make your eyes water and burn.

CONTROLLING AIR POLLUTION FROM AUTOMOBILES Photochemical smog occurs mostly in sunny areas where there is considerable automobile traffic. Controlling the emissions from 200 million mobile sources in the United States is difficult, but several things can be done.

One idea is to burn fuel more completely by using a leaner fuel mixture (less fuel, more air). This would put less CO and hydrocarbons into the air, but it would cause some loss in power and would increase the nitrogen oxide emissions. Burning a rich mixture (with less air) causes the

---

CHEMISTRY SPOTLIGHT FROM PINEAPPLES TO PHOTOCHEMICAL SMOG TO CLEANER AIR

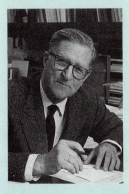

Figure 12.11 Arie J. Haagen-Smit. (Courtesy of California Institute of Technology)

Air pollution first became a serious problem in Los Angeles in the early 1940s. By 1947 strict laws banned all outdoor burning, and electrostatic precipitators (see Figure 12.7) were required for all incinerators. Although particulate emissions were sharply reduced, the air still had a yellow-brown haze and irritated people's eyes. Officials and scientists thought that sulfur dioxide was the culprit.

In the early 1950s, Arie J. Haagen-Smit (Figure 12.11), a chemist at the California Institute of Technology, accidentally solved the mystery. He was trying to isolate the main ingredient in the aroma of pineapples. When he collected the gaseous products given off by pineapple, he smelled ozone (O_3). Noticing that the smog was bad that day, he suspected that the ozone came from the air instead of from the pineapple. So he collected a sample of air, and he again smelled ozone. Analyzing the air sample, he found that it also contained hydrocarbons, aldehydes, and other organic compounds.

Haagen-Smit spent the next year doing research. He established that nitrogen oxides (released primarily by car engines) interacted with hydrocarbons, other air components, and sunlight to produce the irritating mixture of chemicals we now call photochemical smog (see Figure 12.9).

Realizing that air-pollution control was more than a scientific problem, Haagen-Smit became active politically. As a member of the state Motor Vehicle Pollution Control Board and later as head of the Air Resources Board, he worked to see that tough air pollution–control laws were enacted and enforced in California. He also contributed significantly to the first national air-quality laws that were passed in 1970.

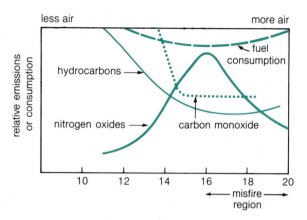

Figure 12.12 Increasing the air-to-fuel ratio (that is, reacting more air with a given amount of fuel) in the internal combustion engine changes engine performance and the pollutants emitted.

reverse situation—low nitrogen oxide levels and high CO and hydrocarbon emissions. Figure 12.12 shows how we trade one form of pollution for another when we change the air-to-fuel ratio.

A more promising answer is a *catalytic converter*. This device mixes engine exhaust with outside air in a chamber containing catalysts to oxidize carbon monoxide and unburned hydrocarbons:

$$2CO + O_2 \longrightarrow 2CO_2$$

$$\text{hydrocarbons} + O_2 \longrightarrow CO_2 + H_2O$$

These catalytic converters reduce the emissions of CO and hydrocarbons but have little effect on nitric oxide. Scientists are working to develop catalysts that reduce nitric oxide (NO) to nitrogen gas (N_2). But they haven't yet found a satisfactory combination of catalysts that reduces NO while also oxidizing hydrocarbons and CO. Lead ruins any of these catalysts, so cars with catalytic converters must use unleaded gasoline.

Another answer is to modify the internal combustion engine. Engines that run at lower temperatures, for example, emit less NO but more CO and hydrocarbons. Engines also could be modified to burn fuels that pollute less. Studies show that burning oxygen-containing fuels such as methanol, ethanol, and a particular ether (called methyl *tert*-butyl ether) would significantly reduce the emissions of carbon monoxide and ozone-forming hydrocarbons.

LEAD POLLUTION: SOLVING A PROBLEM Lead levels have been increasing throughout the world, even in remote areas. Greenland ice, which has layers preserved from as long ago as 800 BC, reveals the history of lead accumulation in the environment (Figure 12.13). Notice the rapid rise between 1940–1967, when cars began burning leaded gasoline.

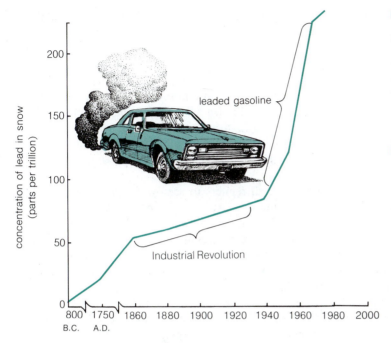

concentration of lead in snow
(parts per trillion)

leaded gasoline

Industrial Revolution

The main sources of lead are paint chips (from paint made before 1976), air, and water that circulates through leaded pipes. Most of the tiny particles of lead in the air come from burning leaded gasoline in cars; industrial smelters are another source. As you might suspect, urban air has more lead than rural air, and urban dwellers have more lead in their blood than people living in rural areas.

Although the effects of low lead levels aren't fully known, high concentrations cause nerve, brain, and kidney damage. The adults at risk include workers in lead smelters and battery plants and people who do extensive soldering. Young children, however, are the main victims because they cannot excrete the lead they inhale fast enough to keep it from reaching dangerous levels in their bodies. In addition, lead harms children while their brains are still developing; much of this damage is permanent.

A major source of lead in the air is **tetraethyl lead**, $Pb(C_2H_5)_4$, an additive that raises the octane rating of gasoline:

$$CH_3CH_2-\underset{\underset{\displaystyle CH_2CH_3}{|}}{\overset{\overset{\displaystyle CH_2CH_3}{|}}{Pb}}-CH_2CH_3$$

tetraethyl lead (TEL)

In an internal combustion engine, lead reacts with oxygen in the air to form lead(IV) oxide:

$$Pb + O_2 \longrightarrow PbO_2$$

Lead(IV) oxide is a solid that fouls an automobile's spark plugs, valves, pistons, and exhaust system. So leaded gasoline has other additives (such as 1,2-dichlorethane and 1,2-dibromoethane) that convert PbO_2 into $PbBr_2$, $PbCl_2$, and $PbBrCl$, which vaporize and pass out of the exhaust system. This is good for the engine but not for people who breathe the air.

In 1985 the EPA ordered refiners to remove 90 percent of the lead in gasoline by the end of the year. Automobile manufacturers redesigned engines so that cars could run on unleaded gasoline. The petroleum industry modified its refining processes to provide a larger proportion of aromatic hydrocarbons (which have higher octane values) to replace tetraethyl lead in fuel mixtures. The additional refining raises the price of gasoline, and some of the aromatic compounds used are suspected or known to cause cancer.

But the changes have made a significant difference. The EPA reported that the concentration of lead in the air decreased by 19 percent in 1987 and had dropped 88 percent in the past ten years (since 1978). Now the risk of serious lead poisoning comes mostly from drinking water that flows through leaded plumbing or from (children) eating chips of old, leaded paint.

12.4 Effects on Materials and Living Things

Because they have large concentrations of cars and factories, cities normally have more air pollution than do rural areas. But air pollutants respect no boundaries. Winds spread pollutants to the countryside and to other downwind urban areas. For example, SO_2 from tall smokestacks in Great Britain and Europe changes into sulfuric acid as it travels to Sweden and Norway, falling out and harming trees and aquatic life in those countries.

DAMAGE TO PROPERTY AND PLANTS   Air pollutants damage many common materials. Soot and grit soil statues, buildings, cars, and clothing, costing hundreds of millions of dollars each year for cleaning and maintenance. Sulfuric acid, sulfur dioxide, nitrogen oxides, nitric acid, and some particulates corrode metals such as steel, iron, and zinc. Sulfuric acid and ozone (O_3) attack and fade rubber, leather, paper, paint, and fabrics such as cotton, rayon, and nylon.

Acids attack statues and building materials. Granite and sandstone, which are mostly silicon compounds, are quite resistant to sulfuric acid and nitric acid in the air. Much more vulnerable, however, are limestone, marble, mortar, and slate, which are made of calcium carbonate ($CaCO_3$). They react with acids such as sulfuric acid:

$$CaCO_3 + H_2SO_4 \longrightarrow CaSO_4 + CO_2 + H_2O$$

The CO_2 escapes into the air, leaving behind a crumbly mass of calcium sulfate ($CaSO_4$) where a hard marble surface used to be. Calcium sulfate then dissolves in water and eventually washes away, leaving the building or statue pitted. Some of the world's finest historical and artistic monuments are deteriorating. The Parthenon in Athens, India's Taj Mahal, and many other stone and marble art treasures show visible signs of damage due to air pollution.

Plants are another target. Sulfur oxides, nitrogen oxides, ozone, and PANs damage the leaves of crop plants and trees. Crop losses in the midwestern United States amount to about $5 billion per year. In the Sudbury area of Ontario, Canada, sulfur dioxide from nickel smelters essentially destroyed a forest area extending 8 km (5 miles) downwind and damaged trees and plants as far as 30 km (19 miles) away. Sulfur oxide emissions from a copper smelter at Ducktown, Tennessee, did similar damage (Figure 12.14). Ozone, or a combination of ozone and sulfuric acid, has damaged trees in the Appalachian Mountains from Georgia to New England and has harmed more than one-third of all the trees in West Germany.

DAMAGE TO HUMAN HEALTH   Air pollution harms humans and other animals in a number of ways (Figure 12.15). The most vulnerable people are the very young, the old, the poor (who often live in highly polluted areas), and those already weakened by heart and lung diseases.

Figure 12.14 Fumes from a copper smelter at Ducktown, Tennessee, killed the forest that once flourished on this land. (Tennessee Valley Authority)

Figure 12.15 Possible effects of air pollution on the human body.

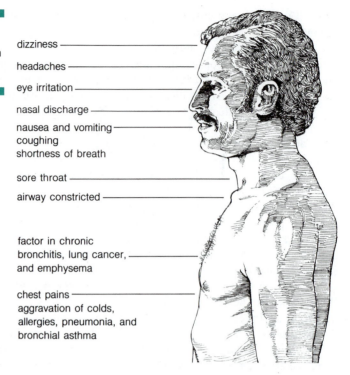

dizziness

headaches

eye irritation

nasal discharge

nausea and vomiting
coughing
shortness of breath

sore throat

airway constricted

factor in chronic
bronchitis, lung cancer,
and emphysema

chest pains
aggravation of colds,
allergies, pneumonia, and
bronchial asthma

One villain is ozone (O_3) in photochemical smog. Inhaling ozone causes coughing, shortness of breath, and nose and throat irritation and aggravates chronic diseases such as asthma, bronchitis, and emphysema. Excessive smoking and high levels of sulfur dioxide and sulfuric acid can lead to *chronic bronchitis*, a persistent inflammation of the lungs.

Smoking and air pollution also cause *emphysema* (Figure 12.16, center), a condition in which a person cannot expel most of the air from the

Figure 12.16 Magnified cells in the lung tissue of a normal person (left) and the lungs of people having emphysema (center) or lung cancer (right). (Massachusetts Audubon Society)

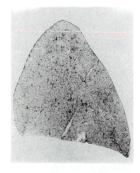

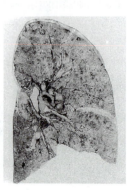

lungs. The lungs enlarge and become less efficient. Breathing gets harder, walking becomes painful, and running is impossible. Eventually the victim may die of suffocation or heart failure.

Some air pollutants cause lung cancer. Smoking is the number one cause, but lung cancer also has been linked to other air pollutants including asbestos, radioactive plutonium-239, polycyclic aromatic hydrocarbons (such as 3,4-benzopyrene), automobile exhaust, and particulate metal substances such as beryllium, chromium, and nickel.

Another culprit is carbon monoxide (CO). This colorless, odorless gas binds to hemoglobin in our blood (Chapter 21) 200 times more strongly than does oxygen. When CO ties up 5 to 20 percent of our hemoglobin, we don't get enough oxygen to our cells. Drivers, for example, can experience headaches, fatigue, and impaired judgment as they breathe higher levels of carbon monoxide during rush-hour traffic. Smoking adds even more carbon monoxide to their lungs and blood.

ACID DEPOSITION Droplets of sulfuric acid and nitric acid dissolve in rain and snow and fall out of the atmosphere. Particles of sulfate and nitrate salts also dissolve in rainwater or simply drop out of the air as particles. All this fallout is known as **acid deposition**, commonly called **acid rain**.

Regions rich in materials such as limestone, $CaCO_3$, and dolomite, $CaMg(CO_3)_2$, can resist acidification because their minerals neutralize acids. Two reactions are

$$2HNO_3 + CaCO_3 \longrightarrow Ca(NO_3)_2 + H_2O + CO_2$$

$$2H_2SO_4 + CaMg(CO_3)_2 \longrightarrow CaSO_4 + MgSO_4 + 2H_2O + 2CO_2$$

Lakes and thin soils in some areas, however, are vulnerable to acid deposition because they lack these alkaline materials (Figure 12.17).

Besides damaging property and plants (as discussed above), acid deposition has the following harmful effects:

— Kills fish, aquatic plants, and microorganisms in lakes and streams.

— Causes and aggravates human respiratory diseases.

— Dissolves lead from water pipes into drinking water and dissolves mercury and other toxic metals stored in the sediments of lakes and rivers (Figure 12.18).

According to the National Academy of Sciences, damage from acid deposition in the United States costs at least $6 billion a year.

Because it dissolves CO_2 and produces carbonic acid (H_2CO_3), rainwater is normally acidic and has a pH value of about 5.6. But precipitation falling on much of eastern North America has a pH of 4.0 to 4.2.

Figure 12.17 Areas in the United States where lakes and streams are especially vulnerable to acid deposition because they have low concentrations of alkaline substances that can neutralize acids. (Data from Environmental Protection Agency)

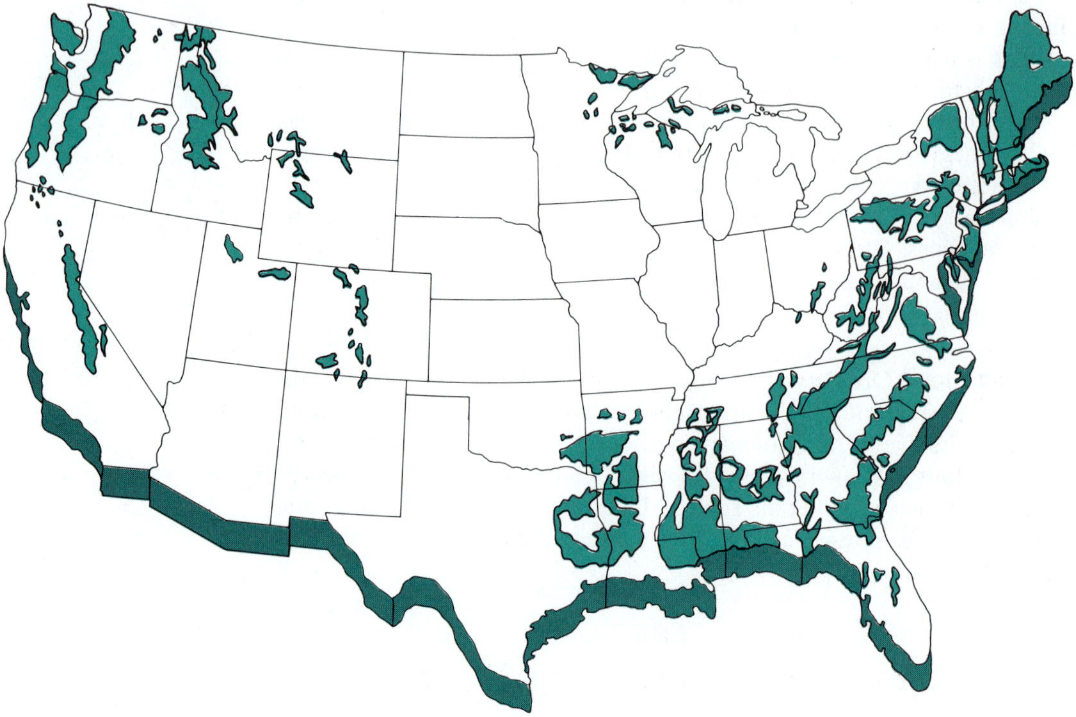

This is thirty to forty times more acidic than precipitation that fell there several decades ago. Much of the airborne acid comes from coal- and oil-burning power and industrial plants in Ohio, Indiana, Pennsylvania, and neighboring states.

Airborne acids often travel elsewhere. Most of Canada's acid deposition comes from the United States. Three-fourths of the acid deposition in Norway, Switzerland, Austria, Sweden, Finland, and the Netherlands blow in from other industrialized countries in Europe. In 1985 the Soviet Union and twenty-one European countries signed an agreement to reduce their sulfur oxide emissions. But the United States has declined to do so, citing scientific uncertainties about the harmful effects of such emissions.

In some areas, chemical treatments have restored soil or lakes that have become too acidic. Adding lime (CaO) or limestone ($CaCO_3$) neutralizes the acidity. For example,

$$2HNO_3 + CaO \longrightarrow Ca(NO_3)_2 + H_2O$$

These treatments, however, are expensive and add minerals such as $Ca(NO_3)_2$ to the water and soil.

12.5 Global Effects: Depleting Ozone and Warming the Earth

ARE WE DEPLETING THE OZONE LAYER? Ozone (O_3) is a villain at ground level, but it becomes a hero up in the stratosphere, 20 to 50 km (12 to 31 miles) above the earth. This **ozone layer** shields life on earth by absorbing nearly all of the sun's harmful UV radiation.

Ozone forms in the stratosphere when light from the sun splits oxygen molecules into oxygen atoms; these atoms then can react with oxygen (O_2) molecules to form ozone. This complex series of reactions can be summarized as

$$O_2 \xrightarrow{\text{UV radiation}} \underset{\text{oxygen atoms}}{2O}$$

$$O + O_2 \xrightleftharpoons{\text{UV radiation}} O_3$$

The second reaction goes in the reverse direction when ozone absorbs high-energy UV light.

As a result of these reactions, ozone constantly forms and breaks down. A dynamic equilibrium (Section 7.4) keeps a fairly steady concentration of ozone in the stratosphere through the overall reaction

$$3O_2 \rightleftharpoons 2O_3$$

Our protective ozone layer is diminishing. The most dramatic losses—up to 50 percent of the ozone—occur seasonally over Antarctica, especially in September to November. Smaller losses over the Arctic were discovered in 1988. That same year, National Aeronautics and Space Administration (NASA) scientists reported a drop of about 5 percent in the average worldwide ozone levels since 1979.

What is depleting our ozone layer? A major cause is chlorofluorocarbons (CFCs), also known as Freons (Section 8.5). CFCs are fairly nonreactive compounds used widely in spray cans, refrigerators, and air conditioners. Once in the air, many CCl_2F_2 (Freon-12) and other CFC molecules do not react until they encounter high-energy UV radiation in the stratosphere. This encounter generates products such as Cl atoms.

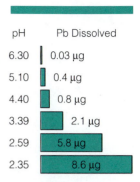

pH	Pb Dissolved
6.30	0.03 µg
5.10	0.4 µg
4.40	0.8 µg
3.39	2.1 µg
2.59	5.8 µg
2.35	8.6 µg

Figure 12.18 More acidic water (lower pH values, shown in color) dissolves greater amounts of the 31.7 µg of lead (Pb) in 1 g of Mississippi River sediments.

One example is

$$CCl_2F_2 \xrightarrow{\text{\textit{UV radiation}}} CClF_2 + \quad Cl$$

Freon-12   *chlorine atom*

Because it needs another valence electron to become stable, a chlorine atom is highly reactive and can catalyze the breakdown of ozone. One reaction is

$$Cl + O_3 \longrightarrow ClO + O_2$$

Figure 12.19 shows how the presence of ClO correlates with the reduction of O_3 in the "ozone hole" over Antarctica.

According to the National Academy of Sciences, each 1 percent loss of ozone would increase the amount of UV radiation reaching the earth by about 2 percent. More UV radiation would mean more skin cancer, more cataracts, more intense photochemical smog, damage to many animals, and decreased yields from crops such as corn, rice, and wheat.

An international treaty, signed by fifty-nine nations, went into effect in 1990. By the end of this century, participants must ban the use of CFCs. The United States and some other countries no longer allow CFCs in spray cans, but Freons are still widely used in refrigeration and air-

Figure 12.19 A NASA airplane flying toward Antarctica measured concentrations of ozone (O_3, color) and ClO on September 16, 1987. As the plane reached the ozone hole, O_3 levels decreased sharply as ClO levels increased.

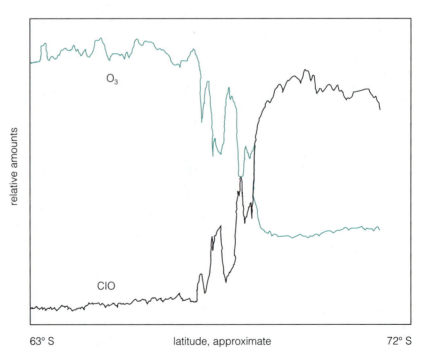

conditioning units (which keep CFCs sealed inside cooling coils). CFCs also are used as foaming agents in polyurethane and other materials. Now chemical companies are developing substitutes for CFCs in these products.

GLOBAL WARMING Solar radiation absorbed by the land and water is radiated back toward space as longer-wavelength infrared (IR) radiation, or heat energy (Figure 12.20). But gases in the atmosphere—especially CO_2 and water vapor—absorb some of the outgoing heat and radiate it back to earth.

The atmosphere acts like the glass in a greenhouse or a car window, which lets visible light enter and depart freely but keeps IR heat from escaping. Thus, the warming effect of the atmosphere is sometimes called the **greenhouse effect**. Without these IR-absorbing molecules in the air, the earth would be an average of about 30°C (54°F) colder than it is now.

We can do little to change the amount of water in the atmosphere, but we are changing the amount of other gases that help warm the earth. For example, CO_2 levels have risen 10 percent since 1958, and methane levels are increasing about 1 percent per year (Figure 12.21). Carbon dioxide (CO_2) accounts for about 50 percent of the greenhouse effect, methane (CH_4) and CFCs for 15 to 20 percent each, and nitrous oxide (N_2O) for about 6 percent.

Carbon dioxide goes into the air when we burn fossil fuels. Green plants and trees remove some CO_2 by photosynthesis, but we are cutting down and burning plants faster than they regrow. The increases of methane come from several sources: anaerobic bacteria producing CH_4 from plant material in rice paddies, wetlands, and the stomachs of cattle and sheep; coal mines; and natural gas leaks from pipelines and during oil production.

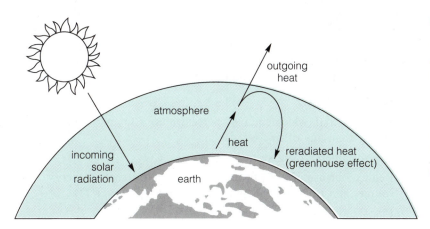

Figure 12.20 The greenhouse effect. Incoming solar radiation is absorbed by the earth and radiated back into the air as heat. Atmospheric gases absorb outgoing heat and radiate some of it back to earth.

Figure 12.21 Concentrations of carbon dioxide (left) and methane (right) in the air are increasing, thus increasing the greenhouse effect.

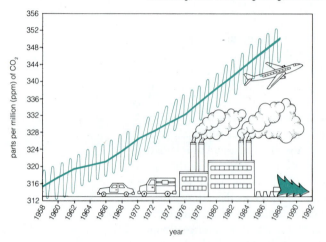

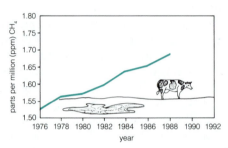

Scientists disagree about what effect continued increases will have on the earth's climate. Many forecast an average increase of 1.5 to 4°C (3 to 7°F) in the next 50 to 100 years. The warming is projected to be greater in the polar regions, where it might melt polar ice and raise the level of the oceans several feet. Temperature changes of even a few degrees are significant. The most recent ice age, 18,000 years ago, occurred at an average global temperature only 5°C colder than it is now.

Is the earth actually getting warmer? The average temperature has risen only 0.5°C since 1860 (Figure 12.22). Several periods of ups and downs have occurred since then, but now the trend is upward. The 1980s were the warmest decade during that period and included six of the ten individual warmest years, including the warmest (1988). In addition, a study in 1989 showed that sea levels are rising slightly—about 2.5 mm a year.

Most climate experts agree that global warming will continue for at least several decades because of the gases already in our atmosphere. Major efforts to slow the warming could include:

— Stop using CFCs.

— Burn less coal (which emits 60 percent more CO_2 per unit of energy than any other fossil fuel).

— Use scrubbers in smokestacks to remove CO_2.

— Reduce our use of fossil fuels by developing alternative energy sources and by using energy more efficiently (Chapter 11).

— Plant and retain more trees.

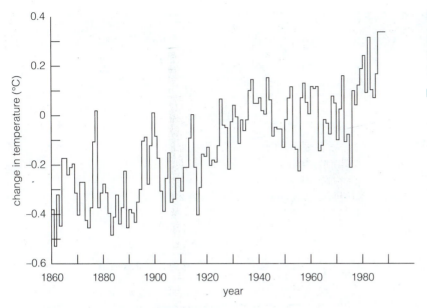

Figure 12.22 Average global temperature changes since 1860.

12.6 A Summary of Air Pollution and Its Control

Table 12.5 summarizes information about air pollutants and our prospects for controlling their emissions. The technology for short-range control of many pollutants is available, and it is being implemented with moderate to high costs.

The air in most cities is improving. Although the air quality in rural areas may be slipping by small degrees, it still remains much cleaner than city air and well below the maximum allowable pollution levels.

Two of the more difficult pollutants to control are nitrogen oxides and ozone. In 1988 one out of three people in the United States lived in cities that often exceeded the ozone levels considered safe. Data from the EPA, however, show that the quality of U.S. air is improving in terms of carbon monoxide, sulfur oxides, particulates, and (most dramatically) lead (Figure 12.23).

Two of the most serious unsolved problems are acid deposition and the greenhouse effect. In both cases, part of the problem is political because the emissions of one country may cause more harm elsewhere. As for acid deposition, the offending countries could pay more money to clean up their emissions. But much of the benefits would go to their neighbors. And a warmer global temperature due to the greenhouse effect could actually benefit a few countries, while harming others.

TABLE 12.5 SUMMARY OF AIR-POLLUTION PROBLEMS AND CONTROLFEASIBILITY

Pollutant	Residence Time	Area Affected	Technological Feasibility	Economic Cost
Particulates	Hours to days	Local and regional	Good	Moderate
Sulfur dioxide, sulfuric acid	4 to 8 days	Local and regional	Fair	High
Carbon monoxide	2 to 3 months	Local and regional	Fair	High
Nitrogen oxides, nitric acid	3 to 4 days	Local and regional	Poor	High
Photochemical oxidants	Hours to days	Local and regional	Fair	High
Freons	Many years	Regional and global	Good by banning Freons	Moderate
Carbon dioxide	2 to 4 years	Regional and global	Very poor	Very high
Heat	Variable	Local, regional, and global	Very poor	Very high

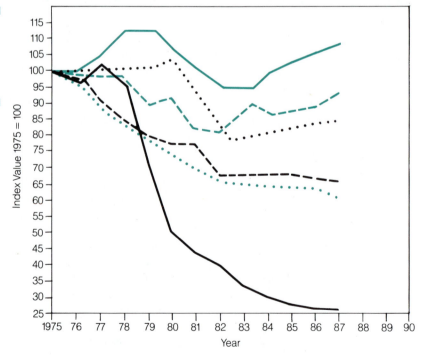

Figure 12.23 Trends in U.S. outdoor air quality for six pollutants, 1975–1987. (Data from EPA)

Suspended particulate matter ·················

Nitrogen oxides ——————

Lead ———————

Sulfur dioxide – – – –

Ozone – – – – – –

Carbon monoxide ············

Nevertheless, it is clear that chemistry and technology are important tools to improve our air quality. The experience of the past two decades shows us that much remains to be done; it also shows us that much can be done.

ERIK P. ECKHOLM   Air pollution can no longer be addressed as simply a local urban problem.

Summary

Air pollution is air that contains chemicals or heat in a high enough concentration to do harm. Indoor air may be as harmful to breathe as outdoor air. Major types of air pollutants by mass are carbon monoxide, nitrogen oxides, sulfur oxides, particulates, and hydrocarbons and other organic compounds.

Primary pollutants in industrial smog are particulate matter and sulfur dioxide (SO_2). Secondary pollutants include sulfur trioxide SO_3; sulfuric acid H_2SO_4; and ammonium sulfate $(NH_4)_2SO_4$. The main sources are fossil fuels. Ways to reduce these emissions include electrostatic precipitators, scrubbers, and burning less fossil fuels.

Primary pollutants in photochemical smog are carbon monoxide (CO), nitric oxide (NO), and hydrocarbons. In the presence of sunlight, secondary pollutants such as PANs, ozone (O_3), and aldehydes form. Automobile emissions are a major source of photochemical smog.

Air pollutants cause respiratory and other damage in humans. Acid deposition harms plants, aquatic life, and building materials. Chlorofluorocarbons (CFCs) reduce the ozone layer in the stratosphere, increasing the damage from high-energy UV radiation; one effect is increased skin cancer. Increased amounts of CO_2, methane (CH_4), CFCs, and nitrous oxide (N_2O) in the air cause the greenhouse effect, which warms the earth.

Terms for Review

After completing this chapter, you should know and understand the meaning of the following terms:

acid deposition (p. 355)

acid rain (p. 355)

air pollution (p. 335)

greenhouse effect (p. 359)

industrial smog (p. 342)

ozone layer (p. 357)

PANs (p. 348)

photochemical smog (p. 346)

primary pollutant (p. 343)

scrubber (p. 344)

secondary pollutant (p. 343)

smog (p. 336)

tetraethyl lead (p. 351)

thermal inversion (p. 341)

Questions

Odd-numbered questions are answered at the back of this book.

1. Distinguish between the troposphere and the stratosphere in terms of distance from the earth's surface.

2. Which of the first eight substances listed in Table 12.1 are increasing in concentration in the atmosphere?

3. What is the most abundant air pollutant by mass?

4. What are the main sources of radon-222 in indoor air?

5. Why do mobile homes often have poorer air quality than conventional homes?

6. Explain the term *thermal inversion* in words and with a diagram. How frequently do thermal inversions occur in your area?

7. What are the major sources of SO_2 in industrial smog?

8. List the names and formulas of secondary pollutants formed from SO_2.

9. Write chemical equations for the formation of secondary pollutants formed from SO_2.

10. List three ways to reduce the emissions of particulate matter.

11. How do scrubbers work? What do they do?

12. Despite better technology to remove SO_2 from smokestacks in the next decade, what factors will work against a sharp decrease in sulfur oxide levels in the air?

13. Why do cities with photochemical smog typically have a brownish haze?

14. List three secondary pollutants that commonly form in photochemical smog.

15. Is ozone (O_3) a desirable or undesirable air component? Explain.

16. Catalytic converters are the *least* effective in removing from automobile exhaust (a) nitric oxide, (b) hydrocarbons, or (c) carbon monoxide.

17. To maintain engine performance at high altitudes, would you adjust your carburetor to provide a richer or leaner mixture? Explain. With this adjustment, which

exhaust pollutants would probably decrease and which would increase?

18. What are the advantages and disadvantages of removing tetraethyl lead from gasoline?

19. Which air pollutant harms humans primarily by binding to hemoglobin in the blood?

20. Write the names and formulas of the two most abundant acids in acid deposition.

21. Which regions of the United States are the least able to neutralize acid deposition?

22. What are the major sources of airborne chlorofluorocarbons?

23. What region of the world seasonally has the *least* protection from ozone in the stratosphere?

24. Which gases in the atmosphere are most responsible for the greenhouse effect?

25. Suppose the greenhouse effect causes the earth's temperature to rise by 1.7°C by the year 2040. How much of an increase would this be in Fahrenheit degrees?

26. List three ways to slow global warming by the greenhouse effect.

27. Larger amounts of certain pollutants reach the air from natural, rather than human, activities. Yet the portion caused by human activities often does more harm. Explain.

Topics for Discussion

1. A number of countries blame much of their acid deposition on air pollution coming from other countries. What action can and should such countries take to protect themselves?

2. Should older coal-burning power plants be required to install scrubbers? What are the advantages and disadvantages of doing this?

3. One of the major sources of indoor air pollution is cigarette smoking. What actions have been taken in your state to reduce this problem, and what further action do you favor? Why?

4. What is the air quality where you live? What are the main sources of air pollution? How often does air pollution exceed federal standards? How often does an inversion occur? What steps have been taken locally to improve air quality?

5. Simulate an air-pollution hearing at which automobile manufacturers request a three-year delay in meeting air-pollution standards set for the coming year. Assign three members of the class as members of a decision-making board and other members as the president of an automobile manufacturing company, two lawyers for that company, two government attorneys representing the EPA, a public health official, and two citizens (one opposing and one favoring the proposal). Then discuss the implications of the final ruling of the board.

CHAPTER 13

Water Resources and Water Pollution: Good to the Last Drop?

General Objectives

1. How is water recycled on earth? How is it used, and what is the water supply in various parts of the world?

2. What are the major physical properties of water, and how are these related to the chemistry of water?

3. What are the major types of water pollutants? What are their key sources and effects?

4. Why is the amount of oxygen (O_2) gas dissolved in water so important? How can this oxygen be depleted?

5. How do acids, salts, and toxic metals pollute water, and how can these problems be controlled?

6. What is groundwater, and what are the major ways it can be polluted?

7. What are the main ways to treat wastewater to improve its quality?

Water is our most abundant and important chemical. It makes up about two-thirds of your body weight and covers 71 percent of the earth's surface. Although you could go about seventy days without food, you could last only a few days without water. In fact, you need to take in about 2 liters (L) of water a day to stay alive.

Water in your body dissolves nutrients, transports them to other organs, and removes waste products. Plants use water to carry out photosynthesis, which supplies food for us and other living things. Water also helps maintain our climate, making the earth hospitable to life.

But in many parts of the world, people don't have enough water, or the water they have is unfit for certain uses. Droughts bring famine and starvation, while polluted water spreads disease, damages crops, and kills aquatic life. This chapter examines the crucial role of water in our lives and how we can improve the quality of our water.

13.1 Water Resources: Will There Be Enough?

WORLD WATER RESOURCES The world's water supply in all forms (vapor, liquid, and ice) is enormous. If we could distribute it equally, every person on earth would have about 280 billion L (74 billion gal).

But only about 0.003 percent of the water is readily available for our use (Figure 13.1). About 97 percent of the earth's water is in oceans and is too salty for drinking, growing crops, and most industrial purposes. Most fresh water, however, is unavailable because it lies too far under the earth's surface or is tied up in glaciers, polar ice caps, the atmosphere, or the soil. This leaves the equivalent of only $\frac{1}{2}$ teaspoon of usable fresh water for every 100 L (26 gal) of water on earth.

367

100 liters (26 gallons)

Total Water
100%

3 liters (0.8 gallon)

fresh water
3%

0.5 liter (0.5 quart)

Available
fresh water
0.5%

0.003 liter
(1/2 teaspoon)

Usable
fresh water
0.003%

Figure 13.1 Only a tiny fraction of the world's water supply is available as fresh water for human use.

Even this tiny fraction amounts to an average of 8.4 million L (2.2 million gal) for each person on earth. Furthermore, water continuously purifies itself in the natural **hydrologic cycle** (Figure 13.2)—as long as we don't pollute it faster than it is replenished.

The water you use comes from two sources: surface water and groundwater. *Surface water* flows in streams and rivers and is stored in natural lakes, wetlands, and reservoirs. Surface water entering rivers and freshwater lakes is called *runoff*. This water source is renewed fairly rapidly (two to three weeks) in areas with average precipitation.

Water also seeps into the ground. Most of it evaporates or is taken up by plants, but some slowly percolates deeper into the earth to renew underground reservoirs. This **groundwater** makes up about 95 percent of the world's freshwater supply.

DO WE HAVE A WATER SHORTAGE?   Although the average amount of water per person seems to be enough, many areas of the world, including many parts of the United States, have water shortages. Figure 13.3 shows how many major urban centers in the United States are located in areas that already have inadequate water or are projected to have water deficits by the end of the century. Worldwide, scarcity and droughts afflict at least eighty countries, mostly in Asia and Africa.

In much of the world, however, the primary hazard is drinking contaminated water. In 1983 the World Health Organization (WHO) estimated that 61 percent of the people living in rural areas of lesser developed countries and 26 percent of the people in urban areas did not have access to safe drinking water (Figure 13.4). WHO estimated that at least 5 million people die each year from cholera, dysentery, diarrhea, and other preventable waterborne diseases.

Water use varies considerably worldwide. Almost 80 percent of the fresh water people in the United States use is for agriculture (primarily irrigation) and to cool electric power plants (Figure 13.5); in a less industrialized country such as China, a higher percentage of water is used for agriculture alone. Most of the water in U.S. homes is used for bathing, washing hands, and flushing toilets (Figure 13.6).

13.2 Water: A Unique Molecule

PROPERTIES OF WATER   If you searched for substances that have extreme or unusual physical properties, you might expect to find a list of exotic chemicals. But, surprisingly, often your search would lead you to water—the most familiar chemical of all. Water is indispensable to life because of its unique properties.

Figure 13.2 Simplified diagram of the hydrologic cycle.

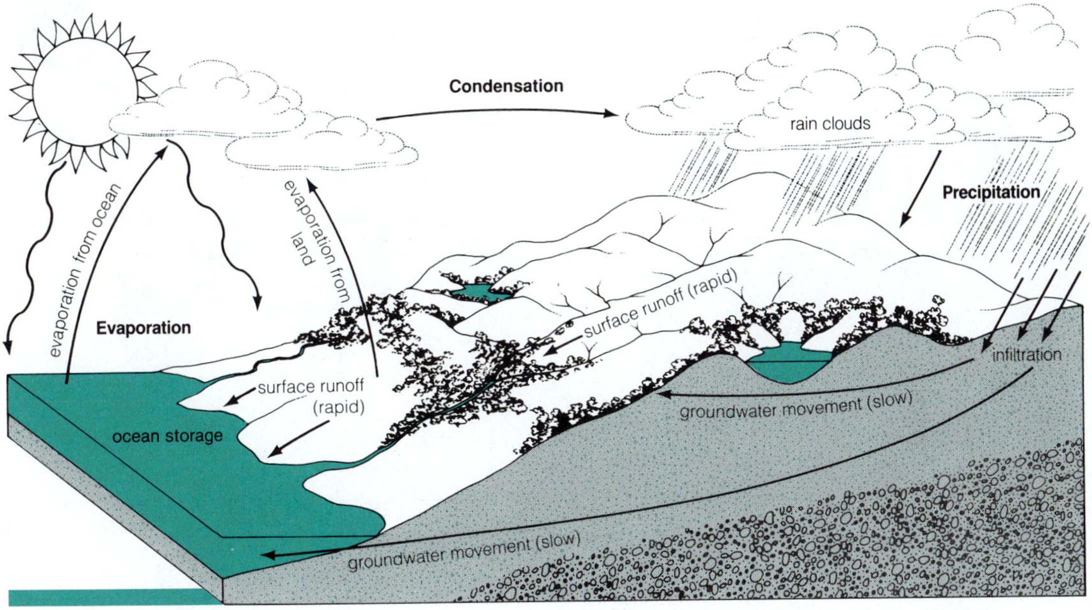

One such property is **heat capacity**, the measure of how much heat is required to raise the temperature of a given mass of a substance 1°C. Water has the highest heat capacity of any common liquid or solid. It takes more heat to warm 1 g of liquid water by 1°C than it takes to warm 1 g of other materials by the same amount (Table 13.1).

The high heat capacity of water prevents its temperature from changing rapidly. As a result, large bodies of water warm and cool slowly. This helps keep our climate moderate and protects living things from the shock of abrupt temperature changes. Because water absorbs heat so well, it also is used to remove heat from power plants and other industrial processes (see Figure 13.5).

Water also is an excellent heat conductor and has a very high **heat of vaporization**: the amount of energy it takes to vaporize a given quantity of a substance at its boiling point. It takes 540 cal to vaporize 1 g of water already at 100°C. In contrast, the heat of vaporization is only about 73 and 204 cal per gram for mercury and ethanol, respectively, at their normal boiling points. These properties further give water its remarkable ability to absorb and store heat.

Each day the sun's heat vaporizes about 12,000,000,000,000 (1.2×10^{13}) L of water. Because water has such a high heat of vaporization, the vapor contains a vast amount of energy. As it changes back into liquid

Figure 13.3 Present and projected water-deficit regions in the United States compared with present metropolitan regions with populations greater than 1 million. (Data from U.S. Water Resources Council and U.S. Geological Survey)

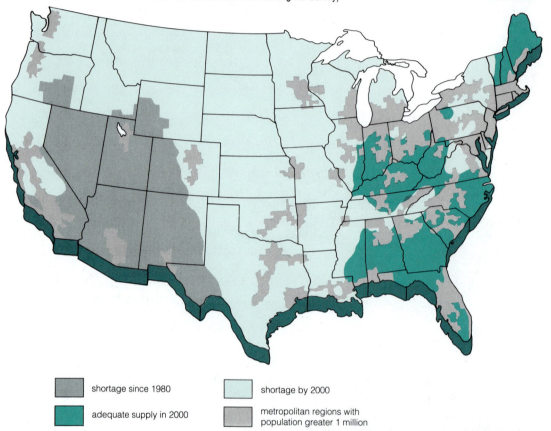

shortage since 1980

shortage by 2000

adequate supply in 2000

metropolitan regions with population greater 1 million

Figure 13.4 These children in Lima, Peru, are scooping up their drinking water from a puddle because the nearby public pump is inadequate. (United Nations)

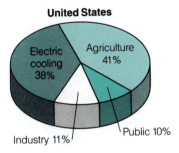

United States

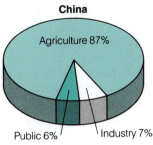

China

Figure 13.5 Use of water in the United States and China. (Data from Worldwatch Institute and World Resources Institute)

Figure 13.6 Domestic uses of water in the United States. (Data from U.S. Geological Survey)

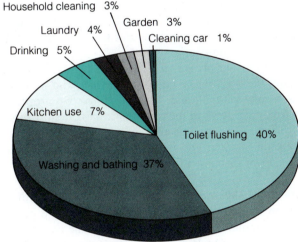

water, the vapor releases energy as heat over land and water. This warms coastal and other areas throughout the world.

Your body also takes advantage of water's high heat of vaporization. Each gram of water that vaporizes from your skin carries away more than 500 cal of heat. So by perspiring, you remove large amounts of heat while losing relatively little water. This keeps you from getting too hot or dry on warm days.

Another unusual property of water is its density (mass per unit of volume). Almost all substances are the most dense when they are solids. But water is different; its maximum density is at 4°C, at which temperature it is still a liquid. As it cools below 4°C, liquid water becomes less and less dense, finally freezing at 0°C. Thus, ice floats on water and in your cola.

Because of this remarkable property, water freezes from the top down instead of from the bottom up. Otherwise lakes and rivers would freeze

TABLE 13.1 HEAT CAPACITIES PER GRAM FOR SOME COMMON SUBSTANCES	
Substance	Calories/ Gram
Water	1.00
Ethyl alcohol	0.58
Ethyl ether	0.53
Olive oil	0.47
Aluminum	0.22
Carbon (graphite)	0.17
Iron	0.11
Copper	0.09
Mercury	0.03

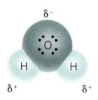

Figure 13.7 The water molecule is bent. The unequal distribution of electrons, or electrical charge, makes water molecules polar.

solid in the winter and kill most higher forms of aquatic life. But because water expands when it freezes, it can break pipes, crack engine blocks (which is why you use antifreeze), and fracture streets and rocks. The first time water freezes inside a living organism is the last time, for the water does irreparable damage to cells and other structures as it expands.

Water also has unusually high freezing and boiling points (Section 5.2). If it behaved as other molecules of a similar weight, water would be a gas instead of a liquid at normal temperatures. But if it were a gas, the earth would have no oceans, lakes, rivers, plants, or animals. All life depends on water being unusual.

Liquid water also is a superior solvent. Because it can dissolve a wide variety of polar and ionic materials, water carries nutrients throughout the tissues and organs of plants and animals, is a good cleanser, and removes and dilutes water-soluble wastes. But this also means it can carry many pollutants in solution.

Two other important properties are water's high **surface tension**, the force that causes the surface of a liquid to contract, and **wetting**, the capability to coat a solid with a liquid. Together, these two properties enable water to rise from tiny pores in the soil up into thin, hollow tubes such as the stems of plants. Certain molecules in soaps and detergents decrease the surface tension of water and increase its wetting ability, making it a better cleaning agent (Section 15.1).

MOLECULAR STRUCTURE AND HYDROGEN BONDING   Water's extraordinary properties come from the structure of its molecule. Water is a bent molecule (Figure 13.7), with polar covalent bonds between the oxygen and hydrogen atoms. Because oxygen is more electronegative than hydrogen, the oxygen part of the molecule carries a partial negative charge while the hydrogens have a partial positive charge. This makes water a polar covalent molecule (Section 4.4).

The negative end of one water molecule attracts the positive end of another molecule by hydrogen bonds (Section 5.1). This attraction is especially strong for water molecules because of their small size and the large partial charges caused by oxygen being so much more electronegative than hydrogen. These hydrogen bonds are like a glue that makes it harder for water molecules to move away from each other. So water is a liquid, not a gas, at room temperature.

Strong hydrogen bonds account for many of water's unique properties. These bonds make it harder for water molecules to move faster (or slower), so water has a high heat capacity and doesn't warm (or cool) as readily as most other materials. This is also why so much energy is needed for water molecules in the liquid state to break free from each other and vaporize. For example, gasoline (which is heavier, nonpolar, and has no hydrogen bonding) has an average heat of vaporization only one-seventh that of water.

As water cools, molecules move slower and pack closer together until they reach maximum density at 4°C. Most liquids continue to contract as they cool, finally, reaching their maximum density as a solid. But at temperatures below 4°C, water molecules begin to form rigid, open, hexagonal networks, as hydrogen bonds hold the molecules in place (Figure 13.8). The molecules are farther apart than at 4°C, so the ice that forms at 0°C is less dense and floats on liquid water. You can see in the hexagonal patterns of snowflakes (Figure 13.9) a sign of how hydrogen bonds hold water molecules together in ice.

Hydrogen bonds also give water its wetting ability and its relatively high surface tension. Water coats (wets) polar and ionic solids to which it can form hydrogen bonds. Water has high surface tension because its molecules at the boundary of air and water are attracted sideways and inward (but not upward) to neighboring water molecules (see Figure 15.4). This tension exists only at the surface; water molecules within the body of the liquid have no net pull because they are attracted equally in all directions.

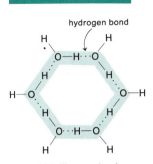

Figure 13.8 Water molecules attracted to each other by hydrogen bonds (dotted lines) form a hexagonal pattern (color) in ice.

13.3 Major Water Pollutants: Types, Sources, and Effects

WATER POLLUTION   Water is such an effective solvent that it often dissolves and carries pollutants. **Water pollution** occurs when water is contaminated to the extent that it cannot meet water-quality standards or cannot be used for a specific purpose. Water that is too polluted to drink may be

Figure 13.9 Snowflakes have a hexagonal pattern because of the way water molecules hydrogen bond in ice (see Figure 13.8). (National Oceanic and Atmospheric Administration)

satisfactory for washing steel or cooling a power plant. Water too polluted for swimming may not be too polluted for boating or for fishing.

So tradeoffs and different value judgments arise in defining what constitutes water pollution. Another complication is that thousands of different chemicals (many in trace amounts) are used commercially each year and are discharged into water. Determining the exact amount of each chemical in the water and its effects on humans and other living things is very difficult and expensive.

TYPES AND SOURCES OF WATER POLLUTION   The major sources of water pollution arising from human activities are industry, agriculture, domestic wastes, mining, and construction. These activities discharge a wide variety of chemicals into water. Table 13.2 lists seven of the major types of water pollutants and their effects.

Degradable organic wastes include domestic sewage, animal manure, decaying plants, and industrial wastes (especially from oil refining, food processing, tanning, textile making, and paper making). Bacteria usually consume these wastes if enough oxygen is dissolved in the water. We examine these pollutants further in Section 13.4.

Disease-causing agents such as bacteria and viruses enter the water from human and animal wastes. They can spread diseases such as typhoid fever, cholera, dysentery, infectious hepatitis, and polio. Typhoid and other disease-causing bacteria are under control in most U.S. water supplies. But chlorination, the most common way to disinfect water, is less

TABLE 13.2 TYPES OF WATER POLLUTANTS AND THEIR EFFECTS

Pollutant	Residence Time	Area Affected	Effects
Degradable organic wastes	Days	Local	Aesthetic; depletes oxygen and thus kills aquatic life
Disease-causing agents	Days to months	Local, regional	Damages humans and other animals
Inorganic chemicals	Months to years	Local, regional	Damages property, plants, and animals
Synthetic organic chemicals and oil	Days to years	Local, regional	Damages humans, other animals, and plants
Plant nutrients	Decades	Local, regional	Aesthetic; depletes oxygen and thus kills aquatic life
Radioactive substances	Days to years	Local, regional	Damages humans and other animals
Heat	Days to weeks	Local	Depletes oxygen and thus kills aquatic life

Figure 13.10 Fish killed by pesticide runoff in Arrowhead Lake, North Dakota. (Nelius B. Nelson, U.S. Fish and Wildlife Service)

effective against viruses such as those that cause hepatitis. Hikers often carry filtration devices (which filter out microbes) or oxidizing agents (which kill microbes) to treat water before drinking it.

Inorganic chemicals are the vast array of acids, bases, salts, and metals that reach our waters from mining and manufacturing processes, irrigation, oil fields, and acid deposition (Section 12.4). We discuss several specific examples in Section 13.5.

Synthetic organic chemicals include organic pesticides, herbicides, plastics, detergents, and industrial chemicals and wastes. Some of the more than 700 synthetic organic chemicals found in trace amounts in U.S. water supplies are toxic (Figure 13.10). The EPA reported in 1987 that the most common toxic pollutants in water included polychlorinated biphenyls (PCBs; Section 14.5) and pesticides such as DDT, chlordane, and dieldrin (Section 17.2). We know little about the environmental effects of many other synthetic organic compounds. The most effective strategy is to minimize their entry into our waters and find short-lived or biodegradable substitutes for them. For example, a vigorous program in the early 1970s to restrict the use of pesticides in the Great Lakes areas has had a positive effect (Figure 13.11).

Oil pollution is becoming an increasingly serious problem. Much of this pollution comes simply from the disposal of lubricating oil from machines and automobile crankcases. Although it has now largely been cleaned up, the Cuyahoga River near Cleveland, Ohio, was so contaminated with oil and other wastes that in the 1960s it caught fire (Figure 13.12). Communities now have programs to collect and reprocess used oil and grease.

Spills from oil tankers are another problem. Aromatic hydrocarbons (Section 8.2) in oil, such as benzene, are toxic to fish and shellfish, but most of these chemicals evaporate within a few days. Some marine birds,

Figure 13.11 Restricting or banning the use of chemicals such as PCBs (left) and dieldrin (right) in the Great Lakes Basin in the early 1970s produced lower concentrations of these toxic chemicals in herring gull eggs from Lake Ontario colonies. (Data from Great Lakes Water Quality Board)

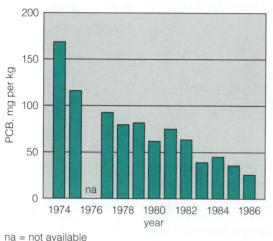
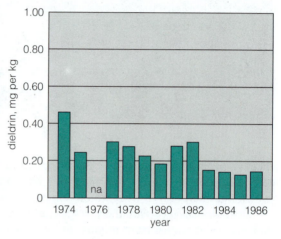

na = not available

Figure 13.12 The oil-polluted Cuyahoga River in Cleveland, Ohio, on fire in the 1960s. (*The Plain Dealer*, Cleveland, Ohio)

particularly diving birds, die when oil destroys the natural insulating properties of their feathers (Figure 13.13). Being nonpolar, oil compounds also dissolve in the fatty tissues of fish and shellfish, making them unfit to eat.

The worst spill in U.S. waters occurred in 1989 in Alaska when the *Exxon Valdez* hit submerged rocks on a reef. The tanker released about 42 million L (11 million gal) of oil. The oil slick killed more than 34,000 birds, more than 1000 sea otters, and unknown numbers of fish and shellfish. Solutions include regulating oil tankers (and offshore oil wells) more strictly, developing better ways to clean up oil spills, and wasting less oil.

Plant nutrients in water include nitrogen, phosphorus, and other substances that help aquatic plants grow. They come from the runoff of fertilizers and animal manure, phosphate-containing detergents (Section 15.4), and effluents from industries and wastewater-treatment plants. We discuss these pollutants further in Section 13.4.

Radioactive wastes include thorium-230, radium-226, radon-222, cesium-137, iodine-131, and strontium-90. These radioisotopes occur naturally deep in the earth and get into surface water from geothermal wells, from various steps in the nuclear fuel cycle (Section 11.4), and from using radioisotopes in medicine, industry, and research (Section 3.6). The safety systems to prevent them from reaching our air, water, and food supplies have worked fairly well.

According to the second law of thermodynamics (Section 7.2), heat is an inevitable by-product when work is done. It is hardly surprising then that power and industrial plants send vast amounts of *heat* into streams, lakes, and oceans. Excess heat drives dissolved oxygen out of water. The effect is like heating a carbonated beverage without a lid; the gas (CO_2) molecules become so energetic that they soon escape from the forces of attraction holding them in solution.

At warmer temperatures, metabolic reactions—like all chemical reactions—go faster (Section 7.3). So heat causes aquatic organisms to consume oxygen (metabolize) faster, while driving oxygen out of the water. For some organisms, this combination is fatal.

The harmful effects of heat can be minimized by

— Using less electricity.

— Limiting the number of power and industrial plants that can discharge heat into the same body of water.

— Discharging heated water at a point away from the fragile shore zone.

— Dissipating heat into the air by using cooling ponds, canals, or large cooling towers (Figure 13.14)

Figure 13.13 A seabird coated with crude oil from an oil spill. Such birds will die unless the oil is removed by washing it with a detergent solution. (Photo Burr Heneman, courtesy Center for Marine Conservation)

Figure 13.14 Cooling towers for the Rancho Seco nuclear power plant near Sacramento, California. Compare the size of the towers with the power plant and automobiles. Each tower is more than 120 m (400 ft) high and could hold a baseball field in its base. (Sacramento Municipal Utility District)

13.4 Dissolved Oxygen: Underwater Breath of Life

DISSOLVED OXYGEN AND BIOCHEMICAL OXYGEN DEMAND   One of the most useful indicators of water quality in a river, lake, or stream is its **dissolved oxygen (DO)** content. Because O_2 is a nonpolar molecule (Section 4.4), only small amounts can dissolve in polar solvents such as H_2O. For example, at 20°C and normal atmospheric pressure, only 9 mg of oxygen can dissolve in 1 L of water. One liter of water weighs 1 million mg, so we can express the number of milligrams of a substance per liter of water as **parts per million (ppm)**. Figure 13.15 shows the correlation between water quality and its ppm of DO.

Figure 13.15 Water quality and dissolved oxygen (DO) content.

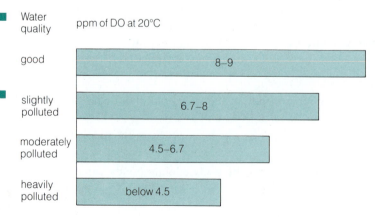

Water quality	ppm of DO at 20°C
good	8–9
slightly polluted	6.7–8
moderately polluted	4.5–6.7
heavily polluted	below 4.5

Most aquatic plants and animals need oxygen to metabolize their food and generate energy. Thus, the amount of DO in any body of water helps determine what can live there. Fish need the highest levels of DO, invertebrates require lower levels, and bacteria need the least. Water with less than 6 ppm of DO cannot support a diverse and balanced aquatic community.

Bacteria and other decomposers help break down degradable organic wastes in the water. Those that require O_2 are called **aerobic organisms;** those that do not are called **anaerobic organisms**. Table 13.3 lists the typical products from each type of decomposition. Aerobic organisms produce water, carbon dioxide, and nitrate, phosphate, and sulfate ions, which are neither offensive nor toxic in moderate amounts. But when DO levels are too low, anaerobic bacteria take over. They produce toxic and foul-smelling substances such as hydrogen sulfide (H_2S), ammonia (NH_3), flammable methane (CH_4) (*swamp gas*), and phosphine (PH_3).

Water usually loses its DO as a result of having too much degradable organic wastes. When aerobic bacteria degrade all this material, they quickly deplete the DO:

$$\text{organic wastes} + O_2 \xrightarrow{\text{bacteria}} CO_2 + H_2O$$

The amount of DO needed to consume such waste is called the **biochemical oxygen demand (BOD)**. It measures (in ppm) how much DO is

---

TABLE 13.3 DECOMPOSITION PRODUCTS IN WATER WITH AND WITHOUT DISSOLVED OXYGEN (DO)

Element	Decomposition Products in Aerobic Water (Dissolved O_2 Present)	Decomposition Products in Anaerobic Water (Dissolved O_2 Absent)
Organic C $\longrightarrow$	CO_2	CH_4 (flammable)
Organic N $\longrightarrow$	NO_3^-	NH_3 + amines (foul odor)
Organic S $\longrightarrow$	SO_4^{2-}	H_2S (poisonous, rotten-egg odor)
Organic P $\longrightarrow$	PO_4^{3-}	PH_3 and other phosphorus compounds (some with unpleasant odors)
Organic H $\longrightarrow$	H_2O	CH_4, H_2O, NH_3, H_2S

consumed by the organic material after a five-day incubation period in the water at 20°C.

Water is considered seriously polluted when its BOD equals or exceeds 5 ppm. A value of 5 means that decomposing all organic wastes in the water would decrease its DO from, for example, 9 ppm to 4 ppm. Thus, a high BOD value means large amounts of organic wastes in the water, which depletes its oxygen. This shifts the decomposition reactions from aerobic to anaerobic, giving polluted water its offensive odors and killing some of its living things.

EUTROPHICATION   Another way to deplete DO is to overload water with plant nutrients, especially phosphate (PO_4^{3-}) and nitrate (NO_3^-) ions.

Eutrophication (Greek for "well nourished") is a natural process by which water gradually becomes enriched in these nutrients. Usually this takes thousands to millions of years. But acid deposition and pollution from wastewater-treatment plants, industries, and runoff of fertilizers and animal wastes can cause the same effect in just a few decades.

When phosphate and nitrate levels rise, rooted plants—such as water chestnuts and water hyacinths—and floating algae undergo population explosions (blooms) until they cover much of the water's surface with a greenish scum. During the day, these blooms add oxygen to the upper layer of water through photosynthesis. But when these organisms die, they fall to the bottom, decompose, and thus deplete the DO. Then trout, whitefish, and many other deep-water fish die of oxygen starvation, leaving the water to fish such as perch and carp, which need less oxygen and live in the upper layers. The bottom water becomes foul and almost devoid of animals, as anaerobic bacteria spew out their smelly products (see Table 13.3).

RIVERS AND LAKES: A COMPARISON   Water pollution and potential remedies for it differ markedly for rivers and lakes. Because they flow, most rivers recover fairly rapidly from some forms of pollution—especially from heat and pollutants that deplete oxygen (Figure 13.16). Just downstream from where large amounts of oxygen-demanding wastes are added, the DO decreases sharply. But further downstream, as the flowing water dissolves O_2 from the air, DO levels may return to normal. How long this takes depends on the river's size, flow rate, and volume of incoming wastes. Slow-flowing rivers and streams take longer to recover.

Reducing the input of oxygen-depleting wastes can make a remarkable difference. The Thames River in England was little more than a flowing anaerobic sewer in the 1950s. Now, thanks to strenuous efforts to clean up the river and to stop polluting it, DO levels have returned to normal. Commercial fishing for salmon and other species is thriving, and

Figure 13.16 Dissolved oxygen (solid) versus biochemical oxygen demand (dashes). Depending on flow rates and the amount of pollutants, rivers recover from oxygen-demanding wastes and heat if given enough time.

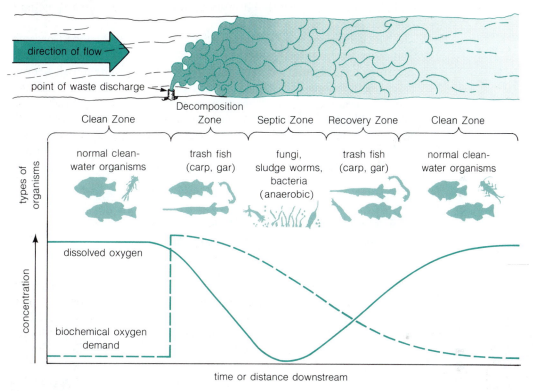

many waterfowl and wading birds have returned to their former feeding grounds.

In contrast to rivers, lakes have little flow. As a result, lakes are less able to replenish their dissolved oxygen and less able to get rid of their pollutants. Although the flushing time of a river can be measured in weeks, in lakes it may take as long as 100 years. Lakes share with rivers the problems of fish kills from toxic chemicals, but they are more vulnerable to slowly degradable pollutants that accumulate in the water and aquatic organisms.

The major problem of lakes, particularly shallow ones near urban or agricultural centers, is accelerated eutrophication. The Great Lakes provide an example of the problem and of what can be done. For decades billions of gallons of untreated sewage, industrial wastes, and agricultural runoffs flowed into the Great Lakes, killing fish, contaminating water supplies, and forcing many bathing beaches to close. The hardest hit were Lake Erie and Lake Ontario, which were relatively shallow and small. By 1970 massive algal blooms choked off oxygen to the bottom two-thirds of Lake Erie, killing nearly all of its native fish and bringing sport and commercial fishing to a halt.

The solution was to reduce the flow of plant nutrients and organic and other wastes into the lakes and to clean them up. Since 1972 these measures have slowed the eutrophication rate in the Great Lakes, even reversing it in some areas. Wastewater treatment was improved, and the input from major U.S. and Canadian industries into the Great Lakes was greatly reduced. As a result, commercial fishing is making a comeback, and swimming beaches that were closed for thirty years have reopened and are crowded with people during the summer. By 1988 only 8 of 516 swimming beaches remained closed because of pollution.

13.5 Inorganic Pollutants: Acids, Salts, and Toxic Metals

Figure 13.17 Acid mine drainage from an abandoned coal mine in West Virginia pollutes a nearby stream. (U.S. Department of Agriculture, Soil Conservation Service)

ACIDITY   Recall (Section 12.4) that sulfuric acid and nitric acid enter rivers and lakes through acid deposition from the atmosphere. Acids also come from industries such as mining and lumber, which use acids and discharge them into the water. This acidity can damage crops, kill fish and other aquatic organisms, and corrode materials such as metal boats, bridges, piers, and plumbing systems.

Another source is abandoned mines, especially coal and copper mines rich in sulfur compounds. As rainwater or groundwater seeps through these mines, the sulfur undergoes a series of reactions, ultimately forming sulfuric acid, which enters nearby rivers and streams (Figure 13.17). Although abandoned mines can be sealed, tight seals are difficult to achieve. Another strategy is to divert acid drainage (from mines or industries) away from nearby bodies of water and neutralize it (Section 6.1) with a base such as lime (CaO):

$$\underset{\substack{\text{sulfuric acid}}}{H_2SO_4} + \underset{\substack{\text{lime} \\ \text{(a base)}}}{CaO} \longrightarrow \underset{\substack{\text{calcium} \\ \text{sulfate}}}{CaSO_4} + H_2O$$

But liming lakes is expensive and must be repeated as long as acid continues to enter the water. Another drawback is that neutralizing acids produces large amounts of salts (such as calcium sulfate), which can also make the water unfit for certain purposes.

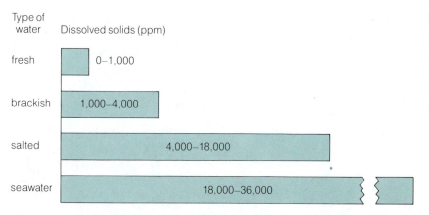

Figure 13.18 Classification of water according to its salinity or total concentration of dissolved ionic solids.

SALINITY About 97 percent of the water in the world is saline. **Salinity** is a measure of the concentration of all dissolved ionic salts. Major sources of salinity in fresh water include mines, industrial wastes, fertilizers, fallout of airborne pollutants, and highways treated to melt ice in the winter. Besides making the water unfit to drink, salinity can destroy freshwater fish and other organisms. Figure 13.18 summarizes the different types of saline water.

A major cause is irrigation. As irrigation water flows repeatedly over and through the ground, it dissolves salts—including those from fertilizers used on the land. As the saline water spreads over the soil, some of the salt stays on the land as water evaporates. This buildup reduces crop yields and leaves fields glistening with salt, looking like fields of freshly fallen snow (Figure 13.19). As it passes through irrigated cropland between northern Colorado and southern Arizona, the Colorado River's salinity increases twentyfold.

Removing salt from water is called **desalination**. One method is **distillation**, in which salt water is heated to evaporate (and then condense) the fresh water, leaving the salts behind. In **reverse osmosis**, high pressure forces salt water through membranes whose pores let water molecules, but not the salt ions, pass through.

Desalination requires energy to remove the widely scattered (high-entropy) ions because of the second law of thermodynamics (Section 7.2). Another problem is how to dispose of all salt that is collected. Because of the energy cost, desalinated water is much more expensive than is water from a tap. Most analysts project that desalinated water will never be cheap enough to be a major source of fresh water worldwide.

In 1985 desalination plants produced only about 0.006 percent of the world's water and only 0.4 percent of the water used in the United States. Most of the plants are small and serve coastal cities in arid regions where water from any source is expensive.

Figure 13.19. Because of poor drainage, white salts have replaced crops that once bloomed in heavily irrigated Paradise Valley, Wyoming. (U.S. Department of the Interior, Bureau of Reclamation)

TOXIC METALS: CADMIUM AND MERCURY   Many of our activities—mining, burning coal, and industrial processes—release toxic metal compounds into air and water. As much as 50 percent of the toxic metals in Great Lakes waters are now thought to come from airborne metals. In the United States, the largest emissions are lead, cadmium, mercury, and arsenic. The major airborne sources are motor fuel combustion (for lead and cadmium), coal combustion (for all four), and municipal waste incineration (for cadmium, mercury, and lead).

Cadmium (Cd) is widely used in plastics, paints, and nickel–cadmium batteries (Section 6.4) and in electroplating metals to prevent corrosion. It is also a contaminant in phosphate fertilizers and mining wastes. About 4 to 8 percent of the cadmium we eat or drink stays in our bodies, mostly in the liver and kidneys, and takes a year or more to expel. Our bodies retain 40 to 50 percent of the cadmium we inhale, and for most people the main source is tobacco smoke.

In 1955 Japanese physicians reported that some people living along the Zintsu River had high levels of cadmium because the water was contaminated by industrial and mining wastes. The victims had excruciating pain in their joints, and their bones slowly weakened as they lost calcium; even standing or coughing could break their bones. Between 1955–1968, several hundred people became sick and at least 100 people died.

Mercury (Hg) enters water as fallout from the air and from discharges by mercury-using industries. In terms of total mass, however, two of the largest sources are natural: mercury vaporizing from the earth's crust and huge stores in bottom sediments in the ocean. As a result, it is dangerous to eat large amounts of tuna, swordfish, and other large ocean species because they may contain high levels of mercury.

Figure 13.20 shows three major forms of mercury and how they change form. Metallic mercury is dangerous when inhaled into the lungs; it is less dangerous when swallowed because most of the mercury passes out of the body in a few days. Inorganic mercury salts are dangerous at high concentrations and cause intestinal and kidney damage. But normally our bodies excrete these salts in the urine before they accumulate to high levels.

An organic form called methyl mercury (CH_3Hg^+) is the most toxic. It can remain in the body for months; can attack the central nervous system, kidneys, liver, and brain tissue; and can cause birth defects. Under acidic conditions, anaerobic bacteria dwelling in bottom mud convert elemental mercury and mercury salts into methyl mercury. Most waters apparently aren't acidic enough for this to happen, but the acidification of lakes by acid deposition may aggravate this problem.

One tragic episode occurred in the late 1950s when 52 people died and 150 suffered serious brain and nerve damage from mercury discharged into Minimata Bay, Japan, from a nearby chemical plant. Most of the victims in this seaside area ate the mercury-contaminated fish three times a day. Another tragedy occurred in 1972 when Iraqi villagers who

Figure 13.20 Some chemical forms of mercury and how they may be transformed.

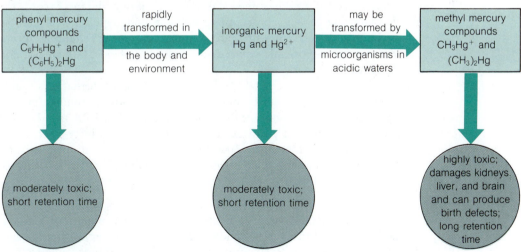

had received a large shipment of seed grain fumigated with methyl mercury fed it to their animals and used it to bake bread instead of planting it. Reportedly, 459 people died, and 6530 were injured.

Mercury poisoning is mostly a local problem that arises when people eat food contaminated with methyl mercury from seeds or food from waters containing too much methyl mercury. Two steps being taken to solve this problem are to use other compounds to fumigate seeds and to have industries recover most of their mercury before discharging wastes into the water. These technologies already exist.

13.6 Groundwater Pollution: Contaminating Our Drinking Water

A GROWING PROBLEM   One of the growing water-pollution problems is the contamination of groundwater. This portion of the hydrologic cycle (see Figure 13.2) provides drinking water for one out of two people in the United States and 95 percent of those in rural areas.

Until recently, groundwater was considered safe from contamination; pollutants that didn't vaporize would presumably be bound by the soil, and bacteria would decompose them before they percolated down to the groundwater. But it turns out that this self-cleaning system can be overloaded. In some areas such as parts of Florida, the soil is porous and enables pollutants to reach the groundwater faster than they can be decomposed. And in some areas, the amount of pollutants exceeds the decomposing capacity of a limited underground population of aerobic bacteria, which operate with scarce supplies of oxygen. Yet another problem is synthetic organic compounds such as pesticides, which bacteria don't readily decompose.

The EPA estimated in 1985 that 1 to 2 percent of the nation's usable groundwater supply had become polluted. The most common contaminants are pesticides and nitrates from fertilizers. The EPA has detected seventy-four different pesticides in groundwater from thirty-eight states. In the mid-1980s more than 1000 Florida wells were shut down because their water contained high levels of a now-banned pesticide (ethylene dibromide, EDB) used to kill nematode worms.

CONTROLLING GROUNDWATER POLLUTION   Groundwater pollution is essentially irreversible because groundwater is replenished very slowly through the hydrologic cycle. Even if further pollution stopped today, our groundwater would continue to carry contaminants for several decades. Another complication is that locating and monitoring groundwater pollution is expensive—up to $10,000 per well being monitored.

Because of these difficulties, the only effective long-range solution to protect groundwater is to prevent contaminating it in the first place. The two largest sources of groundwater pollution are underground storage tanks and hazardous waste dumps. Another important source is the agricultural use of pesticides and fertilizers.

One step being taken is to require by 1998 new safety features on underground gasoline tanks. In addition, new tanks installed after 1993 must have a leak-detection system, and action must be taken immediately when leaks are discovered. Environmentalists call this action too little, too late. Independent operators and local petroleum suppliers who own such tanks say the new rules will drive them into bankruptcy.

13.7 Cleaning Wastewater: Chemical and Ecological Approaches

WASTEWATER TREATMENT Sewage and waterborne industrial and agricultural wastes can be treated in several ways. Wastewater from individual homes may go into *septic tanks*, which trap large solids and use bacteria to decompose organic wastes. In small communities, the water may go into *wastewater lagoons* where solids settle out and wastes are degraded by air, sunlight, and bacteria. In industrialized countries, however, wastewater in most urban areas flows through a network of sewer pipes to centralized *wastewater-treatment plants*.

Primary wastewater treatment is a mechanical process that uses screens to filter out debris such as sticks, stones, and rags. The water then goes into a sedimentation tank where suspended solids settle out as sludge (Figure 13.21). Adding chemicals such as aluminum sulfate produces aluminum hydroxide, $Al(OH)_3$, which settles out and carries with it most of the bacteria and suspended matter:

$$Al_2(SO_4)_3 + 3Ca(OH)_2 \longrightarrow 2Al(OH)_3 + 3CaSO_4$$

| aluminum sulfate | calcium hydroxide | aluminum hydroxide (insoluble) | calcium sulfate |

Anaerobic bacteria decompose the solid material. The water is then chlorinated to kill bacteria before it leaves the plant. Altogether, primary treatment removes about 60 percent of the solid material and about one-third of degradable organic wastes.

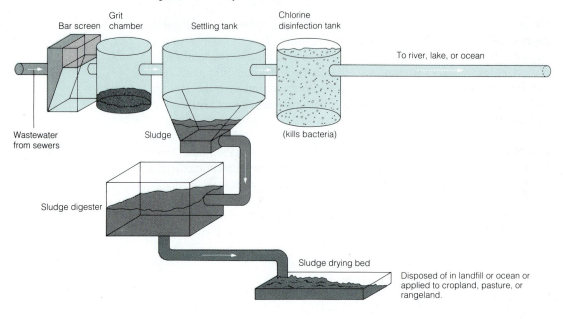

Figure 13.21 Primary wastewater treatment.

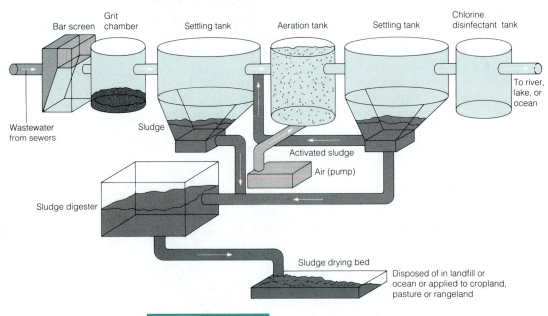

Figure 13.22 Secondary wastewater treatment.

Secondary wastewater treatment includes all steps in primary treatment, plus a step using aerobic bacteria to break down degradable organic wastes (Figure 13.22). Some treatment plants use trickling filters where bacteria degrade sewage as it seeps through a bed of stones. Others use an activated sludge process in which air and bacteria-rich sludge are bubbled through the wastewater to increase bacterial decomposition.

Combined primary and secondary treatment leaves in the water about 3 to 5 percent of the oxygen-demanding waste, 3 percent of suspended solids, 50 percent of the nitrogen (mostly as nitrates), 70 percent of the phosphorus (mostly as phosphates), and 30 percent of most toxic metal compounds and synthetic organic compounds. Virtually none of the long-lived radioisotopes and nondegradable organic substances such as pesticides are removed.

Tertiary (or **advanced**) **wastewater treatment** is a variety of specialized chemical and physical processes that remove pollutants still left after primary and secondary treatment (Figure 13.23). Tertiary treatment is rare

Figure 13.23 Tertiary (advanced) wastewater treatment. This diagram shows several different types of advanced treatment. Often only one or two of these processes are used to remove specific pollutants in a particular area.

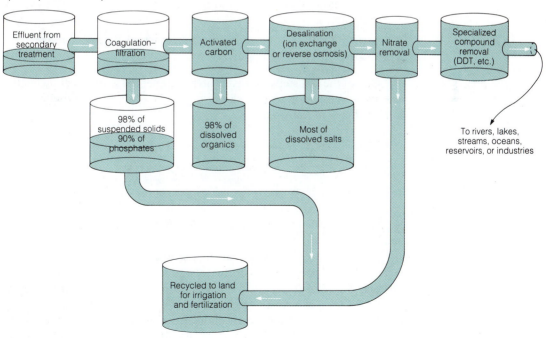

(except in Scandinavian countries) because many methods are still experimental and the additional treatment is expensive—twice as much to build the plant and up to four times as much to operate as secondary treatment costs. Common methods include

— Precipitation (coagulation–filtration) to remove nearly all suspended solids, 90 to 95 percent of the phosphates, and certain toxic metal ions. Lime (CaO) or aluminum sulfate can be added to form insoluble calcium phosphate or aluminum phosphate, respectively. Simplified equations are

$$3Ca^{2+} + 2PO_4^{3-} \longrightarrow \quad Ca_3(PO_4)_2$$
insoluble precipitate

$$Al^{3+} + PO_4^{3-} \longrightarrow \quad AlPO_4$$
insoluble precipitate

Because $AlPO_4$ dissolves in acidic water, the pH of the wastewater affects how efficiently this treatment removes phosphates.
— Reverse osmosis (Section 13.5) to remove dissolved organic and inorganic materials.
— Ion exchange (Section 15.2) to remove dissolved inorganic compounds.
— Adsorption onto activated carbon or other surfaces to remove dissolved organic compounds.

One of the last steps in wastewater treatment is to disinfect the water. The usual method is to add chlorine gas (Cl_2), which reacts with water to form hypochlorous acid:

$$Cl_2 + H_2O \longrightarrow \quad HCl \quad + \quad HClO$$
chlorine   hydrochloric   hypochlorous
acid   acid
(a bleach)

HClO is an oxidizing agent (Section 6.3) that kills bacteria in the water and removes color.

The main problem with chlorine is that it also reacts with organic materials in wastewater to form tiny amounts of chlorinated hydrocarbons. Some of these, such as chloroform ($CHCl_3$) and carbon tetrachloride (CCl_4), can cause cancer. Although the benefits of using Cl_2 far outweigh the risks, the concern about trace amounts of chlorinated hydrocarbons has triggered a search for an alternative treatment.

One of the leading contenders is ozone (O_3) gas, which is used in parts of Europe to disinfect water. Ozone is more expensive than chlorine but is more effective against viruses. Ozone is not very soluble in water, however, and this limits its ability to disinfect water.

During the last two decades, wastewater treatment has improved considerably in the United States. By 1988 87 percent of the country's wastewater-treatment plants, which handle 95 percent of all municipal wastewater, used at least secondary treatment. In addition, 80 percent of industry was in compliance with the permits limiting what they could discharge into rivers, streams, and lakes. The EPA reported in 1987 that 75 percent of U.S. streams and lakes are clean enough for fishing and swimming.

ECOLOGICAL WASTE MANAGEMENT   Some scientists argue that we should return to the land treated wastewater that carries plant nutrients such as phosphates and nitrates. Another potentially rich source of nutrients is the dried sludge collected at wastewater-treatment plants.

A problem, however, is that both sludge and treated wastewater may still contain bacteria, viruses, toxic metals, and hazardous, synthetic organic chemicals. Returning these materials could contaminate the land and the food grown there.

Sludge and wastewater can be cleaned by methods such as those used in tertiary treatment, but this is expensive. In some areas, wastewater and sludge are put on land not used for crops, livestock, or drinking. Examples include forests, surface-mined lands, golf courses, and highway medians.

Even primary, secondary, and tertiary treatments cannot remove 100 percent of all water pollutants. Such a goal is impractical, far too expensive, and it fails to take advantage of natural mechanisms to purify water. The task is to keep our water clean and to help nature do it for us.

WILLIAM ASHFORTH   Born in a water-rich environment, we have never really learned how important water is to us. . . . Where it has been cheap and plentiful, we have ignored it; where it has been rare and precious, we have spent it with shameful and unbecoming haste. . . . Everywhere we have poured filth into it.

Summary

The earth's water supply is renewed through the hydrologic cycle. People in many parts of the world experience water shortages or drink contaminated water.

Water has many unusual properties. Compared to other substances of a similar molecular size and weight, water has a high heat capacity, high heat of vaporization, high freezing and boiling points, and high surface tension. Water is in its densest form as a liquid, not a solid. These properties come from the strong hydrogen bonds between water molecules.

Seven major types of water pollutants are degradable organic wastes, disease-causing agents, inorganic chemicals (acids, salts, and toxic metals), synthetic organic chemicals and oil, plant nutrients (phosphates and nitrates), radioactive substances, and heat.

Dissolved oxygen (DO) is a measure of the quality of water. When aerobic bacteria degrade organic wastes, they consume DO; the amount of O_2 consumed under standard conditions is the biochemical oxygen demand (BOD). Plant nutrients indirectly deplete dissolved oxygen, especially from lakes, by the process of eutrophication.

Acids are another pollutant, and much of the acidity in water comes from acid deposition from the atmosphere. Salinity is a measure of dissolved ionic salts in water. Salinity can be reduced by distillation or reverse osmosis, but it is expensive. Two important toxic metals in water are cadmium (Cd) and mercury (Hg); a third, lead (Pb), is discussed in Chapter 12.

A growing problem is pollution of groundwater, especially from underground storage tanks, hazardous waste dumps, and runoff of agricultural chemicals.

Primary wastewater treatment removes about 60 percent of solid material and about 33 percent of degradable organic wastes from water. Secondary treatment removes about 95 percent of those materials but leaves significant amounts of nitrates, phosphates, toxic metals, and synthetic organic compounds. Tertiary (advanced) treatment removes additional pollutants, but at considerable expense.

Terms for Review

After completing this chapter, you should know and understand the meaning of the following terms:

aerobic organism (p. 379)

anaerobic organism (p. 379)

biochemical oxygen demand (BOD) (p. 379)

desalination (p. 383)

dissolved oxygen (DO) (p. 378)

distillation (p. 383)

eutrophication (p. 380)

groundwater (p. 368)

heat capacity (p. 369)

heat of vaporization (p. 369)

hydrologic cycle (p. 368)

parts per million (ppm) (p. 378)

primary wastewater treatment (p. 387)

reverse osmosis (p. 383)

salinity (p. 383)

secondary wastewater treatment (p. 389)

surface tension (p. 372)

tertiary (advanced) wastewater treatment (p. 389)

water pollution (p. 373)

wetting (p. 372)

Questions

Odd-numbered questions are answered at the back of this book.

1. What are the major domestic uses of water in the United States?

2. What are some of the major diseases carried by microorganisms in water?

3. Explain, on a chemical basis, why water is a liquid at room temperature while the substances from which it forms—hydrogen (H_2) and oxygen (O_2)—are gases.

4. Would ice float more readily on liquid water that is 4°C or on liquid water that is 0°C? Explain.

5. Would water form hydrogen bonds if it were a linear, rather than bent, molecule? Explain.

6. On a hot day, the concrete around a swimming pool gets much warmer than the water in the pool. Explain.

7. Why do snowflakes have symmetrical, six-sided shapes?

8. If it were possible, would it be desirable to have lakes containing pure, distilled water? Why?

9. After hikers add oxidizing agents to natural water samples to disinfect it, they need to wait a certain length of time before the water is safe to drink. Do you think colder water would require a longer or a shorter waiting period? Explain.

10. Criticize the following statements:
    a. The largest use of water in U.S. homes is for bathing and washing.
    b. Eutrophication is an inexorable, irreversible process.
    c. Rivers and lakes have similar water-pollution problems.
    d. Chlorinated water is safe to drink.

11. Aromatic hydrocarbons are among the most toxic ingredients in an oil spill. What chemical structure do aromatic hydrocarbons have? (Review Section 8.2 if necessary.)

12. What are the two most common ions that are plant nutrients? Write a chemical formula for each combined with sodium.

13. Distinguish between aerobic and anaerobic decay. Why is anaerobic decay usually less desirable?

14. Suppose a sample of water had a dissolved oxygen (DO) level of 8.3 ppm and a biochemical oxygen demand (BOD) of 3.1 ppm. What would the DO level be if the organic wastes were degraded by aerobic bacteria in a closed container? Use Figure 13.15 to rate the quality of the resulting water.

15. How does heat affect the amount of oxygen that can dissolve in water? Do goldfish have a better chance of surviving in cool or warm water?

16. How does heat affect the amount of salts that can dissolve in water? Explain.

17. Explain, in terms of the second law of thermodynamics (Section 7.2), why desalination requires energy.

18. What are the major sources of mercury in water? Can the problem be solved by stopping industrial emissions of mercury into the water?

19. What are the main pollutants that enter groundwater because of agricultural activities?

20. Which of the following do *not* occur in primary wastewater treatment? (a) Chlorination, (b) activated sludge, (c) sedimentation, (d) sludge digestion, and (e) reverse osmosis.

21. Which tertiary (advanced) wastewater-treatment methods may remove toxic metal compounds?

22. Secondary wastewater treatment is especially effective in removing which of the following? (a) Phosphates, (b) toxic metals, (c) suspended solids, (d) degradable organic wastes, and (e) nitrates.

23. Ozone (O_3) molecules are bent and polar, whereas oxygen (O_2) molecules are linear and nonpolar. Though neither is highly soluble in water, which one is more soluble? Why?

Topics for Discussion

1. What water-supply problems, if any, exist in your city or area today? What do you think the situation will be in the year 2000? Why?

2. Some people have speculated that forms of life based on ammonia (NH_3) or hydrogen sulfide (H_2S) instead of water could exist elsewhere in space. Could those substances work as substitutes for water? Why?

3. To what extent is water pollution linked to air pollution? Cite specific examples.

4. What type of wastewater treatment is used in your community? Should all cities and towns be required to install tertiary treatment plants by the year 2000? Why?

5. Should wastewater-treatment plants use ozone or chlorine to disinfect water? What information is the most important to you in answering this question?

PART IV

Consumer Chemistry

RENE DUBOS   In science as in other human activities, the speed of progress is less important than its direction. Ideally, knowledge should serve understanding, freedom, and happiness rather than power, regimentation, and technological development for the sake of economic growth. Scientists must give greater prominence to large human concerns when choosing their problems and formulating their results. In addition to the science of things, they must create a science of humanity.

CHAPTER 14

Synthetic Polymer Products: The World of Plastics

General Objectives

1. What are polymers, and what two major types of chemical reactions are used to prepare them?

2. What are the structures of polyethylene and vinyl plastics, and how are these products made and used?

3. What are the structures of synthetic rubber polymers?

4. What are the structures of Dacron, nylon, and polymers such as Bakelite? How are they made and used?

5. What are the structures and uses of silicone polymers?

6. What are some environmental problems with the widespread use of plastics?

We've all had enough experience in the marketplace to conclude that anything imitation or artificial will likely be cheap and inferior. Silk flowers cannot match the beauty of a garden bouquet; plastic upholstery takes a back seat to leather; and costume jewelry pales in the presence of diamonds. Nothing counterfeit surpasses the real thing.

From these valid observations, however, many people make the mistake of attaching the same connotation to synthetic materials, made by human ingenuity. *Synthetic* refers to a process; it tells nothing about quality. Far from being low-grade impostors for products of nature, many synthetic substances are in every way indistinguishable from their natural counterparts. Some are actually better. And a growing number have properties and uses that cannot be approximated by anything from nature.

This chapter examines synthetic materials made from very large molecules, called polymers. These products include foam (Styrofoam) ice chests, plastic food wrap (Saran wrap), plastic pipes and beverage cartons, coated (Teflon) frying pans, and sprays to keep hair in place.

14.1 Polymers: Really Big Molecules

NATURAL AND SYNTHETIC POLYMERS A **polymer** (from Greek roots meaning "many parts") is a giant molecule (macromolecule) made by linking together many small, repeating molecular units known as **monomers**. A polymer is like a train, with the individual cars being the monomer units. The repeating units can consist of the same monomer or a mixture of different monomers. In the latter case, they form a **copolymer** (Figure 14.1).

Nature has been forming polymers for billions of years. These natural polymers, or *biopolymers*, include many of the molecules of life—cellulose,

Figure 14.1 The formation of a polymer from one type of monomer (top) and a copolymer from two different monomers (bottom). The copolymer shown is only one of many possibilities.

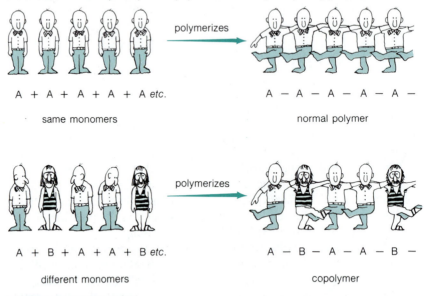

A + A + A + A + A *etc.*

same monomers

A — A — A — A — A —

normal polymer

A + B + A + A + B *etc.*

different monomers

A — B — A — A — B —

copolymer

starch, nucleic acids (DNA and RNA), and proteins. Materials such as leather, wool, cotton, wood, and rubber are also natural organic polymers.

The first synthetic polymers were designed to replace natural substances such as ivory for billiard balls, hemp for rope, and rubber for tires. During the last fifty years, however, the polymer industry has produced a remarkable array of entirely new materials and products. This happened mostly because of three factors:

1. A plentiful and cheap supply of fossil fuels (primarily petroleum and natural gas; Section 8.3) to provide raw material for monomers.
2. Extensive chemical research that provided information about molecular structures of polymers, their physical and chemical properties, and the reactions that form them.
3. The development of machines and large-scale technological processes for making and converting polymers into consumer products.

SOME BASIC TERMS   Before examining specific polymers, we need to become familiar with a few general terms. Synthetic polymers include elastomers, fibers, and plastics. **Elastomers**, like rubber, are elastic; stretch them, and

Figure 14.2 Production of Saran wrap. (Courtesy Dow Chemical Co.)

they will spring back to their original shape. **Fibers** are polymers that have high tensile strength; that is, they don't break or deform when pulled along the length of the material. **Plastics**, such as polyethylene and Saran wrap, are polymers that can be molded or made into sheets or other shapes (Figure 14.2).

Polymers can be linear or branched (Figure 14.3). In a linear polymer, the monomer units join end-to-end to produce a continuous, very long chain. Branched polymers, on the other hand, have chains extending out from other chains, like branches coming out from a tree trunk. In linear polymers, individual molecules (like tree trunks with all limbs sawed off) can pack close together, making the material relatively dense and rigid. In contrast, molecules of branched polymers (like tree trunks with all limbs still connected) cannot pack as closely together and thus produce a softer material with a lower density (Figure 14.4).

Individual polymer molecules, depending on their functional groups (Section 8.4), can attract each other by dipole–dipole or hydrogen bonds (Section 5.1). But they are held even more firmly in place when they join together with covalent bonds, which are called **cross-links** (see Figure 14.3). Controlling the extent of branching and cross-linking are just two ways that polymer chemists can develop products with desired properties for specific uses.

We also classify polymers according to what happens to them during heating. Some, such as polyethylene and nylon, soften and then regain their original texture when they cool; these are **thermoplastic polymers**. Others get harder and more rigid when heated and stay that way after cooling; these are **thermosetting polymers**. This effect is due to heat causing

a linear-chain polymer

a branched-chain polymer

cross-linked polymer chains

Figure 14.3 Some different structural possibilities for polymers.

Figure 14.4 Linear polymers (left) have different properties than branched-chain polymers (right).

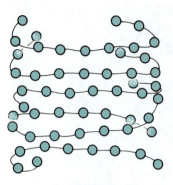

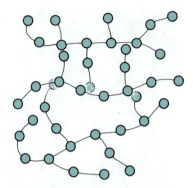

linear polymer
(high melting point,
fairly hard, rigid, and tough)

branched-chain polymer
(no definite melting point,
soft and flexible)

permanent cross-links to form and join the individual polymer molecules into larger, more rigid networks.

14.2 Addition Polymers: Polyethylenes, Vinyls, Rubber, and Paint

POLYMERIZATION REACTIONS   Two general classes of polymers, based on the way they form from their monomer units, are addition polymers and condensation polymers. We discuss condensation polymers in Section 14.3.

Addition polymers form from monomer units that join together without removing any atoms. In other words, monomer units add together, with all atoms in the monomer molecules remaining in the polymer. It is like a group of people joining hands to form a long, flexible chain.

The monomers used to make addition polymers typically have a double bond between two carbon atoms:

$$\diagdown \!\!\! C = C \!\!\! \diagup$$

Under conditions of high temperature and pressure and in the presence of a catalyst, two such unsaturated monomers can combine so that the double bond is eliminated:

$$\diagdown \!\!\! C = C \!\!\! \diagup \ + \ \diagdown \!\!\! C = C \!\!\! \diagup \ \longrightarrow \ \text{---} C \!\!-\!\! C \!\!-\!\! C \!\!-\!\! C \text{---}$$

More monomer units add at either end of the resulting molecule to form long chains (see Figure 14.1).

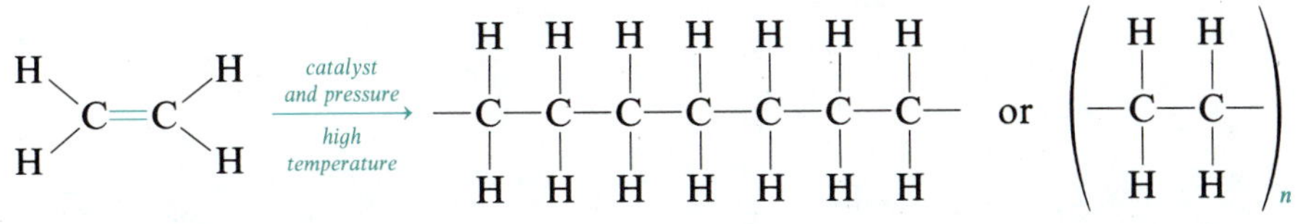

ethylene monomer   polyethylene   repeating
                                                                    unit formula

Actually, carbon atoms join at a slight angle to form a kinked structure, so the carbon skeleton of polyethylene has the following shape:

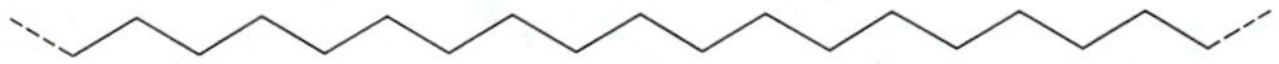

Polyethylenes are inexpensive and, as thermoplastic polymers, can be softened and molded into almost any shape, extruded into fibers or filaments, and precipitated or blown into films. Polyethylene products vary from building materials and electrical insulation to toys and packaging materials.

By changing reaction conditions such as temperature, pressure, and catalysts, polymer chemists can control the average length and branching in a chain as it forms. This in turn changes the properties of polyethylene. At moderate temperatures and pressures, linear chains predominate. The resulting high-density polyethylenes (HDPEs) have great strength and rigidity (see Figure 14.4). They are used for bottles, film, and pipe.

Higher reaction temperatures and different catalysts produce more branching in the molecules. The resulting low-density polyethylenes (LDPEs) are softer, more flexible polymers that can be used for wire and cable coverings, squeeze bottles, plastic bags, films, toys, and many other household articles. LDPE, however, softens at relatively low temperatures.

VINYL ADDITION POLYMERS: THE EFFECTS OF SIDE GROUPS   Many other addition polymers can be made by replacing one or more hydrogen atoms on the ethylene monomer. Table 14.1 lists the structures and uses of many of these polymers.

When one hydrogen is replaced, the rest of the molecule is known as a *vinyl group* (Figure 14.5). The products that form when these monomers polymerize are **vinyl polymers**. For example, substituting a methyl group

TABLE 14.1 SOME ADDITION POLYMERS

Monomer (Common Name)	Monomer Formula	Polymer Formula and Name	Trade Names	Uses
Ethene (ethylene)	H₂C=CH₂ *(H H / C=C / H H)*	—C—C—C—C—C—C— (polyethylene)	Polythene	Bags, films, toys, molded objects, containers
Propene (propylene)	*(H H / C=C / H CH₃)*	—C—C—C—C—C—C— (polypropylene)	Prolene, Vectra, Herculon	Bottles, films, indoor-outdoor carpets, bucket seats, fishing nets
Chloroethene (vinyl chloride)	*(H H / C=C / H Cl)*	—C—C—C—C—C—C— (polyvinyl chloride)	PVC, Koroseal	Floor tiles, phonograph records, raincoats, garden hoses, packaging, car tops
1,1-Dichloroethene (vinylidene chloride)	*(H Cl / C=C / H Cl)*	—C—C—C—C—C—C— (polyvinylidene chloride)	Saran	Food wrap, container liners
Tetrafluoroethene (tetrafluoroethylene)	*(F F / C=C / F F)*	—C—C—C—C—C—C— (polytetrafluoroethylene)	Teflon	Pan coatings, tape, electrical insulation, gasket bearings
Styrene	*(H H / C=C / H C₆H₅)*	—C—C—C—C—C—C— (polystyrene)	Styrofoam Lustrex	Food and drink coolers, insulation, plastic furniture, piping, building material, insulation

TABLE 14.1 CONTINUED

Monomer (Common Name)	Monomer Formula	Polymer Formula and Name	Trade Names	Uses
Acrylonitrile	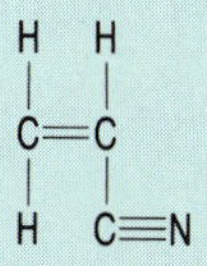 H H C=C H C≡N	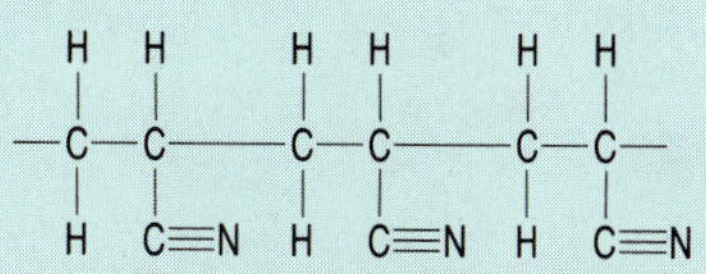 H H H H H H —C—C——C—C——C—C— H C≡N H C≡N H C≡N *polyacrylonitrile*	Acrilan, Creslan, Orlon	Rugs, fabrics, draperies, yarn
Methyl methacrylate	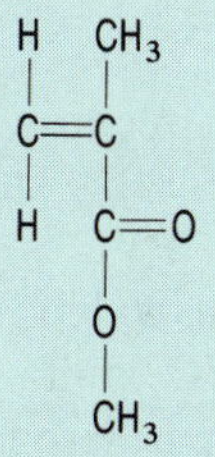 H CH₃ C=C H C=O O CH₃	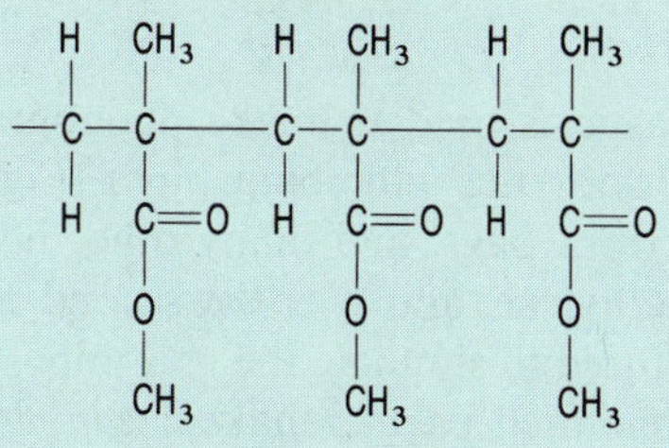 H CH₃ H CH₃ H CH₃ H CH₃ —C—C——C—C——C—C— H C=O H C=O H C=O H C=O O O O O CH₃ CH₃ CH₃ CH₃ *polymethyl methacrylate*	Lucite, Plexiglas, Perspex	Windows, coatings, latex paints, molded items such as combs

for a hydrogen in ethylene gives a new monomer, propylene. Its vinyl polymer, polypropylene, is used to make indoor-outdoor carpets, heavy-duty upholstery, hinged boxes, and other articles (Figure 14.6):

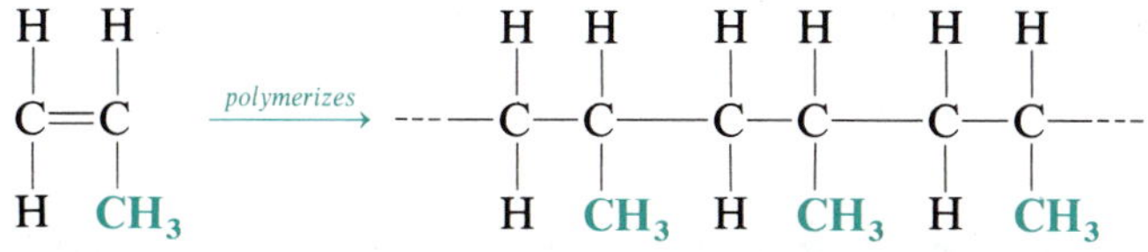

H H   *polymerizes*   H H H H H H
C=C → ----C—C——C—C——C—C----
H CH₃   H CH₃ H CH₃ H CH₃

propylene
monomer

polypropylene

Figure 14.5 The vinyl functional group (in color) bonded to some element other than hydrogen (X).

or

CH₃ CH₃ CH₃ CH₃ CH₃ CH₃ CH₃ CH₃ CH₃

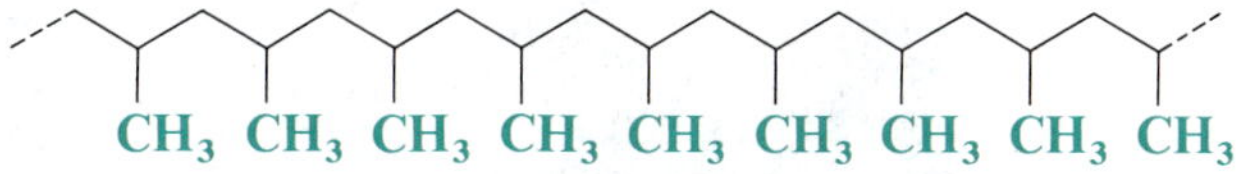

When a chlorine atom replaces a hydrogen in ethylene, the monomer is vinyl chloride. It polymerizes to yield a plastic known as polyvinyl chloride (PVC):

Figure 14.6 The exceptional strength and flexibility of polypropylene allow it to be used in automobile storage batteries. (Courtesy Delco Remy)

$$
\begin{array}{ccc}
H & H \\
| & | \\
C & = & C \\
| & | \\
H & Cl
\end{array}
\quad \xrightarrow{\text{polymerizes}} \quad
\text{---} \begin{array}{cccccc}
H & H & H & H & H & H \\
| & | & | & | & | & | \\
C & C & C & C & C & C \\
| & | & | & | & | & | \\
H & Cl & H & Cl & H & Cl
\end{array} \text{---}
$$

vinyl chloride monomer   *polyvinyl chloride (PVC)*

or

This plastic is used to make phonograph records, garden hose, plastic bottles, floor tile, plumbing pipes (Figure 14.7), sheathing for electrical wiring, trash bags, and many other items. PVC is about as inexpensive as polyethylene, and it is the second most widely used plastic. Colored and textured to simulate the real thing, PVC "imitation leather" products have done their part to make *cheap plastic* a derogatory term.

If both hydrogen atoms on one carbon atom of ethylene are replaced with chlorine atoms, the resulting monomer is vinylidene chloride. Its polymer is polyvinylidene chloride, or Saran, used in Saran wrap (see Figure 14.2):

$$
\begin{array}{ccc}
H & Cl \\
| & | \\
C & = & C \\
| & | \\
H & Cl
\end{array}
\quad \xrightarrow{\text{polymerizes}} \quad
\text{---} \begin{array}{ccccccc}
H & Cl & H & Cl & H & Cl & H \\
| & | & | & | & | & | & | \\
C & C & C & C & C & C & C \\
| & | & | & | & | & | & | \\
H & Cl & H & Cl & H & Cl & H
\end{array} \text{---}
$$

vinylidene chloride monomer   *polyvinylidene chloride (Saran)*

Figure 14.7 Rigid polyvinyl chloride (PVC) used as pipe is lightweight, resistant to chemicals, and easy to install. (Courtesy B. F. Goodrich)

Replacing all four hydrogen atoms on ethylene with fluorine atoms produces the monomer tetrafluoroethylene (see Table 14.1). You probably know its polymer, polytetrafluoroethylene, by the trade name Teflon. The sheath of fluorine atoms encases the carbon backbone and protects it from chemical attack, making Teflon more dense, heat-resistant, and chemically inert than polyethylene. Teflon provides a nonstick coating for cooking utensils because its surface is not wetted by grease or water.

Substituting a phenyl (benzene) group for a hydrogen atom in ethylene yields styrene, which polymerizes into the transparent plastic, polystyrene:

styrene monomer

polystyrene

Polystyrene is used in many household articles including plastic furniture, small radios, wall tile, refrigerator parts, and picnic utensils. Polystyrene formed by adding a chemical that produces gas bubbles within the plastic is called Styrofoam. Probably the most recognizable plastic, Styrofoam is in insulating linings for buildings and picnic coolers, packing cushions for shipping, hinged cartons for fast-food items, and inexpensive cups for hot drinks.

CHEMISTRY SPOTLIGHT   THE DISCOVERY OF TEFLON

In 1938 Du Pont chemists were trying to produce new fluorine compounds to use as refrigerants. In one experiment, they stored tetrafluoroethylene gas in a cold cylinder that happened to contain small amounts of O_2 gas as an impurity. When they opened the cylinder, they discovered a waxy, white, slippery solid that was heat-resistant and insoluble in almost all solvents. The material was a polymer of tetrafluoroethylene, which we now know as Teflon.

The first Teflon-coated cookware went on sale in 1960. But the Teflon didn't stick to the pans' metal surfaces and peeled badly. That problem was solved in 1962. Now Teflon is used not only for nonstick cookware but also for such things as Gore-Tex jogging suits, chainsaw blades, specialized tubing, and even the underpinnings of the Statue of Liberty. In medicine, Teflon is used in artificial veins and arteries, such as the aorta, that have been implanted in many thousands of patients.

Replacing one of the hydrogens with a nitrile (cyanide) group (—C≡N) produces the acrylonitrile monomer, which polymerizes to form polyacrylonitrile (see Table 14.1). These polymers can be spun into various trade name acrylic fibers such as Orlon, Acrilan, Creslan, and Dynel, most of which are copolymers that contain a mixture of acrylonitrile (35 to 85 percent) and vinyl chloride monomers.

Another vinyl polymer, polyvinylpyrrolidene (PVP) is widely used in hair sprays. These sprays are essentially solutions of a polymer such as PVP in an easily evaporated solvent. After being sprayed on hair, the solvent vaporizes, leaving a plastic film that holds hair in place:

vinylpyrrolidene monomer   *polyvinylpyrrolidene (PVP) polymer*

Acrylic polymers are made from acrylic acid

$$CH_2{=}CH{-}\overset{\overset{\displaystyle O}{\|}}{C}{-}OH$$

and its derivatives. They include polymethyl methacrylate (Lucite, Plexiglas; see Table 14.1), for "unbreakable" windows and contact lenses, and copolymers of acrylonitrile and vinyl chloride (Acrilan, Orlon), which are spun into fibers and used as clothing fabrics.

ELASTOMERS   Rubber is an addition polymer of isoprene that occurs naturally in tropical rubber trees:

isoprene monomer   *polyisoprene or natural rubber (5000 to 8000 monomer units)*

Its long, flexible polymer chains can coil and uncoil, making rubber an elastomer.

During World War II, German and American chemists developed several synthetic rubbers that could overcome the war-caused shortages of natural rubber. Since then, chemists have synthesized a wide variety of elastomers with different properties. Today, less than one-fourth of everything we call rubber in the United States is natural rubber; the rest is various types of synthetic rubber. In fact, rubber itself (polyisoprene) can be synthesized in a form that is more chemically pure than the natural product. It is one example of a synthetic substance that is better than its natural counterpart.

The first synthetic rubber produced in the United States was neoprene. Substituting a chlorine atom for a methyl group on the isoprene monomer produces chloroprene, which polymerizes to form polychloroprene, or neoprene:

chloroprene monomer

neoprene (polychloroprene) rubber

Neoprene resists aging and doesn't dissolve in oils, gasoline, and other organic solvents. These qualities make it useful for products such as garden hoses, shoe soles, gloves, adhesives, highway joint seals, gaskets, and automotive and conveyor belts.

Most synthetic rubbers are copolymers. About one-third of the synthetic rubber produced in the United States is SBR (styrene–butadiene rubber), which is made from 1,3-butadiene and styrene monomers. The ratio of butadiene to styrene in the polymer chain is normally 3 to 1:

1,3-butadiene   *styrene*

styrene–butadiene rubber (SBR)

The single largest use of both natural and synthetic rubbers is in tires. A typical tire, however, is only about 45 percent rubber; the rest is carbon black for color, belts (steel, polyester, or nylon) for strength, oils and softeners for consistency, and sulfur for vulcanization (see the "Chemistry Spotlight").

PAINTS   Paints have three main ingredients: (1) a *pigment*, which provides the color and helps shield against high-energy ultraviolet radiation from the sun; (2) a *solvent*, which keeps the paint easy to spread and evaporates readily; and (3) a *binder*, which becomes solid on exposure to the air.

A paint pigment has to have the right color and be able to hide the original color of the painted surface. Most paints contain either titanium

CHEMISTRY SPOTLIGHT   CHARLES GOODYEAR'S ACCIDENTAL DISCOVERY OF VULCANIZATION

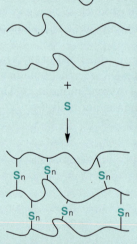

Charles Goodyear had spent many years trying to find a way to make rubber stronger and less sticky. One day in 1839, he was carrying out an experiment with sulfur and rubber. Some of the mixture accidentally fell on a hot stove and formed a charred rubber blob—much like a pancake. Goodyear noticed that the rubber pancake was tough, but not sticky or brittle.

Goodyear used this accidental discovery to develop the process called *vulcanization*—named after Vulcan, the Roman god of fire. Later work showed that treating rubber with sulfur at about 140°C causes sulfur-to-sulfur cross-links to form (Figure 14.8), making the rubber more rigid and resistant to wear.

Figure 14.8 The vulcanization process cross-links rubber molecules with sulfur (color). The subscript *n* indicates a variable number of sulfur atoms.

dioxide (TiO_2) or zinc oxide (ZnO) as their primary pigments. Paints no longer use white lead ($Pb(OH)_2 \cdot 2PbCO_3$) for this purpose because lead is toxic. Each of the primary pigments is white and extremely opaque. The desired color is made by blending the primary pigment with colorants, which usually are inorganic compounds such as orange cadmium sulfide (CdS), red iron oxide (Fe_2O_3), or green chromium oxide (Cr_2O_3).

The solvent, sometimes called the *carrier*, keeps paint ingredients suspended in an easily spreadable mixture. Once the paint is applied, however, the solvent must evaporate quickly so that the paint does not stay wet forever. For many years, oily hydrocarbon liquids were the only practical solvents. But now synthetic polymer ingredients have allowed water to become an effective paint solvent; painters can thin their paint and clean their equipment with plain or soapy water.

This advance to convenient water-based paints came because of improved binders. Oil-based paints use unsaturated drying oils (such as linseed oil), which polymerize in air to form a tough, solid coating as the paint dries. Water-based latex paints, however, use synthetic polymer binders—tiny droplets of polystyrene, polymethyl methacrylate, or a polystyrene–butadiene copolymer suspended in water.

As long as the paint is wet, you can clean up latex paint with soapy water. But once the water evaporates, the microscopic globules of polymer combine to form a continuous film of paint that no longer dissolves in water. In fact, dried latex paints are more washable than oil-based paints. They also are more resistant to damage from sunlight, and they last longer as exterior coatings.

14.3 Condensation Polymers: Dacron, Nylon, and Thermosetting Polymers

POLYESTERS: DACRON   **Condensation polymers** form when monomer units bond together and eliminate a small molecule such as water, an alcohol, or hydrogen chloride (HCl). Forming these polymers is like a group of people having to remove their gloves, staple the matched pairs together, and throw them aside before being able to join hands in a chain.

A condensation polymer with regular, repeating ester linkages is a **polyester**. Recall (Section 8.6) that a carboxylic acid reacts with an alcohol to form an ester and water. Polyesters are copolymers made from a monomer with two carboxylic acid groups and a monomer with two alcohol groups. An alcohol group in one monomer reacts with a carboxylic acid group in the other monomer, forming an ester and eliminating water.

One of the most common polyesters is Dacron. It forms by the reaction

terephthalic acid monomer   *ethylene glycol monomer*

Dacron (a polyester copolymer)   *water*

Figure 14.9 Dacron fibers being spun. (Courtesy Du Pont Co.)

Dacron fibers (Figure 14.9) are used widely in clothing and fabrics. In medicine, Dacron is coated onto artificial heart valves and artery segments that replace damaged blood vessels.

When this polyester is spun into a thin film, it is called Mylar, a common material for computer diskettes, recording tape, and packaging for frozen food. Another, more complex polyester known as Lexan is used for break-proof windowpanes, bullet-proof vests, and astronauts' helmets because it is as clear as glass but almost as tough as steel.

If one or both monomers contain additional functional groups, those groups can cross-link the polyester strands. Cross-linked polyesters, known as *alkyd resins*, are hard and tough. They are used in lacquers, paints, and molded fiberglass objects such as boats and furniture.

POLYAMIDES: NYLONS   Condensation polymers with repeating amide linkages are called **polyamides**. Recall (Section 8.7) that a carboxylic acid reacts with an amine to form an amide and eliminate water. Polyamides are copolymers made by linking monomers having two carboxylic acid groups with monomers having two amino groups. Figure 14.10 shows how the most important polyamide, nylon, forms from its monomers.

Nylon is very versatile. It can be woven into fabrics or molded into various shapes. Its strength comes from hydrogen bonds between amide groups of nearby molecules (Figure 14.11). Because its polar groups are already tied up in hydrogen bonds, nylon doesn't readily form hydrogen bonds with water; this makes nylon easy to dry. By varying the length of the carbon chain in monomer units, chemists can modify the properties of nylon. Some types, for example, are hard enough to use as gears in machinery and are very resistant to wear.

Nylon and other polymers such as Dacron and Orlon can be spun into fibers, which can then be twisted into threads. Here the long molecules lie stretched out along the direction of the fiber, providing considerable strength during stretching. The extensive side-to-side bonding between

Figure 14.10 The polymerization of nylon-66.

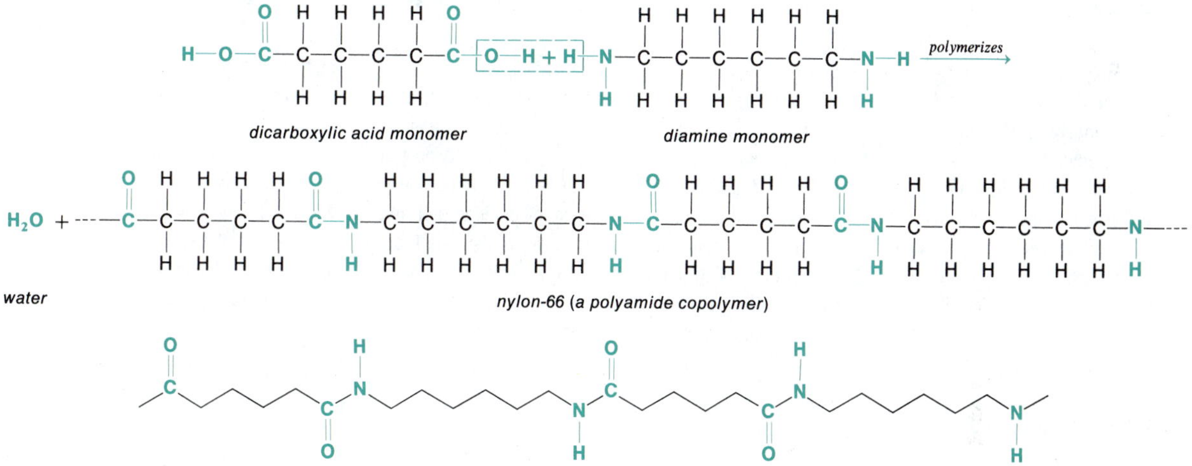

dicarboxylic acid monomer   diamine monomer   *polymerizes* →

water   nylon-66 (a polyamide copolymer)

Figure 14.11 Hydrogen bonds (shown in color) between nylon chains.

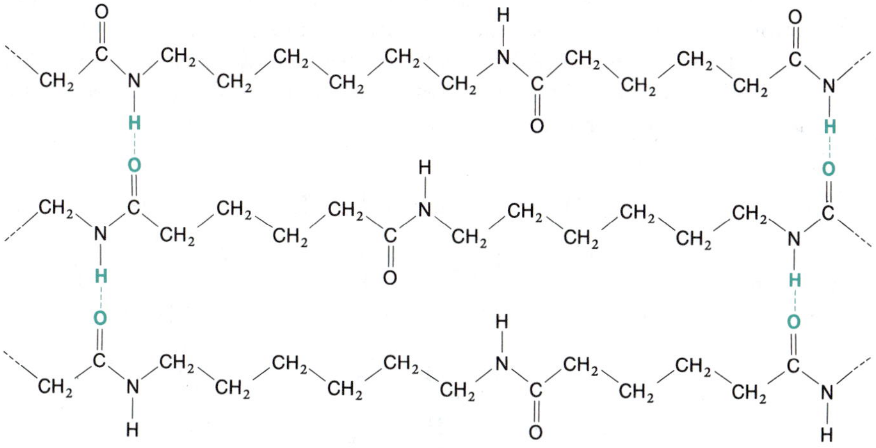

molecules keeps the molecules from slipping past each other. This makes nylon useful for such things as fishing line and tire cord.

THERMOSETTING POLYMERS   After thermoplastic polymers such as nylon, polyethylene, and vinyls melt, they regain their original texture when they cool. But a number of materials—especially certain condensation polymers—harden permanently when they melt. These thermosetting

CHEMISTRY SPOTLIGHT

THE SYNTHESIS OF NYLON

Beginning in the 1920s, chemists at the Du Pont Company tried to develop a polymer to replace silk or cotton. Their first attempts produced polyesters that were unstable in water and melted at warm (about 70°C) temperatures. Finally, in 1935 they synthesized a polyamide from six-carbon dicarboxylic acid and six-carbon diamine monomer units. The product was called nylon-66.

World War II cut off U.S. supplies of silk and hemp (for rope) from the Far East, increasing the need for substitute materials. Nylon was rushed into production, and it became a major material for parachutes, rope, and other military articles. Nylon also became a glamorous new fabric for clothing and other products. In 1940 the first nylon stockings went on sale. Within four days, nearly all the 4 million pairs that had been made were sold. A new era was born.

The Du Pont chemist most responsible for discovering nylon was Wallace Carothers. In 1937, depressed by feelings of failure, he committed suicide.

polymers get rigid because heating causes strong cross-links to form between polymer molecules.

The first thermosetting plastics were phenol–formaldehyde resins, developed in 1909 and sold under the trade name Bakelite. They form in a complex reaction (involving several intermediate chemicals) between phenol and formaldehyde, with water being split out and driven off by heat as the polymer sets:

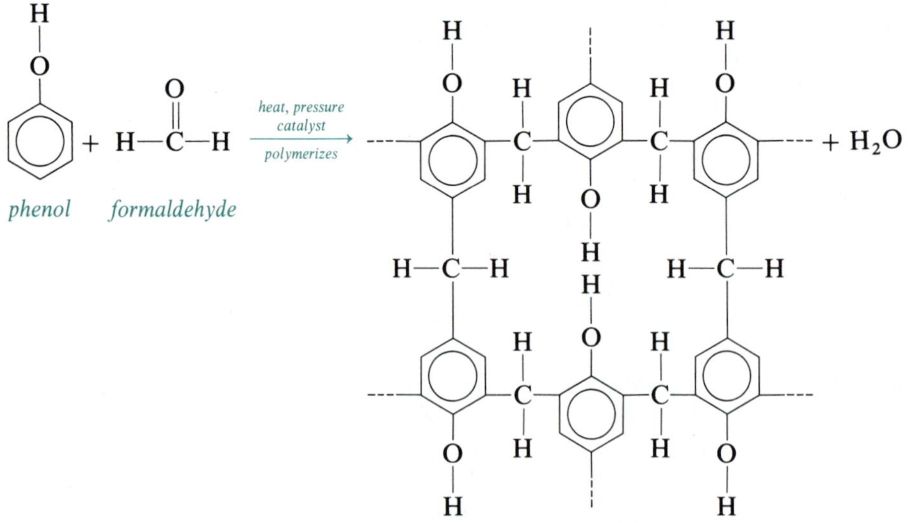

phenol   *formaldehyde*

phenol–formaldehyde polymer
Bakelite

Its stiff, rigid chains and multiple cross-links make this polymer very hard, rigid, and heat-resistant. It is widely used for radios, television circuit boards, plywood adhesive, electrical insulation, buttons, cookware handles, and distributor caps.

Using melamine (Figure 14.12) instead of phenol produces thermosetting melamine–formaldehyde resin, which has the trade name Melmac. Its main use is high-quality, everyday tableware that is dishwasher-safe and nearly unbreakable. It also is one of several polymers used to make permanent-press fabrics. Fabrics impregnated with a suitable polymer, such as melamine–formaldehyde resin, have their fibers cross-linked and fixed in their original, pressed shape. Then when the cloth is washed or worn, its individual fibers cannot slip over one another to put lasting wrinkles into the fabric.

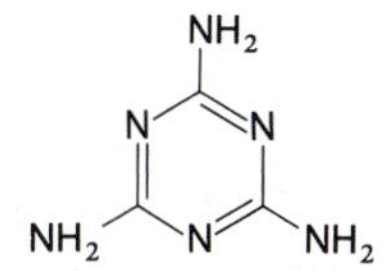

Figure 14.12 Melamine, a monomer unit for Melmac.

Polyurethanes are thermosetting (and condensation) polymers made from monomers with two isocyanate (—N=C=O) groups and monomers with two or more alcohol (—OH) groups (such as ethylene glycol or glycerine):

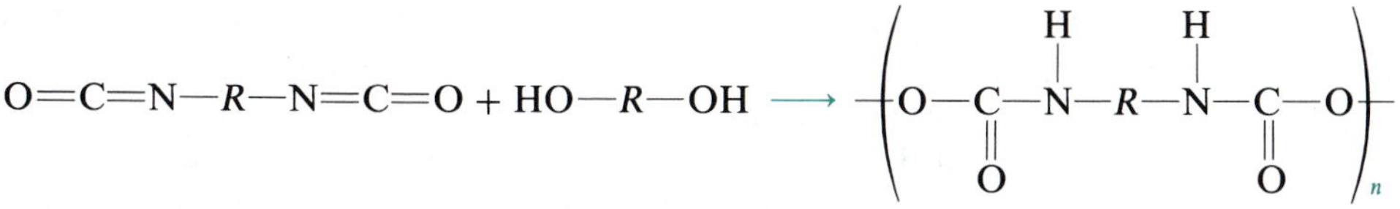

a di-isocyanate *a dihydroxy alcohol* *a polyurethane*

A blowing agent can be added to produce tiny gas bubbles in the plastic mass so that it becomes a rigid, quick-setting foam. These polyurethanes are known as "foam rubber." Polyurethanes are also used in stretch fabrics such as Spandex and Lycra, which are both rigid and elastic.

Epoxies are another family of thermosetting polymers. One of their monomer units has an oxygen atom in a three-membered ring bonded to two carbon atoms. One example is

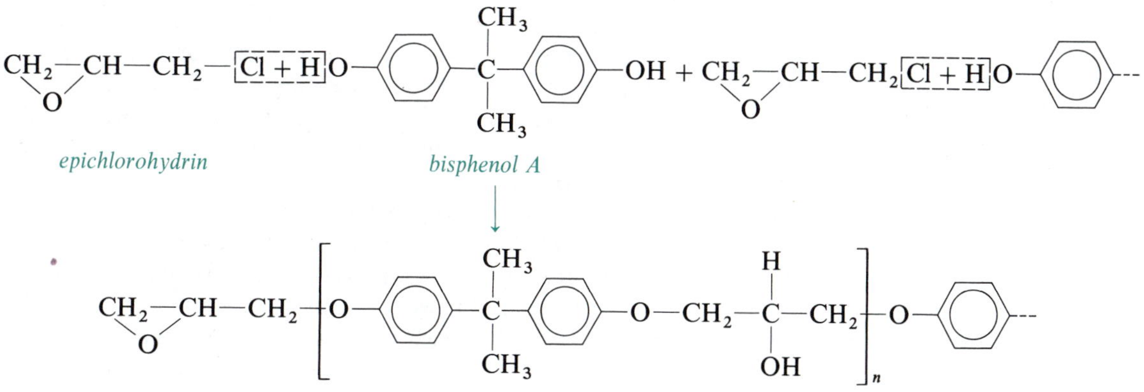

epichlorohydrin *bisphenol A*

an epoxy polymer

Epoxy polymers are extremely tough. They stick tightly to a variety of surfaces, and they are excellent insulators of electricity. Epoxy glues are usually sold in a pair of tubes whose contents must be mixed just before use. One tube contains the epoxy polymer and the other a curing agent that brings about the irreversible hardening of this type of thermosetting plastic. Once hardened, it holds two surfaces together so tightly that the seam is often stronger than the fastened materials themselves. Epoxy polymers also are used for tooth fillings and dentures.

14.4 Other Polymers: Silicones

SILICONES   All polymers discussed up to now have been organic polymers; that is, all contain a carbon backbone. In their search for polymers with a wide range of heat-resistant and structural properties, however, chemists have also produced polymers with noncarbon backbones.

The element most capable of replacing carbon as a backbone material in polymers is silicon (Si), which appears below carbon in the same group on the periodic table. Unlike carbon, silicon cannot form long chains of its atoms bonded to each other. But silicon can form long chains of alternating silicon and oxygen atoms. This is the basis for silicone polymers.

Chemists have synthesized **silicone polymers** (also known as *siloxanes*) from dihydroxysilicone monomer units:

silicone alcohol monomers   *a silicone polymer*

Depending on the average chain length and the degree of cross-linking, silicone polymers can be oils, greases, waxes, or solids. Because they aren't chemically reactive and don't decompose at high temperatures or thicken at low temperatures, silicone oils are effective lubricants in engines. Other oils are used in cosmetics, car waxes, and waterproof coatings. One kind of silicone oil, mixed with chalk, is known as Silly Putty.

A variety of silicone rubbers can be made by cross-linking fairly long silicone chains. Silicone rubber stays flexible at freezing temperatures and is more stable than other rubbers when exposed to heat, oxidizing agents, and other reactive chemicals. Like silicone oils, silicone rubber is useful in aircraft and space vehicles, which experience a wide range of temperatures. Some types are also used for electrical insulation, gaskets, O-rings, or as skin substitutes for burn victims (Figure 14.13). Plastic surgeons frequently implant silicone material to improve the appearance of the ears, chin, nose, cranium, breasts, or other body parts.

Besides organic and silicon-based polymers, chemists have made a few polymers containing boron, phosphorus, sulfur, and other elements. For example, boron–boron and boron–nitrogen polymers may have even greater heat resistance than silicone polymers. Most of these other polymers, however, are still in the experimental stage of development.

14.5 Polymers and the Environment: A Price to Pay

TOXIC MATERIALS   Polymers often contain additives to make them more useful for consumer products. For example, Styrofoam and polyurethane in the form of "foam rubber" contain foaming agents to provide bubbles as the polymer hardens; this makes the product light and an effective heat insulator. Chlorofluorocarbons (CFCs) are widely used as foaming agents, but their use is decreasing because they deplete the earth's protective ozone layer (Section 12.5).

Other additives include coloring agents, reinforcing materials, and plasticizers. Plasticizers make plastics—especially vinyls—softer and more pliable. One problem is that plasticizers eventually vaporize, leaving the polymer brittle and cracked. Another problem is that the plasticizers, once released into the environment, may be toxic.

A major problem comes from polychlorinated biphenyls (PCBs), a mixture of about seventy different but closely related compounds (Figure 14.14). PCBs have been used as insulating fluids in electrical transformers and in the production of plastics, paints, rubber, adhesives, sealants, and other products. Being nonpolar, PCBs don't dissolve in water and accumulate in fatty tissues in the body. Tests have shown that PCBs cause liver, kidney, stomach, and skin damage in laboratory animals. And recently PCBs were found to cause birth defects in humans (Section 21.5).

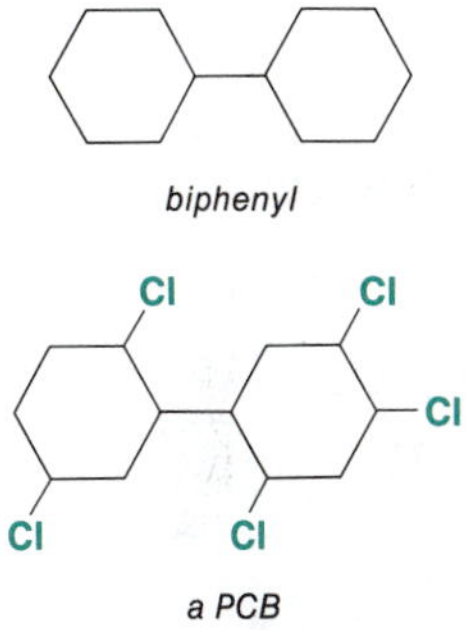

By law PCBs are no longer produced in the United States. But occasional accidents and indiscriminate dumping in landfills, fields, sewers, and roadsides have spread PCBs throughout the environment. Traces have been found all over the world in soil, surface and groundwater, fish, and human breast milk and fatty tissues—even in Arctic snow.

An alternative plasticizer is dioctyl phthalate (Figure 14.15) and related compounds. Though not approved for use with foods, these plasticizers aren't very toxic at single doses. But their long-term, accumulative effects are largely unknown.

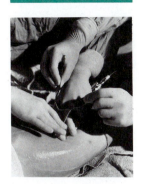

Figure 14.13 Membranes made of silicone polymer are used as a skin substitute for burn victims during their recovery. (Leonard Kamsler/ Medichrome)

biphenyl

a PCB

Figure 14.14 Biphenyl and one of many polychlorinated biphenyls (PCBs).

Figure 14.15 Dioctyl phthalate, a plasticizer.

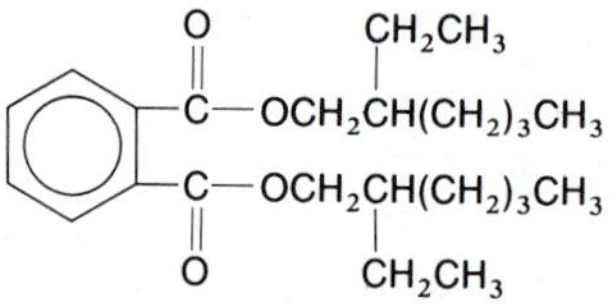

Another toxic material is vinyl chloride (see Table 14.1), the monomer used to make polyvinyl chloride (PVC). This was discovered when an unexpectedly high number of factory workers exposed to vinyl chloride developed liver cancer. New safety measures have greatly reduced workers' exposure to this cancer-causing agent. Unlike its monomer, PVC is nontoxic.

POLLUTION   Can you imagine going back to a world without plastics? Our homes would be missing much of their roofing, wall coverings, furnishings, and plumbing. The tires, mats, trim, upholstery, seats, grilles, instrument panels, and steering wheels would be absent from our cars. Electrical appliances and electronic gadgets would be gutted. Most products we buy would lack their packaging. And without synthetic fibers and textiles in our clothes, many of our outfits would be threadbare indeed.

The United States consumed 42 billion kg (19 billion lb) of plastics in 1970 and 106 billion kg (48 billion lb) in 1985 and will use 170 billion kg (76 billion lb) a year by the year 2000, according to estimates by the Society of the Plastics Industry. We use plastics mostly for packaging (30 percent), building and construction (23 percent), consumer products (8 percent), and such things as electrical components, transportation, coatings, and adhesives.

Where does all of this material go? Each person in the United States generates an average of about 40 kg (90 lb) of plastic trash per year. Plastics are manufactured to be durable, so they will remain unchanged in landfills for centuries. Beverage six-pack holders have a life expectancy in seawater of 450 years.

Figure 14.16 Seal muzzled by plastic. (Courtesy of the Center for Marine Conservation)

The Entanglement Network, a coalition of environmental groups, estimates that discarded plastic kills 2 million sea birds and 100,000 marine mammals each year. Seals get tangled in plastic fishing nets; fish and turtles eat plastic bags that look like jellyfish; and birds get tangled in plastic and eat plastic pellets, mistaking them for fish eggs (Figure 14.16). The plastic often lodges in the digestive tract of these animals, blocking further digestion.

Many polymers pollute the air when they burn. When buildings or furniture that contain plastics burn or when we intentionally burn plastics to dispose of them, we often observe a thick, stinky smoke. Some of the gases are toxic. As you can see from their structures (see Table 14.1), burning PVC produces caustic hydrogen chloride (HCl) gas, while burning acrylonitriles (such as Orlon) produces deadly hydrogen cyanide (HCN) fumes.

We have several ways to reduce pollution by plastics. One way is to produce less plastic by recycling and reusing it. In addition, plastics may become limited because they are petrochemicals (Section 8.3); as our supply of petroleum dwindles in the next century, it will become more difficult and expensive to produce plastics.

Another option is to make plastics that decompose in the environment. One strategy, for example, is to add cornstarch to certain plastics; when bacteria ingest the cornstarch, the plastic disintegrates within a few years. Beverage companies are experimenting with specially treated polyethylene bottles and six-pack rings that decompose within about four months when exposed to ultraviolet light.

In the near future, we are likely to use more, not less, plastic materials. Thus, it is increasingly important to find ways to deal with the growing problem of plastic wastes.

JOHN UPDIKE

In Praise of $(C_{10}H_9O_5)_x$

My tie is made of terylene [Dacron];

   Eternally I wear it,

For time can never wither, stale,

   Shred, shrink, fray, fade or tear it . . .

Summary

Polymers are very large molecules made by joining many small molecules, called monomers. Polymer molecules may be branched or linear and may be connected by cross-links; this affects their properties.

Addition polymers form when monomer units join together without the removal of any atoms. Polyethylene, polyvinyl chloride (PVC), polypropylene, polystyrene, and various synthetic rubbers are examples. Condensation polymers form when monomer units join together with the elimination of a small molecule, often water. Nylon, polyesters, and various thermosetting polymers are examples. Silicones are polymers based on silicon, not carbon.

Some ingredients in the production of plastics are toxic and accumulate in the environment. Disposing of plastic wastes is a problem.

Terms for Review

After completing this chapter, you should know and understand the meaning of the following terms:

addition polymer (p. 400)

condensation polymer (p. 409)

copolymer (p. 397)

cross-link (p. 399)

elastomer (p. 398)

fiber (p. 399)

monomer (p. 397)

plastic (p. 399)

polyamide (p. 410)

polyester (p. 409)

polymer (p. 397)

silicone polymer (p. 414)

thermoplastic polymer (p. 399)

thermosetting polymer (p. 399)

vinyl polymer (p. 401)

Questions

Odd-numbered questions are answered at the back of this book.

1. Distinguish among and give examples of (a) a monomer, (b) a natural polymer, and (c) a synthetic polymer.

2. Which of the following are copolymers? (a) Nylon, (b) polypropylene, (c) Orlon, and (d) Dacron.

3. Distinguish between (a) addition polymerization and (b) condensation polymerization. Give an example of each.

4. Compare and contrast high-density polyethylene (HDPE) and low-density polyethylene (LDPE) in terms of (a) method of preparation, (b) structure, (c) properties, and (d) common uses.

5. Which of the following monomers could be used to prepare addition polymers?

a. $H-\overset{\overset{\displaystyle H}{|}}{C}=\overset{\overset{\displaystyle H}{|}}{C}-\overset{\overset{\displaystyle H}{|}}{C}=\overset{\overset{\displaystyle H}{|}}{C}-H$

b. $H-\overset{\overset{\displaystyle H}{|}}{\underset{\underset{\displaystyle H}{|}}{C}}-\overset{\overset{\displaystyle H}{|}}{\underset{\underset{\displaystyle H}{|}}{C}}-O-H$

c. CH_2-OH
   $\overset{|}{C}H-OH$
   $\overset{|}{C}H_2-OH$

d. $HO-\overset{\overset{\displaystyle O}{\|}}{C}-\langle\bigcirc\rangle-\overset{\overset{\displaystyle O}{\|}}{C}-OH$

e. $CH_3CH_2-NH_2$

f. $H_2N-CH_2CH_2-NH_2$

g. $H_2N-\overset{\overset{\displaystyle H}{|}}{C}=\overset{\overset{\displaystyle H}{|}}{C}-NH_2$

6. For the appropriate monomer units in Question 5, write the formula (as in Figure 14.1) for the addition polymer each would form.

7. Would you expect Teflon to be less reactive chemically than nylon? Explain.

8. Would you expect polypropylene or polystyrene to be more flexible? Explain.

9. What is the polymer used in acrylic clothes fibers?

10. Give four examples of vinyl addition polymers.

11. For both of the following addition polymers, identify the monomer from which it was synthesized (see page 419):

a.
$$\cdots-CH_2-\underset{\underset{CH_3}{|}}{\overset{\overset{CH_3}{|}}{C}}-CH_2-\underset{\underset{CH_3}{|}}{\overset{\overset{CH_3}{|}}{C}}-CH_2-\underset{\underset{CH_3}{|}}{\overset{\overset{CH_3}{|}}{C}}-CH_2-\cdots$$

b.
$$\cdots-CH_2-\underset{\underset{F}{|}}{\overset{\overset{F}{|}}{C}}-CH_2-\underset{\underset{F}{|}}{\overset{\overset{F}{|}}{C}}-CH_2-\underset{\underset{F}{|}}{\overset{\overset{F}{|}}{C}}-CH_2-\underset{\underset{F}{|}}{\overset{\overset{F}{|}}{C}}-CH_2-\cdots$$

12. Which of the following are elastomers? (a) Polyvinyl chloride (PVC), (b) neoprene, (c) natural rubber, and (d) Plexiglas.

13. The process of using sulfur to cross-link rubber is called _____.

14. Name the three principal components of paints. Which of the three involves polymers? How have polymers made modern paints longer-lasting and easier to use?

15. Which of the substances in Question 5 could be used as a monomer to synthesize a condensation polymer?

16. Which substance in Question 5 could be a monomer unit for producing a polyester? A polyamide?

17. You buy some clothing made of polyester fiber. Explain the meaning of *polyester* in terms of chemical structure and reaction.

18. What is nylon-66?

19. What is a silicone polymer? How are such polymers formed?

20. Compare the general polymer structures of silicone oils and silicone rubber.

21. Examine Table 14.1. Which polymers, when burned, could give off hydrogen chloride (HCl) gas? Hydrogen cyanide (HCN) gas?

22. What is the function of PCBs as plasticizers? What are the problems with using PCBs?

23. Examine Figure 14.14. Draw the structures for two other compounds that could be classified as PCBs.

Topics for Discussion

1. Discuss three major reasons for the rapid growth of the synthetic polymer industry. Do you think rapid growth will continue during this century? Why or why not? Which types of growth should be encouraged or discouraged?

2. Should we reserve a significant portion of the world's fossil fuel resources exclusively for the production of polymers? Why or why not? What are some possible consequences of this decision for your own life?

3. List all the synthetic plastic, fiber, and rubber products you use each day (or those found in your room, home, or kitchen). List others you use occasionally. Don't forget packaging. Now classify each item as "essential," "nice but not necessary," or "frivolous, wasteful, and unnecessary."

4. Should we restrict the use of plastics to those that are readily biodegradable or recyclable? Why or why not?

CHAPTER 15

Laundry Products: Getting the Dirt Out

General Objectives

1. What are surfactants, and how do they help remove dirt from clothes, dishes, skin, and other items?

2. How is soap made, and why isn't it still widely used in the United States for washing clothes?

3. What are the major ingredients in modern laundry detergents, and what is the function of each of these chemicals?

The most widely used washday chemical is water. Laundry detergents, bleaches, fabric softeners, enzyme presoaks, water conditioners, and all other kinds of laundry products taken together amount to only a small fraction of the chemicals used to wash clothes. It is water that wets fabric, lifts out dirt, and carries grime down the drain. Without water, washing clothes would be impractical.

But trying to get clothes clean with water alone would be equally frustrating. Although a good solvent for many substances, water is not very effective for dissolving common laundry soil. Clothing could be rinsed in water for days and still remain dirty.

The combination of laundry products and water makes a clean wash possible. In particular, certain compounds called surfactants, or surface-active agents, in laundry detergents and other laundry products allow water to remove otherwise insoluble dirt from fabrics. Other ingredients are added mainly to assist the surfactant–water mixture in its cleaning. Thus, the study of wash-water chemistry, the subject of this chapter, is the study of surfactant solutions and how they work.

15.1 Surface-Active Agents: Removing Dirt and Grease

RINSING IS NOT ENOUGH Nearly everyone washes clothes in water because it is the cheapest, most abundant solvent available. It would be nice if water were the most effective solvent as well, but it is not. Water alone dissolves only a few common soils such as pancake syrup and perspiration. Simple rinsing with water does not remove most washday grime. But with the assistance of certain additives called surfactants, or surface-active agents, water becomes a very effective cleaner. Surfactant–water mixtures lift out ground-in dirt, remove greasy smudges, and carry away unsightly stains. A small amount of surfactant in water entices many substances, otherwise insoluble, to mix readily in water. Were it not for surfactants, the torrents of water that flood our washing machines would be ineffective.

SURFACTANTS AND SOLUBILITY   What are surfactants, and how can they change the solubility of dirt and stains? To answer these questions, we must first recall some principles of solubility (Section 5.3).

Solutions form only when solute molecules have intermolecular forces that are similar to those of the solvent molecules—*like dissolves like*. If the forces are very different, solute molecules stay to themselves and do not intermingle with solvent molecules. Water molecules, for example, attract one another by hydrogen bonding (see Figure 5.6), which is among the strongest of intermolecular forces (Section 5.1). The only things that dissolve in water therefore are substances with fairly strong intermolecular forces themselves—ionic compounds, such as salts, acids, and bases, and polar covalent compounds, such as alcohols and sugars; and not even all of these dissolve. Because nonpolar substances, such as hydrocarbons, have only weak London forces of attraction between their molecules (Figure 5.3), they cannot associate with water on a molecular basis.

The solubility of organic compounds in water is interesting because the various functional groups (Section 8.4) differ so widely in their polarities. Figure 15.1 shows a few examples. Ethane, a hydrocarbon with no functional groups, is nearly insoluble in water. But ethyl alcohol, which is ethane with a hydroxyl (—OH) group replacing one hydrogen, is polar enough to dissolve in water in all proportions. A hydroxyl group by itself, however, cannot make substances soluble that have large amounts of nonpolar hydrocarbon material. Butyl alcohol (C_4H_9OH), for example, is only partially soluble in water, and cetyl alcohol ($C_{16}H_{33}OH$) is even less soluble in water than is ethane. In fact, cetyl alcohol has so much hydrocarbon material that it is very soluble in oil (composed of hydrocarbons) despite its polar hydroxyl group.

Figure 15.1 The effects of functional groups (in color) and hydrocarbon content on solubility in water.

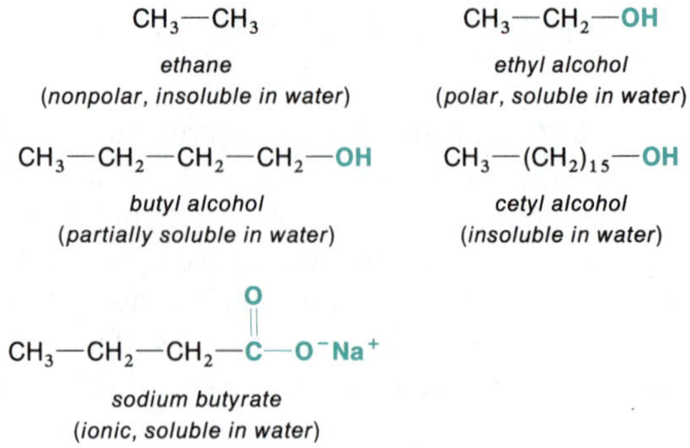

$CH_3—CH_3$
ethane
(nonpolar, insoluble in water)

$CH_3—CH_2—OH$
ethyl alcohol
(polar, soluble in water)

$CH_3—CH_2—CH_2—CH_2—OH$
butyl alcohol
(partially soluble in water)

$CH_3—(CH_2)_{15}—OH$
cetyl alcohol
(insoluble in water)

$$CH_3—CH_2—CH_2—\overset{\overset{\textstyle O}{\|}}{C}—O^-Na^+$$
sodium butyrate
(ionic, soluble in water)

If the functional group is ionic instead of just being polar, the substance has an even greater tendency to be water-soluble (and oil-insoluble). For instance, sodium butyrate, the salt of a carboxylic acid, is more soluble in water than is butyl alcohol, which has the same amount of hydrocarbon material but is polar instead of ionic.

With these solubility ideas fresh in mind, we can now talk about what surfactants are.

SURFACTANTS: MOLECULES WITH SPLIT PERSONALITIES Let's look at sodium palmitate, the sodium salt of the fatty acid known as palmitic acid (Figure 15.2). Sodium palmitate is to cetyl alcohol what sodium butyrate is to butyl alcohol. Sodium palmitate is a typical soap. Figure 15.2 also shows another surfactant, linear alkylbenzene sulfonate (LAS), which is widely used in laundry detergents. Notice that both surfactants in the figure consist of

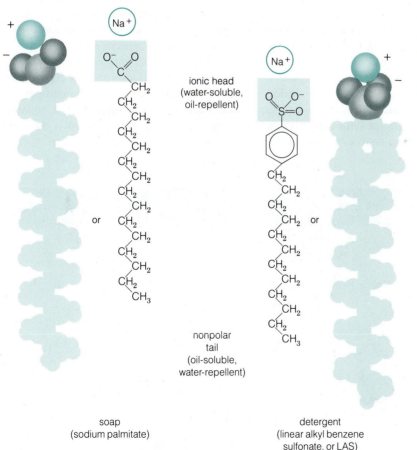

Figure 15.2 Typical surfactants have an ionic or polar end and a nonpolar end.

ionic head
(water-soluble,
oil-repellent)

nonpolar
tail
(oil-soluble,
water-repellent)

soap
(sodium palmitate)

detergent
(linear alkyl benzene
sulfonate, or LAS)

two parts, each with contradictory solubilities: a water-soluble, ionic end (the head) and a nonpolar, oil-soluble end (the tail).

surfactant head
(water-soluble, oil-repellent)

surfactant tail
(oil-soluble, water-repellent)

These opposite properties prevent the entire molecule from dissolving well in either oil or water. Nevertheless, the head dissolves in water and pulls the molecule part way into the water. Likewise, the tail portion dissolves in oil. But in either case, the soap tends to avoid the body of the solvent and remains at the surface (Figure 15.3). Substances that concentrate at the surface of a solvent are called **surface-active agents**, or **surfactants**.

Surfactants must possess a water-soluble, oil-repellent head and an oil-soluble, water-repellent tail. Many combinations are possible, but the tail section of a laundry surfactant typically resembles those shown in Figure 15.2: a hydrocarbon chain (with or without a benzene ring) of twelve to eighteen carbons. Carbon chains of this size are long enough to be fully insoluble in water, but not long enough to pull a strongly water-soluble head into an oil solution. The head section of a surfactant may take any one of dozens of structures. Some are ionic, carrying either a negative charge (anionic) or a positive charge (cationic). Others are polar but have no charge (nonionic). Table 15.1 shows an example of each type.

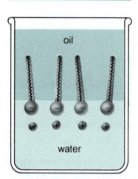

Figure 15.3 Oil and water do not mix. A surfactant remains at the surface separating them, with its nonpolar tails in the oil layer and the polar or ionic heads in the water.

TABLE 15.1 SURFACTANTS WITH DIFFERENT WATER-SOLUBLE HEADS

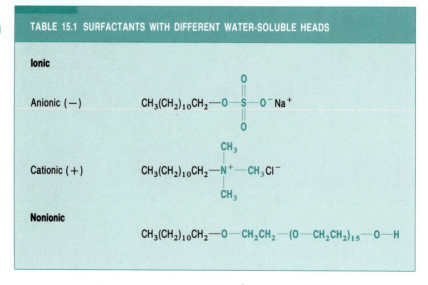

Ionic

Anionic (−) $CH_3(CH_2)_{10}CH_2—O—\overset{\displaystyle O}{\underset{\displaystyle O}{\overset{\|}{\underset{\|}{S}}}}—O^-Na^+$

Cationic (+) $CH_3(CH_2)_{10}CH_2—\overset{\displaystyle CH_3}{\underset{\displaystyle CH_3}{N^+}}—CH_3Cl^-$

Nonionic

$CH_3(CH_2)_{10}CH_2—O—CH_2CH_2—(O—CH_2CH_2)_{15}—O—H$

Because they concentrate at the surface of a liquid and nowhere else, surfactants have four special properties. When mixed with a liquid, surfactants (1) reduce the liquid's surface tension, (2) increase its tendency to wet hard surfaces, (3) cause foaming, and (4) promote detergent action.

SURFACE TENSION   Every molecule is attracted to the other molecules that surround it. Molecules located within the body of a liquid are attracted evenly on all sides and are not pulled in any one direction (Figure 15.4). Surface molecules, on the other hand, have no molecules above them. Thus, they feel a net force of attraction called **surface tension** and are pulled closer together into the liquid (Figure 15.4). Because of surface tension, a liquid takes the shape that gives it as little surface area as possible.

The surface of a liquid behaves as if it were under a tension and had a thin skin, or membrane, stretched across it. The surface tension of water causes many familiar phenomena. Rain falls in the form of droplets because, in that shape, water has the least amount of surface area. Similarly, two raindrops on a window pane tend to coalesce into one so that their surfaces are minimized. Water glasses can be filled above the brim without overflowing. A carefully placed steel sewing needle can float on water, and certain insects can flit around on water without sinking despite being denser than water.

When a surfactant mixes with a liquid, the surfactant takes the place of surface molecules (Figure 15.5). Each surface molecule replaced in this way becomes just like any other molecule in the body of the liquid and no longer feels the unbalanced forces that produced the surface tension. Although surfactant units on the surface attract each other, these forces are generally so small that the surface tension decreases dramatically. A glass cannot be overfilled with a water-surfactant mixture; a water bug has trouble staying afloat on one; and hard surfaces become entirely wet and droplet-free.

WETTING   Everyone knows that if you fall in the water, you get wet. Your skin becomes covered with moisture. But not everything gets wet in water. The newly waxed hood of a car, for example, stays mostly dry in the rain.

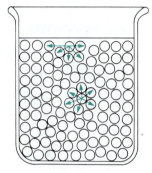

Figure 15.4 Molecules in the interior of a liquid are attracted in all directions. Molecules on the surface, however, are attracted downward and closer together to minimize the surface area of the liquid. This is surface tension.

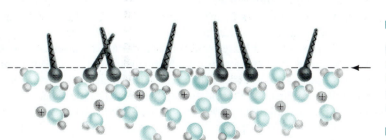

Figure 15.5 When surfactant units take the place of the liquid's molecules at the surface, surface tension is reduced.

The water that hits it beads up into droplets and leaves most of the hood without a coating of water.

Wetting is the coating of a solid with a liquid. How wet the solid gets depends on the forces between molecules of the solid and those of the liquid. Because of this, we can talk about wetting in the same way we talked about solubility: *Like wets like.* Wetting occurs only when surface molecules of the solid have attractive forces similar to those of surface molecules of the liquid. A polar liquid such as water cannot coat a solid surface that cannot strongly attract polar molecules. On a nonpolar surface, for example, water molecules hydrogen bond to each other, causing the water to form into droplets that do not wet the surface (Figure 15.6, top). Besides the wax coating on your car, common examples of water-repellent surfaces include bathroom tile, kitchenware, eating utensils, and many fabrics.

When a surfactant is added to a liquid, however, surface molecules are replaced, and the forces of attraction at the surface are altered (see Figure 15.5). The liquid can then coat solids because the liquid surface is more compatible with the solid surface (see Figure 15.6, bottom). We might even say that a surfactant makes the liquid wetter. Surfactant–water mixtures wet glassware with a sheet of liquid that does not spot the glass. They soak into clothes faster, and they do not bead up on the hood of a car.

FOAMING If enough surfactant is added to a liquid, every possible surface molecule of the liquid is replaced with surfactant. If the liquid remains perfectly still after this, additional surfactant doesn't interact with it. The slightest agitation or splashing, however, exposes fresh surface molecules that are immediately replaced with surfactant. This fresh surface remains on the outside of the liquid in tiny hollow domes, or bubbles, supported inside and out by the surfactant. We see this effect as foam, or suds.

DETERGENT ACTION The most important property of a surfactant, at least for laundry purposes, is its detergent action, which gives it the ability to remove soil from clothes.

Soil is often a mixture of oily, fatty, or hydrocarbonlike materials. Whether composed of mud, dirt, grit, clay, cooking oil, or meat fat, soil is almost always insoluble in water. This means that rinse water alone cannot wash it away. But with a surfactant in the water, matters are different.

Oil-soluble tails of surfactant molecules bury themselves in any greasy material present, and water-soluble heads orient themselves toward the water (Figure 15.7). Many thousands of surfactant units pack in around the dirt and grease, piercing it like pins in a pin cushion. This releases the dirt and grease from the fabric and breaks it into smaller pieces, each surrounded by a shell of negative charges from the protruding water-soluble heads (Figure 15.7). Repulsive forces between these like-charged particles

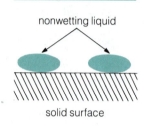

nonwetting liquid

solid surface

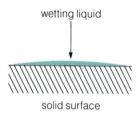

wetting liquid

solid surface

Figure 15.6 A nonwetting liquid (top) forms beads of liquid on a solid surface. A wetting liquid (bottom) such as a surfactant spreads out and covers a solid surface.

soap or
detergent
surfactant
units

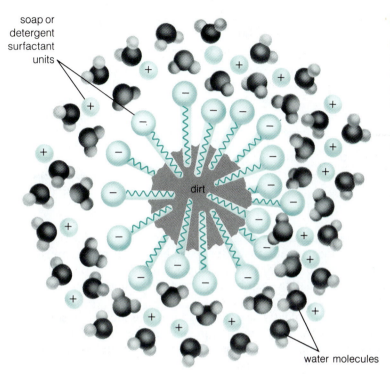

dirt

water molecules

keep them from coming together to reform larger particles. As a result, dirt particles stay suspended in water until they are rinsed away.

15.2 Soap and Hard Water: Ring Around the Collar

THE EARLIEST SURFACTANT   Soap is the most familiar surfactant because it has been in use longer than any other. The early Egyptians used an earthy alkali as a washing aid; the Pilgrims and pioneers laundered with home-made lye soap; and the twentieth-century U.S. population purchases a billion pounds of soap a year.

Soap is made by reacting a strong base, typically sodium hydroxide (NaOH), with triglycerides (Section 9.3) in animal or vegetable fats (Figure 15.8). The reaction, called **saponification**, produces glycerol and sodium salts of the fatty acids and is like the reverse of the reaction for the formation of a fat or triglyceride (see Figure 9.9). After glycerol is removed, the remaining mixture of sodium salts of fatty acids is **soap**.

Figure 15.8 The saponification reaction, in which soap is made by the reaction of a strong base with animal or vegetable fat.

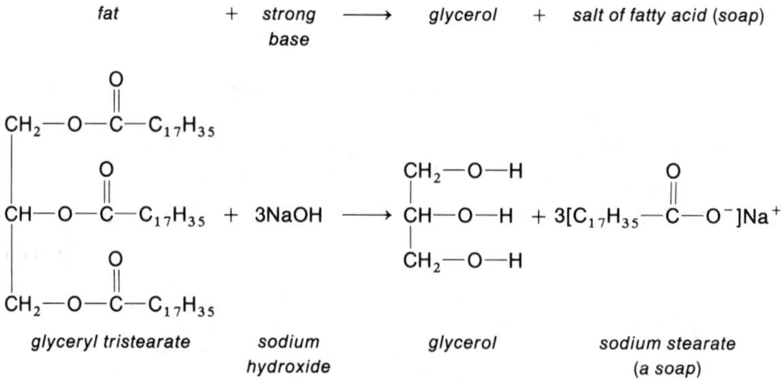

$$\text{fat} \quad + \quad \underset{\text{base}}{\text{strong}} \quad \longrightarrow \quad \text{glycerol} \quad + \quad \text{salt of fatty acid (soap)}$$

$$
\begin{array}{c}
\overset{\displaystyle O}{\overset{\|}{\text{CH}_2\text{—O—C—C}_{17}\text{H}_{35}}} \\
\overset{\displaystyle O}{\overset{\|}{\text{CH—O—C—C}_{17}\text{H}_{35}}} + 3\text{NaOH} \longrightarrow \begin{array}{c} \text{CH}_2\text{—O—H} \\ \text{CH—O—H} \\ \text{CH}_2\text{—O—H} \end{array} + 3[\text{C}_{17}\text{H}_{35}\overset{O}{\overset{\|}{\text{—C—O}^-}}]\text{Na}^+ \\
\overset{\displaystyle O}{\overset{\|}{\text{CH}_2\text{—O—C—C}_{17}\text{H}_{35}}}
\end{array}
$$

glyceryl tristearate   sodium hydroxide   glycerol   sodium stearate (a soap)

A soap may contain any or all of the sodium salts shown in Table 15.2, plus salts of other, unsaturated fatty acids. No commercial soap is anywhere near 99 and 44/100 percent pure anything—at least not in the chemical sense. Floating soap bars are made by whipping air into them to make them less dense than water.

Potassium salts of fatty acids, analogous to the sodium salts in Table 15.2, are also soaps. They are used in shaving creams and liquid-soap preparations. Tincture of green soap, often used in hospitals for cleaning skin wounds, is a solution of potassium soaps in alcohol.

TABLE 15.2 SOME SUBSTANCES FOUND IN SOLID SOAPS

Name	Source	Number of Carbons	Formula
Sodium laurate	Coconut oil	12	$CH_3(CH_2)_{10}\overset{O}{\overset{\|}{C}}\text{—O}^-\,Na^+$
Sodium myristate	Palm oil	14	$CH_3(CH_2)_{12}\overset{O}{\overset{\|}{C}}\text{—O}^-\,Na^+$
Sodium palmitate	Palm oil	16	$CH_3(CH_2)_{14}\overset{O}{\overset{\|}{C}}\text{—O}^-\,Na^+$
Sodium stearate	Beef tallow	18	$CH_3(CH_2)_{16}\overset{O}{\overset{\|}{C}}\text{—O}^-\,Na^+$

Regardless of its composition, any soap is a good surfactant. Each salt in Table 15.2 has a water-soluble, ionic head and a water-repellent, hydrocarbon tail. Thus, compared to pure water, soapy water has less surface tension, more wetting ability, a greater tendency to form suds, and better cleaning power. These properties make soap one of the best cleansing agents available.

THE HARD WATER PROBLEM   One property of soap, however, led to a sharp decline in its use for washing clothes since the 1940s. Whenever soap mixes with **hard water** (water that contains dissolved calcium, magnesium, or iron salts), it combines with the metal ions of these salts (Ca^{2+}, Mg^{2+}, Fe^{2+}) to form an insoluble soap curd:

$$2CH_3(CH_2)_{14}-\overset{\overset{\displaystyle O}{\|}}{C}-O^-Na^+ + Ca^{2+} \longrightarrow [CH_3(CH_2)_{14}-\overset{\overset{\displaystyle O}{\|}}{C}-O]_2Ca + 2Na^+$$

sodium palmitate	*(Mg^{2+} or Fe^{2+})*	*calcium palmitate*
(a soap)		*(insoluble in water)*

This reaction causes an unsightly gray precipitate to form in wash water, and it removes the soap from solution so it can't suspend oil and dirt. Soap scum leaves a ring around the bathtub when the water is drained and accounts for the "ring around the collar" and other soiling of clothes washed with soap in hard water. Old-fashioned, ringer-type washers squeezed this gray scum from clothes and yielded a fairly clean wash. Modern, spinning-type washing machines, however, cannot handle soap curd. They force the curd-containing rinse water right through the clothes, leaving the curd embedded in the fabric. Thus, soap is not very useful in hard water.

SOFT WATER   One alternative is to use **soft water**—water that does not contain large amounts of Ca^{2+}, Mg^{2+}, or Fe^{2+} ions. Some parts of the United States have water that is naturally low in these ions, but most of the nation has hard water (Figure 15.9). Here water must be treated if Ca^{2+}, Mg^{2+}, and Fe^{2+} ions are to be removed.

In a typical home water softener, water passes through a tank containing zeolites. *Zeolites* are complex sodium aluminum silicates that have an open structure, with negatively charged aluminosilicate frameworks (represented by Z^-) and sodium ions in the openings of the frameworks. Zeolites strongly attract metal ions (Ca^{2+}, Mg^{2+}, Fe^{2+}) in hard water and bind them to open sites on their surface. Bound ions displace sodium ions and remain in the resin (Figure 15.10). This replacement of one group of ions in water for another is called **ion exchange**. The water that flows out of a home water softener has an increased concentration of sodium ions but very low levels of ions that would create soap curd.

Figure 15.9 Typical hardness of public water supplies in the continental United States. (Courtesy U.S. Geological Survey)

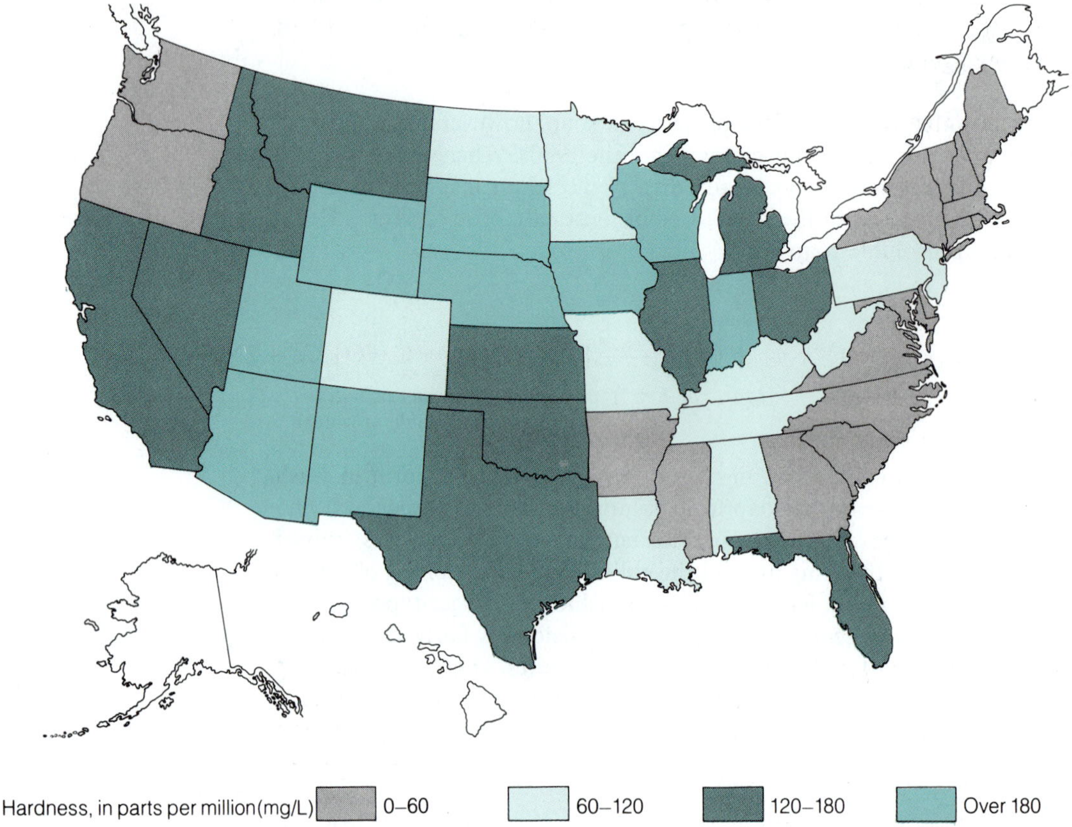

Hardness, in parts per million(mg/L)   0–60   60–120   120–180   Over 180

After a day or two of use, zeolites stop working as ion exchangers because they become saturated with mineral ions. They are then recharged by flooding them with a concentrated solution of sodium chloride. This strips zeolites of mineral ions and replaces them with sodium ions, at which point the tank can soften another couple of days' worth of water. This cycle can be repeated indefinitely.

Yet even in soft water, soap curd forms from minerals carried by the dirty wash water. About the only way to avoid this problem is to use a surfactant other than soap. These surfactants are key components of synthetic detergents that have largely replaced soap for washing clothes in the United States and in most industrialized nations.

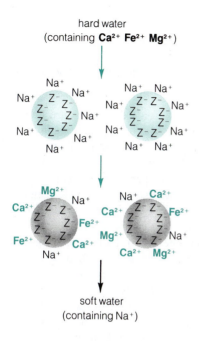

hard water
(containing **Ca²⁺ Fe²⁺ Mg²⁺**)

soft water
(containing Na⁺)

Figure 15.10 How an ion-exchange water softener works. Ca^{2+}, Mg^{2+}, and Fe^{2+} in hard water bind to the negatively charged zeolite (Z^-) resin. This releases Na^+, which is removed in the softened water.

15.3 Synthetic Laundry Detergents: Replacements for Soap

SYNTHETIC DETERGENTS The first synthetic surfactants that worked in hard water were developed in the 1930s. But it was not until 1946, with the introduction of Proctor & Gamble's Tide, that they could begin to compete with the cleaning power of soap. Since then, in most industrialized nations, all-purpose synthetic detergents have largely replaced soap for washing clothes and dishes.

These products contain three main types of ingredients:

— Surfactants
— Builders such as phosphates or water-softening agents
— Other ingredients such as bleaches and brighteners

The most important ingredients, of course, are surfactants. They actually do the cleaning. Yet surfactants constitute no more than 20 percent by weight of any powdered detergent, and sometimes they amount to as little as 8 percent; the water-softening agents, called builders, make up to 50 percent of the product. Heavy-duty liquid detergents, however, are 30 to 40 percent surfactants and contain no builders. Other ingredients in laundry products range from bleach, borax, and brighteners to corrosion inhibitors, processing aids, and antiredeposition agents.

Three types of surfactants are used in laundry products: *anionic*, *cationic*, and *nonionic* (see Table 15.1). Each has advantages and disadvantages.

ANIONIC DETERGENTS   The most widely used anionic detergents are linear alkylbenzene sulfonates (LAS; see Figure 15.2) and alkyl sulfonates (see Table 15.1). The water-soluble head, which is the salt of a sulfonic acid, is negatively charged (or anionic). The oil-soluble alkyl tail is somewhat shorter than soap's, containing nine to fifteen carbons. The presence or absence of a benzene ring makes little difference in the surfactant's cleaning power, but substances that have it are easier and cheaper to manufacture.

The cleaning, foaming, and wetting properties of alkylbenzene sulfonates and soaps are similar. But alkylbenzene sulfonates have one significant advantage: They react with minerals in hard water to form a water-soluble product. In other words, these detergents don't produce scum or any other fabric-fouling film when used in hard water.

One of soap's disadvantages, however, also occurs with alkylbenzene sulfonates: Hard-water minerals still destroy the cleaning action of surfactants. When calcium, magnesium, or iron ions react with them, the resulting salts, though soluble, are no longer surfactants. This is one reason why they must be coupled with builders in order to be effective.

CATIONIC DETERGENTS   The water-soluble head of a surfactant may be positively charged (or cationic), as shown in Table 15.1. These kinds of surfactants, however, find little use as laundry-cleaning agents. They are so attracted to the negatively charged clothes that they cannot lift out and remove dirt. But this very property makes them appropriate as antistatic agents and as fabric softeners. Because they also can kill some germs, cationic detergents are widely used in hospitals, especially as surgical scrubs.

NONIONIC DETERGENTS   A substance need not ionize in solution to be a surfactant. Any functional group that is sufficiently polar can act as the water-soluble head of a surfactant. Such is the case with the major nonionic detergents, ethoxylated linear alcohols, which contain many polar oxygen atoms (see Table 15.1).

Their lack of charge enables nonionic detergents to avoid water-hardness reactions and makes them especially effective in removing non-charged, oily soils. They also produce fewer suds and thus have less tendency to clog washing machines with foam. The quarter-cup liquid detergents rely heavily on nonionic surfactants for their effectiveness.

Being without a charge is also their liability. Because clothes tend to develop a negative charge in the wash, anionic surfactants are repelled and are able to keep soil suspended in wash water. Nonionic surfactants, however, are less able to do this and thus allow dirt to redeposit in the clothes.

BIODEGRADABILITY   Surfactants do not stop being surfactants when the wash cycle is over and they are rinsed down the drain. They continue to cause foaming, wetting, and reduced surface tension until they are decomposed somehow. The billions of pounds of surfactants we use every year could become an enormous environmental problem if they accumulated in the waterways. The bacteria that abound in water, however, can decompose many surfactants by breaking down the hydrocarbon end. Such surfactants are called **biodegradable**.

The earliest synthetic detergents were not biodegradable. The cheapest and easiest surfactants to make had a *branched* hydrocarbon tail rather than a *linear* one. Prior to 1966, the most widely used surfactant in laundry detergents was sodium alkylbenzene sulfonate (ABS) with a formula like

$$CH_3CHCH_2CHCH_2CHCH_2CH \overset{}{\underset{\underset{CH_3\quad CH_3\quad CH_3\quad CH_3}{}}{}} \!\!-\!\!\bigcirc\!\!-\!\!\overset{\overset{O}{\|}}{\underset{\underset{O}{\|}}{S}}\!\!-\!\!O^-Na^+$$

Figure 15.11 Foam on a creek caused by branched-chain, nonbiodegradable detergents used prior to 1966. (U.S. Department of Agriculture, Soil Conservation Service)

This surfactant has several branched hydrocarbon groups (shown in color), in sharp contrast to modern surfactants such as LAS (see Figure 15.2).

Bacteria cannot break down branched surfactants, so during the 1960s, many waterfalls, rivers, sewage-disposal plants, and septic tanks across the country contained masses of foam (Figure 15.11). In fact, it was common to see certain lakes covered with foam or to turn on a water faucet and get suds. Now all surfactants have linear hydrocarbon tails and are biodegradable. If suds form in the streams and rivers, they are not caused by modern detergents.

15.4 Builders: Better Cleaning

DETERGENT TEAMWORK   Modern surfactants cannot equal the cleaning performance of soap without the help of certain water conditioners called **builders**. Both mineral ions from hard water and hydronium ions from excess acidity reduce the solubility of surfactants in water. Builders protect the surfactant from the effects of these ions, and they stabilize the surfactant-coated soil once it has been lifted from the fabric so that it can be rinsed away without being redeposited.

PHOSPHATES   The most widely used builder is sodium tripolyphosphate (STPP), which dissolves in water to yield a mixture of sodium ions and tripolyphosphate ions:

$$Na^+ \qquad {}^-O-\overset{\overset{\displaystyle O}{\|}}{\underset{\underset{\displaystyle O^-}{|}}{P}}-O-\overset{\overset{\displaystyle O}{\|}}{\underset{\underset{\displaystyle O^-}{|}}{P}}-O-\overset{\overset{\displaystyle O}{\|}}{\underset{\underset{\displaystyle O^-}{|}}{P}}-O^-$$

sodium   tripolyphosphate ion (TPP)
ion

Tripolyphosphate (TPP) ions help soften water and increase the cleaning action of synthetic detergents. The negatively charged TPP ions attract calcium, magnesium, and other mineral ions in hard water, encasing the offending ions so that they cannot reach the surfactant. As the salt of a weak acid, TPP ions also react with hydronium (H_3O^+) ions and raise the pH of the wash water; the surfactant, also the salt of a weak acid in most cases, works much more effectively in such alkaline solutions. TPP ions further add to the surfactant's cleaning power by dispersing and suspending inorganic clay-type soils.

STPP is produced quite cheaply from the abundant deposits of calcium phosphate rock in the earth's crust. It is nontoxic to humans, animals, and plants; mild to the skin and mucous membranes; and harmless to machinery and equipment. Because of its low cost, safety, and effectiveness, nearly a billion pounds of STPP are used in laundry detergents throughout the United States each year.

WATER POLLUTION   A disadvantage of STPP and other phosphate builders is that when they are washed down the drain and discharged into rivers and lakes, they become plant nutrients that accelerate the eutrophication process (Section 13.4). Phosphates fertilize the excessive growth of undesirable blue-green algae. This depletes the oxygen dissolved in the water, destroying fish and encouraging growth of bacteria that produce foul-smelling and toxic chemicals such as hydrogen sulfide (H_2S).

Local legislators have taken action to solve this problem. In about 25 percent of the United States, phosphates were banned from detergents altogether. In other areas, detergent products that once were nearly half STPP (which amounts to 12.3 percent elemental phosphorus) were limited to a third (8.7 percent P) or a quarter (6.1 percent P). Most laundry products now list on the label their percentage of phosphorus.

PHOSPHATE SUBSTITUTES   With these limitations on the phosphorus content in detergents, manufacturers have tried to find suitable substitutes. Several environmentally safe replacements are currently in use, but none is as effective as STPP.

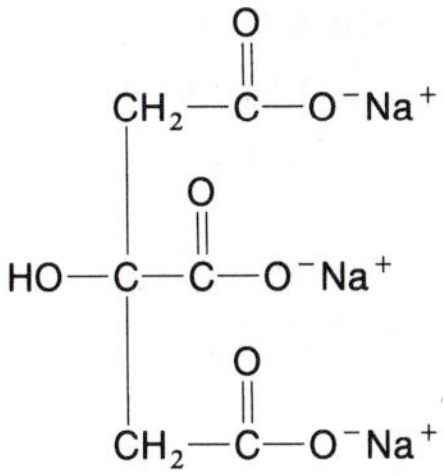

Figure 15.12 Some nonphosphate builders.

sodium citrate

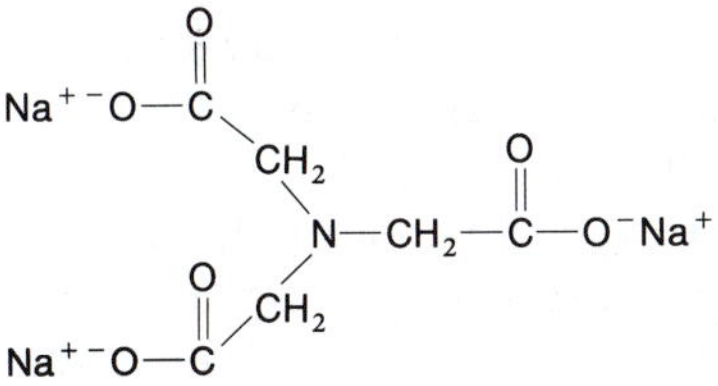

sodium nitrilotriacetate (NTA)

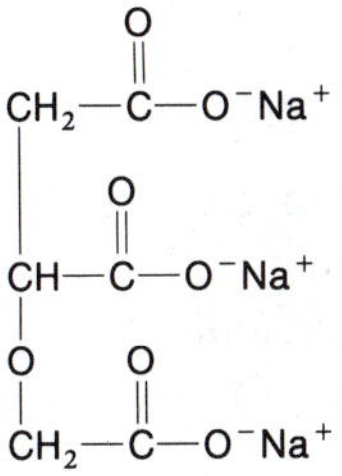

sodium carboxymethyloxysuccinate

$2Na_2O \cdot 2Al_2O_3 \cdot 4SiO_2 \cdot 2H_2O$

zeolite
(sodium aluminosilicate)

The two most widely used substitutes are sodium citrate and zeolite (Figure 15.12). Sodium citrate enjoys a long-established safety record as a natural ingredient in foods. It works well in keeping wash water alkaline but is much weaker then STPP in softening water. Zeolite is an effective softener; in fact, it is used in home water softeners (see Figure 15.10). But it is insoluble and can be used only in powdered detergents.

Possible future substitutes for STPP include trisodium nitrilotriacetate (NTA), sodium carboxymethyloxysuccinate, and sodium carboxymethyltartronate. NTA (see Figure 15.12) was used for several years in the late 1960s. But its ability to solubilize toxic metals in rivers and lakes raised questions about its safety, so NTA was voluntarily taken off the market. Those questions are still not resolved. The other two compounds have been patented by large manufacturers and are still in the testing stages.

15.5 Other Detergent Ingredients: Making Detergents Better

BLEACHES   When a fabric is cleaned by a surfactant–water mixture, a tiny residue of soil is left behind. The amount is so insignificant that it would not bother even the most fastidious homemaker—except when the residue is highly colored. Then it is called a *stain*.

Bleaches remove the color in stains by oxidation (Section 6.4). It is a fortunate fact of nature that oxidized states of most stains are colorless. Hence, bleaches are oxidizing agents that do not remove stains; they simply make stains invisible.

Bleaches have typically contained sodium hypochlorite (NaOCl) as the oxidizing agent (Section 6.4). But problems with residual chlorine and its odor have increased the use of sodium perborate:

$$Na_2 \begin{bmatrix} H{-}O & O{-}O & O{-}H \\ & B & & B \\ H{-}O & O{-}O & O{-}H \end{bmatrix} \cdot 6H_2O$$

sodium perborate (a laundry bleach)

Sodium perborate decomposes in warm or hot water to give hydrogen peroxide (H_2O_2), which is responsible for the bleaching action. These bleaches are mild enough oxidizing agents so that they don't alter the color of permanent dyes in textiles. Thus, they improve the laundry's whiteness while being color-safe.

AUXILIARY BUILDERS   Sodium tetraborate, $Na_2B_2O_7 \cdot 10H_2O$, or borax, obtained from the desert dry lakes of southern California, acts as an auxiliary builder. Not only does borax increase the pH of the wash water, but it also removes any calcium and magnesium ions by forming insoluble solids with them.

BRIGHTENERS   The appearance of the fabric is a critical part of its cleanliness—especially for white fabrics, where even the tiniest amount of residual yellow turns the color off-white. The old-fashioned remedy for this was *bluing*, a blue dye added to the wash so that any yellow color would be absorbed. Virtually all modern detergents, however, contain whiteners called *optical brighteners*. Instead of taking visual color away, they actually add to the fabric's brightness.

White fabrics reflect almost all visible and ultraviolet (UV) components of light, although our eyes cannot see UV light. Certain complicated organic molecules, because of the arrangement of their double bonds between carbon atoms, can absorb UV light and re-emit it as visible blue. One optical brightener is

stilbene–triazine (an optical brightener)

Our eyes see more light reflecting off treated fabric than shines on it. So these optical brighteners really do make a fabric "whiter than white."

ANTIREDEPOSITION AGENTS   When a surfactant fails to suspend removed soil in wash water, dirt redeposits before the rinse cycle. After repeated washings, these deposits build up until clothes look dull and dingy. To prevent this, most detergents contain sodium carboxymethylcellulose, a cottonlike substance that absorbs into fabric and repels surfactant-removed soil so that none returns.

ENZYMES   Television commercials relentlessly tell viewers that the worst kind of dirt is greasy dirt. Hydrocarbon-based soil of this sort may be repulsive, but it is the very kind of grime that surfactants handle best. Much more difficult to remove are stains caused by protein-based substances— such as blood, mucus, milk, and eggs. These are often insoluble in both water and oil, and they tend to lock themselves onto the fibers of the fabric. The surfactants in detergents can't remove them.

For this reason, enzyme catalysts (Section 9.4) are added to some detergent products. These enzymes split (hydrolyze) the protein-based soils into simpler substances that the surfactant can remove. They do this job effectively. One minor concern, however, is that enzyme detergents, if spilled on the skin, may cause irritation and rashes by degrading some of the skin protein.

MISCELLANEOUS INGREDIENTS   Many substances that don't have direct cleaning functions also appear in detergents. Sodium silicate, for example, helps prevent corrosion of the washing machine. Other ingredients help keep the product homogeneous, and fragrances and colorings make it more appealing to the senses. An immense amount of chemistry separates us from the days when people stirred boiling pots to make alkali soap.

ARTHUR D. CAMPBELL   Changing back to soaps for washing clothes would not only create an unfillable demand for natural fats—it would require equipping home laundries with new machines.

Summary

Soaps and synthetic detergents are surfactants that help water remove nonpolar soils from fabrics and other materials. Surfactants do this by reducing water's surface tension, by increasing wetting and foaming, and by promoting the suspension and removal of dirt and grease. To suspend nonpolar soils in water, surfactants have a water-insoluble nonpolar end and a water-soluble ionic or polar end.

Soaps are salts of fatty acids. They are effective surfactants but form an insoluble scum in hard water. Soaps have been largely replaced as laundry agents by three types of synthetic detergents—anionic (negatively charged), cationic (positively charged), and nonionic (uncharged).

Besides surfactants, laundry products contain builders, which help soften water, increase its pH (alkalinity), and help keep soils suspended. Other ingredients include bleaches, optical brighteners, enzymes, corrosion inhibitors, and agents to prevent redeposition of removed dirt.

Terms for Review

After completing this chapter, you should know and understand the meaning of the following terms:

biodegradable (p. 433)

builder (p. 433)

hard water (p. 429)

ion exchange (p. 429)

saponification (p. 427)

soap (p. 427)

soft water (p. 429)

surface-active agent (p. 424)

surface tension (p. 425)

surfactant (p. 424)

wetting (p. 426)

Questions

Odd-numbered questions are answered at the back of this book.

1. Why is ethyl alcohol (C_2H_5OH) soluble in water when cetyl alcohol ($C_{16}H_{33}OH$) is not?

2. What structural features must a surfactant have?

3. Identify which of the following would act as surfactants:

a. $Na^+ \ ^-O-\overset{\overset{\displaystyle O}{\|}}{C}-CH_2CH_3$

b. $CH_3(CH_2)_{15}-\overset{\overset{\displaystyle O}{\|}}{C}-O^-Na^+$

c. $Na^+ \ ^-O-\overset{\overset{\displaystyle O}{\|}}{\underset{\underset{\displaystyle O}{\|}}{S}}-O-(CH_2)_2CH_3$

d. $CH_3(CH_2)_{11}-OH$

e. $CH_3(CH_2)_{17}-\overset{\overset{\displaystyle CH_3}{|}}{\underset{\underset{\displaystyle CH_3}{|}}{N^+}}-CH_3Cl^-$

4. Name the four special properties of a surfactant.

5. Comparing surfactant–water mixtures to plain water, identify which of the following statements are true: (a) The mixture forms droplets more easily on a smooth surface; (b) the mixture tends to be more sudsy; (c) the mixture dissolves more oily materials; (d) dust floats better on the mixture; (e) the mixture has more difficulty soaking into fabrics.

6. Describe how a surfactant (a) reduces surface tension, (b) increases wetting, and (c) causes suds and foam.

7. Explain how a surfactant can break up oily and greasy dirt into microscopic droplets and suspend them in water.

8. What is soap?

9. What is the difference between hard water and soft water? How can hard water be softened?

10. How does soap interact with hard water?

11. Why isn't soap widely used in the United States for washing clothes?

12. What are the two principal types of ingredients in a modern laundry product?

13. Which of the substances in Question 3 could be classified as (a) anionic detergent, (b) cationic detergent, or (c) nonionic detergent?

14. Why are anionic surfactants more effective than cationic detergents for washing clothes?

15. What are the advantages and disadvantages of a nonionic surfactant?

16. What is a biodegradable surfactant, and why is it desirable?

17. What is a builder, and what is its purpose in a laundry detergent?

18. Name three ways that sodium tripolyphosphate (STPP) assists a surfactant to get clothes clean.

19. What environmental problem do phosphate builders pose? How have legislators and manufacturers acted to solve this problem?

20. How do bleaches work? Do they physically remove a stain from clothes?

21. What is the purpose of optical brighteners in a detergent? In what sense do they make clothes "whiter than white"?

Topics for Discussion

1. To help reduce the damaging growth of blue-green algae in lakes and waterways, legislators in many states have placed limits on the amount of phosphates in detergents. But fertilizers continue to introduce more phosphates into the water than detergents ever did (Chapter 17). Should launderers be required to use inferior low-phosphate or phosphate-free detergents while farmers are allowed to use all the phosphate-rich fertilizer they please? Why or why not?

2. Most people in the United States wash their clothes in spinning-type washing machines using modern detergents that soften the water chemically. Ideally, would it be better to use ringer-type washers with soap instead? Would ion-exchange water softeners on all machines be better still? Explain.

3. Would you be willing to use less effective laundry detergents if it would help protect the environment? What kinds of additives do you think could be omitted from laundry products? Why?

CHAPTER 16

Personal Products: Taking Care of Our Teeth, Skin, and Hair

General Objectives

1. What is tooth decay, how can it be prevented, and what are the key chemical ingredients in toothpastes?

2. What is the nature of skin, and what are the major ingredients and functions of skin cleansers, moisturizers, and acne treatments?

3. How do sunscreens and sunglasses help prevent skin and eye damage, and what are the active ingredients in instant-tanning preparations?

4. What are the major ingredients of perfumes, colognes, antiperspirants, and deodorants?

5. What are the key ingredients and functions of shampoos, hair conditioners, and dandruff preparations?

You've probably never given it much thought, but all the products we use for personal grooming and hygiene involve chemistry. Those products—together with our hair, skin, and teeth—are made of chemicals in the form of atoms, ions, and molecules. But you may not realize that all personal products such as toothpastes, skin cleansers, moisturizers, shampoos, sunscreens, and antiperspirants operate according to the basic principles of chemistry that you have learned.

16.1 Teeth, Tooth Decay, and Toothpastes: Clean and Healthy

WHAT ARE TEETH?   Except for your eyes, your mouth and teeth are probably your most striking facial features. Sparkling white teeth are especially attractive, and they are likely to be the most healthy as well. All this makes your choice and effective use of toothpaste important.

The main part of a tooth (Figure 16.1) is a tough, bony substance called *dentin*. Covering the exposed outer portion of the dentin is material called *enamel*, the hardest substance in your body. Enamel can withstand all the mechanical stresses of biting and chewing. Only the heaviest of blows can cause it to crack or chip.

Both enamel and dentin consist of a crystalline lattice (Section 4.3) of calcium (Ca^{2+}) ions, phosphate (PO_4^{3-}) ions, and hydroxide (OH^-) ions. This substance, called hydroxyapatite, has the formula $Ca_5(PO_4)_3OH$. Fibrous protein (Section 9.4) fits in the spaces between the ions. This network of ions of hydroxyapatite makes teeth hard and rigid, whereas protein provides springiness and toughness.

441

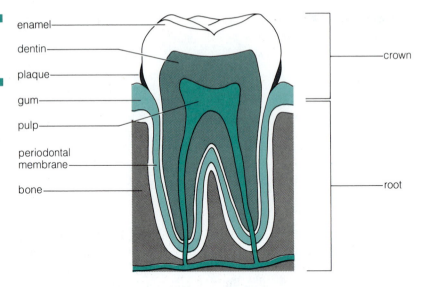

Figure 16.1 Major parts of a tooth.

enamel

dentin

plaque

gum

pulp

periodontal membrane

bone

crown

root

Teeth form by the process of **mineralization**—the deposit of calcium, phosphate, and hydroxide ions in the form of hydroxyapatite. Dissolving these ions in saliva is **demineralization**. The enamel on teeth is always dissolving to a tiny extent, forming ions in solution. At the same time, however, some of these ions are recombining to deposit enamel back on teeth. As long as mineralization and demineralization occur at the same rate, there is a state of dynamic equilibrium (Section 7.4) between these two opposing reactions, and no net loss of enamel results:

$$Ca_5(PO_4)_3OH \; \underset{mineralization}{\overset{demineralization}{\rightleftharpoons}} \; 5Ca^{2+} + 3PO_4^{3-} + OH^-$$

TOOTH DECAY AND GUM DETERIORATION   Tooth decay (dental caries) and gum deterioration (periodontal disease) result when demineralization exceeds the rate of mineralization. Severe tooth decay leads to such a large loss of enamel and dentin that the tooth either disintegrates or must be extracted. Decay is the leading cause of tooth loss before the age of thirty-five. After that age, tooth loss comes mostly from gum disease, which slowly destroys the gums, connective tissue, and bone that support teeth in their sockets.

Look again at the demineralization–mineralization equation. Anything that shifts the position of the dynamic equilibrium to the right results in a loss of enamel and (if allowed to proceed far enough) a loss in dentin. According to Le Châtelier's principle (Section 7.4), any process (other than the reverse reaction) that removes calcium, phosphate, or hydroxide ions from the system causes the equilibrium position to shift toward the right.

This shift occurs when acids are present. An acid–base reaction (Section 6.1) occurs when acid molecules provide hydronium (H_3O^+) ions that react with hydroxide (OH^-) ions in hydroxyapatite:

$$H_3O^+ + OH^- \longrightarrow 2H_2O$$

When this happens, OH^- is removed to become H_2O, and a new equilibrium position becomes established, with less enamel than before. Calcium and phosphate ions diffuse out of the enamel and are washed away by saliva. The missing enamel forms pits, or cavities, in your teeth (Figure 16.2), and you then suffer tooth decay.

Decay is a slow process, usually requiring months to occur, so only H_3O^+ ions having long and continuous contact with your teeth can begin to cause cavities. But your mouth, with its abundant moisture, warmth, and food in the form of sugars, is a paradise for acid-producing bacteria to stick to your teeth. Unless you clean you teeth throughly by brushing, flossing, and rinsing after eating, colonies of these bacteria can build up on your teeth in a matter of hours. These white or off-white deposits, consisting of about 70 percent bacteria, are **plaque**. The bacteria in plaque thrive on sugars, especially sucrose (Section 9.2), and turn them into various carboxylic acid (Section 8.6) products. The normal pH of saliva is about 6.8 (see Figure 6.1), but plaque-produced acids can decrease the pH to 5.5 or less, causing a loss of enamel.

Wherever plaque persists, decay begins. Plaque flourishes in out-of-the-way cracks and crevices between your teeth and near your gums. There the plaque can absorb minerals and harden into **tartar**, a tough crystalline substance consisting mainly of calcium phosphate, $Ca_3(PO_4)_2$; calcium carbonate, $CaCO_3$; and organic substances. Here hydronium ions get the uninterrupted time they need to dissolve enamel.

Plaque and tartar also cause gums to deteriorate. Bacterial products inflame the gums, and the gums then produce a number of chemicals to destroy the bacteria. If present in sufficient quantities over a long enough period, these chemicals can also destroy the gum tissue and fibers that hold teeth in place. The gums then begin to shrink away from the teeth.

The chief culprit in both of these dental diseases is sugar, mostly in the form of sucrose. Eskimos living on their natural sucrose-free diet of animal fat and protein have almost no cavities; when they switch to a westernized diet, their incidence of tooth decay rises sharply. The length of exposure is important, too. For example, sugar in caramels, which cling to the teeth, causes more tooth decay than the same amount of sugar in soft drinks, which remain in the mouth only briefly. And people who eat sugary snacks between meals tend to develop more cavities than those who consume sugar only during meals.

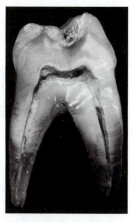

Figure 16.2 A molar tooth with decay in a fissure. (Len Barbiero/Medichrome)

USING FLUORIDES TO COMBAT TOOTH DECAY   Limiting sucrose in your diet is an obvious way to combat tooth decay, but it is not the only one. Fluoride (F^-) ions inhibit the demineralization of teeth by converting up to 30 percent of the hydroxyapatite in enamel into fluoroapatite:

$$Ca_5(PO_4)_3OH + F^- \longrightarrow Ca_5(PO_4)_3F + OH^-$$

hydroxyapatite   *fluoroapatite*

Fluoride ions fit better in the apatite lattice than do the slightly larger hydroxide ions. This leads to a more stable crystal that is about 100 times less soluble in acids than is hydroxyapatite. When fluoridated enamel dissolves in saliva, few if any hydroxide ions are generated—just calcium,

CHEMISTRY SPOTLIGHT   FLUORIDATED DRINKING WATER

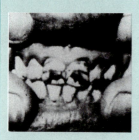

Figure 16.3 Although fluoride-ion concentrations of 2 mg/L or higher in drinking water can discolor tooth enamel, it also reduces tooth decay. (Courtesy National Institutes of Health)

In 1916 Frederick S. McKay, a young dentist in Colorado Springs, Colorado, found that *mottling*, or discoloration (Figure 16.3), on the teeth of many of his patients was caused by something in the local drinking water. He also observed that people with mottled teeth had very few cavities.

It was not until 1931, however, that Harry V. Churchill, a chemist at ALCOA, discovered that high concentrations of fluoride ions in drinking water caused mottling. Later studies showed that noticeable mottling and protection against tooth decay take place when each liter of drinking water contains 2 mg or more of fluoride. Studies also showed that when the water contains 1 mg/L of fluoride ions no mottling is observed, but the decay protection remains.

In 1942 the Public Health Service found that children, living in communities where the water naturally contained 1 mg/L of fluoride, had about 60 percent fewer cavities than those in communities with fluoride-free water. This led to the first large-scale artificial fluoridation program when Grand Rapids, Michigan, added enough sodium fluoride (NaF) to its drinking water to bring the fluoride concentration to 1 mg/L. Today about half of the U.S. population drinks water that has been fluoridated naturally or artificially at a fluoride concentration of 1 mg/L.

For several decades, some people have opposed water fluoridation. They consider it a form of compulsory mass medication, whose long-term effects are not fully known. They have also claimed that even at 1 mg/L it could be a contributing factor to cancer, heart disease, allergies, kidney disease, sterility, Down's syndrome, and impaired bone development. However, several studies, including one by the World Health Organization, have found no such hazards at the 1-mg/L concentration. Expense is not a major issue. For every dollar spent on fluoridation, between $30 and $50 are saved in treating tooth decay.

phosphate, and fluoride ions. Plaque-produced hydronium ions have little affinity for any of these ions, so little demineralization occurs.

Fluoride ions may also help prevent decay by inhibiting certain enzymes, found in plaque bacteria, that catalyze the conversion of sugars to organic acids in the first place. They may also inhibit the formation of sticky polysaccharides that promote the adhesion of bacteria to enamel surfaces. Fluoride even helps reverse decay in young children by increasing the mineralization of tooth enamel.

People in the United States have access to fluoride ions in fluoride-containing toothpastes and mouthwashes, in dentist-prescribed fluoride drops and tablets, in concentrated fluoride gels applied in mouth trays by dentists, and in drinking water. Thanks to fluorides and better dental hygiene, tooth decay has declined 50 percent among all age groups in the United States during the past fifteen years.

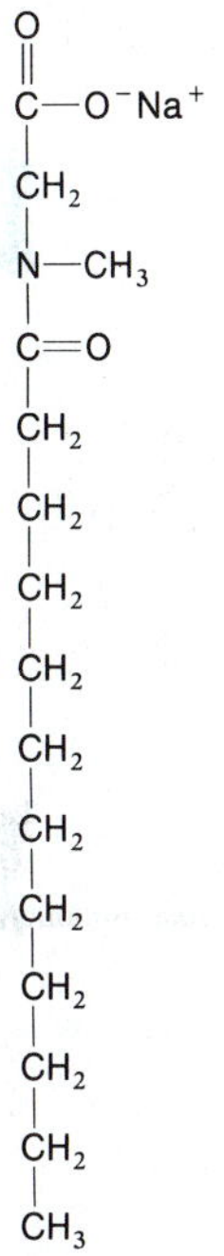

WHAT'S IN TOOTHPASTE?   The main purpose of any toothpaste, gel, or powder is to help remove plaque from teeth. In addition, toothpastes can provide fluoride, help prevent the formation of tartar, and freshen breath.

To accomplish their primary aim, all toothpastes contain cleaning and polishing agents known as *abrasives*. These give teeth their shine by scouring the enamel with a hard substance that has been finely powdered. More than half of the toothpastes use some form of silicon dioxide (SiO_2) as their abrasive. Various calcium compounds—including chalk ($CaCO_3$), calcium monohydrogen phosphate ($CaHPO_4$), and calcium pyrophosphate ($Ca_2P_2O_7$)—are also common. Each substance is hard enough to scratch off plaque deposits. But only calcium compounds are softer than and hence harmless to enamel; SiO_2 has to be specially processed so that it does not mar the surface of teeth.

Toothpastes containing sodium pyrophosphate ($Na_4P_2O_7$) can prevent tartar from building up by interfering with the formation of crystalline solids (tartar) in plaque. But none of the abrasives can dislodge tartar once it has formed. Having a dentist or hygienist scrape it off is the only way to remove it.

Another target for toothpastes is breath odor. Besides plaque, bacteria in your mouth can cause bad breath, so some toothpastes—particularly the gels—contain ingredients that kill these bacteria. Two such compounds are sodium N-lauroyl sarcosinate (Figure 16.4) and sodium lauryl sulfate (see Figure 16.16). Compounds such as these also act as surfactants (Section 15.1) that help clean teeth and produce the foam we expect from a toothpaste.

About 80 percent of the toothpastes sold in the United States contain fluoride compounds at approximately the level of 0.1 percent fluoride. The most common forms are stannous or tin(II) fluoride, SnF_2; sodium monofluorophosphate (MFP), Na_2PO_3F; and sodium fluoride, NaF.

Figure 16.4 Sodium N-lauroyl sarcosinate.

Putting fluoride in toothpaste presents some technical problems, however. A typical tube of toothpaste sits on the shelf for six months or more before it is purchased. In that many months, the reactive fluoride can find a number of ways to become deactivated. One way is to form insoluble calcium fluoride (CaF_2) by reacting with the abrasive. Therefore, not every toothpaste claiming to contain fluoride can provide it in its active F^- ion form when you brush.

Current formulations that do deliver active fluoride contain sodium fluoride (NaF) with the SiO_2 abrasive, stannous fluoride (SnF_2) with the $Ca_2P_2O_7$ abrasive, and sodium MFP (Na_2PO_3F) with just about any abrasive. The MFP ions release fluoride ions when they react with water in saliva:

$$PO_3F^{2-} + H_2O \longrightarrow H_2PO_4^- + F^-$$

Each of these combinations has been clinically tested. People using them showed anywhere from 13 to 44 percent fewer cavities than did people using identical toothpastes without fluoride.

Recognizing an effective toothpaste is easy. The American Dental Association's (ADA) Council on Dental Therapeutics examines toothpaste claims. When a product meets their exacting standards, they award it their seal of acceptance. Such accepted toothpastes do work, and their labels are not shy about saying so.

16.2 Skin-Care Products: Clear and Supple

YOUR SKIN   If you are a typical college-age (eighteen to twenty-two years old) student, you are in the prime of life, and your skin (Figure 16.5) is at its healthiest. You are beyond the days of oily skin and acne, and you have yet to see the time of dry skin and wrinkles. All you need to do is keep things this way.

The top layer of skin is the **stratum corneum** (Figure 16.5), a protective covering of dead cells. As these dead cells wear away, they are replaced by live cells from below. **Sebaceous glands** (Figure 16.5) exude oily sebum, which coats the stratum corneum and helps maintain moisture.

All layers of skin, plus hair and fingernails, are made of **keratin**, the sturdiest protein in your body. Unlike most internal proteins, which become denatured and useless outside their carefully controlled environments, keratin can withstand the rigors of all outdoors. Wide ranges of heat or cold, acidity or alkalinity, sunlight or darkness, and moisture or drought have little effect on keratin.

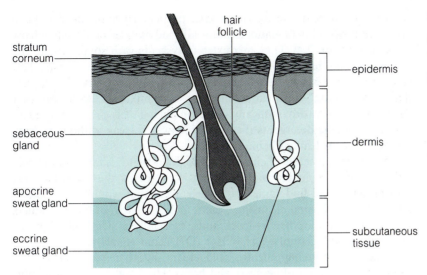

Figure 16.5 Cross section of skin.

But skin protein is not indestructible. Extremes in any one of these conditions can overwhelm even keratin. Skin can become dirty, dried, cracked, irritated, or diseased if you don't give some care to it. Fortunately, all sorts of cleaners, moisturizers, and medications are available to help you with this care.

SKIN CLEANSERS   To remove the unsightly or unpleasant substances that always get onto skin, you have three choices: you can rinse them off, tissue them off, or dust them off.

Because water and oil do not mix by themselves, rinse water alone cannot remove oily soil from skin. A surfactant like soap is needed for this purpose. One end of the surfactant is drawn to oil-like molecules, while the other end is attracted to water molecules. Thus compelled by its very nature, each surfactant unit attaches itself between water and oil, lifting the soil from skin and mixing it with the water (Sections 15.1 and 15.2).

One problem with soap is its tendency to be alkaline in water. An equilibrium forms between soap and water to produce the fatty acids and sodium hydroxide from which the soap was synthesized:

$$CH_3(CH_2)_{16}\overset{O}{\overset{\|}{C}}-O^-Na^+ + H_2O \rightleftharpoons CH_3(CH_2)_{16}\overset{O}{\overset{\|}{C}}-O-H + Na^+OH^-$$

soap	*fatty acid*	*lye*
(sodium stearate)	*(stearic acid)*	*(sodium hydroxide)*

Sodium hydroxide (lye or caustic soda) is particularly harsh on your skin. To make soap milder, manufacturers take advantage of Le Châtelier's principle (Section 7.4). They mix extra fatty acids with soap, driving the position of equilibrium to the left and reducing the amount of harsh alkali that forms. This is why many bar soaps on the market—even ones that claim to be "pure"—are surfactant–fatty acid mixtures. Bath bars that do not have the word *soap* anywhere on their labels are composed of modern synthetic detergents (Section 15.3), which clean without causing bathtub rings or leaving behind a murky film.

Although surfactants help, water is so polar that it is far from the best solvent for dissolving dirt and oily grime. Why not use the "like-dissolves-like" principle and find some less polar solvent that would work better? This is the idea behind cleansing creams and oils. Most of these products are mixtures of nonpolar hydrocarbons, especially mineral oil or petroleum jelly. Cetyl alcohol—$CH_3(CH_2)_{15}OH$—and squalane (Figure 16.6) are other common ingredients. Cold cream, another widely used skin cleanser, is a mixture of beeswax and borax. Unlike water, all these nonpolar solvents dissolve skin soils easily. You can then wipe off the resulting solution with a tissue, and your skin is clean.

You can also use powders, such as talc, for cleaning. Spongelike, the individual grains of the powder physically absorb the dirt and oil. Dusting off the powder rids your skin of the grime.

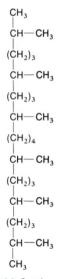

Figure 16.6 Squalane.

MOISTURIZERS   Your skin encloses and protects your body. It is nearly impervious, keeping in all vital fluids and keeping out contaminants. The dead cells of the stratum corneum—the outer layer of skin—are like a brick wall, forming a secure perimeter about your body.

The mortar surrounding these keratin bricks is an oily mixture that can suspend up to six times its weight in water. Water can come either from the live cells under the stratum corneum or from the external world. When the coating has soaked up enough water, the underlying keratin becomes pliable, and your skin feels soft and supple. Dry skin occurs when the moisturizing mixture becomes parched. No longer swollen by water, keratin fibers revert to their intrinsically rough and scaly form. Chronic dryness can lead to cracking of the stratum corneum, and exposure of the less resistant cells below can bring on irritation and infection.

Moisturizing products help your skin increase and maintain its water content. One way is to attract water from the outside. Ingredients that do this are **humectants**. The other way is to coat your skin with a waterproof layer that prevents water from escaping. Components that do this are **emollients**. All commercial moisturizers are combinations of humectants and emollients.

To be a humectant, a compound must have the same water-attracting properties as the head of a surfactant (see Table 15.1). It must be a polar

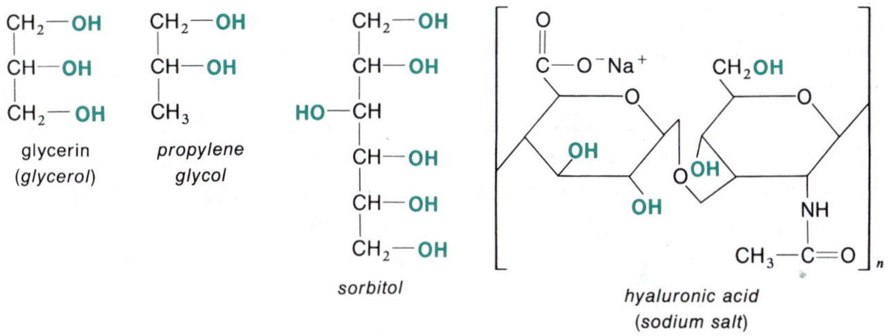

CH_2—OH $\quad$ CH_2—OH $\quad$ CH_2—OH

CH—OH $\quad\quad$ CH—OH $\quad\quad$ CH—OH

CH_2—OH $\quad$ CH_3 $\qquad$ HO—CH

glycerin $\qquad$ propylene $\qquad$ CH—OH
(glycerol) $\qquad$ glycol

$\qquad\qquad\qquad\qquad\qquad$ CH—OH

$\qquad\qquad\qquad\qquad\qquad$ CH_2—OH

$\qquad\qquad\qquad\qquad$ sorbitol

hyaluronic acid
(sodium salt)

Figure 16.7 Some humectants. Water-attracting hydroxyl (alcohol) groups are in color.

molecule, typically containing a number of oxygen atoms. The most common humectants are glycerin, propylene glycol, and sorbitol—alcohols with multiple hydroxyl groups (Figure 16.7). The sodium salt of hyaluronic acid (Figure 16.7), a natural humectant in skin, is a featured ingredient in some expensive skin creams.

Emollients must be as water-resistant as the nonpolar end of a surfactant. A wide variety of hydrocarbons, fatty acids, triglycerides, and other nonpolar compounds can serve as barriers to prevent water from escaping from your skin. Petroleum products (mineral oil, petroleum jelly), animal oils (mink, lanolin), vegetable oils (avocado, sesame), common oils (soybean, wheat germ), exotic oils (jojoba, aloe vera), natural oils (sweet almond, safflower), synthetic oils (caprylic triglycerides, glyceryl trioctanoate), and numerous others all have the same major function—to form a waterproof layer that feels smooth and slick on your skin.

Healthy skin does its own moisturizing. Sebaceous glands below the stratum corneum secrete an oily mixture of fats and waxes called *sebum* that coats the outer layer and acts as an emollient. But moisturizers are needed when age has slowed the natural moisturizing process, when work and weather have chapped portions of your skin, or when diseases such as psoriasis and eczema occur.

ACNE Ducts for sebaceous glands are located at pores and hair follicles (see Figure 16.5) all over your body—especially on the face, back, and chest—so that sebum can provide its normal moisturizing action. But one of the side effects of puberty is an increased production of sebum together with more keratin in the ducts. Inevitably, the greater flow through more constricted channels leads to blockages. And when sebum gets backed up underneath the skin, it becomes the whiteheads and blackheads of acne that many teenagers experience.

Because clogged pores are the problem (not hygiene or chocolate or other old wives' tales), effective acne medication must free up the sebum passageways. Unlike many skin-care products, substances that treat acne are considered drugs and must be approved by the Food and Drug Administration. Of the active ingredients approved for this purpose, all work by irritating the skin. The aggravation causes skin cells to dry up and slough off more rapidly, and it loosens any debris blocking the ducts. This deliberate drying is just the opposite of moisturizing. During treatment then, it is wise to wash off the natural emollient your skin continues to produce and to avoid using artificial moisturizers.

One of the most effective acne medications is benzoyl peroxide:

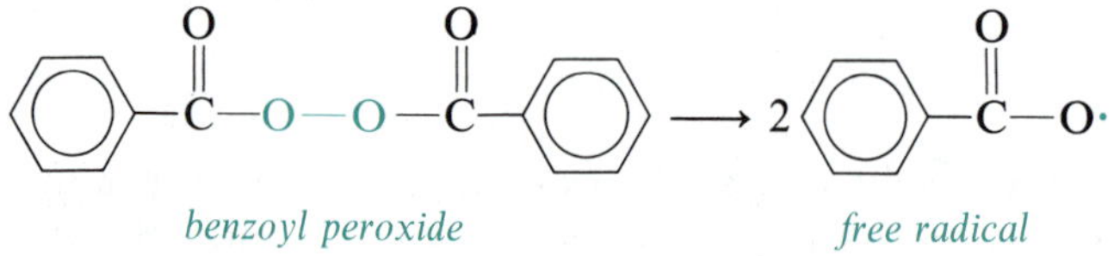

benzoyl peroxide   free radical

Like other peroxides, benzoyl peroxide contains oxygen atoms that have been unable to attract the usual number of electrons. As a result, it is very reactive and readily breaks apart at the oxygen–oxygen bond to form fragments with one unpaired electron (shown as a dot in its formula). Molecular fragments with an odd number of electrons, such as these, are known as **free radicals**. They are even more reactive than the benzoyl peroxide itself—so much that they attack and destroy the relatively inert substances of the stratum corneum, causing irritation.

Free radicals also kill the ever-present acne bacteria that tend to infect oily pimples. Infection causes the scarring and disfigurement that characterize tragic cases of acne. Furthermore, through a process that is not understood, free radicals seem to moderate the excessive production of sebum. Thus, benzoyl peroxide attacks acne on many fronts.

Another substance used to treat acne is retinoic acid (Figure 16.8), a form of vitamin A sold under the trade name Retin-A. Retinoic acid irritates the skin and causes epidermal cells to multiply faster, causing dead cells to be shed faster. Besides treating acne, this action seems to smooth wrinkles in the skin, thus producing a more youthful appearance. This has greatly increased the demand for this drug, whose long-term effects are not known. In the United States, retinoic acid is available by prescription only, and only for treating acne.

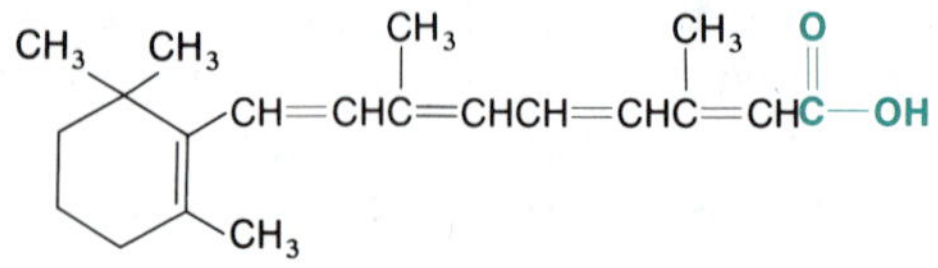

Figure 16.8 Retinoic acid. The difference between this structure and that of vitamin A (Figure 18.7) is in color.

16.3 Sun-Protection Products: Tan and Smooth

ULTRAVIOLET LIGHT   Light from the sun is more than what you see with your eyes. Invisible gamma rays, X rays, ultraviolet (UV) light, infrared (IR) light, and radio waves are all part of sunlight (Figure 16.9). Some of these forms of radiation are wholesome; others are harmful. All, however, interact in some way with the molecules of your body. IR light, for example, is less energetic than the deepest red in a rainbow. It carries just the proper energy to cause vibrations in the molecules at or near the surface of your body, and you sense these vibrations as warmth. Gamma rays and X rays have enough energy to uproot electrons and strip them away from molecules. If these potent radiations were not filtered out by the upper atmosphere, they would be deadly. Of the sunlight that does reach the earth's surface, only UV light is more energetic than visible light. Packing enough power to ionize molecules or to move electrons in atoms from one energy level to another (Section 2.4), UV light has both good and bad effects on your body.

The major beneficial action is the synthesis of vitamin D from a steroid—7-dehydrocholesterol—in your skin (Section 18.2). Vitamin D

Figure 16.9 The sun radiates a wide range of energies with different wavelengths. Because much of this radiation is either reflected or absorbed by the earth's atmosphere, mostly moderate- to low-energy radiation actually reaches the earth's surface.

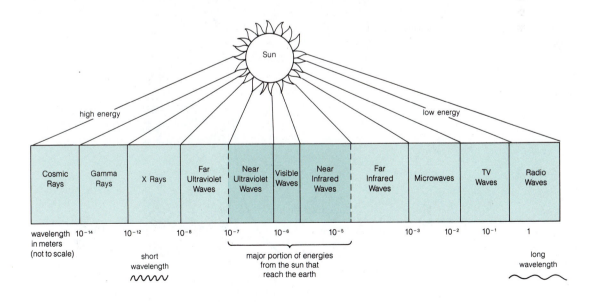

(Figure 18.7) regulates the body's use of calcium and phosphorus to make bones and teeth strong, and its synthesis in skin by UV light is a major source of this vitamin. Apart from this benefit, however, UV radiation is not very healthful. Exposure to UV light can cause wrinkles, age spots, and even skin cancer. Much of the worn and weathered skin conditions of old age come from excessive exposure to UV radiation. Once the skin becomes thick and leathery from too much UV radiation, it cannot be restored to its original youthful appearance.

SUNBURNS AND SUNTANS Most of the sun's UV light that reaches earth has a wavelength in the range 300 to 400 nm (nm, nanometer = 10^{-9} m). This is sometimes subdivided into UV-B (about 300 to 320 nm) and UV-A (320

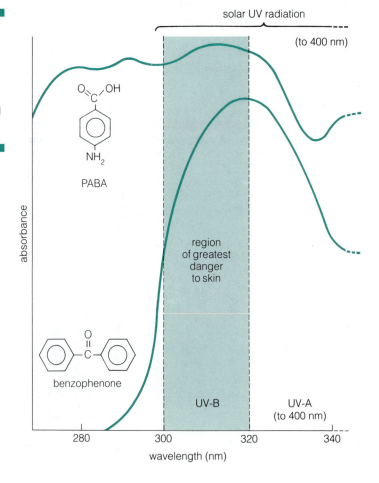

Figure 16.10 Solar UV light reaching earth is mostly in the range 300 to 400 nm. PABA and benzophenone, two sunscreens, absorb the UV light that is most harmful to the skin.

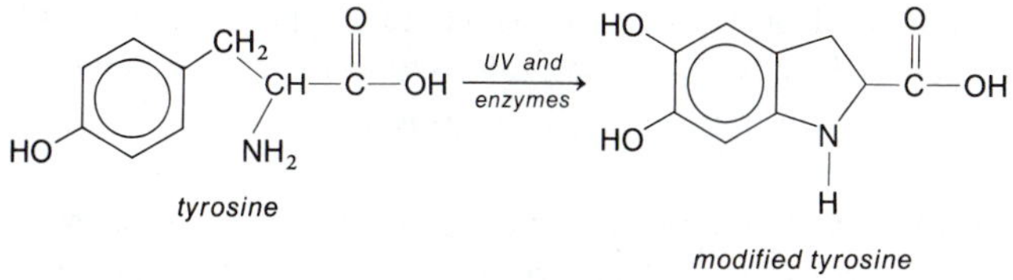

Figure 16.11 UV radiation in sunlight or sunlamps activates enzymes that modify the amino acid tyrosine.

to 400 nm) radiation (Figure 16.10). At higher wavelengths (about 400 to 800 nm), the light becomes visible to our eyes.

To control the dosage of UV light that penetrates skin—enough to make vitamin D but not enough to cause permanent aging—the body reacts in two ways, depending on the type and quantity of UV radiation. Sudden high levels cause skin to burn. This sunburn sets off the body's standard warning signal: pain. Redness and inflammation of affected tissues make it uncomfortable to prolong the exposure.

Steady UV light at lower levels causes a more subtle reaction. The radiation activates enzymes that modify tyrosine, an abundant amino acid in the skin protein (Figure 16.11). Many modified tyrosine molecules interconnect into giant molecules, collectively known as **melanin** (Figure 16.12). Brown in color, melanin is the pigment of the skin that determines how dark-complexioned a person is. The UV-stimulated production of more melanin results in a suntan, which is really part of the body's defense against more UV damage. The larger and deeper the melanin molecules become, the darker their color. And the deeper the brown, the more UV radiation they absorb. Thus, tanning is a signal that increased amounts of UV light are reaching the skin and that measures are being taken to combat the increased risk of skin cancer and prematurely aged skin.

Most dermatologists advise against seeking any kind of a suntan. They reason that tanning should alert a person to possible danger, like a blown fuse or a dashboard warning light. But many people regard a

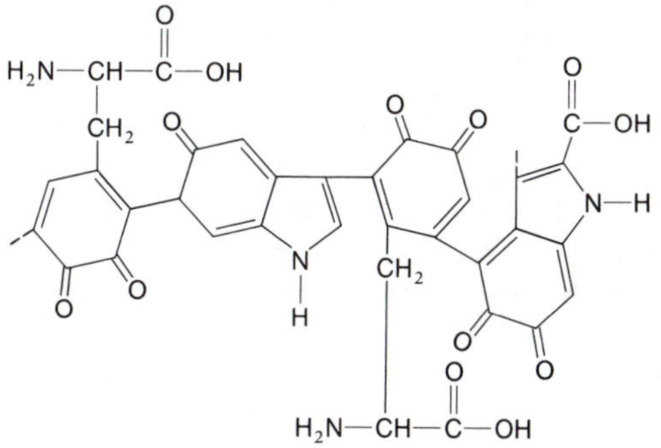

Figure 16.12 Melanin, the pigment of your skin.

tan as desirable—a symbol of health and leisure time. The brisk sales of sun products and the popularity of tanning salons attest to this. Knowing some chemistry can help you make an informed decision.

SUNSCREENS Molecules of each substance have a particular set of energy levels in which their electrons reside (Section 2.4). Consequently, each substance has a characteristic spectrum of energies that it absorbs to enable its electrons to move to higher energy levels. A substance that absorbs UV light from the sun thus can protect skin against UV light that causes sunburns and aged skin. Such a compound is a **sunscreen**.

The easiest sunscreens to formulate were those that absorb or reflect all light, visible or invisible. Zinc oxide (ZnO) and titanium oxide (TiO_2) have long been used for this purpose. Ointments containing these oxides are so intensely white, they are opaque. They are also messy and garish.

More attractive are sunscreens that absorb dangerous UV radiation but transmit visible light. These appear colorless and transparent but are just as protective. The most widely used are derived from para-aminobenzoic acid (PABA; see Figure 16.10), salicylic acid (see Figure 20.3), cinnamic acid, and benzophenone (see Figure 16.10). Acids react with alcohols to form esters (Section 8.6), and it is often esters (or other derivatives) of these acids that function as sunscreens. Likewise, benzophenone sunscreens often contain benzophenone with various groups attached to its benzene rings. By absorbing (and thus removing) the dangerous UV radiation from sunlight, these compounds prevent both sunburns and suntans in proportion to their concentration on the skin.

You need to choose the right sunscreen for your complexion. The more fair-skinned you are, the stronger the protection you need. And because little energy is needed to maintain an existing tan, using products with high protection after tanning is a prudent practice. All sunscreen products are labeled with a sun-protection factor (SPF)—a number between 2 and 15—to help you with the choice. The higher the number, the greater the protection. A product with SPF 4, for example, provides four times the skin's natural sunburn protection. Some specialty products advertise SPF ratings of 30 and even more, but the extra protection beyond SPF 15 may not be significant for most people.

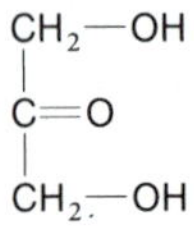

SUNLESS, QUICK TANS For some people, the threat of radiation damage, the risk of sunburn, and the boredom of sunbathing may be too high a price to pay for a genuine suntan. But they might still want the tanned look. Artificial tanning substances such as dihydroxyacetone and muconic aldehyde (Figure 16.13) both form brownish complexes with skin protein. The results vary from person to person, and they may not prove satisfactory to everyone. In any case, the application of chemistry widens the options.

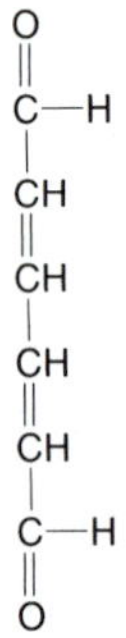

dihydroxyacetone

muconic aldehyde

Figure 16.13 Two artificial tanning substances.

SUNGLASSES Just as sunscreens protect the skin by absorbing UV light, materials in sunglasses can absorb UV (and visible) light from the sun. Although the data are not conclusive, some ophthalmologists believe that UV-B may cause cataracts, a condition in which the lens of the eye becomes cloudy or opaque. And UV-A may harm cells in the retina of the eye.

Dark and tinted sunglasses screen out more visible light, but not necessarily more UV light. Sunglasses typically absorb 95 percent of UV-B and 60 to 92 percent of UV-A light, while special-purpose glasses can screen out 99 percent of UV light. A simple rating system, similar to the SPF values, is being developed for sunglasses.

16.4 The Chemistry of Good Scents

PERFUMES AND COLOGNES Throughout history people have used chemicals that give off a pleasing fragrance. Most of the essential oils used in perfumes and colognes come from natural sources—rose, jasmine, violet, peppermint, rosemary, and many others. These oils are mixtures of alcohols, ethers, aldehydes, ketones, hydrocarbons, esters, and other compounds. Figure 16.14 shows the structures of a few major ingredients in a few oils.

Commercial perfumes and colognes consist of a blend of essential oils (up to 200), a solvent, and a fixative that slows evaporation and thus helps fragrances last longer. Essential oils, which provide fragrance, are 20 to 40 percent of the material in perfumes but only 3 to 5 percent in colognes. The most common solvent is ethanol (or an ethanol–water mixture), which comprises 60 to 80 percent of a perfume and 80 to 90 percent of a cologne. Natural fixatives include civetone (from the civet cat; see Figure 16.14), castor (from beaver), musk (from deer), and ambergris (from sperm whale).

Scientists have identified seven primary odors—camphorous, musky, floral, pepperminty, ethereal, pungent, and putrid. To cause an odor, a

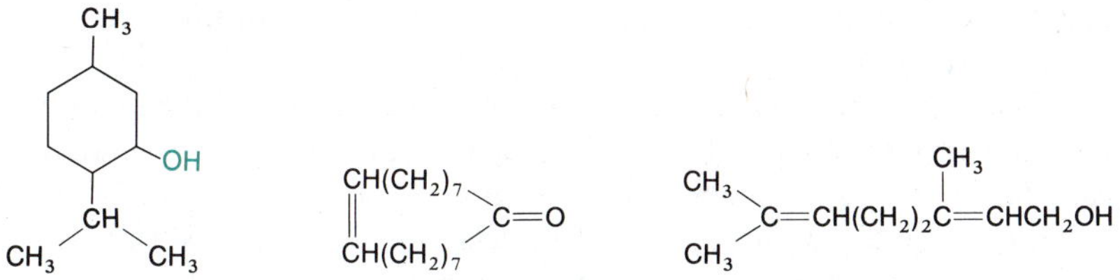

menthol
(in oil of peppermint)

civetone
(from civet cat)

geraniol
(in rose oil)

Figure 16.14 Some natural substances in perfumes and colognes.

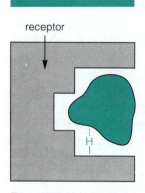

Figure 16.15 Molecules that impart a pepperminty aroma typically fit into a wedge-shaped receptor in the nose and can form hydrogen bonds.

substance vaporizes and binds to one of the seven types of receptor sites in the nose. The size, shape, and other chemical features (such as unsaturation or the ability to form hydrogen bonds) of the molecule determine which type of receptor it binds to and which type of aroma it produces. Pepperminty molecules, for example, can form hydrogen bonds and have a wedge shape (Figure 16.15).

ANTIPERSPIRANTS AND DEODORANTS   We also use chemicals to mask or prevent unpleasant body odors and sweat. There are two kinds of sweat—eccrine and apocrine. **Eccrine sweat**, produced in eccrine sweat glands (see Figure 16.5) on almost all parts of the skin, is the cooling mechanism of your body. Whenever exercise or environment threatens to raise your temperature, eccrine sweat is exuded onto skin to evaporate. Evaporation, being endothermic, takes away excess heat energy so that your body temperature remains fairly constant. Besides water, eccrine sweat contains some organic compounds and salts but does not produce offensive odors.

   Apocrine sweat, however, is a different story. Apocrine glands terminate in hair follicles (see Figure 16.5) at only a few places on your body—your underarms being one of those locations. Your nervous system activates these glands, which secrete liquid in proportion to the stress you feel. Although mostly water, about 1 percent of apocrine sweat consists of fat, cellular fragments, and bacteria. When exposed to the air, bacteria begin to flourish, producing smelly compounds and hence body odor.

   There are five ways products can combat this body odor:

1. Inhibit the production of apocrine sweat
2. Prevent the sweat produced from reaching the open air on the skin
3. Kill offending bacteria in the exposed sweat
4. Decompose foul-smelling substances the bacteria create
5. Mask odors with more pleasant fragrances

Clearly, the most effective actions are at the top of the list.

   The federal government requires that manufacturers reveal the general action of their product. If it works by Methods 1 or 2 above, then it can be called an *antiperspirant*. If it works by any of the others, it must be called a *deodorant*. Some products with combinations of ingredients can claim to be both.

   The active ingredient in most antiperspirants is one of the aluminum chlorohydrates, $Al_2(OH)_5Cl$ or $Al_2(OH)_4Cl_2$, or a zirconium–aluminum salt. These are water-soluble ionic compounds that produce Al^{3+} ions in solution. Aluminum ions bind to the ducts of sweat glands, shrinking the openings and forming an aluminum–keratin complex that plugs up many ducts. The flow of perspiration is reduced or, for some glands, prevented altogether. In addition, aluminum chlorohydrates kill bacteria in the apocrine sweat that does reach the skin. This pore-clogging action cannot be used by everyone. Because sebum glands open up in the same places the apocrine glands do, both can get obstructed. For certain susceptible people, rashes (sort of an underarm acne) can develop.

Deodorants, which have ingredients to kill bacteria and absorb, decompose (by oxidation), or mask odors, are alternatives for people who are unable to use antiperspirants. Mouthwashes are essentially oral deodorants that work in a similar way. Besides providing a pleasing aroma, they include ingredients such as alcohols (which kill bacteria by dehydrating them) and various phenols (which kill bacteria by denaturing their proteins).

16.5 Hair-Care Products: Shampoos and Conditioners

Most of your body systems are maintained automatically. Damage is repaired, chemical imbalances are corrected, and waste is removed with no conscious effort on your part. But your hair is not one of those systems. Made entirely of keratin, every strand of hair is dead. If any hair shaft becomes dry, cracks, or loses its softness or pliability, your body has no direct way of restoring it; deciding when and how to clean, style, or repair your hair is entirely up to you. The answers, however, come from some of the chemical principles you already know.

SHAMPOOS Shampoos are more than just hair cleansers. If cleanliness were the only goal, any heavy-duty laundry detergent would do a superb job. But shampoos must also help keep hair healthy, soft, and shiny. These additional requirements call for a specialized product.

Your hair, being all keratin, has many of the same requirements as your skin. In particular, it needs sebum as an emollient to soften it and give it natural body and luster. Every hair follicle has its own sebaceous gland for this purpose (see Figure 16.5). But sebum needs to be present in the optimum amount. With too little sebum, your hair is dry and strawlike; with too much, it is greasy and matted. Therefore, shampoos must be able to wash away the greasiness without removing the shine. They do this with mild surfactants (Section 15.1) that have only limited cleaning ability. Sodium lauryl sulfate (Figure 16.16) is the most widely used surfactant in shampoos. It helps you keep that "Goldilocks" quantity of sebum on your hair: not too much, not too little, but just right.

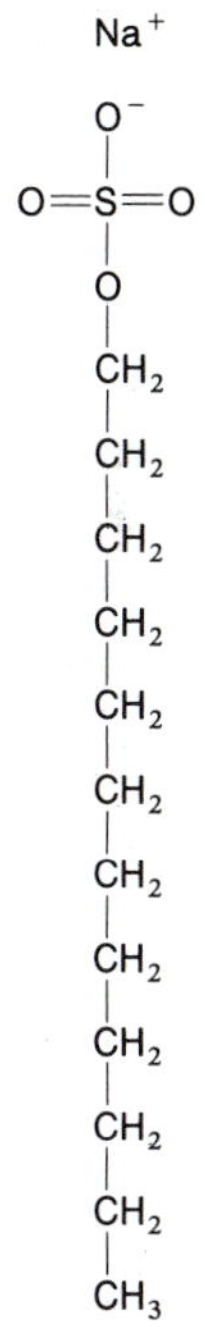

Na^+

O^-

$O=S=O$

O

CH_2

CH_2

CH_2

CH_2

CH_2

CH_2

CH_2

CH_2

CH_2

CH_2

CH_2

CH_3

Figure 16.16 Sodium lauryl sulfate.

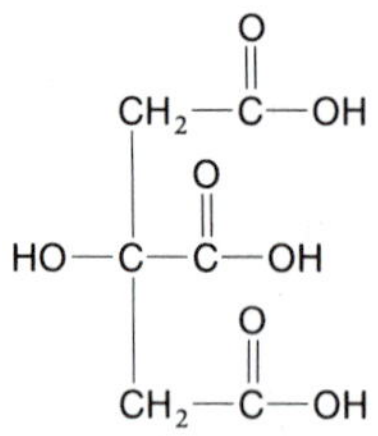

citric acid

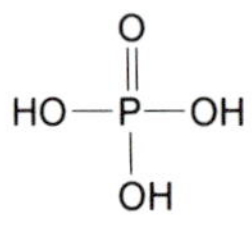

phosphoric acid

Figure 16.17 Two acids used in shampoos.

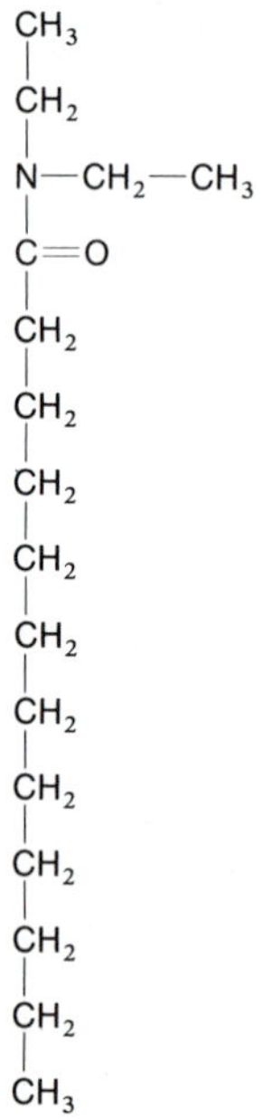

Figure 16.18 Lauramide diethylamine, a foaming agent and thickener.

PRACTICE EXERCISE

Some shampoos have different formulations for people with naturally dry hair or naturally oily hair. What levels of surfactants go into those products?

People with oily hair need more cleaning power in their shampoo, so oily-hair formulations contain more surfactant. Dry-hair shampoos have less surfactant.

SOLUTION

Harsh conditions can damage hair. Extremes in acidity or alkalinity can cause your hair's protein to denature (Section 9.4) and decompose. Hair needs a pH between 4 and 6—that is, slightly on the acid side of neutral—to achieve its maximum wet strength. Because most surfactant–water mixtures are strongly alkaline, typically with pH values of 10 or more, shampoos often contain acids to lower the pH. The most common are citric acid (the same compound that gives tartness to citrus fruits) and phosphoric acid, a mild acid often found in soft drinks (Figure 16.17). So many people are uneducated in chemistry that manufacturers advertise their products as "nonalkaline" or "pH-controlled" or even "acid-balanced," but they don't dare say that their shampoos are acidic.

The price of shampoo is higher than it needs to be because of those uneducated consumers. Each shampoo is filled with unnecessary ingredients including foaming agents (such as lauramide diethylamine; Figure 16.18) to make rich lathers, moderators to help the foaming agents work, and thickeners (such as lauramide diethylamine and sodium chloride) to give the runny liquids a richer texture. But the performance of the shampoo is not raised by any of these additives—only the price.

CONDITIONERS   Besides cleanliness and shininess, a number of other qualities may be desirable in hair. If you are like most people, you appreciate hair that is easy to comb (no tangles), is free from damage (no split ends), and is never unruly (no fly-aways). Most of all, you probably like the fullness and manageability of hair with body. That is why conditioners are on the market.

Like other proteins, the molecules of hair are made of twenty different types of amino acids joined together. Some of these amino acids (aspartic acid and glutamic acid; see Table 9.3) have free carboxylic acid groups that tend to donate protons; others (for example, lysine; see Table 9.3) have free amino groups that are bases and tend to accept protons. Thus, hair has built-in acid–base properties. It has more acidic groups than basic ones, so at a pH higher (more alkaline) than 3.8 (a pH value between 4 and 6 is typical), hair has a net negative charge (Figure 16.19). This static charge causes strands of hair to repel one another, causing wild, fly-away hair that is difficult to style.

One function of a conditioner then is to supply positively charged ions to neutralize the negative charge. Most conditioners do this with ionic substances in which one or more amino groups is electrically positive:

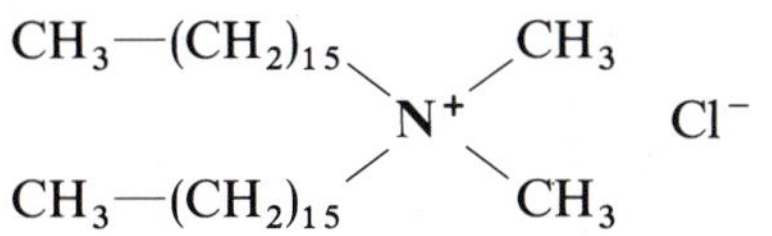

$$CH_3-(CH_2)_{15} \diagdown \quad \diagup CH_3 \atop N^+ \qquad Cl^- \atop CH_3-(CH_2)_{15} \diagup \quad \diagdown CH_3$$

Your hair ceases to be charged once these amino compounds bind to it with ionic bonds.

Long-chain hydrocarbon groups in the conditioner also serve other functions: They replace the shine-producing coating removed by shampoos; they act as an oil-like lubricant between hair strands to minimize tangles; and they add thickness to the hair, contributing to its body. On the negative side, however, these molecules can build up on hair and make it limp.

Swimming, sunning, and styling take their toll on your hair. The outer layer of protein can get roughened or broken. The ends can become frayed, like a rope. In severe cases, whole strands of hair can split in two. And all this damage can detract from your appearance. This is the most difficult problem for a conditioner to handle because the damage is not uniform; each strand of hair can have its own unique defect. Fortunately, your hair's inner core has a different amino acid composition from that of its outer layer and tends to develop a greater negative charge. Thus, damage that exposes the inner core creates a site that attracts more conditioner. In other words, the positively charged amine compounds in a conditioner tend to flock toward places where they are needed the most.

Most conditioners also contain protein fragments to help repair damage. Derived from animal hides and hoofs, the protein is not quite the same as your own. However, like plaster on a wall, it serves to fill in the cracks and dents. The fragments are polar molecules that are attracted to the more negative (and damaged) parts of your hair. As these protein segments bind to the hair's own protein fibers, split ends recombine, rough spots smooth out, and hair gets extra body. Conditioners also may include oils (such as lanolin, glycol stearate, and wheat germ oil) to act as sebum substitutes, carbohydrates (such as honey, beer, and aloe) to act as

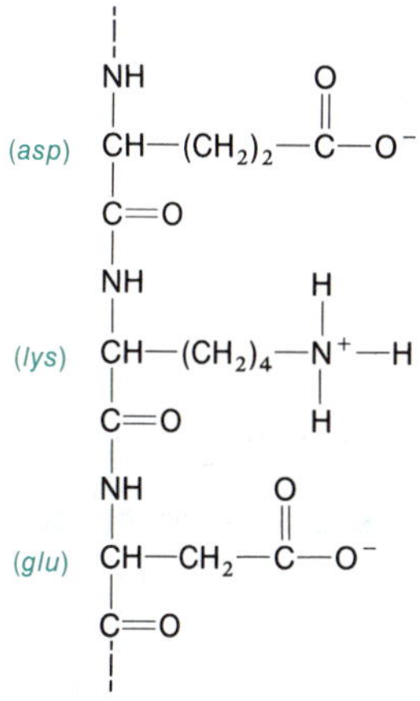

Figure 16.19 Part of a keratin molecule with aspartic acid (asp), lysine (lys), and glutamic acid (glu) in the ionic forms they assume at pH 4 to 6.

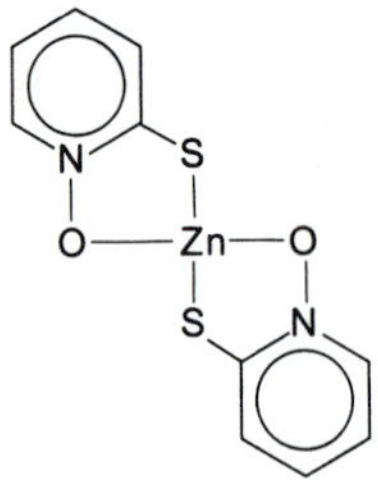

Figure 16.20 Zinc pyrithione.

humectants, and many other substances (such as vitamins and botanicals) that are generally of little consequence.

DANDRUFF   Like any other part of your skin, the stratum corneum of the scalp is made of dead cells that have migrated to the surface (see Figure 16.5). It normally takes twenty to thirty days for this migration to occur, after which the cells slough off individually into your hair, almost imperceptibly. When a person has the abnormality called dandruff, however, the migration takes only seven to ten days and ends with cells being shed in large clumps or flakes.

This unsightly flaking can be controlled in two ways. The first method is to slow the runaway migration of skin cells. The most popular dandruff shampoos work in this way. Their active ingredients are either selenium sulfide (SeS_2) or zinc pyrithione (Figure 16.20). The other antidandruff technique is to break up the flakes into insignificant pieces. Ingredients for this purpose include elemental sulfur (S) and salicylic acid (Figure 20.3). Because antidandruff materials aren't very soluble, shampoos containing them are opaque instead of clear.

Hair care is up to you, and much of it consists of applying chemical principles. Thus, chemistry really can make you more attractive.

MICHAEL AND IRENE ASH

Cosmetic use today is a product of the twentieth-century technology explosion. The . . . future potential for endless consumer products has naturally included cosmetics, and probably as long as our culture continues to grow more complex and specialized so too will the cosmetics market.

Summary

Demineralization of teeth produces decay. The process is stimulated by acids and is inhibited by fluoride (F^-) ions. Toothpaste provides abrasives to clean teeth, antibacterial agents, and usable forms of fluoride.

Skin cleansers may consist of surfactants, nonpolar solvents, or absorbent solids. Emollients prevent water evaporation from skin, whereas humectants attract water to skin. Acne is treated with substances that irritate skin and cause cells to slough off more rapidly. Sunscreen products (and sunglasses) absorb harmful UV radiation from the sun and thus protect skin (and eyes) from damage.

Perfumes and colognes consist of compounds with pleasing fragrances, a solvent, and a fixative. The aroma depends on the ability of a substance to bind to the appropriate receptor in the nose. Antiperspirants block apocrine sweat from reaching the skin's surface, whereas deodorants combat the odor resulting from such sweat.

Hair shampoos contain mild surfactants (for cleaning) and acids to neutralize alkalinity. Conditioners contain ingredients that bind to hair to repair damage, minimize tangling and fly-aways, and provide greater body. Antidandruff agents slow the flaking rate from the scalp or break the flakes into smaller pieces.

Terms for Review

After completing this chapter, you should know and understand the meaning of the following terms:

apocrine sweat (p. 456)

demineralization (p. 442)

eccrine sweat (p. 456)

emollient (p. 448)

free radical (p. 450)

humectant (p. 448)

keratin (p. 446)

melanin (p. 453)

mineralization (p. 442)

plaque (p. 443)

sebaceous gland (p.446)

stratum corneum (p. 446)

sunscreen (p. 454)

tartar (p. 443)

Questions

Odd-numbered questions are answered at the back of this book.

1. Write a balanced chemical equation for the decay of tooth enamel.

2. Explain, in terms of Le Châtelier's principle, how acids promote tooth decay.

3. Identify which of the following fluorine compounds can deliver active fluoride (F^-) ions in toothpaste: (a) NaF, (b) CaF_2, (c) PO_3F^{2-}, and (d) SnF_2.

4. How do fluoride (F^-) ions protect your teeth from decay?

5. What is an emollient? A humectant?

6. What is sebum? What does it do for your skin?

7. Identify which of the following two compounds would be the better emollient for your skin and explain why: (a) $CH_3(CH_2)_2OH$ or (b) $CH_3(CH_2)_{14}OCH_3$.

8. What are the effects of too much sebum and too little sebum on your skin?

9. The following five materials can be used to clean your skin. Classify the active ingredient of each as a surfactant, a solvent, or an absorbant: (a) petroleum jelly, (b) mineral oil, (c) corn starch, (d) dish detergent, and (e) bubble bath.

10. How does a sunscreen absorb UV light?

11. What dangers does sunbathing pose to skin?

12. What does an SPF value of 10 mean?

13. Under what conditions should a person choose a sunscreen with a low SPF? A high SPF?

14. In comparison with visible light, UV radiation has a _____ (shorter or longer) wavelength and a _____ (higher or lower) energy.

15. Why do darker-tinted sunglasses not necessarily provide greater eye protection against the sun's radiation?

16. What is the main function of ethanol in colognes?

17. What is the main chemical difference between perfumes and colognes?

18. Distinguish between an antiperspirant and a deodorant.

19. What does aluminum chlorohydrate do in an antiperspirant?

20. What is hair made of?

21. What are the effects of too much sebum and too little sebum on hair?

22. At pH 4 to 6, hair typically has a _____ (positive or negative) electrical charge.

23. Identify which of the following features of a shampoo might make a difference to the cleanliness, appearance, or health of your hair:
    a. Pearls sink slowly through the liquid.
    b. Its pH is 5.2.
    c. It contains protein.
    d. It contains eggs and beer.
    e. You can see through the liquid.
    f. It has a fresh herbal fragrance.
    g. Your scalp tingles when it is applied.
    h. It has a mild surfactant.

24. What is the main function of positively charged amines in hair conditioners?

Topics for Discussion

1. Do you favor tighter or looser government regulation of personal products such as the ones in this chapter? Why?

2. An effective acne-treatment drug used on the skin was found to increase the risk of birth defects in children born to pregnant women who used the product. The federal government allowed the drug to be used for treating acne, but required warnings to potential users. Do you favor this approach? Why?

3. The Council of Dental Therapeutics of the ADA has approved several brands of toothpaste. Are there valid reasons for using other brands?

4. What information do you need on a product's label? Look at the labels of personal products (such as toothpaste, soap, moisturizer, shampoo, deodorant, and sunscreen) that you use. What are the functions of the ingredients listed?

5. Before toothpastes, baking soda ($NaHCO_3$) was widely used for cleaning teeth because it has good abrasive properties. For much of that time, dental science was not advanced enough to take into account the effects of the bicarbonate (HCO_3^-) ions' acid–base properties. Consult Chapter 6 to determine whether bicarbonate acts as an acid or base in water and tell what side effects that might have on teeth. Are they beneficial or harmful?

CHAPTER 17

Food-Growing Products: Using Fertilizers and Pesticides

General Objectives

1. What major chemical nutrients do plants need, and how do they obtain these nutrients naturally?

2. What major types of organic fertilizers and commercial inorganic fertilizers are used to increase plant growth?

3. What are the major types and properties of pesticides?

4. What are the advantages and disadvantages of using synthetic chemical pesticides?

5. What are some alternative ways to control pests?

Each day our world has 200,000 new people to feed. United Nations population experts estimate that the world's population—5.2 billion people in 1987—will increase 50 percent by the year 2010. Yet all the plants and animals growing in the wild, as numerous as they are, cannot supply enough food for everyone. The only reason billions of people can survive is the huge production of food from agriculture. Even so, more than 10 million people a year die from starvation (lack of food), malnutrition (lack of nutritious food), or related diseases.

How will we feed everyone? Two important ways are (1) to increase food production by using fertilizers and (2) to reduce food loss (Figure 17.1) by using pesticides and other methods. We examine these two factors in this chapter.

17.1 Fertilizers: Feeding the Plants

THE FOOD CHAIN Despite the dominion we enjoy over other living things, humans cannot claim priority in the food chain. Food, our source of substance and energy, comes ultimately from the light of the sun and the simple molecules of the earth. Powered by sunlight, plants take carbon dioxide and water and convert them into carbohydrates and oxygen through *photosynthesis*:

$$6CO_2 + 6H_2O \xrightarrow{sunlight} C_6H_{12}O_6 + 6O_2$$
$$\textit{glucose}$$

Plants also take nitrogen, phosphorus, potassium, and other minerals from the soil and use them to synthesize products such as proteins. By 465

Figure 17.1 Throughout recorded history, locusts have devoured wild and cultivated plants used to feed people. (United Nations Food and Agriculture Organization)

contrast, our bodies have little ability to make these essential ingredients from scratch. Indeed, we can stay alive only by eating other organisms that supply us with scores of complex, ready-made, organic compounds.

Plants are self-sufficient. All they need are sun, soil, water, and air. They mature, propagate, and die on their own. Then with the help of bacteria and fungi, they decompose, regenerating nutrients for other plants.

The food cycle for animals and humans is merely a branch of the plants' cycle (Figure 17.2); it cannot operate independently. Our place in the world is that of a permanent dinner guest, totally dependent on plant-kingdom hosts to prepare our meals. Vegetation on earth is so vast and resilient that it can accommodate the entire wild-animal kingdom and many of us people as well. But it cannot sustain a world population of 5 billion such hangers-on. Survival requires that we be paying guests, enhancing the natural food cycle with an enormous agricultural effort.

PLANT NUTRITION   Plants produce glucose during photosynthesis and then convert glucose into other necessary materials. They make about 90 percent of their dry material in this way. Stimulating plants to grow then is largely a matter of assisting photosynthesis.

Besides the carbon, hydrogen, and oxygen used in the main reaction, plants need at least thirteen other elements to carry out photosynthesis (Table 17.1). Plants absorb all these essential elements from soil in ionic form. Plants require relatively large amounts of certain elements, especially nitrogen, phosphorus, and potassium; these are called *macronutrients. Micronutrients* are just as vital, but they are needed only in trace amounts.

Generalizing about the kinds of nutrients a plant needs is easy; generalizing about the amounts needed is not. Each variety of plant has its own nutritional needs. Leafy plants, for example, need more nitrogen and

Figure 17.2 The basic components of the food cycle. Solid arrows represent cyclical movement of chemicals within the system, and open arrows represent the one-way flow of energy through the system.

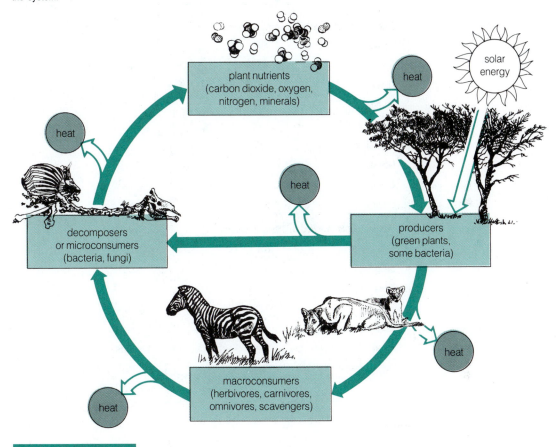

magnesium than do others; sugar beets require an amount of boron that would be toxic to soybeans; alfalfa has a higher potassium requirement than does grass. In the wild, the quality of soil is as important as the climate in determining which plants will grow well. In agriculture, however, we can use fertilizers to provide optimum amounts of the nutrients to grow a particular crop.

ORGANIC FERTILIZERS Erosion, leaching, and crop harvesting remove plant nutrients from soil. **Organic fertilizers**—which consist of animal manure,

TABLE 17.1 CHEMICAL NUTRIENTS NEEDED BY PLANTS

Name	Obtained from	Primary Chemical Form
Macronutrients (Needed in Large Amounts)		
Carbon	Air	CO_2
Hydrogen	Water	H_2O
Oxygen	Air and water	CO_2, H_2O
Nitrogen	Soil	NH_4^+, NO_3^-
Phosphorus	Soil	$H_2PO_4^-$, HPO_4^{2-}, PO_4^{3-}
Potassium	Soil	K^+
Magnesium	Soil	Mg^{2+}
Calcium	Soil	Ca^{2+}
Sulfur	Soil	SO_4^{2-}
Micronutrients (Needed in Trace Amounts)		
Chlorine	Soil	Cl^-
Copper	Soil	Cu^{2+}
Zinc	Soil	Zn^{2+}
Iron	Soil	Fe^{2+}, Fe^{3+}
Manganese	Soil	Mn^{2+}
Boron	Soil	$H_2BO_3^-$, $B(OH)_4^-$
Molybdenum	Soil	Mo^{6+}, MoO_4^{2-}

green manure (plant stalks, wastes, and other crop residues), and compost—can restore many of these nutrients.

Animal and green manure improve soil structure, increase nitrogen content, and stimulate the growth and reproduction of soil bacteria and fungi. Animal manure is particularly useful on rotation crops such as corn, cotton, potatoes, and cabbage. Historically the cheapest of all fertilizers, animal manure is now very expensive because of the cost of transporting it from animal-raising areas to crop-raising areas.

Green manure is fresh, green vegetation plowed into the soil to make its organic matter available to the next crop. It may consist of weeds, grasses, or clover that have taken over a field or legumes such as alfalfa or soybeans intentionally grown to build up soil nitrogen.

Compost is a rich natural fertilizer usually made by piling up layers of carbohydrate-rich plant wastes (such as cuttings and leaves), protein-rich animal manure, and topsoil. The soil provides microorganisms that decompose the other layers into ready-to-use plant nutrients.

COMMERCIAL INORGANIC FERTILIZERS **Commercial inorganic fertilizers** are made from rock deposits of nitrates, phosphates, and other minerals or by synthesizing ammonia (NH_3) from N_2 and H_2 (Section 5.4). These fertilizers are a concentrated source of nutrients that are cheaper and easier to store and apply than are organic fertilizers. But they contain only the specific ingredients formulated by humans.

Nitrogen is the most widely applied fertilizer ingredient, for soil is more likely to be deficient in nitrogen than in anything else. Although air is 78 percent N_2, plants cannot use nitrogen in this form. Instead, plants must absorb nitrogen from soil in an oxidized form such as nitrate (NO_3^-) ions or a reduced form such as ammonium (NH_4^+) ions. Certain leguminous plants (beans, peas, clover, and alfalfa) have bacteria in their root nodules that convert N_2 directly into a usable form of nitrogen. But most plants have to get their nitrogen from usable forms left in soil by other plants or animals. If soil lacks this nitrogen, the deficiency can be corrected by using nitrogen fertilizers such as liquid or gaseous ammonia, NH_3; urea, $CO(NH_2)_2$; sodium nitrate, $NaNO_3$; ammonium nitrate, NH_4NO_3; ammonium sulfate, $(NH_4)_2SO_4$; or mixtures of these.

Phosphorus is another essential plant nutrient, especially for fruits. Plants absorb phosphorus from the soil as phosphate ions, especially HPO_4^{2-} and $H_2PO_4^-$. The main source of phosphorus fertilizers is apatite (rock phosphate), which is a combination of $Ca_3(PO_4)_2$ and CaF_2. But treating it with phosphoric acid (H_3PO_4) or sulfuric acid (H_2SO_4) converts the insoluble $Ca_3(PO_4)_2$ into a water-soluble fertilizer, $Ca(H_2PO_4)_2$:

$$Ca_3(PO_4)_2 + 4H_3PO_4 \longrightarrow 3Ca(H_2PO_4)_2$$

insoluble   *water-soluble*

$$Ca_3(PO_4)_2 + 2H_2SO_4 \longrightarrow Ca(H_2PO_4)_2 + 2CaSO_4$$

insoluble   *"superphosphate"*
                                more soluble

In recent years, however, this type of phosphate fertilizer has been superseded by $(NH_4)_2HPO_4$, produced by reacting rock phosphate with ammonia.

Potassium completes the trio of key macronutrients. Among its many functions, potassium helps maintain the structural integrity of plant cells and is necessary for the catalytic activity of many plant enzymes. Plants absorb and use potassium only in the K^+ ionic form, so any potassium salt works as a fertilizer. The most common forms are potassium chloride (KCl), potassium sulfate (K_2SO_4), and potassium nitrate (KNO_3).

Fertilizer labels carry a set of three numbers to indicate the amounts of each key macronutrient—N, P, and K. The first number tells the percentage of nitrogen by mass in the fertilizer. The second and third numbers represent the percentage by mass of phosphorus (as P_2O_5) and potassium (as K_2O), respectively. In comparing a 12–42–8 fertilizer with a 35–10–10 fertilizer, we can immediately see that the first is especially rich in phosphorus (making it good for tomatoes and various fruits); the second is high in nitrogen, which makes it effective for lawns and house plants.

Although they have greatly increased food production, commercial inorganic fertilizers have some disadvantages. They reduce the natural production of usable nitrogen in the soil, so in later years farmers have to use ever-larger amounts. Commercial fertilizers also make the soil less porous, reducing its uptake of oxygen and fertilizer. Furthermore, fertilizers entering rivers and lakes cause plants and algae to grow, thus accelerating the process of eutrophication (Section 13.4).

Another concern is cadmium (Cd), a natural component of phosphate rock that accumulates in soil and water where phosphate fertilizers are used. Because cadmium is highly toxic (Sections 13.5 and 21.4), many countries are restricting the amount of cadmium that can occur in commercial fertilizers.

SOIL ACIDITY AND ALKALINITY   The soil's acidity or alkalinity affects its ability to support certain types of crops. For example, wheat, spinach, peas, corn, and tomatoes grow best in slightly acidic soils. Potatoes and berries flourish in very acidic soils, whereas alfalfa and asparagus thrive in neutral soils.

In Section 6.1, we saw that an acidic solution contains more hydronium (H_3O^+) ions than hydroxide (OH^-) ions and has a pH less than 7, whereas a basic or alkaline solution contains more OH^- ions than H_3O^+ ions and has a pH greater than 7. Much of a soil's acidity comes from the reaction between water and CO_2 gas in the air to produce hydronium (H_3O^+) ions:

$$CO_2 + 2H_2O \rightleftharpoons HCO_3^- + H_3O^+$$

Acidity also comes from acid rain (Section 12.4), from the decomposition of organic matter in soil, and from the reaction of water in soil with fertilizers containing ammonium (NH_4^+) and phosphate (in the form $H_2PO_4^-$) ions:

$$NH_4^+ + H_2O \rightleftharpoons NH_3 + H_3O^+$$
$$H_2PO_4^- + H_2O \rightleftharpoons HPO_4^{2-} + H_3O^+$$

When soils are too acidic for a desired crop, acidity can be partially neutralized by adding lime (ground-up limestone, $CaCO_3$):

$$CaCO_3 + H_3O^+ \rightleftharpoons Ca^{2+} + HCO_3^- + H_2O$$

Because lime also causes organic matter in the soil to decompose faster, manure or other organic fertilizers are often added along with lime to help keep the soil fertile.

Soil can become too alkaline in areas such as the semiarid valleys of the western United States, where there is little rain to dissolve and remove calcium and other alkaline compounds in soil. In areas that are well drained, irrigation water can eventually leach away the alkalinity.

Other areas are treated by adding sulfur to soil. Soil bacteria convert sulfur into sulfuric acid (H_2SO_4), which then neutralizes the alkalinity.

17.2 Pesticides: Decreasing the Competition

PESTS AND PESTICIDES   A **pest** is any unwanted organism that directly or indirectly interferes with human activity. The interference can range from flies and ants being a nuisance at a picnic to mosquitoes transmitting malaria in the tropics. It also can mean a life-or-death battle for food. Each year pests consume or destroy about 45 percent of the world's food. Throughout history, for example, locusts (see Figure 17.1) have ravaged wild and cultivated food crops. A 3.5-g locust can eat its own weight each day. A swarm may contain 1 billion locusts, and dozens of swarms may exist during a plague.

   Pesticides are chemicals that kill organisms that humans consider undesirable—insects (*insecticides*), weeds (*herbicides*), rodents such as mice and rats (*rodenticides*), fungi (*fungicides*), and others. If we could design an ideal pesticide, we would want it (1) to be safe and inexpensive to use, (2) to kill only the target pest and be harmless to other organisms, (3) to

Figure 17.3 The heads of these pyrethrum flowers being harvested in Kenya, Africa, are ground into a powder and used as insecticides or converted to other pyrethroid insecticides. (United Nations Food and Agriculture Organization)

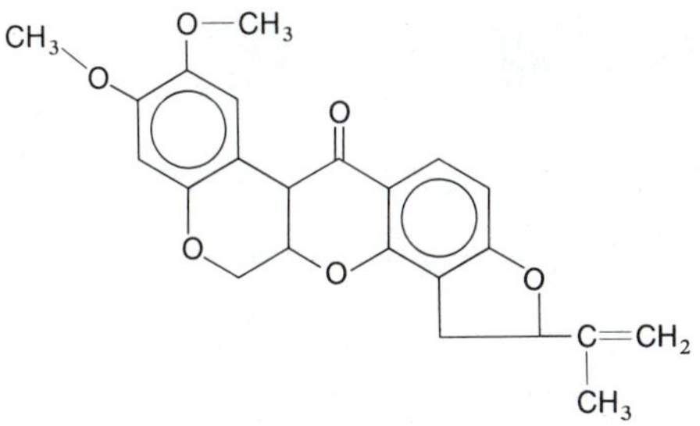

Figure 17.4 Rotenone.

not allow the target pest to develop resistance, and (4) to break down quickly in the environment into harmless chemicals.

INSECTICIDES For thousands of years, plants have naturally produced insect-killing chemicals. For example, more than 2000 years ago the Chinese used pyrethrum, a substance occurring in chrysanthemums that is still used as an insecticide (Figure 17.3). Caffeine (see Figure 20.18), another example, kills mosquito larvae, tobacco hornworms, and milk-weed bugs. Other natural insecticides include nicotine from tobacco, rotenone (Figure 17.4) from the tropical derris plant, and garlic and lemon oils.

Before 1940 the most common synthetic insecticides were ionic compounds made from toxic metals or metalloids such as arsenic, lead, mercury, copper, and zinc. They are not commonly used today because they are toxic to people and animals. In addition, some of these compounds remain in soil for 100 years or more, thus preventing new plants from growing.

An insecticide revolution began in 1939 when Paul Müller, a Swiss chemist, discovered that DDT (*di*chlorodiphenyl*tri*chloroethane), a chemical known since 1873, was a powerful insecticide. In 1948 Müller received the Nobel Prize in medicine for his discovery.

Now chemists have developed many synthetic organic chemicals to kill insects, weeds, rodents, and other pests. Today about 600 active ingredients have been formulated into more than 50,000 pesticide products in the United States. In 1985 an average of 2 kg (4.4 lb) of pesticides per person were used in the United States. The three main groups of synthetic organic insecticides are chlorinated hydrocarbons, organophosphates, and carbamates (Table 17.2).

Chlorinated hydrocarbons contain a functional group of one or more chlorine atoms attached to a carbon atom (Figure 17.5). Also notice in this figure that these insecticides typically have one or more carbon rings. These compounds kill insects by altering the balance of sodium (Na^+)

TABLE 17.2 THREE COMMONLY USED TYPES OF INSECTICIDES

Characteristic	Chlorinated Hydrocarbons	Organophosphates	Carbamates
Examples	DDT, aldrin, dieldrin, endrin, heptachlor, toxaphene, lindane, chlordane	Malathion, parathion, methyl parathion, diazinon, TEPP, mevinphos, chlorpyrifos	Carbaryl, propoxur, carbofuran, aldicarb, methomyl
Major use	Broad-spectrum insecticides kill a wide variety of target and nontarget organisms	Broad- and narrow-spectrum insecticides; a few fungicides and herbicides	Broad- and narrow-spectrum insecticides, fungicides, and herbicides
Action on pests	Nerve poisons that cause convulsions, paralysis, and death	Nerve poisons that deactivate an enzyme involved in transmitting nerve impulses	Nerve poisons that deactivate an enzyme involved in transmitting nerve impulses
Persistence	High (two to fifteen years)	Low to moderate (normally one to twelve weeks, but some can last several years)	Usually low (days to weeks)
Animal toxicity	Relatively low for humans, but high for some animals	Very high for humans and other animals	Low to high for humans and other animals

ions and potassium (K^+) ions in the nerve cells that are necessary for normal nerve functioning.

In the 1950s and 1960s, DDT and other chlorinated hydrocarbon insecticides were widely used on insects such as the gypsy moth, codling moth, corn earworm, and spruce budworm. The compounds are inexpen-

Figure 17.5 Some chlorinated hydrocarbon insecticides.

DDT

methoxychlor

aldrin

dieldrin

lindane

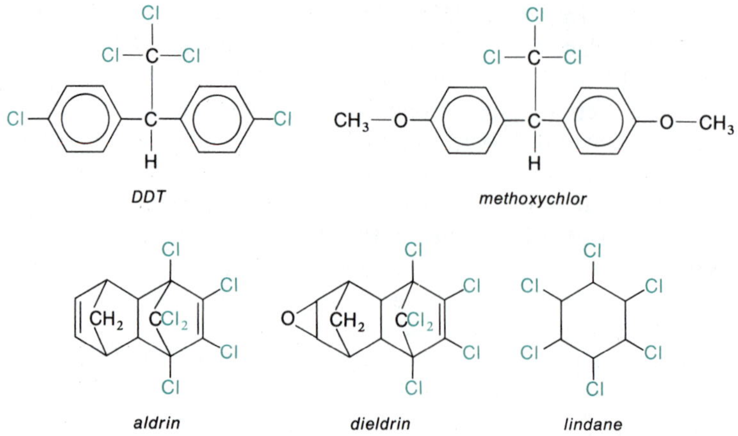

Figure 17.6 Some organophosphate insecticides.

malathion

methyl parathion

mevinphos (Phosdrin)

chlorpyrifos (Dursban)

dichlorvos (Vapona)

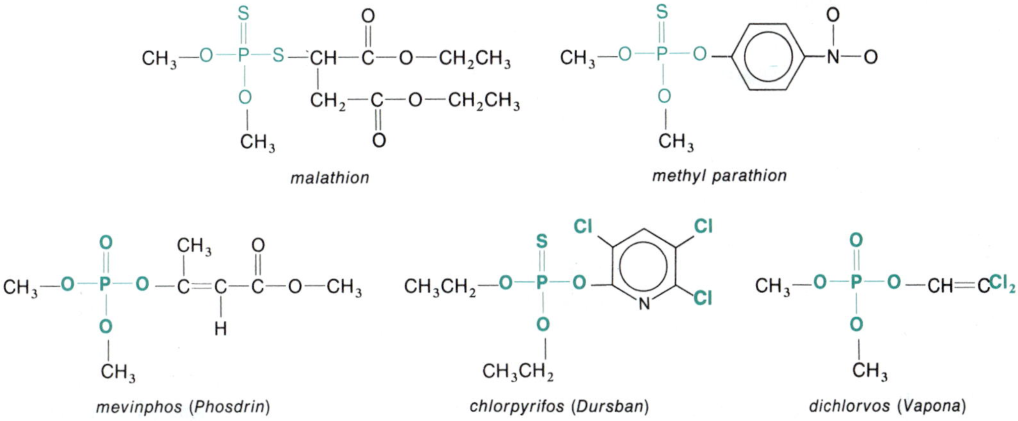

sive to use, kill a variety of pests, and stay in the environment a long time (often several years) while continuing to kill pests (see Table 17.2). But these properties also produced serious side effects (Section 17.3). As a result, in 1973 the EPA banned the use of DDT and in the late 1970s restricted the use of other chlorinated hydrocarbon pesticides.

The decreased use of chlorinated hydrocarbons has led to a greater use of **organophosphates**, which are organic pesticides that contain the phosphate (PO_4^{3-}) group. Notice in Figure 17.6 that in some cases sulfur replaces one or more oxygen atoms in the phosphate group. As we discuss in more detail in Section 21.3, organophosphates paralyze their victims by inhibiting an enzyme they need to transmit nerve impulses.

Organophosphate insecticides normally break down rapidly in the environment but are much more toxic to people and animals (see Table 17.2) than are chlorinated hydrocarbons. Because of organophosphates' rapid breakdown, farmers apply them at regular intervals so that they stay in the environment almost continuously. They are effective against a variety of home, garden, and livestock pests.

Some insecticides contain more than one functional group. Dichlorvos (see Figure 17.6), for example, contains both chlorine and organophosphate groups. It readily vaporizes and is the active ingredient in Shell No-Pest Strips and in some flea collars.

Carbamates, the third group of synthetic organic insecticides, are derived from carbamic acid (Figure 17.7). As you can see from the structures of various carbamates in the figure, one or more of the hydrogens in carbamic acid may be replaced by organic groups. Carbamates work

Figure 17.7 Carbamic acid
and some carbamate
insecticides.

Figure 17.7 Carbamic acid
and some carbamate
insecticides.

carbamic acid

carbaryl (Sevin)

carbofuran (Furadan)

aldicarb (Temik)

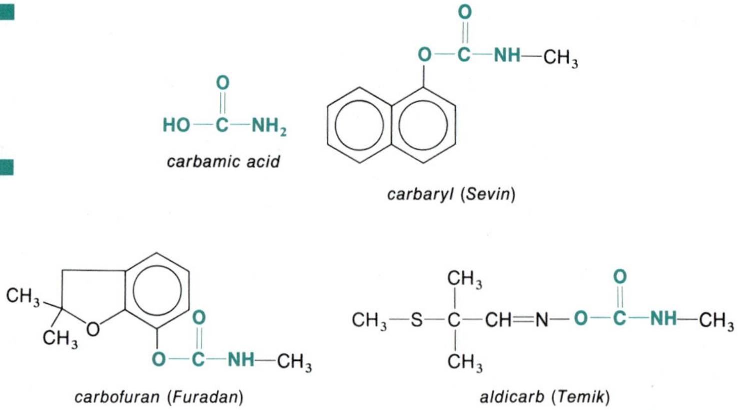

the same way that organophosphates work: They kill insects by disrupting their nervous system.

The use of carbamates also has grown as organochlorine insecticides were banned or restricted. Carbamates tend to have a broad spectrum of action, making them effective as lawn and garden insecticides. They are less toxic to animals than are organophosphates, and they decompose within a few days or weeks in the environment (see Table 17.2).

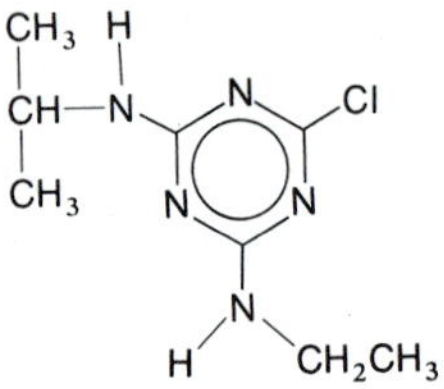

HERBICIDES More than 180 different synthetic organic herbicides are used in the United States. Most are active for relatively short periods (days to weeks). The three main classes, based on their effect on plants, are contact herbicides, systemic herbicides, and soil sterilants.

Contact herbicides kill plant foliage within a few days through direct contact. One class are the triazines, which contain three nitrogen atoms in a six-member ring. These compounds kill plants by interfering with photosynthesis. Atrazine (Figure 17.8), a widely used triazine, is used on corn fields because it kills many weeds but doesn't harm corn plants.

Systemic herbicides are absorbed by the foliage or roots and then travel through the entire plant. One group are phenoxy compounds such as silvex, 2,4-D, and 2,4,5-T (Figure 17.9). Once absorbed, these herbi-

Figure 17.8 Atrazine, a
widely used triazine contact
herbicide.

Figure 17.9 Two important
phenoxy herbicides.

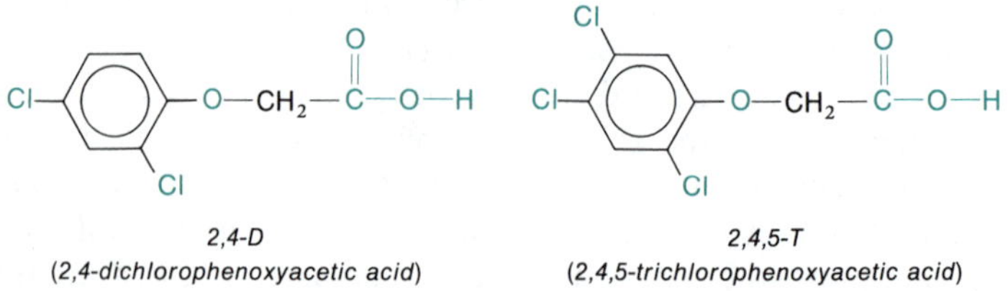

2,4-D
(2,4-dichlorophenoxyacetic acid)

2,4,5-T
(2,4,5-trichlorophenoxyacetic acid)

cides stimulate excessive growth; the plant literally grows itself to death because it cannot get enough nutrients to keep up with its growth.

A mixture of 2,4-D and 2,4,5-T, known as Agent Orange, was used by U.S. military forces in Vietnam to defoliate jungles and to destroy crops. The mixture also contained small amounts of dioxins (see Figure 21.14), which are toxic. Because of the experiences of soldiers and civilians exposed to these herbicides and because of tests done with laboratory animals, the safety of these herbicides to humans is being seriously questioned (Section 21.5). In 1985 the EPA banned the use of 2,4,5-T.

The third type of herbicide, *soil sterilants*, kills plants by destroying soil microorganisms essential to growth. Examples include Treflan (Figure 17.10), Dymid, and Dowpon. Most of these also act as systemic herbicides. Most soil sterilants are active for a few days or weeks, but some remain active for as long as two years.

With herbicides, farmers don't need to till the soil as often to control weeds. This not only reduces soil erosion, but it also means less money spent on energy, fertilizer, and water.

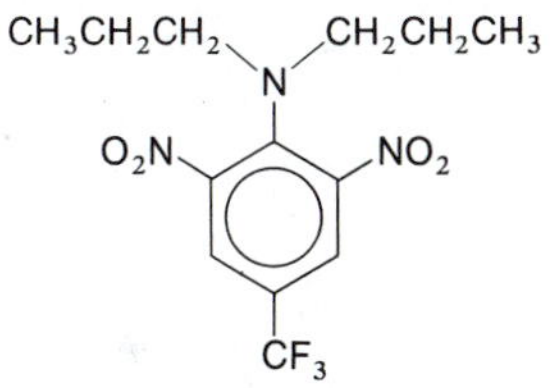

Figure 17.10 Trifluralin (Treflan).

17.3 Pesticides: How Effective Are They?

THE CASE FOR PESTICIDES By increasing the food supply and by helping to control insect-transmitted diseases such as malaria, typhus, and sleeping sickness, DDT and other insecticides have probably saved more lives than any other synthetic chemicals in history.

A dramatic example is malaria, which is carried by the *Anopheles* mosquito. The World Health Organization (WHO) estimates that the extensive use of DDT and other chlorinated hydrocarbon pesticides has freed more than 1 billion people from the risk of malaria and saved the lives of at least 7 million people since 1947. According to WHO, a worldwide ban on these pesticides would lead to large increases in disease, human suffering, and death.

Pesticides have also reduced the price of food and increased its supply (Figure 17.11). The yearly $3 billion that the United States invests in pest control yields about $12 billion in increased crops. Although they also cause about $1 billion in social and environmental damage, pesticides are still a good investment in economic terms. The Department of Agriculture estimates that the extensive use of pesticides and storage facilities in the United States lowers the price of food 30 to 50 percent.

Section 17.4 presents some alternatives to using chemicals to control insects and weeds. But proponents argue that pesticides have several advantages over other approaches: They can control most pests quickly

Figure 17.11 A rat-trapper in an Indian wheat field. In one year, the offspring of one pair of rats can eat enough grain to feed five people. (Courtesy of the Rockefeller Foundation, Marc and Evelyne Bernheim)

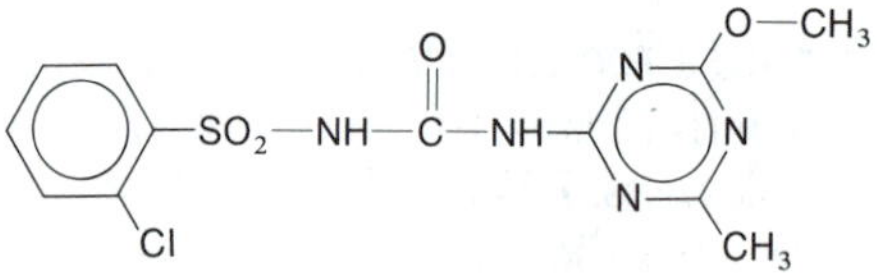

Figure 17.12 Chlorsulfuron (Glean).

and at reasonable cost; they have a relatively long shelf-life; they are easily shipped and applied; and they are safe when handled properly. When genetic resistance occurs, farmers can usually switch to other pesticides.

Indeed, chemical companies are developing pesticides that are more effective while having fewer unwanted side effects. One example is a group of herbicides known as sulfonylureas, which have trade names such as Glean (Figure 17.12) and Oust. Farmers may apply about 500 g (1.1 lb) per acre of conventional herbicides, but with sulfonylureas they may apply as little as 5 g per acre. Furthermore, these herbicides do not leach into groundwater, so they are less likely to spread throughout the environment.

THE CASE AGAINST PESTICIDES   The most serious drawback to using pesticides is that most pest species—especially insects—can develop genetic resistance. When an area is sprayed with a pesticide, most of the pest organisms die. But a few individuals may have genes that make them resistant to the pesticide. Those few will survive and produce offspring that are also resistant. In time then, an ever-greater proportion of the pest population becomes resistant. Within five years or so, most insecticides become ineffective. Plant diseases and weeds also develop genetic resistance, but not as quickly as most insects.

One example is the cotton boll weevil (Figure 17.13), the largest single target of pesticides in the United States. Dieldrin and aldrin (see Figure 17.5) effectively controlled this pest during the late 1940s. But by the early 1950s, weevils had become so resistant that these pesticides could kill only a small percentage of the insects. More than 450 insect species are resistant to DDT. And Colorado potato beetles have defeated a wide array of insecticides; now they become resistant to new pesticides within one season.

Another problem is that most modern insecticides also destroy the target pest's natural predators and parasites. Many times the survivors in the pest population find themselves with few natural enemies and lots of available food. Under these conditions, the pest population poses an even larger threat to crops.

In California's San Joaquin Valley, for example, farmers sprayed cotton crops with heavy doses of the organophosphate insecticide Azodrin to control the cotton bollworm. After three sprayings, so many natural predators had been killed that the cotton bollworms were able to destroy 20 percent of the cotton crop. Scientists at the University of California estimated that if no pesticide had been used, bollworms would have destroyed only about 5 percent of the cotton.

DDT and other long-lasting pesticides can travel far beyond their point of application, and they increase in concentration as they move through food chains. Arctic seals, Antarctic penguins, and Eskimos now have detectable amounts of pesticides in their fatty tissues. What is worse, those amounts are magnified in the higher animals. For example, plankton absorb DDT from the water. Small fish eat thousands of plankton and store thousands of times the plankton's portion of DDT. Big fish eat small fish, and fish-catching birds dine on big fish. The result is, for these birds, a seafood dinner that has been fortified with pesticides millions of times more concentrated than the amount in the water.

This concentration of pesticides can lead to harmful or even deadly doses for animals high on the food chain (Figure 17.14). In the 1950s and 1960s, for instance, the population of robins and other songbirds declined drastically in forests and areas sprayed with chlorinated hydrocarbons. The strange springtime silence provided Rachel Carson with the title of her famous 1962 book, *Silent Spring*, which warned of the harmful effects of pesticides.

Figure 17.13 At one time, about 35 percent of the pesticides used in the United States were used against the cotton boll weevil. (U.S. Department of Agriculture)

Figure 17.14 The population of peregrine falcons in the United States declined between 1950 and 1970 when high levels of DDT accumulated in their bodies. The affected birds produced eggs with shells so thin that many chicks died before they could hatch. (Luther C. Goldman/ U.S. Fish and Wildlife Service)

CHEMISTRY SPOTLIGHT   THE BHOPAL TRAGEDY

The world's worst industrial accident occurred in 1984 at a Union Carbide pesticide plant in Bhopal, India. The plant stored large amounts of methyl isocyanate, a very toxic liquid used to manufacture the carbamate pesticide, aldicarb (see Figure 17.7). Because of poor safety procedures, methyl isocyanate vapors leaked from a storage tank and killed about 2100 people.

The death toll was at least 3300 people, with another 75,000 people injured. The victims suffer from disorders such as blindness, sterility, kidney and liver infections, tuberculosis, and brain damage. The Indian government sued Union Carbide and was awarded $470 million as compensation to be given to the victims.

This tragedy has increased public concern about the safety of pesticide and other chemical plants. In 1985 a gas leak from another Union Carbide aldicarb plant—this one in West Virginia—sent 135 nearby residents to the hospital. This time no one appeared to be injured seriously.

Although shorter-lived organophosphate and carbamate pesticides have partially replaced chlorinated hydrocarbons, the problem of toxicity remains. Each year pesticides—mostly carbamates—kill 20 percent of the honeybee colonies in the United States and damage another 15 percent of the colonies; the reduced pollination causes annual crop losses of at least $135 million. And an estimated 500,000 farm workers, pesticide plant employees, and children worldwide become seriously ill each year from exposure to toxic insecticides, especially organophosphates. The effects on humans of the herbicides 2,4-D and 2,4,5-T are another concern.

17.4 Other Methods of Pest Control: Working with Nature

Although pesticides have significant drawbacks and advantages, they are an important way to help increase the world's food supply and reduce insect-carried diseases. Pesticides, however, are just one of the ways to control pests.

NONCHEMICAL METHODS   For centuries farmers have discouraged pests by rotating crops. Pests feeding on a particular crop then find themselves without food when a different crop is planted. Planting different crops in alternate rows or strips also can help. In addition, farmers can time the planting so that many of the pests die, or their predators become abundant, before the crop ripens.

Another way to control pests is to increase their exposure to organisms that kill them, such as natural predators, parasites, bacteria, and viruses. Ladybugs, for example, kill aphids, and a bacterial agent (*Bacillus thuringiensis*) in powder form kills many strains of caterpillars, mosquitoes, and gypsy moths. Although this approach has the risk that the predators can multiply and become a problem, this method in appropriate situations has saved U.S. farmers an average of $30 per $1 spent.

Insect reproduction can be blocked by using radiation to sterilize large numbers of male insects and then releasing them into an area to mate unsuccessfully with females. This works best in isolated areas and with insects whose females mate only once. In the 1970s, the U.S. Department of Agriculture released huge numbers of radiation-sterilized screwworm flies in the southern states to control this serious cattle pest.

Another strategy is to use genetic engineering (Section 22.1) to make crops resistant to pests or to pesticides. For example, chemical companies have transferred genes into tomato (Figure 17.15), corn, tobacco, and other plants to make them produce toxic compounds that keep away caterpillars or other insect pests.

Crops made resistant to herbicides can grow effectively in fields where the herbicide is used to kill weeds. Companies have now developed strains of cotton, soybeans, tobacco (Figure 17.16), and other plants that are resistant to specific herbicides.

Monsanto Company, for example, is developing crops resistant to its best-selling herbicide glyphosate (Roundup). Glyphosate works by inhibiting an enzyme that helps plants synthesize aromatic amino acids,

Figure 17.15 Untreated tomato plant (right) and a genetically engineered tomato plant (left) after exposure to caterpillars. (Courtesy of Monsanto Company)

Figure 17.16 Genetically engineered (top row) and untreated (bottom row) tobacco plants after exposure to no pesticide (first column) or to pesticide concentration of 0.5 lb (1.1 kg) per acre (second column) or 1.0 lb (2.2 kg) per acre (third column). (Courtesy of Dr. David M. Stalker. *Science Vol. 242*, page 421, 21 October 1988, by Dr. David M. Stalker. Copyright 1986 by the AARS.)

which are needed for proteins (Figure 17.17); if this enzyme action is blocked, plants die. By giving crops genes to make extra amounts of this enzyme, Monsanto has produced crops that can withstand exposure to high levels of glyphosate. Another approach, used with other herbicides, is to make crops resistant by giving them genes to produce enzymes that degrade the herbicide.

CHEMICAL ALTERNATIVES   Pheromones and hormones are chemical alternatives to insecticides. In many insects, a female that is ready to mate

Figure 17.17 Because of its structural similarity to a natural substance called PEP (shown in color), glyphosate blocks the synthesis of aromatic amino acids in weeds, thus killing them.

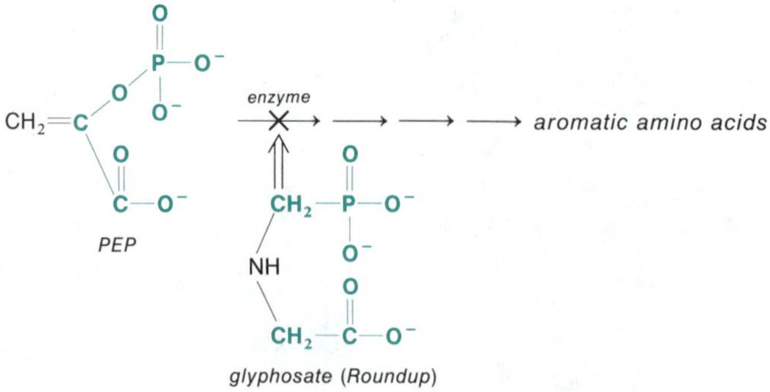

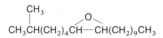

Figure 17.18 The elaborate antennae of the male gypsy moth (bottom) can detect the pheromone (top) emitted by a female gypsy moth. (U.S. Department of Agriculture)

releases into the air a species-specific sex-attractant chemical called a **pheromone**. Males follow this scent to find the females.

Chemists have now identified and synthesized pheromones for more than 400 insect species (Figure 17.18). They can be used in several ways. Spraying an infested area with pheromone prevents males from finding females; the scent is everywhere. Pheromones are an effective way to bait insect traps, luring the male insects to their death. Another strategy is to spray pheromone to attract a predator of the target pest.

Pheromones are effective in small concentrations, break down rapidly in the environment, and are much more species-specific than insecticides. But pheromones don't actually kill insects, and they are expensive. In a few cases, they have proved to be very effective. A vineyard in New York treated with pheromone for the grape berry moth reduced damage to less than 1 percent; untreated areas had 20 percent damage, while two fields treated with the insecticide carbaryl had 18 percent and 2.5 percent damage.

Hormones are chemicals produced in cells that travel through the bloodstream and help regulate an organism's growth and development. Each step in an insect's development is regulated by the timely release of particular hormones. Juvenile hormone, for example, controls early stages of insect development, and molting hormone helps trigger later development. By spraying an area with such hormones, scientists can keep insects from maturing completely, thus making it impossible for them to reproduce (Figure 17.19).

This strategy has been used to control several strains of malaria-carrying mosquitoes. This approach, however, requires having an available hormone for a specific pest and spraying effective amounts at just the right time in the pest's life cycle.

Figure 17.19 Some chemical hormones can keep insects from maturing completely, thus making it impossible for them to reproduce. Compare the normal mealworm (above) with one that failed to develop an adult abdomen after being sprayed with a synthetic hormone. (U.S. Department of Agriculture)

INTEGRATED PEST MANAGEMENT   A number of scientists argue that pest control should be viewed as an ecological problem rather than a chemical one. They argue that pest control involves the interaction of the pest with its total environment rather than with some particular poison only. An integrated pest-management approach considers each crop and its major pests as an ecological system; a control program is developed that uses a variety of biological, chemical, and cultural control methods, with the proper sequence and timing.

Using pesticides only at critical times and rotating the pesticides and the crops can minimize the development of genetic resistance. For example, five years after using synthetic pyrethroids in Australia only during a critical forty-day period of each year, no evidence of genetic resistance has appeared.

More than three dozen integrated pest-management studies over the past thirty years show that these techniques can be successful. They are initially costly, labor-intensive, and complicated. But they can reduce crop losses by half, reduce pesticide and fertilizer use, and produce substantial long-term savings.

G. Y. JACKS AND
R. O. WHYTE

Below that thin layer comprising the delicate organism known as the soil is a planet as lifeless as the moon.

Summary

To grow, plants need certain nutrients in trace amounts (micronutrients) and others in relatively large amounts (macronutrients). Fertilizers replenish the soil with nutrients. Commercial inorganic fertilizers especially provide nitrogen, phosphorus, and potassium. Organic fertilizers are rich sources of carbon and nitrogen.

Pesticides are chemicals that kill undesired organisms such as insects, weeds, rodents, and fungi. Three major chemical classes are the chlorinated hydrocarbons, organophosphates, and carbamates. Because they damage the environment, some pesticides (especially chlorinated hydrocarbons) have been banned or restricted. Herbicides can be classified as contact herbicides, systemic herbicides, or soil sterilants.

Alternative methods of pest control include crop rotation, natural pest predators, sterilizing insects, new genetic strains of crops, pheromones, and insect hormones.

Terms for Review

After completing this chapter, you should know and understand the meaning of the following terms:

carbamate (p. 475)

chlorinated hydrocarbon (p. 473)

commercial inorganic fertilizer (p. 469)

hormone (p. 483)

organic fertilizer (p. 467)

organophosphate (p. 475)

pest (p. 472)

pesticide (p. 472)

pheromone (p. 483)

Questions

Odd-numbered questions are answered at the back of this book.

1. (a) Write a balanced chemical equation for photosynthesis. (b) Then write a balanced chemical equation for the complete oxidation of glucose ($C_6H_{12}O_6$) by humans. [*Hint:* Review Section 5.4.]

2. A fertilizer label is marked 20–15–5. Explain what each number indicates.

3. Identify which of the following elements plants use in large amounts and which they need only in trace amounts: P, Ca, S, N, Mo, Zn, K, Mg.

4. Every plant is surrounded by the plentiful nitrogen present in the air, yet many plants are deficient in nitrogen. Why?

5. Write a balanced chemical equation for the Haber process.

6. Why do we use fertilizers?

7. What are the main sources of organic fertilizers? For which macronutrients are they a rich source?

8. What are the advantages and disadvantages of commercial inorganic fertilizers?

9. If soil is too acidic for growing a particular crop, what can be done to raise its pH?

10. If soil pH is too high, what can be done to lower it?

11. Classify each of the following pesticides as a chlorinated hydrocarbon, an organophosphate, or a carbamate:

a. Cl—⬡—Cl

paradichlorobenzene

b.

$$CH_3,\!\!\diagdown N\!\!-\!\!\bigcirc\!\!-\!\!O\!-\!\!\overset{\overset{\displaystyle O}{\|}}{C}\!-\!\!\overset{\overset{\displaystyle H}{|}}{N}\!-\!CH_3$$

Zectran

c. $CH_3CH_2-O-\overset{\overset{\displaystyle O}{\|}}{\underset{\underset{\displaystyle CH_3CH_2-O}{|}}{P}}-O-\overset{\overset{\displaystyle O}{\|}}{\underset{\underset{\displaystyle O-CH_2CH_3}{|}}{P}}-O-CH_2CH_3$

TEPP

d. $CH_3-\overset{\overset{\displaystyle O}{\|}}{\underset{\underset{\displaystyle F}{|}}{P}}-O-\overset{\overset{\displaystyle |}{CH}}{\underset{\underset{\displaystyle CH_3-\overset{|}{\underset{\underset{CH_3}{|}}{C}}-CH_3}{|}}{}}-CH_3$

GD (a nerve gas)

e.

$$\bigcirc\!\!-\!\!O\!-\!\!\overset{\overset{\displaystyle O}{\|}}{C}\!-\!\!\overset{\overset{\displaystyle H}{|}}{N}\!-\!CH_3$$

$CH_3-\overset{|}{\underset{\underset{CH_3}{|}}{CH}}$

Baygon

12. Which compound in Question 11 is the most likely to remain in the environment for a long time after being used?

13. Examine Figures 17.5 to 17.7. What functional groups are present in molecules of (a) DDT, (b) dichlorvos, (c) malathion, and (d) carbaryl?

14. What functional groups are present in (a) Treflan (Figure 17.10) and (b) Roundup (Figure 17.17)?

15. Explain the difference between a contact herbicide and a systemic herbicide.

16. What are the herbicides in Agent Orange? Why was this product used in Vietnam?

17. What are the advantages and disadvantages of using herbicides and insecticides that break down rapidly in the environment?

18. What were the advantages and disadvantages of DDT? Why was it banned for general use in the United States?

19. Explain how glyphosate (Roundup) kills plants.

20. What substance—carbohydrate, lipid, protein, or vitamin—can a plant synthesize directly when it has been given a new gene? [Review Section 9.5 if necessary.]

21. A substance was recently discovered that inhibits the normal decomposition of juvenile hormone in the larva stage of insect development. Do you think treating larvae with this substance would produce unusually small or large larvae? Why?

22. List several ways that pheromones can be used to control an insect population.

Topics for Discussion

1. Why do some people use commercial inorganic fertilizers when organic fertilizers can do the job?

2. Which is more persuasive—the case for pesticides or the case against? Why?

3. U.S. companies can manufacture and sell pesticides that are banned in the United States to other (mostly poor) nations where these chemicals are not banned. Do you agree with this policy? Why?

4. Do you agree with a recent proposal to permit the use of very effective pesticides if their residues in food pose a cancer risk from a lifetime's exposure of less than one in a million? Why?

CHAPTER 18

Nutrients and Additives in Food Products: Eating to Stay Healthy

General Objectives

1. What are the major types of nutrients you must eat to stay healthy, and what happens to them in your body?

2. What happens if you get too few or too many calories or too little protein?

3. What are some key vitamins and minerals you need to be healthy?

4. How do natural foods differ from synthetic or processed foods, and what are the major types of chemical additives in processed foods?

5. How safe are food additives?

To stay alive, your body is full of chemical reactions—reactions to produce energy for activities, to replace worn-out materials, or to remove waste products. Although it recycles many materials, your body would eventually use up all the reactants for its reactions unless you replenish them. That is why you eat, drink, and breathe. Eating and drinking provide the nutrients you need, while breathing supplies the oxygen to help metabolize those nutrients.

Omnivorous, you can digest food from plants or animals. Although you could get everything your body needs by eating a wide enough variety of these foods, in practice your intake is restricted to foods that taste good and fit with your culture and your health concerns. Famine is virtually unknown among U.S. college students, but malnutrition and overnutrition are not. You can select from a diverse array of foods, but you must choose sufficient proteins, carbohydrates, lipids, vitamins, and minerals from this bountiful supply to keep your body healthy. This chapter shows you how knowing some chemistry can help you make good choices.

18.1 Human Nutrition: Are You What You Eat?

METABOLISM: THE CHEMICAL MACHINERY OF LIFE Five fundamental chemical processes—photosynthesis, biosynthesis, digestion, fermentation, and respiration—keep us alive and healthy. As we have seen (Section 17.1), photosynthesis in plants provides the basic carbohydrate fuel for all food chains. Biosynthesis includes the thousands of chemical reactions that produce the organic molecules needed for life (Chapter 9). **Digestion** is the chemical splitting of food or nutrient molecules into a pool of simpler compounds that can dissolve and be absorbed into body fluids and tissues for use. Fermentation and respiration break down these organic compounds and store their chemical energy for use in the body. Collectively, these five processes are called **metabolism** (Figure 18.1).

NUTRITION   Like other living things, you require several types of substances in your body: carbohydrates (Section 9.2), lipids (fats and oils, Section 9.3), proteins (Section 9.4), nucleic acids (Section 9.5), water, and miscellaneous other substances, particularly vitamins and minerals (Section 18.2). Of these, only nucleic acids can be supplied completely by biosynthesis (see Figure 18.1); all others must be in your diet.

Food also contains roughage, which consists mostly of the indigestible cell walls of plants made of cellulose (Section 9.2). Although roughage is not a nutrient, it provides bulk that helps the digestive system function

Figure 18.1 Summary of metabolism in an animal cell.

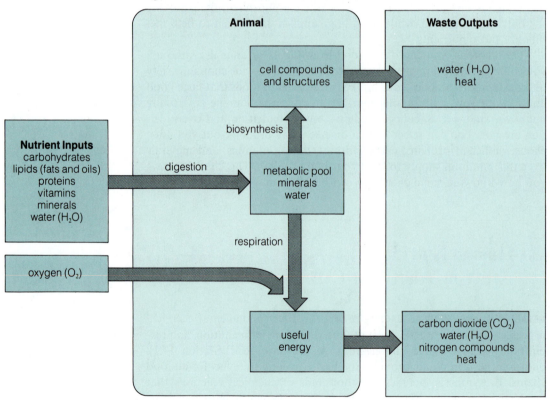

properly and prevents constipation. Roughage also absorbs many materials, including cancer-causing substances, and prevents them from being absorbed into the blood. In this way, dietary roughage may reduce the risk of colon cancer.

Figure 18.2 summarizes how your body digests nutrients in the food you eat. Specific enzymes (Section 9.4) catalyze the breakdown of carbohydrates, lipids, and proteins into simpler substances. This makes the component parts of the nutrients available for other metabolic processes (see Figure 18.1) so that your body can use them for energy (respiration) or reassemble them into the specific molecules your need. A protein in a hamburger patty, for example, has a sequence of amino acids (Section 9.4) that may have been useful to cattle but probably does not match what you need; so you digest the protein into its individual amino acids and then rearrange those amino acids into a different protein, one that your cells need.

CALORIES   The official SI unit of measurement for energy is the joule (J) (Section 1.4), but dieters still use calories to measure food energy. A nutritional **Calorie** (with a capital C, where 1 Cal = 1 kcal = 1000 calories) is nearly 4200 J, and it is more convenient to say that a serving of chocolate pudding has 385 Cal than that it has 1,600,000 J.

You can estimate the calories you need each day to maintain your body weight by multiplying your weight (in lb) by 10 (if you are physically inactive), 15 (if you are moderately active), or 20 (if you are very active)

Figure 18.2 Digestion of various nutrients in food.

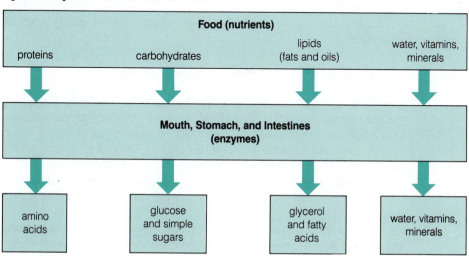

and then subtracting 10 Cal for each year over the age of 35 (to a maximum of 400 Cal).

Estimate the daily caloric needs of a forty-year-old, 140-lb person who is moderately active.

$140 \times 15 = 2100; \ 2100 - 50 = 2050 \ \text{Cal}$

SOLUTION

Getting the necessary energy to you is the top priority for your metabolism. The carbohydrates, lipids, and (to a lesser extent) proteins you eat are the three nutrients you use directly for energy. And if they are not enough, the carbohydrate stores (glycogen), the fat deposits (triglycerides), and the muscle proteins of your body are broken down to provide the necessary calories. Nothing short of death by starvation can prevent you from getting the energy you need.

CARBOHYDRATES   Carbohydrates are the chief source of our energy. In fact, 50 to 60 percent of the calories in a typical U.S. diet come from monosaccharides, disaccharides, and polysaccharides (Section 9.2). For people in less developed nations, carbohydrates provide an even higher percentage of calories.

Each gram of glucose ($C_6H_{12}O_6$) produces 4 Cal of energy when it is oxidized:

$$C_6H_{12}O_6 + 6O_2 \longrightarrow 6CO_2 + 6H_2O + \textbf{energy}$$

Your body normally maintains a blood glucose level of 70 to 100 mg of glucose in each dL (100 mL) of blood. As a result, at any given time you have only about 40 Cal of ready energy available in your blood. Stored in your liver and muscles, however, are 300 to 350 g of glycogen, which can be quickly converted into blood glucose on demand. Thus, carbohydrates can provide about half a day's supply of energy. Additional energy comes mostly from ingesting more food or by breaking down stored fat.

Glucose occurs naturally in most fruits, especially figs, dates, raisins, and grapes. Another monosaccharide, fructose, is also abundant in fruits.

Honey is a mixture of glucose and fructose. Like all carbohydrates, these two monosaccharides supply about 4 Cal/g. Because fructose is the sweetest tasting carbohydrate (see Figure 9.6), food manufacturers can use smaller amounts of fructose than table sugar (sucrose) in food products, making the food just as sweet but with fewer calories. On the other hand, promoting pure crystalline fructose as a replacement for "refined" sugar is misleading because both sugars come from plant sources and are processed similarly.

The main disaccharides in the diet are sucrose and lactose (Section 9.2). A compound of glucose linked to fructose, sucrose must be broken apart by enzymes in the small intestine to be absorbed into the body. Like any other digested sugar, sucrose provides 4 Cal/g. Contrary to widespread beliefs, sucrose is not harmful to your health unless it is taken in excess and displaces other nutrients. Molasses, raw sugar, and brown sugar are all common sources.

Lactose, a disaccharide of glucose and galactose (Section 9.2), is milk sugar. Like sucrose, it must be broken down into monosaccharide form in order to be absorbed from the small intestine. But some adults—especially those in certain ethnic groups such as Thais, Chinese, and blacks—lack the necessary enzyme (lactase) to convert lactose into glucose and galactose. For them, milk products can be unpleasant. Undigested lactose attracts water molecules into the small intestine. The lactose then passes into the colon (large intestine), where bacteria ferment much of it into carbon dioxide and other by-products. The extra intestinal water and gas cause nausea, cramping, pain, and watery diarrhea. Affected individuals can avoid milk and other dairy products that contain lactose, or they can drink milk that has been pretreated with the enzyme lactase to remove the lactose. This changes the milk's taste because the glucose and galactose that form are sweeter than the original lactose (see Figure 9.6).

Most of our carbohydrate intake is in the form of complex polysaccharides (Section 9.2), especially starch found in potatoes, pasta, rice, and corn. We have enzymes (called amylases) to digest these polymers into glucose. Cellulose, another polysaccharide, is the most abundant polymer of glucose in the world. The stems and leaves of all plants are mostly cellulose. But we cannot digest cellulose because we lack the enzymes to break the slightly different links between its glucose units (see Figure 9.7). Cellulose in your diet, such as celery stalks and lettuce leaves, provides bulk, or roughage, but no usable calories. Horses, cows, and goats, on the other hand, have bacterial enzymes in their stomachs to digest cellulose. One idea, still in the experimental stage, is to isolate large amounts of these enzymes and use them to convert cellulose into glucose for human consumption. Nourishing meals from straw and sawdust could mark a giant increase in the world's supply of edible carbohydrates.

FATS AND OILS   The most concentrated sources of energy in food are fats and oils, which are mostly triglycerides (Section 9.3). Being composed mostly of hydrocarbons, digested fats and oils can be more extensively oxidized than can carbohydrates; that is, their molecules can gain more oxygen atoms before they form carbon dioxide and water. This extra oxidation, being exothermic, releases correspondingly more energy. Thus, fats and oils provide about 9 Cal/g, more than twice as much energy as the 4 Cal/g released by the oxidation of carbohydrates (or proteins).

Fats have to be mobilized from their storage in adipose tissue before they can be used for energy, so they are not as quick a form of available energy as glucose or glycogen. You might think of glucose as your energy checking account, glycogen as your energy savings account, and fats as an energy insurance policy. The average adult who is not overweight has enough fat to provide energy for thirty to forty days. Stored fat also gives contour to the body, helps absorb shocks, protects bones and organs from damage, and acts as a heat insulator.

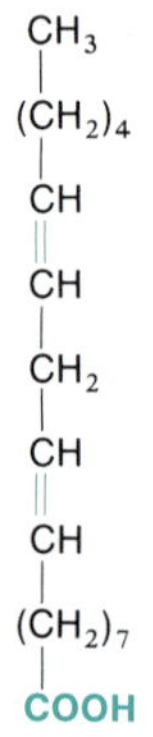

Besides energy, fats and especially oils (see Table 9.2) in your diet provide the essential polyunsaturated fatty acids—linoleic acid (Figure 18.3) and linolenic acid (see Figure 9.7). Itchy, runny rashes on the skin result when these fatty acids are missing. Although the body can synthesize fats from excess dietary glucose—perhaps all too well in many people's experience—it cannot produce polyunsaturated ones. You have to consume them ready-made.

Saturated fats in the diet (see Figure 9.7), on the other hand, have been correlated with excessive amounts of cholesterol in the bloodstream (Section 9.3) and with heart disease. The American Heart Association recommends that your diet provide less than 30 percent of its calories as fat, with saturated and polyunsaturated fats each accounting for no more than 10 percent of your total calories; in addition, you should consume no more than 300 mg a day of cholesterol. Figure 18.4 compares the distribution of calories in a typical cheeseburger/French fries/milk shake meal with a recommended allotment.

$$CH_3$$
$$|$$
$$(CH_2)_4$$
$$|$$
$$CH$$
$$||$$
$$CH$$
$$|$$
$$CH_2$$
$$|$$
$$CH$$
$$||$$
$$CH$$
$$|$$
$$(CH_2)_7$$
$$|$$
$$COOH$$

Figure 18.3 Linoleic acid, one of the two polyunsaturated fatty acids (linolenic acid is the other) required in the diet.

PROTEINS   Proteins are polymer molecules made by linking together various amino acid units (Section 9.4). Although proteins can be stripped of their amino groups and metabolized for energy (yielding the same 4 Cal/g as carbohydrates), their principal uses are for structure and for enzymes.

Your body digests dietary proteins into amino acids (see Figure 18.2); then your cells use the amino acids to synthesize all the different proteins you need. Each cell must have all twenty amino acids present or protein synthesis stops. Some amino acids can be synthesized from other materials, but eight of them (ten in children) cannot be. These eight, called the **essential amino acids**, must be in your diet (Table 18.1). Besides getting

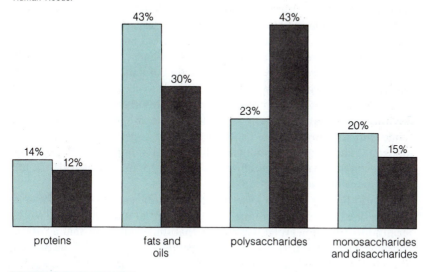

Figure 18.4 The distribution of calories for a typical cheeseburger/fries/shake meal (color) compared to the distribution recommended by the Senate Select Committee on Nutrition and Human Needs.

TABLE 18.1 ESSENTIAL AMINO ACIDS* AND THEIR PERCENTAGE CONTENT IN PROTEINS

	Cow's Milk	Human Milk	Egg (Whole)	Meat	Whole Wheat	Soybean	Corn (Zein)
Arginine[†]	3.5	5.0	6.7	6.6	4.3	7.3	1.7
Histidine[†]	2.7	2.7	2.4	2.8	1.8	2.9	1.3
Isoleucine	6.5	5.2	6.9	4.7	4.4	6.0	7.3
Leucine	9.9	15.0	9.4	8.0	6.9	8.0	23.7
Lysine	8.0	7.2	6.9	8.5	2.5	6.8	0
Methionine	2.4	2.0	3.3	2.5	1.2	1.7	2.4
Phenylalanine	5.1	5.9	5.8	4.5	4.4	5.3	6.2
Threonine	4.7	4.6	5.0	4.6	3.9	3.9	3.5
Tryptophan	1.3	1.9	1.6	1.1	1.2	1.4	0.1
Valine	6.7	5.5	7.4	5.5	4.5	5.3	3.5

* Structures of amino acids are in Table 9.3

[†] Essential for infants only

the right type of protein, you should eat about a gram of protein each day for every kilogram of your body weight. Typically, this amounts to 55 to 70 g (2 to 2.5 oz) of protein per day.

Complete proteins provide all eight essential amino acids. The most convenient of these are animal proteins such as meat, eggs, milk, and cottage cheese. Plant proteins are **incomplete proteins** because they lack sufficient amounts of one or more of the essential amino acids (see Table 18.1). Because proteins are the most expensive foodstuffs, people with little money usually have to survive on a grain diet of limited variety. This sort of diet is deficient in one or more essential amino acids unless plant foods with different essential amino acids are eaten in the proper combinations. Soybeans, for example, are low in methionine, whereas corn is low in lysine and tryptophan (see Table 18.1); eaten together, however, these foods provide adequate amounts of all essential amino acids. Vegetarians must be careful to select combinations of plant foods that provide the full complement of essential amino acids.

World hunger leads to *undernutrition* (insufficient caloric intake) and *malnutrition* (dietary lack of one or more essential nutrients). The World Health Organization (WHO) estimates that undernutrition, malnutrition, and their associated diseases cause an average of about 40,000 deaths *per day*—half of them children under the age of five years.

The two most widespread nutritional diseases are marasmus and kwashiorkor. **Marasmus** (from the Greek "to waste away") occurs when a diet is low in both calories and protein. Most victims are infants under the age of one year in poor families where children are not breast-fed or where there is insufficient food after the children are weaned from breast milk. An infant with marasmus (Figure 18.5, top) typically has a bloated belly, thin body, shriveled skin, wide eyes, and an old-looking face. Diarrhea, dehydration, muscle deterioration, anemia, a ravenous appetite, and possible brain damage also occur. If treated in time with a balanced diet, however, most of these effects can be reversed (Figure 18.5, bottom).

Kwashiorkor (a Ghanian term meaning "displaced child") occurs in infants and very young children (one to three years old) when a younger sibling or other reason causes them to change from mother's milk to a starchy diet based on sweet potatoes, maize flour, wheat, or other cereals. These diets supply enough calories but lack essential amino acids. Children with kwashiorkor have skin sores, swollen tissues, liver degeneration, permanent stunting of growth, mental apathy, and possible mental retardation. Many die because of their susceptibility to diseases such as whooping cough, dysentery, chicken pox, and measles. Like marasmus, kwashiorkor can be cured by a balanced diet if the malnutrition is not prolonged.

Figure 18.5 Most effects of severe protein–calorie malnutrition can be corrected. This two-year-old Venezuelan girl suffered from marasmus (top) but recovered after ten months of treatment and proper nutrition (bottom). (WHO, FAO, Rome, Italy)

OBESITY AND DIETING   While about 10 percent of the people in less developed nations have nutritional deficiency diseases, about 15 percent of the people in more developed countries are overnourished. In the United States, 10 to 15 percent of children and 30 to 35 percent of middle-aged adults are obese, weighing at least 20 percent more than their normal, desirable weight. Eating diets high in calories, saturated fats, salt, sucrose, and processed foods and low in fresh vegetables, fruits, and fiber, these people have an increased risk of diabetes, high blood pressure, stroke, heart disease, intestinal cancer, tooth decay (Section 16.1), and other health problems.

The causes of obesity are complex and not well understood. Inherited genetic factors, metabolic disorders, and psychological conditions may all play significant roles. Experts agree, however, that a major cause of obesity is overeating. When you take in more energy from nutrients than your body uses, you gain weight—about 1 g of fat and 0.5 g of water for every 9 Cal of excess intake. So it takes about 4000 extra Calories for you to gain 1 lb (0.45 kg) of fat plus 1/2 lb (0.23 kg) of water weight.

The most effective way to lose weight is to consume fewer calories than what the body needs, while maintaining healthful quantities of water, carbohydrates, fats, proteins, vitamins, and minerals (Section 18.2). A daily deficit of 1000 Cal through reduced calorie intake, increased energy expenditure, or both produces about a 2-lb (0.9-kg) weight loss per week, the maximum recommended rate. To expect results from anything but this daily, difficult, undramatic process is wishful thinking.

Nevertheless, in their search for a magic way to lose weight, people have used various drugs, undergone surgery, attempted an incredible variety of fad diets, checked into resorts or "fat farms," and begun jogging, aerobics, or other exercise programs. Drugs used in the quest for quick and easy weight loss include (1) appetite suppressants, such as amphetamines or "speed" (Section 20.4); (2) thyroid hormones (Section 19.4), to speed metabolism; (3) diuretics or laxatives, to lose water (which leads to dramatic but temporary weight loss); (4) local analgesics (Section 20.2), such as benzocaine, to numb taste buds; and (5) methyl cellulose, which swells up in the stomach to make it seem full. Surgical methods for weight loss include wiring the jaws shut, surgically removing fat, and stomach staplings and bypasses. Diets include low-calorie, low- or high-carbohydrate, low- or high-protein, low- or high-fat, one food (such as the grapefruit, rice, potato, or water diets), no food (fasting), and countless variations on these. Such methods may or may not produce satisfactory results.

Social pressures to be thin have also driven many people—especially young white women—to eating disorders. *Bulimia* is a cycle of eating binges followed by purging (usually by vomiting) to expel the food before it is metabolized. People with *anorexia nervosa* have an extreme fear of

TABLE 18.2 VITAMINS REQUIRED IN HUMAN NUTRITION

Vitamin	Recommended Daily Allowance [RDA]*	Sources	Possible Deficiency Effects
Fat-Soluble			
A (retinol)	750 μg	Fish-liver oils, butter, egg yolks, green leafy and yellow vegetables, milk	Night blindness, scaly skin, acne
D (cholecalciferol)	10 μg	Fish-liver oils, yeast, liver, fortified milk, egg yolks	Rickets (defective bone formation)
E (α-tocopherol)	4–9 mg	Plant oils, whole-grain cereals, egg yolks, beef liver, wheat germ, green leafy vegetables	Anemia? More rapid aging?
K	65–80 μg	Leafy vegetables, produced by bacteria in the intestines	Hemorrhages, slow clotting of blood
Water-Soluble			
B_1 (thiamine)	1 mg	Cereal grains, organ meats, milk, green vegetables, pork	Beriberi
B_2 (riboflavin)	1–4 mg	Beef liver, milk, eggs, yeast, leafy vegetables	Sores on lips, bloodshot and burning eyes
B_5 (niacin or nicotinic acid)	18 mg	Meats, vegetables, rice, eggs, yeast, whole grains	Pellagra
B_6 (pyridoxine)	2 mg	Eggs, liver, whole grains, milk, fish, legumes	Convulsions in infants; retarded growth; insomnia; eye, nose, and mouth sores
B_{12} (cobalamin)	3 μg	Meats, eggs, liver, seafood, dairy products	Pernicious anemia
Folacin	400 μg	Green vegetables, whole-wheat products	Anemia, diarrhea
Pantothenic acid	7 mg	Cereals, beef liver, kidney	Emotional instability, digestive disorders
Biotin	300 μg	Liver, egg white, produced by bacteria in the intestines	Skin disorders
C (ascorbic acid)	60 mg	Citrus fruits, raw green vegetables (especially tomatoes and green peppers)	Scurvy, low resistance to disease?

* For an adult male in good health

gaining weight. They avoid food and engage in purging when they do eat; for some people, this condition is fatal.

18.2 Vitamins and Minerals: The Little Things That Count

VITAMINS AND DEFICIENCY DISEASES   **Vitamins** are a group of about twenty-one organic compounds that we need in small amounts for good health but cannot synthesize in our bodies (except for vitamin D, which the body can synthesize if exposed to enough sunlight). The federal government has set recommended daily allowances (RDAs) for the amount of each nutrient we should consume for good health (Table 18.2).

The scarcity of food in many parts of the world today leads to millions of cases of vitamin-deficiency diseases each year. In more developed nations such as the United States, a combination of balanced diets, vitamin-fortified foods, and vitamin supplements have greatly reduced the incidence of these diseases. But these disorders still occur in people suffering from chronic alcoholism and in those who go on extreme diets and fail to take vitamin and mineral supplements.

Although people have known for centuries that specific foods can prevent certain diseases, not until the early twentieth century did chemists begin to isolate and identify the actual vitamins in various foods that cured deficiency diseases. For example, centuries ago sailors who traveled for long periods without fresh fruits suffered from weight loss, weakness, slow-healing wounds, skin lesions, vomiting, anemia, loose teeth, and bleeding and sore gums—a condition known as **scurvy**. British sailors were called "limeys" because early in the nineteenth century it was learned that daily rations of lime juice or lemon juice could prevent scurvy. Today we know that vitamin C in fresh citrus juices is the chemical that prevents scurvy.

Liver has long been used to treat **night blindness**. Today we know that this problem is caused by a deficiency of vitamin A. Severe deficiency can cause partial or total blindness, especially in children. In India and other parts of Asia, at least 2 million cases of blindness have been caused by vitamin-A deficiency. Vitamin-A deficiency can also cause the skin to become scaly and subject to acne. Indeed, recall (Section 16.2) that retinoic acid, a form of vitamin A, is used to treat acne.

In the eighteenth century, people used cod-liver oil to treat **rickets**, a disease that causes deformities in children when their bones weaken and bend (Figure 18.6). Now we know that this disease results from a lack of vitamin D, which increases the absorption of calcium from the intestinal

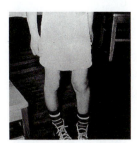

Figure 18.6 Child with bowlegs and knock-knees as a result of rickets. (Centers for Disease Control, Atlanta, Georgia)

tract and regulates the amount of calcium deposited in the bones and teeth. This disease occurs mostly in children living in northern areas where the sun is less effective in synthesizing enough vitamin D in their skin. Milk, an important source of calcium, is now fortified with vitamin D in many countries.

In parts of Asia, people who survive primarily on a diet of polished rice, which is made by removing the outer hulls, often develop **beriberi**. Victims of this disease, shown in 1911 to be caused by a deficiency in vitamin B_1, or thiamine (found in the hulls of rice), suffer from stiff limbs, an enlarged heart, paralysis, pain, loss of appetite, and eventual deterioration of the nervous system. People who eat unpolished rice or polished rice fortified with vitamin B_1 remain healthy.

Pellagra is a disease found in people who lack niacin (vitamin B_5) in the diet. Those people have scaly skin, diarrhea, an inflamed mouth, poor digestion, and an impaired central nervous system (dementia). Pellagra is common among people who eat mostly corn. Corn lacks the amino acid tryptophan (see Table 18.1), which is converted into niacin in the body. This disease can be prevented by adding whole-wheat cereals to the diet, or it can be reversed through niacin therapy.

CLASSES OF VITAMINS   Vitamins have a wide variety of structures and functional groups. As Figure 18.7 and Table 18.2 show, vitamins can be classified as water-soluble or fat-soluble. The fat-soluble vitamins (A, D, E, and K) have fewer polar groups and tend to be bigger molecules with more nonpolar hydrocarbon material; thus, they tend to be insoluble in water and soluble in nonpolar fat molecules. The water-soluble ones (various B vitamins and C) contain more polar groups; this makes them more soluble in water because they form hydrogen bonds with water molecules (Section 5.3).

Food processing, such as peeling fruit or potatoes or removing the outer husk of grains, removes vitamins from the food. High temperatures can break down vitamins such as C and B_1 (thiamine), and cooking foods in water dissolves out some of the water-soluble vitamins. To counter these losses, some foods are supplemented or fortified with vitamins to improve their nutritional value (Section 18.3). Many people also take vitamin supplements to ensure sufficient intake or because they believe that large doses are beneficial. Whether high doses provide better health or just expensive urine remains to be established.

Because vitamins B and C dissolve in water, you rapidly excrete them in urine. In fact, people who take multivitamin or B-complex vitamin pills commonly have urine that is bright yellow because of the extra vitamins, especially B_2 (riboflavin), it contains. Water-soluble vitamins are constantly being flushed from your body, and they are needed in your diet on a daily basis.

Figure 18.7 Structures of fat-soluble and water-soluble vitamins. Polar and ionic groups are in color.

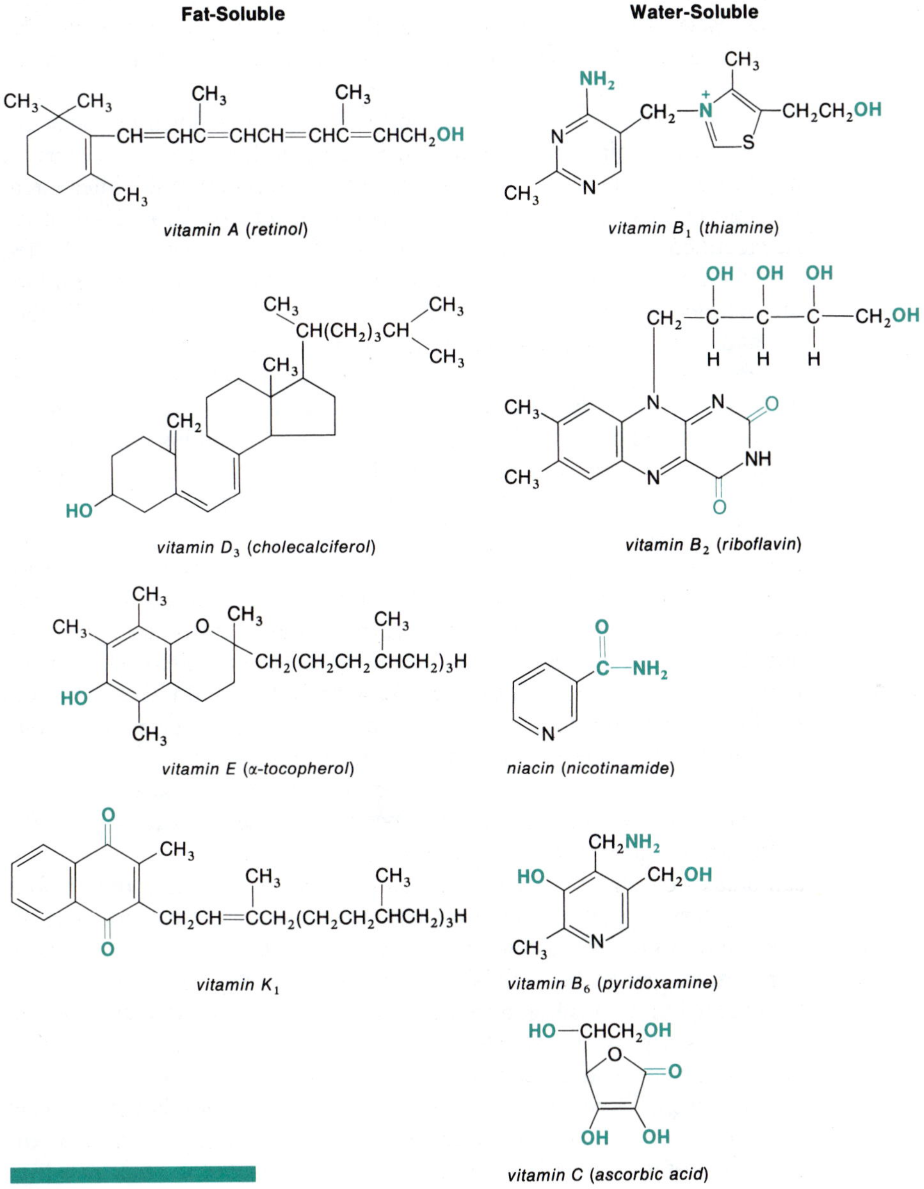

Fat-Soluble

vitamin A (retinol)

vitamin D₃ (cholecalciferol)

vitamin E (α-tocopherol)

vitamin K₁

Water-Soluble

vitamin B₁ (thiamine)

vitamin B₂ (riboflavin)

niacin (nicotinamide)

vitamin B₆ (pyridoxamine)

vitamin C (ascorbic acid)

Because they are nonpolar, vitamins A, D, E, and K can accumulate in fatty tissues. Thus, you don't need them in your diet quite so often. The problem with fat-soluble vitamins arises if you take too much. Instead of excreting the excess vitamins, you store them in your body; and sometimes, especially with vitamins A and D, they can reach toxic levels. Too much vitamin D, for example, causes bonelike calcium deposits in the kidneys.

You can also develop deficiencies (the opposite problem) because the fat-soluble vitamins are so nonpolar. When you digest these vitamins (or any lipid), your liver provides surfactant bile molecules to the intestine to suspend them in water and help absorb them into the blood. When the surfactants do not reach the intestine (due to liver damage or a blocked bile duct, for example), the vitamins simply stay in the intestine until they are excreted. People who cannot absorb the fat-soluble vitamins that they eat and drink develop deficiencies of them.

People often rely on vitamins as a source of extra pep and energy. This is misleading because vitamins provide no calories; only carbohydrates, lipids, and proteins can do that. But many vitamins (especially the B vitamins) help enzymes metabolize the three caloric materials. Only in that indirect sense do they give people the lift they seek.

MINERALS   Your body needs various inorganic compounds to do such things as transmitting electrical impulses in the nerves and muscles, forming bones and teeth, transporting oxygen in blood hemoglobin, and regulating metabolic processes. Most of these **minerals** required by the body are ionic compounds (Section 4.2)—such as sodium chloride, NaCl, and calcium phosphate, $Ca_3(PO_4)_2$—and exist as ions in the body. Figure 18.8 shows some sources and uses for these minerals.

Many physical and mental disorders are linked to deficiencies or excesses of minerals. Scientists are just beginning to learn how many. A few examples are given below.

Traces of zinc are essential for normal cell growth and repair. Zinc also appears to promote immunity to various diseases, to play a role in the operation of more than eighty enzymes, and to take part in the synthesis of DNA, the genetic material of the cell (Section 9.5). Zinc deficiency can cause hair loss, skin sores, and a decreased ability to taste or smell foods.

Iron is a part of hemoglobin protein molecules (Section 9.4) and of many enzymes. Iron-deficiency anemia, which affects about 10 percent of all adult men, 33 percent of all adult women, and more than 50 percent of children of less developed nations, causes fatigue and increases the likelihood of infections. On the other hand, too much iron can cause liver damage and congestive heart failure.

Hemoglobin concentration in the body depends not only on the level of iron but also on the amount of copper present in an enzyme required

Figure 18.8 Some sources and uses of minerals in the human body.

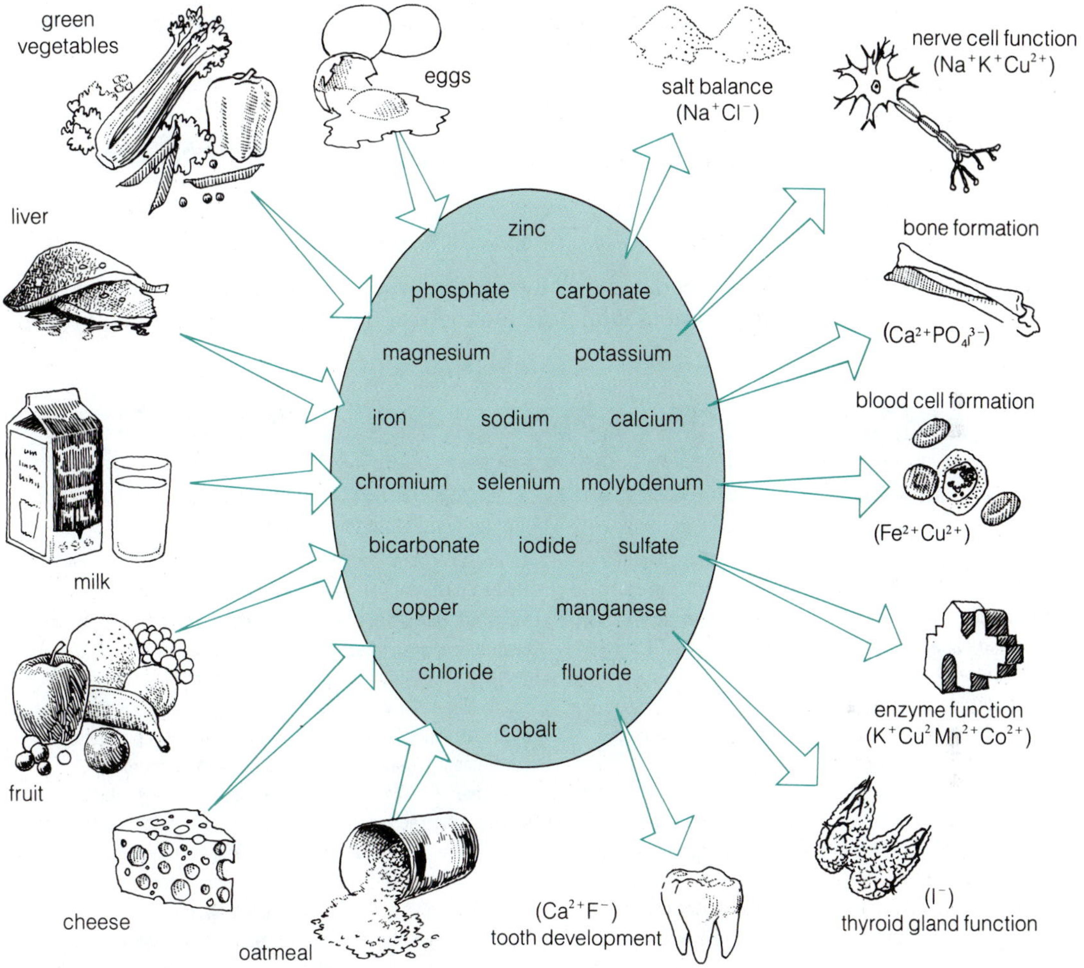

green vegetables

eggs

salt balance (Na^+Cl^-)

nerve cell function ($Na^+K^+Cu^{2+}$)

liver

bone formation

($Ca^{2+}PO_4^{3-}$)

zinc

phosphate   carbonate

magnesium   potassium

iron   sodium   calcium

chromium   selenium   molybdenum

bicarbonate   iodide   sulfate

copper   manganese

chloride   fluoride

cobalt

milk

blood cell formation

($Fe^{2+}Cu^{2+}$)

enzyme function ($K^+Cu^2\,Mn^{2+}Co^{2+}$)

fruit

cheese

oatmeal

($Ca^{2+}F^-$) tooth development

(I^-) thyroid gland function

for hemoglobin formation. The absorption of copper from the diet, in turn, depends on there being a correct amount of another mineral, molybdenum.

Too little iodine can cause goiter, an abnormal enlargement of the thyroid gland in the neck (see Figure 19.17). This can be prevented in most people by adding trace amounts of sodium or potassium iodide (NaI or KI) to table salt, to yield iodized salt.

Many adults, especially women, do not receive enough calcium in their diets. Necessary for healthy bones and teeth, calcium is also essential to transport ions in and out of cells. Whenever the body's supply of calcium runs low, it is replenished from and at the expense of the bones. Thus, long-term calcium deficiency is responsible for a bone-thinning abnormality called *osteoporosis*, which afflicts many elderly people (see the "Chemistry Spotlight").

18.3 Food Additives: How Natural Is Natural?

To the editor of the [Albany, New York] *Times-Union*:

Give us this day our daily calcium propionate (spoilage retarder), sodium diacetate (mold inhibitor), monoglyceride (emulsifier), potassium bromate

---

CHEMISTRY SPOTLIGHT

CAN TAKING TUMS OR OTHER CALCIUM SUPPLEMENTS HELP PREVENT YOUR BONES FROM BREAKING IN OLD AGE?

Relatively few American adults get the RDAs of 800 mg per day of calcium (1500 mg for women after menopause). Many physicians consider this intake necessary to prevent the debilitating loss of calcium from bones known as osteoporosis.

Calcium loss from bones begins after age thirty-five, and with women increases sharply after menopause, mainly because of the sharp drop in the secretion of estrogen by the ovaries. Osteoporosis afflicts 20 million Americans, mostly women over age forty-five, and leads to 1.3 million bone fractures a year. Older women suffer fractures of the hip, ribs, and other bones from simple movements such as bending to pick up a book, rolling over in bed, coughing, sneezing, or stepping off a curb. Many who suffer hip fractures are permanently crippled, and about 40,000 die each year of complications from such fractures—making osteoporosis the leading cause of death among older women. Men have an average of 30 percent more bone mass than women do and thus are less likely to develop severe osteoporosis.

Although most experts advise people to get as much of their daily calcium as possible from food, studies have not yet established that dietary calcium alone is an effective remedy for osteoporosis. Eating three dairy products a day or taking a calcium supplement (equivalent to two tablets of Tums) satisfies the recommended daily intake of calcium. A pint of milk contains 600 mg of calcium, and a cup of ice cream has almost 200 mg. Other sources are the bones in canned salmon, mackerel, and sardines and vegetables such as kale, collard greens, broccoli, tofu, and kidney beans. In addition, the calcium-loss rate can be reduced by curtailing the intake of alcohol, coffee (no more than five cups a day), and salt and by exercising.

Some nutrition experts fear that overreliance on calcium supplements can lead to overdosing. Excessive intake of calcium can cause painful kidney stones and alter the levels of other important minerals such as zinc, iron, and manganese. Some physicians caution against using calcium supplements such as dolomite and bone meal (widely sold in health-food stores) because they may also contain toxic elements such as lead.

(maturing agent), calcium phosphate monobasic (dough conditioner), chloramine T (flour bleach), aluminum potassium sulfate (acid baking powder ingredient), sodium benzoate (preservative), butylated hydroxy-anisole (antioxidant), mono-isopropyl citrate (sequestrant); plus synthetic vitamins A and D.

Forgive us, O Lord, for calling this stuff BREAD.

J. H. Read

TO ADD OR NOT TO ADD: THAT IS THE QUESTION   The letter lists only a few of the 100 or so chemicals that may be added to "enriched" bread. Today at least 2600 different chemicals are added to foods in the United States as we become more and more dependent on convenience and processed foods, which already account for more than half of the food people in the United States buy.

Some additives, such as vitamins and essential amino acids, are valuable nutrients that improve the quality of foods or replace nutrients lost in processing. Adding vitamin D to cow's milk prevents rickets (see Figure 18.6); the addition of iodide ions (I^-) to table salt has virtually eliminated goiter (see Figure 19.17); and since B vitamins (see Table 18.2) have been added to white bread, beriberi and pellagra have become rare in the United States. But when processing techniques remove nutrients, manufacturers may put them back (sometimes only partially), label the product *enriched* or *fortified*, and sell it at a higher price than the unprocessed food.

Additives are used for many other reasons besides nutrition—for instance, as preservatives, emulsifiers, sweeteners, or flavorings or as bleaching, thickening, brightening, or coloring agents (Table 18.3). More than $500 million is spent on additives each year. The average U.S. resident consumes more than 2.3 kg (5 lb) of these nonnutritive additives each year.

HOW SAFE ARE CHEMICAL ADDITIVES?   Nonfood chemicals have been added to food throughout history. Although most additives seem harmless, the rapidly increasing numbers and types of chemicals added to our food have increased concern that a handful may prove harmful. The extremes of this controversy range from "essentially all chemicals in food are bad and we should eat only natural foods" to "there's nothing to worry about because there is no absolute proof that chemical X has ever harmed a human being." As usual, the truth probably lies somewhere in-between. Some additives are necessary, useful, and safe; others are unnecessary or of doubtful safety.

TABLE 18.3 SOME COMMON FOOD ADDITIVES, PROCESSES, AND CONTAMINANTS

Class	Functions
Preservatives	To retard spoilage from bacterial action
Antioxidants (oxygen interceptors or freshness stabilizers)	To retard spoilage of fats by excluding oxygen or by inhibiting the breakdown of fats (rancidity)
Nutrition supplements	To increase the nutritive value of natural food or to replace nutrients lost in food processing
Flavors and flavor enhancers	To add or enhance flavor
Coloring agents	To add color for esthetic or sales appeal or to hide colors that are either unappealing or show a lack of freshness
Acidulants	To provide a tart taste or to mask undesirable aftertastes
Alkalis	To reduce natural acidity
Emulsifiers	To disperse droplets of one liquid (such as oil) in another liquid (such as water)
Stabilizers and thickeners	To provide smooth texture and consistency, prevent separation of components, and provide body
Sequestrants	To tie up metal ions that catalyze oxidation and other spoilage reactions in food, to prevent clouding in soft drinks, and to add color, flavor, and texture
Contaminants	Not deliberately added to foods

Examples	Some Typical Uses
Processes: drying, smoking, curing, canning (heat and sealing), dehydration, freezing, pasteurization, refrigeration. Chemicals: salt, sugar cure, sodium nitrate, sodium nitrite, calcium and sodium propionate, sorbic acid, potassium sorbate, benzoic acid, sodium benzoate, citric acid, sulfur dioxide, sulfite salts	Breads, cheeses, cakes, jellies, chocolate syrups, fruits, vegetables, meats
Processes: sealed cans, wrapping, refrigeration. Chemicals: lecithin, butylated hydroxyanisole (BHA), butylated hydroxytoluene (BHT), propyl gallate	Cooking oils, shortenings, cereals, potato chips, crackers, salted nuts, soups, Pop Tarts, Dream Whip, Tang
Vitamins and essential amino acids	Bread and flour (vitamins, amino acids, and iron), milk (vitamin D), rice (vitamin B_1), cereals
More than 1000 substances including saccharin, monosodium glutamate (MSG), and essential oils such as cinnamon (cinnamaldehyde), banana (isopentyl acetate), vanilla (vanillin)	Ice cream, artificial fruit juices, toppings, soft drinks, candies, pickles, salad dressings, spicy meats, low-calorie foods and drinks (sweeteners), and most processed heat-and-serve foods
Natural color dyes such as cochineal, turmeric; synthetic coal-tar dyes	Soft drinks, butter, cheese, ice cream, breakfast cereals, candies, cake mixes, sausages, fruit
Phosphoric acid (H_3PO_4), citric acid, tartaric acid	Cola and fruit soft drinks, desserts, fruit juices, cheeses, salad dressings, gravies, soups
Sodium carbonate (Na_2CO_3), sodium bicarbonate ($NaHCO_3$)	Canned peas, some wines, olives, coconut cream pie, chocolate eclairs
Lecithin, propylene glycol, monoglycerides, diglycerides, polysorbates	Ice cream, candies, nondairy creamers, dessert toppings, mayonnaise, salad dressings
Vegetable gums (gum arabic), seaweed extracts (agar, algin), dextrin, gelatin	Cheese spreads, ice cream, pie fillings, salad dressings, icings, diet canned fruits, cake and dessert mixes, syrups, instant breakfasts, soft drinks
EDTA (ethylenediamine tetraacetic acid), citric acid, chlorophyll	Soups, desserts, artificial fruit drinks, salad dressings, canned corn and shrimp, soft drinks, beer, cheese, canned frozen foods
DDT; PCBs; compounds of mercury, lead, and other heavy metals; radioisotopes; bacteria from poor hygiene and improper storage and processing; insects	Can occur in a wide variety of foods depending on accidental exposure or improper food processing

In the United States, the Food and Drug Administration (FDA) authorizes what substances (and amounts) can be used as food additives. Under the 1958 Delaney Amendment to the Food and Drug Act, there is zero tolerance for carcinogens; substances found to cause cancer in test animals are banned as food additives.

NATURAL VERSUS SYNTHETIC FOODS   Synthetic additives are not necessarily harmful, and "natural" foods are not necessarily safe. Many synthetic food additives, such as vitamins, citric acid, and sorbitol, also occur in natural foods. Whether natural or synthetic, a chemical is a chemical is a chemical—as long as it is pure. It makes no difference whether you get vitamin C, for example, from oranges or from synthetic vitamin-C tablets.

Similarly, poisoning by a manufactured chemical or by a natural substance is equally dangerous. And nature's own additives include an array of potentially harmful and toxic substances. Eating polar bear or halibut liver can cause vitamin-A poisoning. Lima beans, sweet potatoes, cassava (yams), and sugar cane contain compounds that produce small amounts of deadly hydrogen cyanide (HCN) in the human intestine. Eating cabbage, cauliflower, turnips, mustard greens, collard greens, and Brussels sprouts can cause goiter in susceptible individuals. Certain amines that can raise blood pressure dramatically are found in bananas, various acidic cheeses (such as Camembert), and some beers and wines. Spinach and rhubarb contain oxalic acid ($H_2C_2O_4$), which can precipitate kidney stones as calcium oxalate (CaC_2O_4). In addition, natural foods can be contaminated with food-poisoning bacteria such as *Salmonella* and *Clostridium botulinum* through improper processing, food storage, or personal hygiene.

On the other hand, natural foods may contain trace amounts of beneficial substances not found in purified, synthetic-chemical substitutes. In addition, a chemical that has been thoroughly tested and found to be harmless may interact with another chemical to produce a harmful (or beneficial) effect greater than the effect of either chemical used alone; this is called a **synergistic interaction**. As the number of synthetic chemicals in our food increases, the probability of such interactions also increases. Extensive tests for a single food additive or drug may take up to eight years and cost several hundred thousand dollars. It is impossible, both financially and scientifically, to test for possible synergistic interactions of a chemical with all of the many thousands (or even millions) of other chemicals. Yet, without some synthetic additives—the ones that inhibit deadly botulism bacteria, for example—eating natural foods can be risky.

Neither natural foods nor synthetic additives are free of risks, and neither can claim all the advantages. The more we know about both, the better we can choose the optimum combination for our diets.

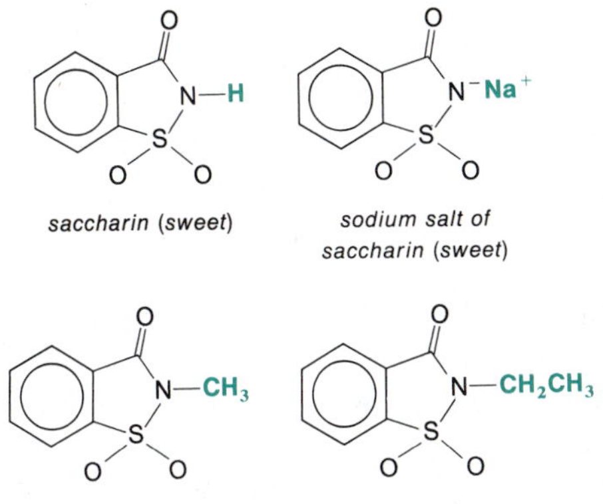

saccharin (sweet)

sodium salt of
saccharin (sweet)

tasteless

tasteless

Figure 18.9 The effect
of molecular structure on
the taste of saccharinlike
molecules.

18.4 Some Controversial Food Additives: How Safe Is Safe?

ARTIFICIAL SWEETENERS   For a number of years, overweight people and diabetics who must restrict their intake of sugar used saccharin, which is 300 to 500 times sweeter than sucrose; 1 or 2 g of it match the sweetness of a whole cupful of table sugar. Although it provides no calories, saccharin leaves a bitter aftertaste. In attempting to find better sugar substitutes, chemists made slight changes in the structure of saccharin (Figure 18.9); those seemingly minor changes produced dramatic changes in taste, but not a better sweetener.

In 1937 Michael Sveda, a chemistry graduate student at the University of Illinois, accidentally discovered a new sweetener that was later marketed by Abbott Laboratories under the name cyclamate (usually as soluble sodium or calcium cyclamate). Sodium cyclamate (Figure 18.10) is about thirty times sweeter than sucrose and has little of the bitter aftertaste of saccharin. Cyclamates were widely used in soft drinks and other foods until they were banned in 1969. The ban occurred because rats injected with large doses of cyclamates showed a high incidence of bladder cancer. Other studies gave conflicting results as to whether cyclamates were weak carcinogens. Nevertheless, cyclamates were banned as additives in accordance with the Delaney Amendment.

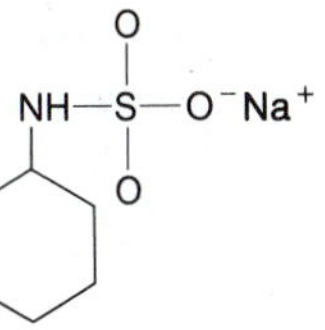

Figure 18.10 Sodium cyclamate.

After the ban, most food manufacturers returned to using saccharin, the lone artificial sweetener remaining on the approved list of the FDA. Since 1977, however, experiments have linked saccharin with cancer in laboratory animals. By law, such evidence should have triggered a ban on saccharin just as it did on cyclamates. But, at the time, that would

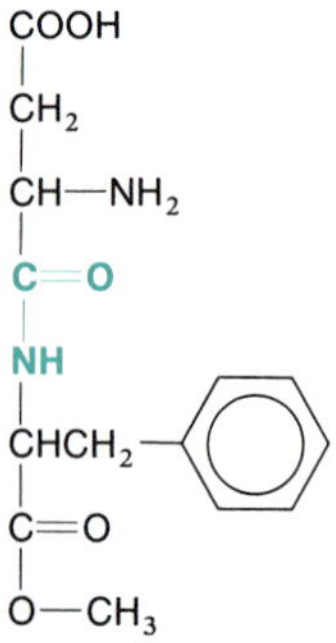

COOH
|
CH$_2$
|
CH—NH$_2$
|
C=O
|
NH
|
CHCH$_2$—⬡
|
C=O
|
O—CH$_3$

Figure 18.11 Aspartame. The amide (peptide) bond is in color.

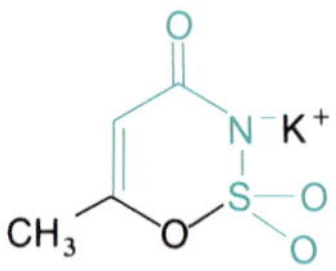

Figure 18.12 Potassium salt of acesulfame. Structural similarities to the sodium salt of saccharin (Figure 18.9) are in color.

have left no artificial sweeteners on the market. Congress, listening to a public outcry against a saccharin ban, imposed a moratorium on a ban that remains in effect at this writing.

In 1981 the FDA approved a new sweetener, aspartame (G. D. Searle's Nutrasweet), after the substance passed the most extensive safety testing of any food additive in history. Aspartame is a methyl ester of the dipeptide between the amino acids aspartic acid and phenylalanine (Figure 18.11). It is metabolized as a protein, yielding 4 Cal/g. However, 1 spoonful does the sweetening of about 180 spoonfuls of table sugar, so very little of it needs to be added to foods. Unlike protein, aspartame also releases methyl alcohol (CH_3OH) as it is digested. At normal consumption levels, however, aspartame provides no more methyl alcohol than do a number of other foods (such as tomato juice) and far less than the amount that can harm the human body. The phenylalanine component of aspartame is a problem for people with PKU (phenylketonuria), who lack an enzyme to metabolize that amino acid. Although these people can handle aspartame as well as they can any other source of phenylalanine, products containing aspartame have a warning label about PKU.

The newest artificial sweetener, approved by the FDA in 1988, is acesulfame potassium (Figure 18.12), known by the trade name Sunnette. Because it is not metabolized in the body, it provides no calories. Acesulfame is about as sweet as aspartame but is more similar in structure to saccharin.

FLAVORS AND FLAVOR ENHANCERS: THE MSG CONTROVERSY   Flavors and flavor enhancers are the largest class of food additives. Of the more than 2000 approved natural and synthetic flavors, at least 1600 are synthetic. Many are esters or aldehydes (Figure 18.13).

Figure 18.13 Some flavorful ester and aldehyde compounds. Most imitation flavors contain a mixture of substances.

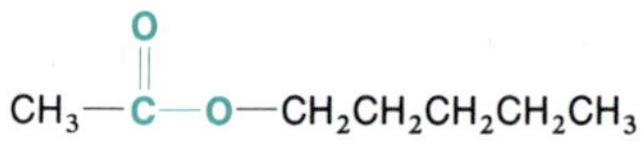

$$CH_3-\overset{O}{\overset{\|}{C}}-O-CH_2CH_2CH_2CH_2CH_3$$
pentyl acetate
(flavor of pears if the C$_5$ chain is straight and of bananas if it is branched)

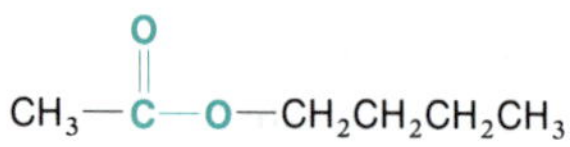

$$CH_3-\overset{O}{\overset{\|}{C}}-O-CH_2CH_2CH_2CH_3$$
butyl acetate
(flavor of raspberries if C$_4$ chain is straight and of strawberries if it is branched)

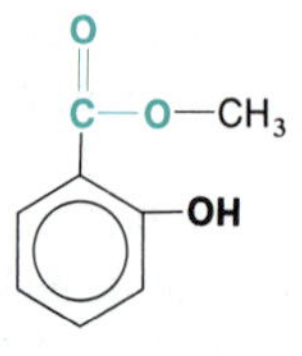

methyl salicylate
(oil of wintergreen)

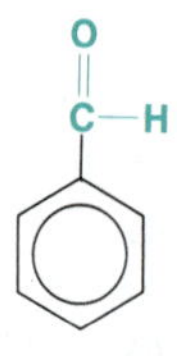

benzaldehyde
(oil of bitter almond)

Most natural flavors are elaborate mixtures of many compounds—sometimes hundreds of them. Usually only a few appear as major components in the mixture. Artificial or imitation flavors are typically blends of those major components only. For example, one imitation cherry flavor contains fifteen different esters, alcohols, and aldehydes. Although its taste is recognizable as "cherry," it lacks the subtle details of real cherry flavor because it lacks all trace compounds in the fruit.

Many flavorings have been used for decades. Most are assumed to be safe but have not been extensively tested for their potential to cause cancer, birth defects, or genetic damage. The government does not require that specific flavorings used in a product be listed (as most other additives must) on the label; their presence is simply indicated by the words *artificial flavoring added.*

Flavor enhancers have little or no taste of their own, but they amplify the flavors of other substances. Perhaps the best-known, most widely used, and most controversial flavor enhancer is monosodium glutamate (MSG), the sodium salt of a naturally occurring amino acid, glutamic acid (Figure 18.14). MSG is added to many processed foods in the United States and is ever-present in soy sauce and Oriental dishes.

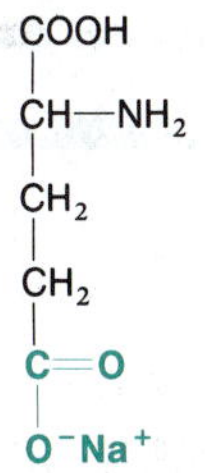

MSG is approved by the FDA, but it is being re-evaluated because of several animal studies and because it seems to be responsible for the "Chinese-restaurant syndrome." Within fifteen to thirty minutes after eating a meal containing large amounts of MSG (especially soups served in Chinese restaurants), susceptible people experience headaches, loss of breath, numbness, burning sensations, chest pains, general weakness, and even fainting spells. The symptoms are often confused with those of a heart attack.

Although most foods do not contain large amounts of MSG, there is concern about potential danger to infants. Baby-food manufacturers voluntarily stopped adding MSG to their products in 1969. MSG was added primarily for the parents' benefit anyway because babies cannot distinguish subtly different flavors.

COLORING AGENTS: COAL-TAR DYES   Artificial coloring agents are another widely used class of additives—not only in foods but also in lipstick, makeup, and shampoo. Even though colors add nothing to the nutritive value of foods, they make food products more attractive to consumers. Manufacturers frequently add colors to restore the natural ones lost in food processing or to give synthetic preparations the colors we expect. For example, oranges (especially Florida oranges), red potatoes, sweet potatoes, and other produce are dyed. In addition, to please consumers, growers add a dye to chicken feed to produce a yellow-skinned chicken.

Originally, many color additives were natural pigments and dyes, but today most artificial colors are synthesized from coal-tar extracts (Section

$$COOH$$
$$|$$
$$CH-NH_2$$
$$|$$
$$CH_2$$
$$|$$
$$CH_2$$
$$|$$
$$C=O$$
$$|$$
$$O^-Na^+$$

Figure 18.14 Monosodium glutamate (MSG).

8.3). They are made by heating coal in the absence of air to produce coke, coal gas, and coal tar (a black liquid). When purified and reacted with other chemicals, coal-tar extracts take on brilliant colors. Most food colors are aromatic compounds, often with double bonds outside the ring(s) (Figure 18.15). These compounds absorb certain wavelengths of visible light and pass on the remaining visible light, which we see as a particular color.

Because some aromatic compounds from coal tar have been shown to cause cancer, the use of coal-tar dyes as coloring agents has been under attack for several decades. Indeed, most of the additives banned by the FDA since 1960 have been coal-tar dyes. About ten colors are approved for use in foods in the United States, but regulations vary widely throughout the world. Although by law the labels of all foods must show that artificial colors have been added, the actual colors often are not listed.

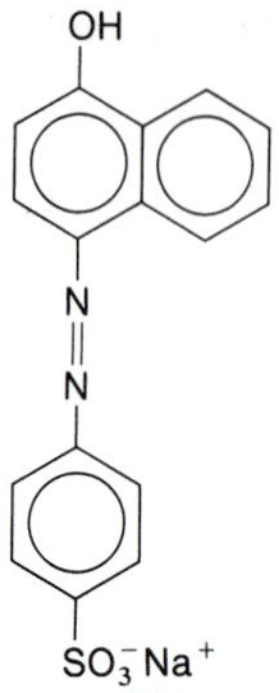

Figure 18.15 Orange No. 1, an approved food dye.

EMULSIFIERS   **Emulsifiers** disperse one liquid as finely suspended droplets in another liquid, thus preventing the liquids from separating into layers. Emulsifiers are like surfactants (Section 15.1), having both a nonpolar region and a polar or ionic region. As a result, emulsifiers serve as surfactantlike bridges to keep nonpolar substances, such as fats and oils, suspended in water.

One emulsifier is glycerol distearate (Figure 18.16). The hydrocarbon chains on the two ester groups make the molecule fat-soluble, whereas the hydroxyl (—OH) group lends water solubility to another portion of the molecule. Emulsifiers are widely used in dairy products and confections to disperse tiny globules of an oil or fatty liquid in water. Up to 10 percent of peanut butter may be emulsifiers.

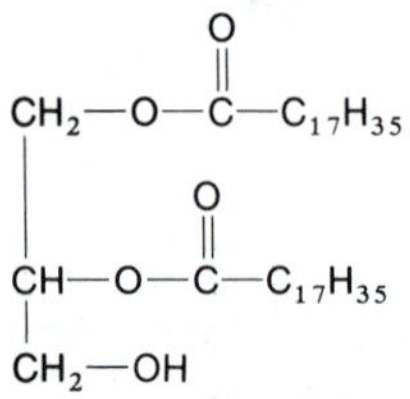

Figure 18.16 Glycerol distearate.

FAT SUBSTITUTES   Most of us consume more fat than is healthy. But now food manufacturers are trying to develop substitutes for fat. Monsanto has patented a process for heating and blending milk or egg protein into tiny spheres that have a smooth, creamy, fatlike texture. The product, called Simplesse, provides just 1.3 Cal/g and might eventually be used in dairy products and oil-based products such as salad dressing, mayonnaise, and margarine. It was approved by the FDA in 1990.

Another fat substitute awaiting action by the FDA is sucrose polyester (Figure 18.17). Known by the trade name Olestra, this large molecule is not absorbed into the blood and thus provides no calories. Unlike Simplesse, which changes texture upon heating, sucrose polyester can be used for cooking. For example, a 220-Cal serving of French fries cooked in vegetable oil would have just 150 Cal—and much less lipid—when fried instead in oil that is 75 percent sucrose polyester.

PRESERVATIVES: NITRATES, NITRITES, AND SULFITES   Microorganisms such as bacteria and molds spoil large amounts of food. One way to prevent this

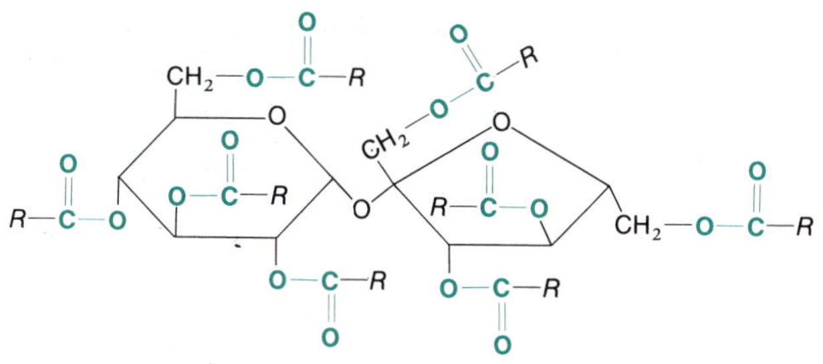

CH₂—O—C—R
R—C—O
O—C—R
O—C—R
CH₂
R—C—O
O—C—R
CH₂—O—C—R

Figure 18.17 Sucrose polyester, a potential fat substitute. The *R* groups are not necessarily the same; all are long hydrocarbon chains in natural fatty acids that are joined here to sucrose by ester linkages (in color).

is to exclude water, which microorganisms need to stay alive. One of the oldest preservative techniques is drying grains, fruits, fish, and meat. Salting and sugar-curing meat and storing fruits in concentrated sugar solutions have also been used for ages. These methods work because a high concentration of salt or sugar attracts water from the microorganisms, thus destroying them. Besides new methods of dehydration, we also use refrigeration and freezing, which slow the rate of chemical reactions so microorganisms cannot grow rapidly.

We also use chemical preservatives (Figure 18.18; see Table 18.3) to keep food safe to eat. Most of these preservatives are safe, and their use has greatly reduced the risk of eating contaminated food. In recent years, however, a few have become controversial. Sodium nitrate ($NaNO_3$) and sodium nitrite ($NaNO_2$) are widely used to give cured meats such as hot dogs, ham, bacon, sausage, and smoked fish a red color and to prevent them from turning an unappetizing gray or gray-brown. Nitrate is converted to nitrite and then to nitric oxide (NO), which reacts with the brown, iron-containing pigments (heme) in muscle and blood to form stable, bright red compounds. More important, nitrates and nitrites prevent the growth of deadly botulism bacteria. In fact, the meat industry says that these are the only chemical preservatives available that give adequate protection against botulism.

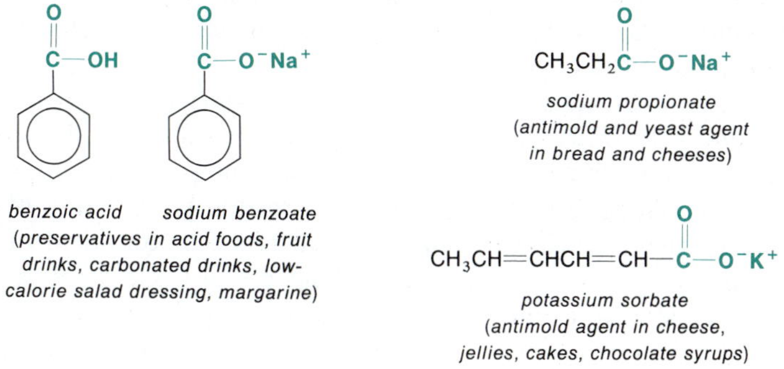

C—OH
C—O⁻Na⁺
benzoic acid
sodium benzoate
(preservatives in acid foods, fruit drinks, carbonated drinks, low-calorie salad dressing, margarine)
CH₃CH₂C—O⁻Na⁺
sodium propionate
(antimold and yeast agent in bread and cheeses)
CH₃CH=CHCH=CH—C—O⁻K⁺
potassium sorbate
(antimold agent in cheese, jellies, cakes, chocolate syrups)

Figure 18.18 Some chemical preservatives used in foods.

The safety of these additives has been questioned because some evidence indicates that nitrites can combine with amines in the stomach to form nitrosamines, which can cause cancer. In the acidic environment of the stomach, nitrite ions first become nitrous acid, which then reacts to form nitrosamines:

$$NO_2^- + \quad H_3O^+ \quad \longrightarrow \quad HNO_2 \; + H_2O$$

nitrite   *hydronium ion*   *nitrous acid*

$$HNO_2 \; + \; \begin{matrix} R \\ \diagdown \\ \diagup \\ R' \end{matrix} N{-}H \; \longrightarrow \; \begin{matrix} R \\ \diagdown \\ \diagup \\ R' \end{matrix} N{-}N{=}O + H_2O$$

nitrous acid   *amines*   *nitrosamine*

The FDA is reviewing the safety of these additives, weighing the possible increased risk of cancer against the danger of botulism if the additives are banned.

Sulfite salts such as sodium metabisulfite ($Na_2S_2O_5$) have been used on produce to prevent mold growth. As many as 1 million people in the United States, however, are allergic to sulfites; their reactions include breathing difficulties, nausea, headaches, and in a few cases even death. In 1986 the FDA withdrew its general approval of sulfites as a preservative, though trace amounts (less than 10 mg/kg of produce) are allowed in foods such as grapes.

ANTIOXIDANTS: BHA AND BHT   Food also can spoil by reacting with oxygen in the air to form products with unpleasant odors or tastes. Lipids and other unsaturated compounds are especially vulnerable, and a familiar example is butter becoming rancid. **Antioxidants** are compounds that prevent or slow the reaction of these food nutrients with oxygen in the air.

Three of the most common antioxidants are butylated hydroxyanisole (BHA), butylated hydroxytoluene (BHT), and propyl gallate (Figure 18.19). They are often added to shortening, vegetable oil, chewing gum, breakfast cereal, potato chips, candy, and other oil-containing products. They are also added to foods containing B vitamins (which can be destroyed by oxidation). Vitamin E, a natural antioxidant, is also effective, but it is much more expensive than synthetic additives.

Extensive studies on animals have not revealed any toxic or reproductive effects from these three additives, but effective long-term studies to determine whether they can cause cancer or genetic damage have not been done. On the other hand, one study showed that relatively high levels of BHT lengthened the average life span of certain rats the equivalent of twenty extra human years. Synthetic antioxidants are just one more example of food additives that present both benefits and drawbacks.

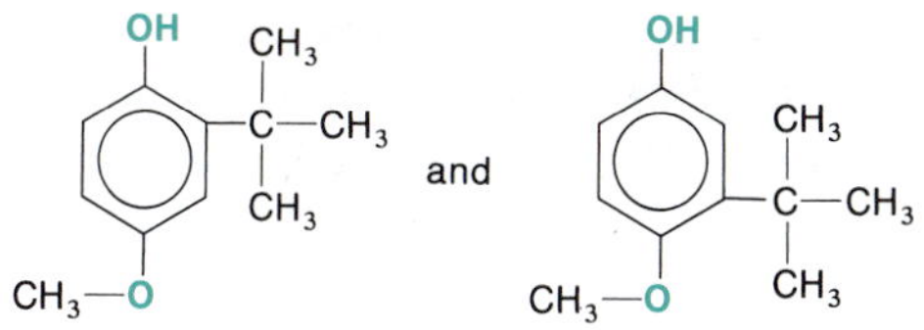

Figure 18.19 Some common antioxidants.

BHA (butylated hydroxyanisole)
(a mixture of two compounds)

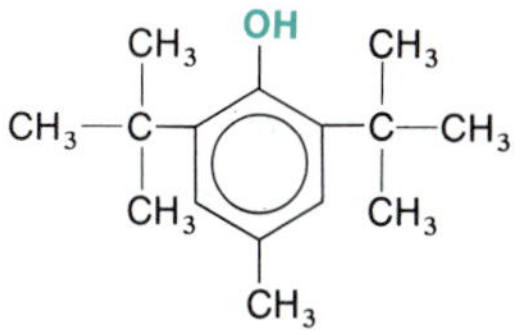

BHT (butylated hydroxytoluene)

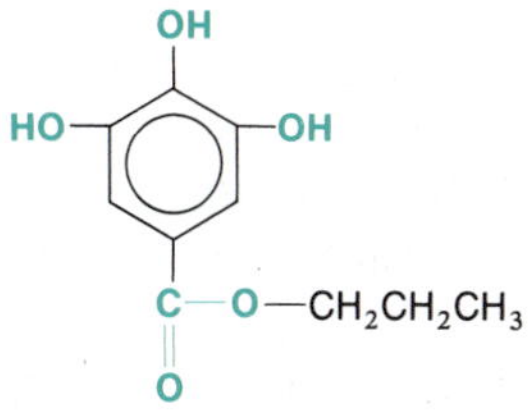

propyl gallate

CHOICES The entire subject of food products—from their production with the aid of fertilizers and pesticides to their consumption in a form combined with nutrients and other additives—can be viewed as an elaborate interplay of advantages and disadvantages. There is no "yellow brick road" that connects all the good and bypasses all the evil. Every food issue is thorny because each has life-and-death implications. Controversy over food products will always be with us. Opinions will continue to range from doomsday hysteria ("food additives cause nothing but hyperactivity and cancer"; "food additives aren't natural, so they are bound to be harmful") to fantasy-world optimism ("if the government approves it, it must be wholesome"; "I've eaten this all my life, and it hasn't hurt me yet"). All we can do is act on the best information available to us, and chemistry is part of that.

An exchange at a U.S. Senate hearing on the National Science Foundation budget:

SENATOR: I hear you biochemists want to adulterate bread by adding vitamin B to it. If the good Lord had wanted vitamin B in bread He would have put it there.

PHILIP HANDLER (President of the National Academy of Sciences): The good Lord did put vitamin B in bread. It was man who took it out in order to make white bread. The scientists who want to put it back are doing God's work.

Summary

We digest and metabolize food to provide the energy and synthesize the materials our bodies need. Our caloric requirements are met by carbohydrates (4 Cal/g), lipids (9 Cal/g), and proteins (4 Cal/g) that we eat; carbohydrates typically supply most of the calories.

Besides calories, we need adequate amounts of certain polyunsaturated fatty acids and amino acids in order to be healthy. Deficiencies or (in some cases) excess intake of vitamins and minerals also can cause disease. Vitamins are classified as fat-soluble or water-soluble, depending on how polar they are.

Food additives serve a wide variety of functions (see Table 18.3). Additives are used to provide food products that are often more appealing to consumers, less likely to be contaminated with microorganisms, and more convenient to use. The FDA judges their safety and regulates their use. Most additives are considered safe, but some present possible risks that must be weighed against their benefits. Examples include artificial sweeteners, monosodium glutamate (MSG), coal-tar dyes, nitrites, and sulfites.

Terms for Review

After completing this chapter, you should know and understand the meaning of the following terms:

antioxidant (p. 514)

beriberi (p. 500)

Calorie (p. 491)

complete protein (p. 496)

digestion (p. 489)

emulsifier (p. 512)

essential amino acid (p. 494)

incomplete protein (p. 496)

kwashiorkor (p. 496)

marasmus (p. 496)

metabolism (p. 490)

mineral (p. 502)

night blindness (p. 499)

pellagra (p. 500)

rickets (p. 499)

scurvy (p. 499)

synergistic interaction (p. 508)

vitamin (p. 499)

Questions

Odd-numbered questions are answered at the back of this book.

1. What is dietary roughage or bulk? Why is it a part of good nutrition?

2. What is metabolism? What bodily processes make up metabolism?

3. Which of the six major types of food nutrients provides the greatest amount of energy per gram?

4. Explain in general what happens to each of the following nutrients when it is eaten: (a) glucose, (b) potato starch, (c) table sugar (sucrose), (d) meat fat, and (e) animal protein.

5. Saliva contains the digestive enzyme amylase. Therefore, if you chew your food longer than usual, you are helping digest (a) fats, (b) simple sugars, (c) protein, (d) cholesterol, or (e) starch.

6. Criticize the statement that most people with poor diets die of starvation.

7. Name three classes of foods that you need in large amounts. Give examples of each.

8. For each of the following foods, tell whether its major ingredient is protein, carbohydrate, or lipid: (a) sweet potatoes, (b) eggs, (c) fish, (d) shortening, (e) carrots, (f) pork, (g) salad oil, and (h) Hostess Twinkies.

9. What is measured in calories? Does any food actually contain calories? What is meant by the statement that a candy bar has 245 Cal?

10. What is the main carbohydrate fuel of your body? Write the overall energy-producing reaction involving this fuel.

11. What structural properties do essential fatty acids possess? Why are they needed in the diet?

12. What are the advantages and disadvantages of fat in your body?

13. All twenty amino acids are needed for protein synthesis. Why are only eight labeled essential?

14. Distinguish among scurvy, beriberi, marasmus, pellagra, rickets, and kwashiorkor. Explain how each can be prevented.

15. What are some foods that contain complete proteins? How can a strict vegetarian get complete protein?

16. Explain the danger of large doses of vitamins A and D.

17. Which vitamins accumulate in the body? Explain why this occurs only for certain vitamins.

18. Why does aspartame have 4 Cal/g while saccharin has none?

19. Besides carbon and hydrogen, most flavoring agents contain the element _____.

20. Classify the food additives listed in the letter at the beginning of Section 18.3 as (a) sweeteners, (b) flavorings or flavor enhancers, (c) emulsifiers, (d) preservatives, or (e) colorants.

21. Examine the structures in Figure 18.18. Draw structural formulas for two related compounds that are also used as preservatives: sorbic acid and potassium benzoate.

22. Why are $NaNO_2$ and $NaNO_3$ added to meats? What is the danger of their presence? What is the danger of their absence?

23. Why are BHA and BHT added to some foods?

Topics for Discussion

1. A common statement is, You are what you eat. In what way(s) is this true? In what way(s) is it not true?

2. Debate the following controversial issues:
   a. BHT, BHA, MSG, coal-tar dyes, and nitrites should be banned as food additives.
   b. Saccharin should be banned as a food additive.
   c. We should eat only natural foods.

3. Classify the food additives listed in the letter at the beginning of Section 18.3 as (a) safe and necessary, (b) safe but unnec- essary, (c) potentially harmful but neces- sary nevertheless, or (d) potentially harmful and unnecessary. Justify your choices.

4. Find a list of ingredients on some food products (such as Sugar Smacks, Mr. Goodbar, or cake mix) and try to deter- mine the function of each additive.

5. According to the Delaney Amendment, a substance cannot be used as a food addi- tive if any amount of it (large or small) causes cancer in any organism. Do you agree or disagee with this law. Why?

PART V

Chemistry and Health

ISAIAH 40:31

They shall rise up with wings as eagles.

They shall run and not be weary; they shall walk, and not faint.

CHAPTER 19

Medical Drugs: Treating Diseases and Preventing Pregnancy

General Objectives

1. What are the sulfa drugs and the antibiotics used to fight infections? What are some problems from the overuse of antibiotics?

2. How are drugs that block metabolic processes used to treat AIDS, cancer, and heart attacks?

3. What drugs are used to treat people who have thyroid gland disorders or diabetes?

4. How are steroid hormones used as oral contraceptives?

Those who long for the "good old days," when life was supposedly simpler and better, know little about the past. Today the average life expectancy in the United States is more than seventy years. In 1850 it was only about thirty-five years, and in 1900 it was about forty-five years. In the nineteenth century, epidemics of yellow fever, smallpox, typhoid fever, cholera, and other infectious diseases were common in U.S. cities. The desolation and human suffering caused by these and other great epidemics of the past are difficult for most of us in the Western world to comprehend.

We take for granted that we will not get an infectious disease when we drink water from a faucet, dine out, or cook processed food from a grocery store. But this is not the case for most people in the world. While the average life expectancy for people in more developed nations is seventy-three years, it is only fifty-eight years for those in less developed nations. Many people in the world's poorer nations drink contaminated water, spend much of their time trying to get enough food to stay alive and healthy, and are susceptible to infectious diseases that rarely afflict more developed nations.

Much of the increase in life expectancy has come from applying chemistry to medicine. Improved sanitation and hygiene have halted the spread of infectious diseases. In addition, scientists have developed a wide array of drugs to combat disease. For example, we now routinely take antibiotics such as penicillin to cure diseases that once were fatal.

This chapter discusses some of the ways chemistry is applied to medicine. We look at drugs that are used to cure infections, fight cancer, control diabetes and other hormonal disorders, and prevent pregnancy.

19.1 Drugs: A Brief History

foxglove
(digitalis)

willow bark
(salicylic acid,
also used to
make aspirin)

periwinkle
(anti-cancer
compounds)

Figure 19.1 Some of the many medicinal plants from which modern drugs have been isolated. (New York Public Library)

THE BEGINNINGS OF CHEMOTHERAPY: USING MOLECULES TO CURE HUMAN ILLS

People have used chemicals to treat illnesses, aches, and pains since at least the dawn of recorded history. Many early drugs were natural animal and plant materials, and some seem a bit strange today. A remedy for blindness, for example, was to mix pigs' eyes, antimony, and honey and pour the concoction in the patient's ear.

But some of the recipes worked, and many contained drugs we now recognize as beneficial. An early Egyptian remedy for crying children contained poppy seeds—a known source of the pain reliever morphine. And a useful cure for night blindness was "liver of ox, roasted and crushed"—a good source of vitamin A.

Until well into the nineteenth century, physicians knew almost nothing about the chemistry of diseases. They also didn't know exactly what was in their potions, which ingredient was active, or why it worked. The drug era began in 1806 when the German pharmacist Friedrich Serturner isolated morphine (Section 20.2) from the opium poppy. Soon other drugs were isolated in pure form from animal and plant materials (Figure 19.1). At last physicians could know exactly what substance and how much of it they were giving their patients, and they could study its effects on the body.

In 1865 the French scientist Louis Pasteur proposed that bacteria cause infectious diseases. During the next three decades, scientists confirmed that microorganisms cause many diseases. In 1867 Joseph Lister discovered that phenol (Figure 8.14) kills microorganisms and thus could be used as a disinfectant to prevent infections resulting from surgery. Although still used to disinfect tables, floors, and other surfaces in hospitals, phenol is not used directly on skin because it causes blistering.

After Lister's discovery, scientists began searching for chemicals that could be taken internally and would kill infectious microbes without harming the body's cells. In 1907 Paul Ehrlich, a German physician and chemist, synthesized the arsenic-containing compound arsphenamine (Salvarsan) (Figure 19.2), and found that it was toxic to the tiny parasite that causes syphilis. Although it wasn't very effective in treating syphilis and it had unwanted side effects, arsphenamine was the first drug to kill harmful organisms without seriously damaging the human body. For this Ehrlich has been called the "father of chemotherapy."

Chemotherapy is the use of chemicals to treat disease. Ehrlich thought of such drugs as "magic bullets" that could destroy the harmful cells while leaving the normal body cells alone. But drugs can do this only if they can exploit some chemical difference between harmful and normal cells. Often this is difficult to do. Indeed, nearly three decades passed following Ehrlich's work before truly effective chemotherapeutic drugs (the sulfa drugs) were discovered.

A drug is usually defined as any substance that can have a noticeable effect on the structure and functioning of the body. But this means very little, for almost any chemical can fit that definition. In this chapter, we consider a **drug** to be any chemical that is used as medicine to treat disease. We focus on drugs that kill infectious organisms or that act on normal chemical reactions (metabolism) in the body. In Chapter 20, we examine substances that act primarily on the nervous system.

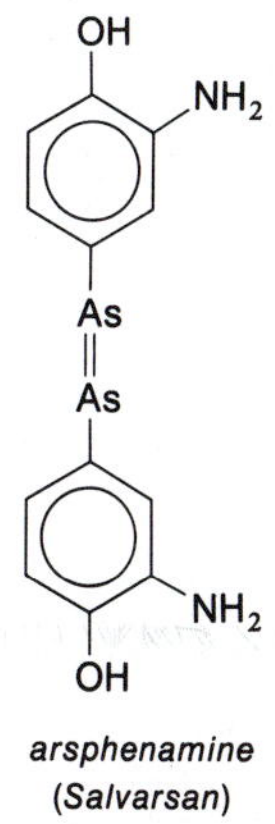

19.2 Sulfa Drugs and Antibiotics: Fighting Infection

SULFA DRUGS In 1932 the German bacteriologist Gerhard Domagk discovered that a red dye called prontosil killed certain infectious bacteria in mice, although it had no antibacterial action when used directly in the test tube. It turned out that the mice (but not the test tubes) were metabolizing prontosil into sulfanilamide (Figure 19.3), which was the actual substance killing the bacteria. Domagk used this drug to cure a deadly infection in a ten-month-old infant, thus launching a revolution in the treatment of infectious diseases.

Sulfanilamide destroyed streptococcal, staphylococcal, and certain other bacteria that caused diseases such as pneumonia, diphtheria, and gonorrhea. But it was ineffective against some disease-causing microorganisms, and it had unwanted side effects such as causing kidney damage at high doses. So chemists did what is now a common strategy in the drug industry: They synthesized new compounds by making structural changes in the sulfanilamide molecule and tested the compounds to see

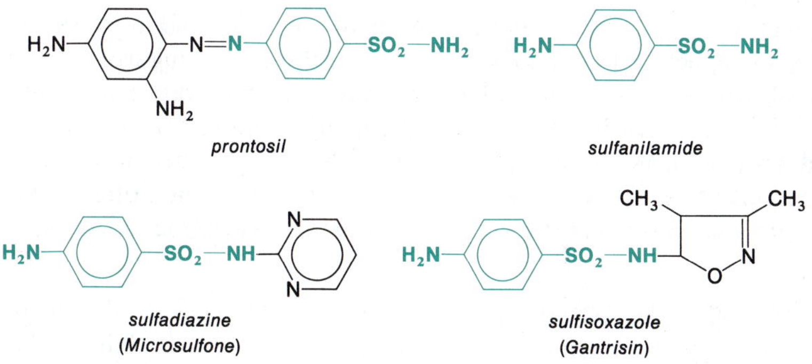

Figure 19.2 Arsphenamine, the first chemotherapeutic drug.

arsphenamine (Salvarsan)

prontosil

sulfanilamide

sulfadiazine (Microsulfone)

sulfisoxazole (Gantrisin)

Figure 19.3 Three sulfa drugs and prontosil. The atoms shown in color show the similarities in structure.

Figure 19.4 Because of its structural similarity to PABA, sulfanilamide prevents bacteria from synthesizing folic acid from PABA.

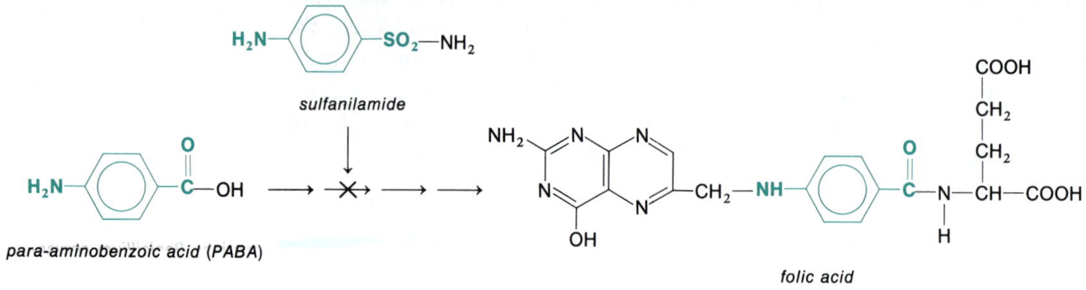

if any were safer or more effective than the original drug. Out of more than 5400 compounds tested, about a dozen proved to be useful drugs.

All the effective drugs contained an $-\overset{\overset{O}{\|}}{\underset{\underset{O}{\|}}{S}}-$ or $-SO_2-$ group and thus are known as **sulfa drugs** (see Figure 19.3).

Sulfa drugs kill bacteria by interfering with their metabolism. All cells need folic acid (a vitamin), and we obtain it from our diets. But certain bacteria cannot transport folic acid into their cells, so they must make their own. Bacteria normally synthesize folic acid from para-aminobenzoic acid (PABA). However, sulfanilamide chemically resembles PABA (Figure 19.4) enough that the bacteria mistakenly try to convert the drug, instead of PABA, into folic acid. It doesn't work, so the bacteria fail to make folic acid and die. Our cells aren't harmed because we don't synthesize folic acid anyway.

Sulfa drugs became the miracle drugs of the 1930s and early 1940s. For example, before 1936 about 95 percent of those suffering from streptococcal meningitis died; sulfanilamide lowered the mortality rate to less than 10 percent. During World War II, many thousands of lives were saved when soldiers and civilians sprinkled "sulfa" into their wounds to prevent infection. Although now largely replaced by antibiotics, sulfa drugs are still used to treat wounds, burns, and urinary tract infections.

PENICILLIN AND OTHER ANTIBIOTICS   **Antibiotics** are chemicals produced by organisms—usually bacteria, molds, or fungi—that kill or inhibit the growth of infectious microorganisms. The first antibiotic was penicillin (see the "Chemistry Spotlight"), which is produced by a *Penicillium* mold (Figure 19.5). Penicillin sharply reduced the incidence and the mortality rate of infectious diseases such as pneumonia, scarlet fever, anthrax, and syphilis.

Figure 19.5 The green mold variety, *Penicillium chrysogenum*, is used to produce almost all of the world's commercial penicillin. (Photo courtesy of Pfizer, Inc.)

CHEMISTRY SPOTLIGHT — THE DISCOVERY OF PENICILLIN

Figure 19.6 Alexander Fleming (1881–1955). (Squibb Corporation)

In the late 1800s, a group of scientists found that a certain mold, a *Penicillium*, killed bacteria that came in contact with it. These important results were ignored and forgotten for more than thirty years.

In 1928 the English bacteriologist Alexander Fleming (Figure 19.6) was studying the growth of bacteria that cause boils and other infections. A stray, airborne spore of a blue-green mold fell onto a culture plate where the bacteria were growing. Many scientists would have discarded the contaminated sample, but Fleming examined the plate and noticed that the bacteria near the mold were being destroyed. Further work showed that the mold contained an antibacterial agent, which he named penicillin because it was found in a *Penicillium* mold. But Fleming couldn't purify the active substance and gave up on the project.

A decade later, two Oxford University scientists, Howard Florey and Ernst Chain, took up the work again and succeeded in isolating a brownish product, penicillin G, from the mold. It cured certain bacterial infections in mice and was first tried on a human in 1941. The patient, a forty-three-year-old London policeman who had a serious case of blood poisoning from a shaving cut, immediately improved and seemed almost recovered after four days. But then the tiny experimental supply of penicillin ran out, the infection worsened, and the man died.

By 1943 penicillin was available for clinical use, and by 1945 there was enough for worldwide use. A *Penicillium* strain discovered on a moldy cantaloupe in a Peoria, Illinois, market proved to be a particularly good source of the new antibiotic. By the end of World War II, penicillin was saving many lives threatened by pneumonia, bone infections, gonorrhea, gangrene, and other infectious diseases.

It was the beginning of a new era in using chemistry to treat disease. For their work, Fleming, Florey, and Chain received the Nobel Prize for medicine and physiology in 1945.

Figure 19.7 Structures of four types of penicillin. Differences from penicillin G are shown in color.

penicillin G
(original penicillin)

penicillin V
(acid-resistant)

ampicillin
(acid resistant)

cloxacillin
(acid- and penicillinase-resistant)

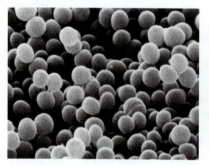
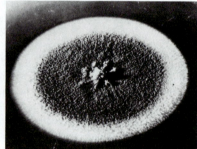

Figure 19.8 Electron micrographs of *Staphylococcus* bacteria in the growing phase before (left) and (right) in a dish after exposure to penicillin. (Left, Science Source/Photo Researchers; right, Squibb Corporation)

The original form (penicillin G) had to be injected because stomach acid destroyed samples taken orally. Another problem was that bacterial strains become resistant to penicillin by producing an enzyme (penicillinase) that destroyed the drug. So scientists chemically modified penicillin to make it even more effective (Figure 19.7). Children often use ampicillin, for example, because it can be taken orally and it kills a wider variety of bacteria than does penicillin G.

All penicillins kill bacteria by preventing them from making normal cell walls, which are complex networks of carbohydrates and protein cross-linked in three dimensions. Penicillin blocks an enzyme that cross-links the cell wall, so the cells become leaky. Substances freely move in and out of the cells, eventually causing them to rupture and die (Figure 19.8). Animal cells don't have cell walls, so they are not harmed.

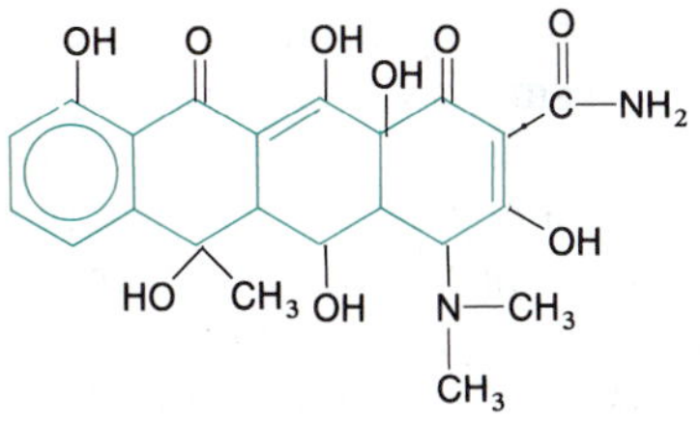

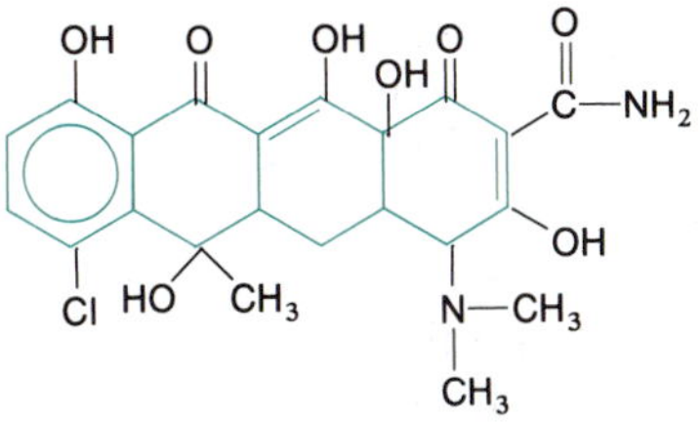

Figure 19.9 Two tetracyclines. Notice their four-ring or tetracyclic structure (shown in color).

Terramycin

Aureomycin

Penicillins, like sulfa drugs, are effective against only certain types of bacteria. So scientists began looking for antibiotics that were effective against a wide variety of disease-causing bacteria. They discovered such broad-spectrum antibiotics as Aureomycin and Terramycin, which are both tetracyclines. As Figure 19.9 shows, all tetracyclines have the same fused four-ring (tetracyclic) structure but differ in the functional groups bonded to this structure. Tetracyclines are among the most widely prescribed drugs in the United States. Table 19.1 lists some antibiotics and their uses.

Antibiotics other than penicillin work in a variety of ways. Many block the synthesis or functioning of DNA in bacteria. Tetracyclines, streptomycin, erythromycin, and several other antibiotics, for example, prevent bacteria from making proteins from their DNA (Section 9.5).

TABLE 19.1 SOME ANTIBIOTICS AND THEIR USES

Antibiotic	Chief Uses
Penicillins	Many types of infections (ear, skin, respiratory, digestive, urinary tract), syphilis, gonorrhea, scarlet fever
Tetracyclines	Many respiratory, digestive, and urinary tract infections; bronchitis; tonsillitis; whooping cough; syphilis; acne; typhus fever
Cephalosporins	Certain ear, throat, skin, and urinary tract infections
Chloramphenicol	Typhoid fever, meningitis, eye and ear infections
Erythromycins	Respiratory, skin, eye, and soft tissue infections; diphtheria
Spectinomycin	Gonorrhea
Gentamycin	Infections of bone, joints, skin, lungs, and abdomen

SOME PROBLEMS WITH ANTIBIOTICS   Although antibiotics have saved millions of lives, they also have disadvantages. First, using broad-spectrum antibiotics (especially tetracyclines) to treat long-lasting infections kills beneficial and harmless bacteria in the digestive tract and causes diarrhea; the destroyed bacteria may then be replaced by more harmful strains. Second, some people are allergic to certain antibiotics. Their reactions may include fever, body rash, and occasionally extreme shock. The Public Health Service estimates that 600 people die every year from allergic reactions to various penicillins.

Third, genetic resistance results from the widespread use of antibiotics. About half of the 14 million kg (30 million lb) of antibiotics produced each year in the United States are used as feed supplements for animals, both to control disease and to increase the animals' growth rate. The Food and Drug Administration (FDA) estimates that 80 percent of the swine, 60 percent of the cattle, and 30 percent of the chickens in the United States are raised with antibiotics in their feed. In an antibiotic-rich environment, it is only the resistant strains of microorganisms that are able to survive and multiply their kind. In time then, an increasing proportion of bacteria are resistant to the antibiotic.

Many resistant microorganisms have developed the ability to produce an enzyme that converts an antibiotic into an ineffective substance. In Mexico where antibiotics are sold over the counter, hospitals report that 20 percent of *Salmonella* strains are resistant to eight or more antibiotics. In the United States, the first case of penicillin-resistant gonorrhea appeared in 1976; nine years later, there were 8500 cases. Some were treated with spectinomycin, which is effective in such cases. But physicians do not want to use spectinomycin widely because this increases the chance that strains resistant to it will develop. Indeed, only three years after using this drug on a regular basis, the first cases of spectinomycin-resistant gonorrhea appeared in 1987 among U.S. servicemen in Korea.

19.3 Antimetabolic Drugs: Treating AIDS, Colds, Cancer, and Heart Attacks

Many drugs work because they structurally resemble a natural substance; when a cell tries to use the drug instead of the natural substance, some normal process is blocked. Drugs that block a metabolic process in this way are called **antimetabolites**. We have already seen that sulfa drugs work in this way, and now we examine other examples.

TREATING AIDS AND OTHER VIRAL INFECTIONS Viruses, unlike bacteria, are parasites that can reproduce only when they enter a real cell. **Viruses** are molecules of DNA or RNA surrounded by a protein coat. Only the nucleic acid (DNA or RNA) enters a cell infected by a virus. The nucleic acid then instructs the cell to synthesize many virus particles. When the cell dies, viruses escape and spread the infection to other cells.

Viruses cause such diseases as AIDS (acquired immune deficiency syndrome), chicken pox, influenza, measles, herpes, and polio. Some of these can be prevented by vaccines, in which the body is exposed to a weakened form of the virus. This enables the body's immune system to build up its defenses so that it can destroy the virus in case of a real infection later. Another approach, especially after the infection begins, is to use drugs.

Several antiviral drugs are antimetabolites that block the infected cells from synthesizing DNA; this keeps viruses from being able to multiply and spread the infection. These drugs block DNA synthesis by chemically resembling the normal bases in DNA (Section 9.5), thus causing the cell to make faulty DNA.

Figure 19.10 shows the resemblance for two important antiviral drugs. Acyclovir is used as an ointment to treat genital herpes. Although

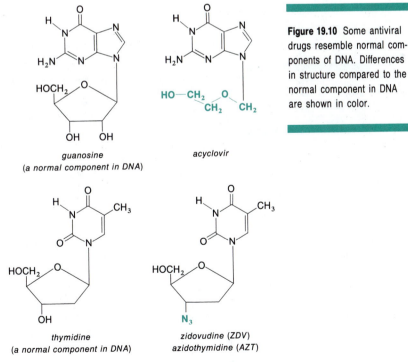

guanosine
(a normal component in DNA)

acyclovir

thymidine
(a normal component in DNA)

zidovudine (ZDV)
azidothymidine (AZT)

Figure 19.10 Some antiviral drugs resemble normal components of DNA. Differences in structure compared to the normal component in DNA are shown in color.

it doesn't cure herpes, acyclovir makes the periodic flare-up of sores less frequent and of shorter duration. Though now largely overshadowed by AIDS, genital herpes afflicts more than 20 million people in the United States.

Zidovudine (ZDV), also called azidothymidine (AZT), is used to treat people who have AIDS. The AIDS virus contains RNA, and the infected cells try to synthesize DNA from the viral RNA. AZT, by masquerading as a normal base in DNA (Figure 19.10), causes the cell to make faulty DNA, thus preventing the virus from multiplying. AZT doesn't cure AIDS, but it alleviates some of the symptoms and prolongs the life of many patients. Current studies are examining the effectiveness of using AZT in combination with other drugs such as acyclovir.

TREATING COLDS   Colds are caused by viruses, but we don't have effective drugs—other than vaccines—to combat those viruses. Instead, we have a wide variety of cold remedies that make us feel better because they treat the symptoms of a cold. Often the remedies contain several ingredients to combat different symptoms. Figure 19.11 shows the structures of a few of the drugs used.

Compounds that treat sore throats include benzocaine or dyclonine, which temporarily numb the throat, and various alcohols (benzyl alcohol, salicyl alcohol) or phenol (or its salt, sodium phenolate), which kill bacteria (but not viruses) on the surface of the throat.

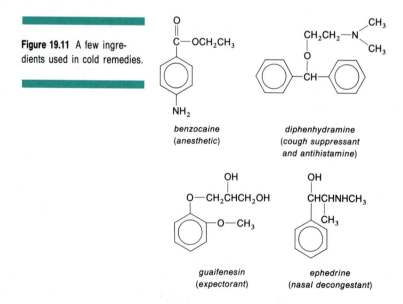

Figure 19.11 A few ingredients used in cold remedies.

benzocaine
(anesthetic)

diphenhydramine
(cough suppressant
and antihistamine)

guaifenesin
(expectorant)

ephedrine
(nasal decongestant)

The most effective cough suppressants, which inhibit the brain's cough center, are codeine (see Figure 20.7), dextromethorphan, and diphenhydramine. The first two are narcotics that can cause addiction if they are abused. Cold remedies also include *expectorants*, which are claimed to loosen mucus so that it can be coughed up. Such compounds include guaifenesin, ammonium chloride (NH_4Cl), and terpin hydrate. Only guaifenesin has been shown to be effective.

When people are exposed to substances to which they are allergic, certain cells secrete histamine, which causes runny noses, nasal stuffiness, and other symptoms. These people are treated with *antihistamines* such as chlorpheniramine, brompheniramine, and diphenhydramine; these antimetabolic drugs block the secretion or actions of histamine. Although many cold remedies contain antihistamines, they are of doubtful value in relieving symptoms due to colds. Of more value are nasal decongestants such as ephedrine (and its derivatives) and phenylpropanolamine, which are effective when used as directed. Some cold remedies also contain a pain reliever such as aspirin, ibuprofen, or acetaminophen (see Section 20.2 and Figures 20.3 and 20.5) to help treat aches and fever.

ANTICANCER DRUGS Antimetabolites also are an important part of cancer treatment. **Cancer** is a group of more than 100 different diseases, all characterized by uncontrolled cell growth and reproduction (Figure 19.12). If not detected and treated in time, many cancerous tumors undergo metastasis; that is, they spread to other parts of the body, making treatment much more difficult.

The major ways to treat cancer include (1) surgery, aimed at removing cancerous growths before they spread to other parts of the body; (2) radiation (as discussed in Section 3.6); and (3) chemotherapy, the use of chemicals to destroy malignant cells. Two major types of anticancer drugs are antimetabolites and cross-linking agents. Both interfere with the rapid growth of cancer cells by blocking the synthesis of DNA needed for new cells.

One drug that does this is the antimetabolite methotrexate, which interferes with the action of folic acid (Figure 19.13). Our cells need folic acid to synthesize the nitrogen bases in DNA. When an antimetabolite prevents this, cells cannot make new DNA and thus cannot multiply. Cancer patients are sometimes given very high doses of methotrexate—doses that would ordinarily be fatal—and after several hours are rescued by large doses of a form of folic acid that counteracts the drug. This limited exposure to toxic levels of methotrexate takes a heavy toll on cancer (and other) cells that normally would multiply during that time.

Other anticancer drugs chemically resemble normal nitrogen bases in DNA and thus prevent cells from synthesizing new, normal DNA. For

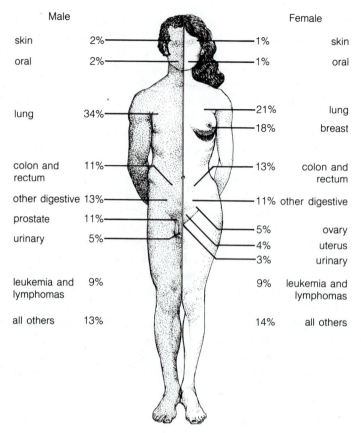

Figure 19.12 Where fatal cancers strike; percentages of all U.S. cancer deaths.

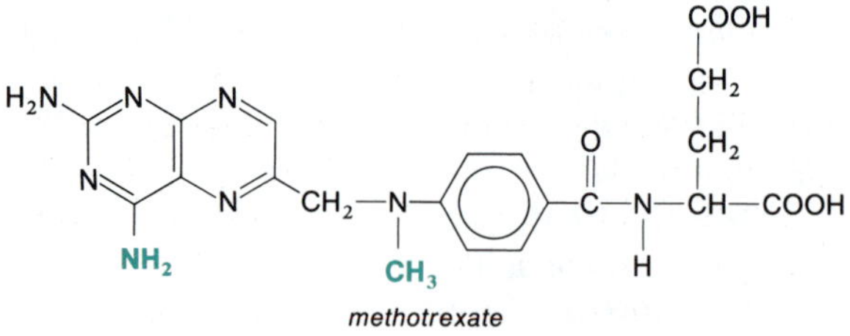

Male				Female
skin	2%		1%	skin
oral	2%		1%	oral
lung	34%		21%	lung
			18%	breast
colon and rectum	11%		13%	colon and rectum
other digestive	13%		11%	other digestive
prostate	11%		5%	ovary
urinary	5%		4%	uterus
			3%	urinary
leukemia and lymphomas	9%		9%	leukemia and lymphomas
all others	13%		14%	all others

Figure 19.13 Structure of the antimetabolite methotrexate closely resembles that of folic acid, shown in Figure 19.4. Structural differences from folic acid are in color.

methotrexate

example, 5-fluorouracil (which resembles the normal DNA base thymine) is used to treat colon, breast, ovarian, and prostate cancer. Methotrexate and 6-mercaptopurine (which resembles the base adenine) have been used for treating acute leukemia.

Another type of anticancer drugs, the **cross-linking agents**, binds to DNA in two locations and prevents it from functioning normally.

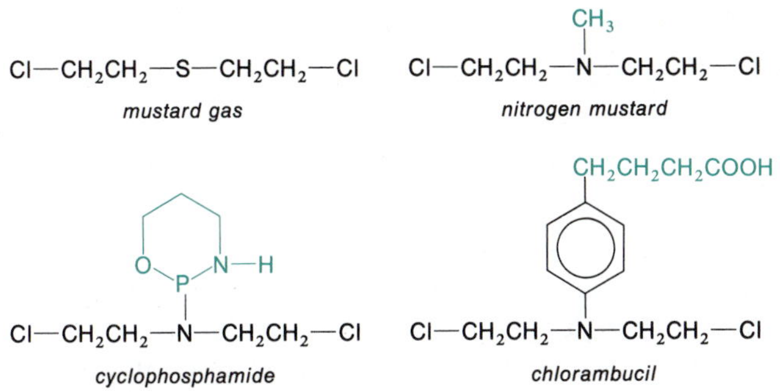

$$Cl—CH_2CH_2—S—CH_2CH_2—Cl$$

mustard gas

$$Cl—CH_2CH_2—\overset{\overset{\displaystyle CH_3}{|}}{N}—CH_2CH_2—Cl$$

nitrogen mustard

Figure 19.14 Mustard gas and nitrogen mustards used to treat cancer. The structural differences are shown in color.

cyclophosphamide

$$Cl—CH_2CH_2—N—CH_2CH_2—Cl$$

$$Cl—CH_2CH_2—\overset{\overset{\displaystyle CH_2CH_2CH_2COOH}{|}}{N}—CH_2CH_2—Cl$$

chlorambucil

Strangely enough, the first known drugs of this type (called *nitrogen mustards*) were closely related to mustard gas, a chemical warfare agent used in World War I. Figure 19.14 shows the structures of mustard gas (which is not effective against cancer) and several cross-linking agents used in treating cancer. Nitrogen mustard, for example, is used to treat Hodgkin's disease (cancer of the lymphoid tissues).

Antimetabolites and cross-linking agents act against all rapidly multiplying cells, not just cancer cells. Thus, they also damage normal cells such as skin, hair, gastrointestinal, and bone marrow, which produces blood cells. This is why cancer patients receiving chemotherapy often have side effects such as fewer blood cells, nausea, vomiting, and hair loss. Through a combination of early detection and improved use of surgery, radiation, and chemotherapy, survival rates for many types of cancer (including skin, testicular, bladder, cervical, breast, and Hodgkin's disease) now are 65 percent or higher. There has been little improvement, however, in the survival rates for people having cancer of the pancreas, esophagus, lung, stomach, or brain.

TREATING HEART ATTACKS Many heart attacks result from cholesterol and other lipids depositing in blood vessels serving the heart (Figure 9.11). As those vessels become clogged, the heart tissue gradually dies because it doesn't receive enough oxygenated blood; eventually, the heart itself fails to work properly.

Lowering blood levels of cholesterol and other lipids is an important way to reduce the risk of heart disease. Low-fat diets, regular exercise, and drugs are effective ways to do this. One such drug is niacin (Figure 18.7), a B vitamin that in large doses reduces blood cholesterol levels. Another is cholestipol, which binds to steroids in the intestine and removes them from the body; cholesterol is then removed from the blood and metabolized to replenish the intestinal supply of steroids.

Another drug, lovastatin (mevinolin), is an antimetabolite that blocks the synthesis of cholesterol in the body (Figure 19.15). Produced by a

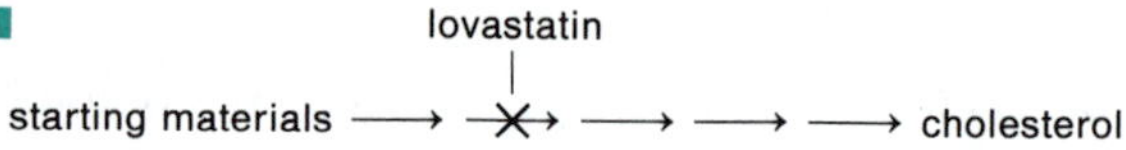

Figure 19.15 Lovastatin blocks cholesterol synthesis in the body.

fungus, lovastatin was approved in 1987 for treating people who are genetically inclined to have high blood cholesterol levels.

Most heart attacks occur when blood clots plug coronary arteries. Drugs that prevent clots are called **anticoagulants**, and several are antimetabolites that resemble vitamin K (Figure 19.16). Our blood needs vitamin K in order to clot; in fact, the *K* comes from an earlier German reference to this substance as a *Koagulations Vitamin*. Thus, antimetabolites that block the action of vitamin K keep the blood from clotting properly.

Anticoagulants are used after surgery or heart attacks to prevent new blood clots from forming and blocking blood vessels. Too much anticoagulant can cause internal bleeding (hemorrhaging), but an overdose can be corrected by giving the patient extra vitamin K. The anticoagulant dicoumarol (see Figure 19.16) was discovered because cows eating spoiled sweet clover (which contains dicoumarol) suffered serious and sometimes fatal internal bleeding. Warfarin is used as a commercial rat poison that causes its victims to hemorrhage to death.

19.4 Treating Hormone Disorders

Certain glands in the body secrete **hormones**, which serve as chemical messengers to produce effects in other parts of the body. Table 19.2 lists

Figure 19.16 Vitamin K$_3$ and two anticoagulants.

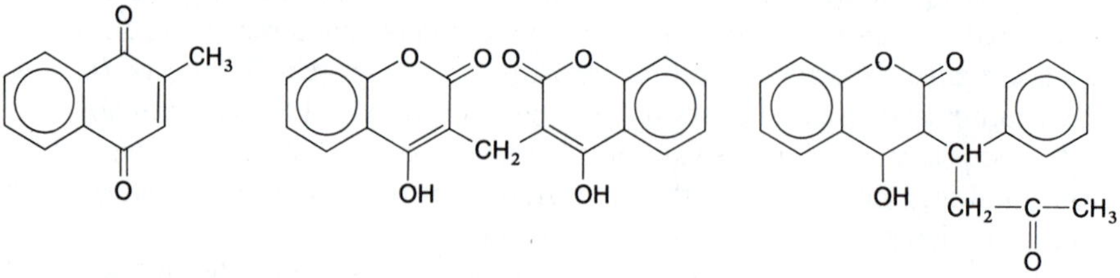

vitamin K$_3$ dicoumarol   warfarin (Coumadin)

TABLE 19.2 SOME IMPORTANT HORMONES

Hormone	Secreted by	Major Function
Thyroxin	Thyroid	Increases rate of metabolic reactions
Insulin	Pancreas (islets of Langerhans)	Decreases blood glucose levels
Glucagon	Pancreas	Increases blood glucose levels
Epinephrine (adrenalin)	Adrenal medulla	Affects many body functions for emergencies
Cortisone and related hormones	Adrenal cortex	Regulates carbohydrate, protein, mineral, and water metabolism
Estrogen	Ovary	Influences development of female sex organs and characteristics
Progesterone	Ovary	Prepares uterus for pregnancy and helps maintain pregnancy
Testosterone	Testis	Influences development of male sex organs and characteristics

several important hormones and their major function. Too much or too little hormone may cause disease, and we consider a few examples here.

THYROID DISORDERS The thyroid gland produces the iodine-containing hormone thyroxin (Figure 19.17, bottom), which stimulates the body's metabolic rate. People with overactive thyroid glands, a condition called **hyperthyroidism**, tend to be slender and very active. In contrast, people with low thyroid activity, a condition known as **hypothyroidism**, tend to be obese and sluggish.

One reason for a deficiency, particularly in poorer countries, is a lack of iodine in the people's diet. Their thyroid glands enlarge to try to compensate for the shortage (Figure 19.17, top), but they just cannot make enough hormone when they lack iodine. The cure for this condition, called *goiter*, is to add iodine to the diet; one way is to add an iodide salt (NaI or KI) to table salt (NaCl), which then is called *iodized salt*. People with other types of hypothyroidism may be treated directly with thyroid hormones.

Hyperthyroidism can be treated in several ways. For reasons not completely understood, giving extra iodide to these patients causes their

Figure 19.17 This person with goiter (top) has an enlarged thyroid gland due to a lack of dietary iodine to produce thyroxin (bottom). (Top, Science Source/Photo Researchers)

thyroid glands to secrete less thyroxin. Two other simple drugs, thiocyanate (SCN^-) and perchlorate (ClO_4^-) ions, block iodide (I^-) from entering the thyroid gland, so less hormone can be made there.

TREATING DIABETES   The pancreas normally secretes insulin (see Figure 22.5), a hormone that helps transport glucose from the bloodstream and into muscle and other cells. If we don't produce enough insulin or if the insulin we produce is not fully effective, high levels of glucose accumulates in the blood—a common sign of **diabetes** (Section 9.2).

There are two types of diabetes. About 10 percent of diabetics have the insulin-dependent type, which usually occurs by age twenty. Because they cannot secrete enough insulin, these diabetics require insulin for treatment. Insulin must be taken by injection because it is a protein and

TABLE 19.3   ORAL HYPOGLYCEMIC DRUGS USED TO TREAT DIABETES

Structure	Trade Name	Duration (hours)
CH_3—⟨ ⟩—SO_2—NH—C(=O)—NH—$(CH_2)_3$—CH_3 *tolbutamide*	Orinase	6–12
Cl—⟨ ⟩—SO_2—NH—C(=O)—NH—$(CH_2)_2$—CH_3 *chlorpropamide*	Diabinese	60
CH_3—C(=O)—⟨ ⟩—SO_2—NH—C(=O)—NH—⟨ ⟩ *acetohexamide*	Dymelor	12–24
CH_3—⟨ ⟩—SO_2—NH—C(=O)—NH—N⟨ ⟩ *tolazamide*	Tolinase	10–14

the digestive tract would destroy it if it were taken orally. As we discuss in Section 22.1, bacteria now produce human insulin for these diabetics.

The other type of diabetes usually occurs later in life, particularly in people who are obese. Here the problem may be a combination of two factors: decreased insulin production and cells having fewer receptors on their surfaces to bind insulin and allow the hormone to work. Symptoms develop gradually and often can be controlled by diet and exercise. But some diabetics also take oral drugs, called **hypoglycemic drugs**, to lower their blood sugar levels. These compounds, which stimulate the pancreas to secrete insulin, vary in their potency and period of effectiveness (Table 19.3).

One complication of diabetes is the formation of cataracts and other eye disorders. A likely reason is that excess glucose in the lens of the eye is converted into sorbitol, which builds up in the lens; this attracts extra water into the lens, causing swelling and initiating the formation of cataracts. Aspirin and ibuprofen (Chapter 20) are drugs that may help relieve this problem by inhibiting the enzyme that synthesizes sorbitol from glucose.

OTHER EXAMPLES OF HORMONE THERAPY The adrenal glands produce several important steroid hormones including cortisone, hydrocortisone (cortisol), and aldosterone (Figure 19.18). Aldosterone causes the kidneys to retain salt and water; the other two hormones (among other things) stimulate the liver to metabolize proteins into glucose and other substances. Cortisone, hydrocortisone, and synthetic compounds (such as prednisone) also help relieve inflammations. They are used to treat skin problems, certain athletic injuries, rheumatoid arthritis, and allergic reactions. Because these compounds may impair growth, children should use them only under a physician's supervision.

Figure 19.18 Three adrenal hormones. Notice the four-ring structure (shaded) characteristic of steroids. Atoms in color show structural differences from cortisone.

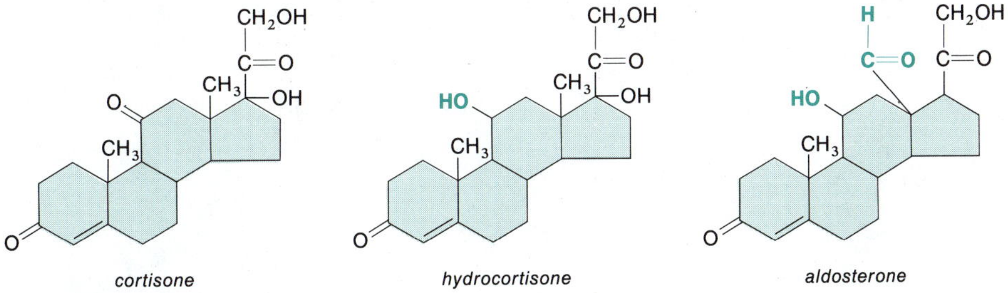

cortisone   hydrocortisone   aldosterone

People with *Addison's disease* have a deficiency of adrenal hormones, often caused by deterioration of their adrenal glands. They experience symptoms such as low blood glucose levels (hypoglycemia) and can be treated effectively with adrenal hormones. People with *Cushing's syndrome*, on the other hand, secrete too much of the adrenal hormones. The cause is usually a tumor of the adrenal glands or of the pituitary or hypothalamus glands in the brain that regulate adrenal secretions. These patients experience muscle weakness, increased infections, and other symptoms due to the excess breakdown of their body proteins. Some patients are treated by surgically removing their adrenal glands, but this is often ineffective.

Another important group of hormones are *prostaglandins*, which are cyclic derivatives of twenty-carbon fatty acids (Section 9.3). Although one of the first known sources was the prostate gland (for which these hormones are named), prostaglandins are produced in most tissues and affect almost every organ system. Some prostaglandins stimulate the contraction of smooth muscles and have been used to induce labor in pregnant women. Other possible uses are to treat hypertension, severe allergic reactions, asthma, and ulcers. One prostaglandin, called prostacyclin, blocks the formation of blood clots and relaxes the blood vessels. Studies are in progress to see whether prostacyclin can help prevent heart attacks.

19.5 Steroids and the Pill: Preventing Pregnancy

STEROID HORMONES   Another important group of steroids is the sex hormones (Figure 19.19), which are secreted by the gonads. Male sex hormones, secreted by the testes, are known as **androgens**. Androgens such

Figure 19.19 Estradiol and progesterone are female sex hormones, and testosterone is a male sex hormone. The four rings shaded in color represent the basic carbon frame common to steroids.

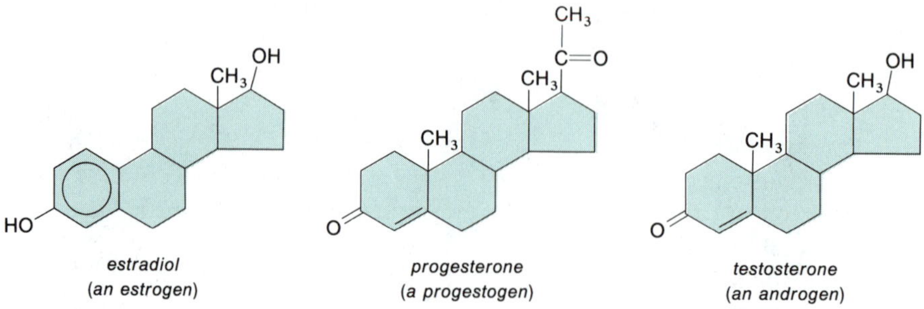

estradiol
(*an estrogen*)

progesterone
(*a progestogen*)

testosterone
(*an androgen*)

as testosterone are responsible for the deepening of the voice, hair-growth patterns such as the beard, development of the sex organs, and other male characteristics. Female sex hormones, produced mainly in the ovaries, are estrogens and progestogens. **Estrogens** such as estradiol control female sexual functions such as breast development and the menstrual cycle. **Progestogens**, particularly progesterone, prepare the uterus for pregnancy and help prevent the release of eggs from the ovaries during pregnancy, thus assuring only one pregnancy at a time.

A major medical development of the 1960s was the worldwide introduction of birth control pills (see the "Chemistry Spotlight"). These pills work by simulating the action of progestogens during pregnancy. By creating a false pregnancy, birth control pills prevent ovulation and thus prevent conception. They also alter the cervical mucus, the lining of the uterus, and the action of the fallopian tubes, further reducing the chances of pregnancy.

Oral contraceptives, when taken as directed, are about 99 percent effective in preventing pregnancy. There are three types: (1) combinations of estrogens and progestogens; (2) sequentially administered estrogens and progestogen–estrogen combinations; and (3) progestogens only. The most common type is the first, which a woman typically takes for twenty-one days and then stops taking for seven days, during which menstruation occurs. The sequential types were removed from the U.S. market in 1976 because of concerns that they may increase the risk of uterine cancer. Figure 19.20 shows the structures of several synthetic steroid hormones.

Side effects of the Pill vary widely and may include nausea, weight gain, headaches, reduced menstrual flow, and depression. Women using the Pill also have an increased risk of clots forming in their blood vessels, which increases their risk of stroke and heart disease—especially if they smoke or have hypertension. But the risk of fatal blood clots is much greater during pregnancy than from using the Pill to prevent pregnancy.

BIRTH CONTROL IN THE FUTURE Scientists are continuing to try to develop better contraceptives. One project is the search for a "morning-after" pill that would prevent pregnancy if taken within twenty-four hours following intercourse. Two effective agents are ethinylestradiol (see Figure 19.20) and diethylstilbesterol (DES), but DES has been linked to an increased incidence of a rare form of vaginal cancer in young women whose mothers used DES.

Once-a-month pills and injections that would be effective for several months are also being developed. For example, injections of progestogens encapsulated in tiny spheres could provide a steady release of the drug for up to six months. Matchstick-size capsules containing progestogen that are implanted under the skin of the forearm can prevent pregnancy for five years. And inserting a vaginal ring that slowly releases progestogen into the uterus provides contraception for three months.

CHEMISTRY SPOTLIGHT   THE BIRTH OF BIRTH CONTROL PILLS

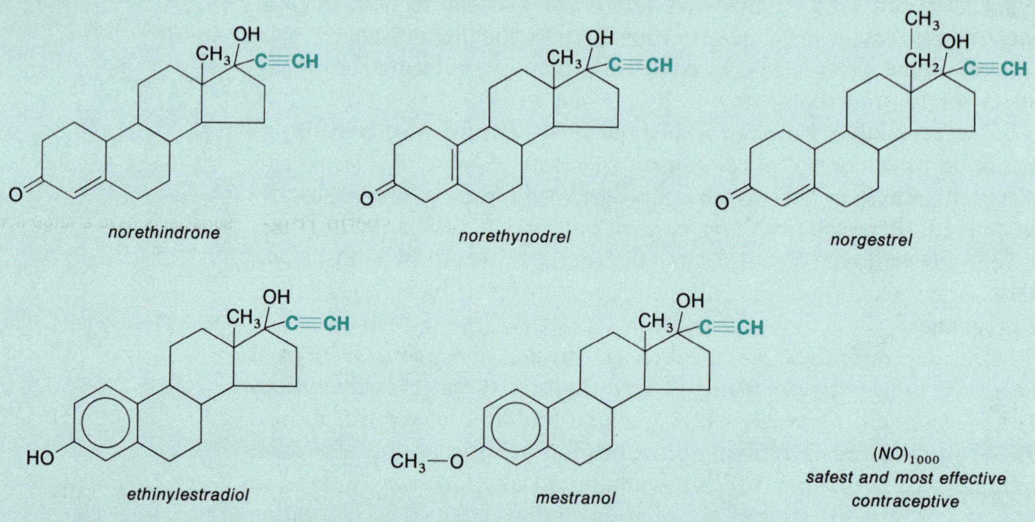

Figure 19.20 Some synthetic progestogens (top row) and estrogens (bottom) used in birth control pills. Notice that they contain an acetylene group (in color) and that the progestogens lack a methyl (—CH₃) group that occurs in progesterone (Figure 19.19).

In the late 1930s, Russell Marker, a chemist at Pennsylvania State College, discovered a way to make progesterone from another steroid called diosgenin. Because large amounts of diosgenin occur in Mexican yams, he wanted to set up a laboratory in Mexico to process the yams and make progesterone. No U.S. drug company would finance his venture, so Marker set up his own company, Syntex. His process brought the price of 1 g of progesterone down from $80 to about $1.

Syntex scientists also pioneered new methods to make other useful steroids. They found a way to remove the —CH₃ (methyl) group from the junction of the first two rings in the steroid structure. Doing this to progesterone produced a synthetic hormone that was far more potent than progesterone itself. The other important discovery, which occurred in Germany during World War II, was that these hormones could be taken orally if an acetylene group (—C≡CH) were bonded to

the five-carbon steroid ring. In 1951 Syntex scientists produced norethindrone (Figure 19.20), a synthetic progestogen with the methyl group removed and the acetylene group attached.

A decade later, two drug companies began marketing birth control pills containing synthetic estrogens and progestogens. Today between 50 and 80 million women worldwide use oral contraceptives, many of which still contain that first synthetic progestogen, norethindrone.

Chemical methods of birth control for men also are being studied. Hormones such as estrogens, progestogens, and testosterone suppress sperm production, but the female hormones have feminizing physical effects and reduce sex drive. Although testosterone is a bit more promising, it may cause heart problems and is inconvenient to use because it must be injected every ten days or so. Another candidate being tested, mostly in China, is gossypol, a yellowish material in cottonseed oil that reduces sperm production after a month or two of use.

Some chemicals also kill sperm directly in the female reproductive tract. Spermicides in foam, cream, or gel are placed near the cervix to destroy any sperm in the vicinity. The spermicide is usually a surfactant (Chapter 15) that helps rupture the cell membranes of the sperm (Figure 19.21). It also can cause minor and temporary irritation while in the cervix.

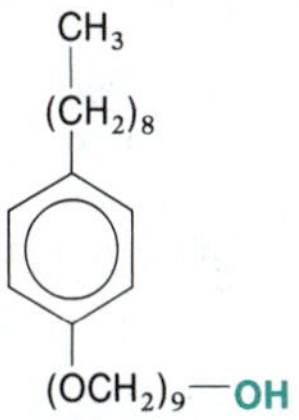

Figure 19.21 Nonoxynol-9, a surfactant used as a spermicide. Recall (Chapter 15) that surfactants have a large nonpolar region and an ionic or polar group (shown in color).

ALBERT EINSTEIN   Concern for humans and their fate must always form the chief interest of all technical endeavors in order that the creations of our minds shall be a blessing, not a curse.

Summary

In this chapter, we define a drug as any chemical used as medicine to treat disease. The first effective drugs were sulfa drugs, which kill bacteria by inhibiting their synthesis of folic acid, a necessary vitamin. Penicillin, an antibiotic, destroys bacteria by interfering with cross-linking in their cell walls. Other antibiotics prevent DNA from functioning normally.

Antimetabolite drugs resemble a normal metabolic substance and thus interfere with its reactions in the body. Antimetabolites are used to treat viral infections and cancer, lower blood cholesterol levels, and prevent blood clotting. Cross-linking agents also are used to treat cancer.

Hormones are secreted by glands and have chemical effects on other parts of the body. Nutrition and drug therapy can treat thyroid disorders. Depending on their type of diabetes, diabetics control their disease with injections of insulin or with a combination of diet, exercise, and hypoglycemic drugs. Birth control pills consist of synthetic progestogens and estrogens that prevent ovulation by inducing a false pregnancy.

Terms for Review

After completing this chapter, you should know and understand the meaning of the following terms:

androgen (p. 538)

antibiotic (p. 524)

anticoagulant (p. 534)

antimetabolite (p. 528)

cancer (p. 531)

chemotherapy (p. 522)

cross-linking agent (p. 532)

diabetes (p. 536)

drug (p. 523)

estrogen (p. 539)

hormone (p. 534)

hyperthyroidism (p. 535)

hypoglycemic drug (p. 537)

hypothyroidism (p. 535)

progestogen (p. 539)

sulfa drug (p. 524)

virus (p. 529)

Questions

Odd-numbered questions are answered at the back of this book.

1. What is the definition of a drug as used in this chapter? Which of the following are drugs according to this definition? (a) Caffeine, (b) KI, (c) LSD, (d) insulin, and (e) birth control pills.

2. How are each of the following thought to be important steps in the history of using drugs to treat disease? (a) Isolation of morphine, (b) use of arsphenamine, and (c) use of sulfa drugs.

3. Are sulfa drugs natural or synthetic compounds? How do they kill bacteria?

4. Examine the structures in Figure 19.3. Which one of the following structural changes to sulfanilamide appears to be the most likely to produce a compound that still works as a sulfa drug? (a) Replace —SO_2— with —CH_2—; (b) in the amino group (—NH_2) next to the ring, replace an H atom with an organic group; or (c) in the amino group next to the SO_2 group, replace an H atom with an organic group.

5. In what way is ampicillin structurally different from penicillin G (see Figure 19.7)? What practical difference does this make?

6. Is penicillin a natural or a synthetic drug? How does it kill bacteria?

7. What are some dangers in the widespread use of antibiotics?

8. Why are antibiotics ineffective in treating colds?

9. What is the function of each of the following in cold remedies? (a) Benzocaine, (b) codeine, and (c) an antihistamine.

10. How does ZDV (AZT) inhibit the action of the AIDS virus? Does it cure AIDS?

11. Which of the two normal components in DNA shown in Figure 19.10 is *not* a normal component in RNA? [Review Section 9.5 if necessary.]

12. What are the two major types of anticancer drugs? How does each type work?

13. What are the most common side effects in cancer patients who are treated with chemotherapy? Why do these side effects occur?

14. What does the drug lovastatin do in the body? Can it be classified as an antimetabolite? Why?

15. One of the following types of drug is used as a rat poison: (a) antibiotic, (b) anticoagulant, or (c) estrogen. Which one is it? Would the drug be poisonous to humans? Why?

16. What is a hormone? Which of the following are hormones? (a) Caffeine, (b) adrenaline, (c) insulin, and (d) glucose.

17. What is goiter, and how is it treated?

18. Why is insulin *not* taken orally?

19. Some diabetics take oral drugs to lower their blood glucose levels. How do these drugs work?

20. There have been murder cases in which it is claimed that the victim was given an overdose of insulin. How might a large dose of insulin be lethal?

21. Which of the following are steroid hormones? (a) Cholesterol, (b) insulin, (c) cortisone, (d) estrogens, and (e) prostaglandins.

22. To which hormones are most birth control pills chemically related? How do these hormones prevent pregnancy?

23. Examine Figure 19.19. What functional group do progesterone and testosterone—but not estradiol—have that accounts for their *-one* suffix? [Review Table 8.9 if necessary.]

24. Why aren't natural hormones used as birth control pills? What are the main chemical differences between the natural hormones and the synthetic hormones used in birth control pills?

Topics for Discussion

1. The *Physician's Desk Reference* (*PDR*), which is available in most libraries, is a good source of information about prescription drugs. Look up the trade name(s), uses, and side effects of the following drugs: methotrexate, meprobamate, chloramphenicol, allopurinol, chlorpropamide, and cyclophosphamide. Which are the safest to use? Why?

2. Some church leaders oppose a "morning-after" contraceptive on the grounds that this is equivalent to abortion. Do you agree or disagree? Why?

3. What are the advantages and disadvantages of reducing the use of antibiotics in animal feeds? What should be the criteria for using antibiotics in these feeds?

4. Many drugs used elsewhere in the world are not approved for use in the United States. Should U.S. drug companies market such drugs in countries where the drugs are allowed? Should U.S. citizens be allowed to use such drugs if they want to? Why?

5. In some situations, physicians will prescribe a placebo for a patient. What is a placebo? Is it appropriate to charge a patient a normal drug price for a placebo? Do you favor any restrictions on the use of placebos? Why?

CHAPTER 20

Chemistry and the Mind: Some Useful and Abused Drugs

General Objectives

1. How are nerve impulses transmitted in the body?

2. What kinds of drugs relieve pain and induce sleep? How do they work?

3. What kinds of drugs are used to relieve tension and anxiety and to treat schizophrenia, depression, hyperactive children, and Parkinson's disease? How do they work?

4. What types of drugs are stimulants and hallucinogens, and how do they work?

5. What are the characteristics of drug abuse, and how can such abuse be treated?

When you read this book, comb your hair, wash your car, or do anything else, your nervous system is working. Nerve signals race between outlying parts of your body and your brain, a 1.4-kg (3-lb) mass of pinkish gray material that has a texture like soft cheese. Your nervous system constantly does things of which you aren't even aware. Besides your conscious actions, your brain monitors and adjusts your heart rate, the diameter of your blood vessels, secretions from your glands, the size of your pupils, and the muscular action of your intestines, bladder, and lungs.

Sometimes the nervous system doesn't work properly. Some people have paralyzed limbs, while others are stricken with various mental diseases. One way to change the workings of the nervous system—for better or for worse—is to use drugs that chemically alter normal nerve and brain activities. In this chapter, we examine some drugs that do this.

20.1 Chemistry and the Nervous System: Sending the Right Messages

WHAT IS A NERVE CELL? Before we can understand how drugs affect the nervous system, we need to know what nerve cells are, how impulses are generated in the cell, and how messages travel from one nerve cell, or **neuron**, to another on their way to target cells such as a gland, cardiac muscle, or skeletal muscle. A typical neuron has a long fiber—the **axon**— at one end, and many branching, antennalike fibers—the **dendrites**—at the other end (Figure 20.1). A nerve impulse travels from the axon of one neuron to a dendrite of a neighboring neuron. But the axon doesn't actually   545

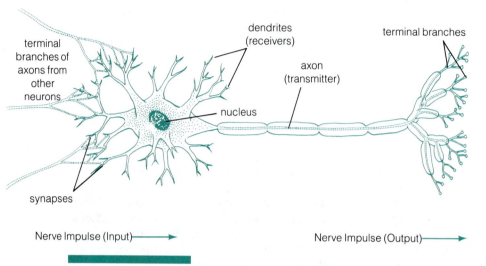

Figure 20.1 A simple neuron, or nerve cell. An electrical impulse travels from the terminal branches of the axon of one neuron across a synapse (gap) to a dendrite of an adjacent neuron.

touch the dendrite of a receiving neuron, so the nerve impulse must somehow cross the tiny gap—a **synapse**—between the axon of one nerve cell and a dendrite of another neuron.

A nerve cell at rest is like a tiny battery waiting to discharge its energy. For a nerve signal to cross the synapse, tiny knobs at the end of the axon have to release chemicals known as **neurotransmitters**. The neurotransmitter crosses the synapse and binds to the dendrite of an adjacent cell, triggering a flow of Na^+ ions into the neuron and K^+ ions out of the neuron. This causes a wavelike change in the membrane's electrical properties that transmits the nerve impulse in less than a thousandth of a second. Once the signal is received, the neurotransmitter must be immediately removed from the dendrite and destroyed so that the neuron can recharge to receive another impulse.

Different neurons use different neurotransmitters. The main neurotransmitters include acetylcholine, dopamine, serotonin, norepinephrine, and gamma-aminobutyric acid (GABA). But recent discoveries have boosted the number of known neurotransmitters to more than sixty, and more probably will be found. Some (such as acetylcholine) excite the receiving cell to send the impulse on to another neuron, whereas others (such as GABA) inhibit the impulse from traveling farther.

One of the most studied neurotransmitters is acetylcholine (ACH). It excites receiving neurons to transmit the impulse and generally has a stimulating effect on our activities. But too much ACH can overstimulate

Figure 20.2 The transmission of a nerve impulse across a synapse. (1) The nerve impulse triggers the release of acetylcholine (ACH) molecules from the axon into the synapse. (2) ACH binds to receptor sites on the dendrite of a nearby neuron and triggers a flow of ions that transmits the nerve impulse. (3) ACH then binds to the enzyme acetylcholinesterase and is deactivated by breakdown to choline and acetic acid.

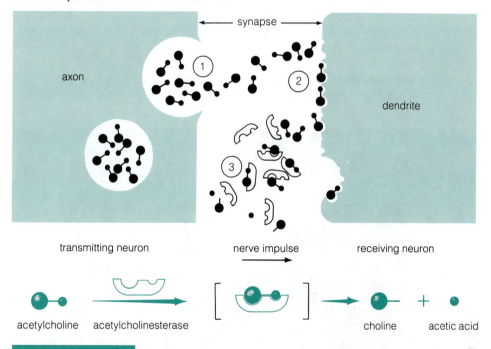

our muscles, glands, and nerves in an uncontrolled way, causing choking, convulsions, paralysis, and even death. To keep this from happening, our cells must destroy ACH on a receiving neuron as soon as the nerve impulse has been transmitted. The enzyme acetylcholinesterase does this by catalyzing the breakdown of ACH to choline and acetic acid (Figure 20.2). Another enzyme eventually converts the choline and acetic acid back into ACH, thus replenishing the supply of this neurotransmitter.

Drugs that alter these nerve signals can affect us in many ways. Some, like nerve gas and organophosphorus pesticides, are poisons that block the breakdown of ACH (Chapter 21). Other drugs mimic or alter the normal action of neurotransmitters and have a variety of effects on the body. Indeed, when we consider that the brain has billions of neurons, each of which can form thousands of synapses to neighboring neurons, with dozens of possible neurotransmitters to cross those synapses, we can see why our brains send and receive an incredible variety of nerve messages.

20.2 Painkillers and Depressants: Relieving Pain and Inducing Sleep

Drugs that relieve pain and induce sleep are among the most important compounds in medicine. Their effects range from localized pain relief to sedation to general anesthesia. But the way these drugs work is known for only a few compounds. In many cases, they slow or block nerve transmission in the body.

MILD ANALGESICS: ASPIRIN IS ASPIRIN IS ASPIRIN   Each year people in the United States spend more than a quarter of a billion dollars for over-the-

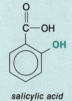

salicylic acid

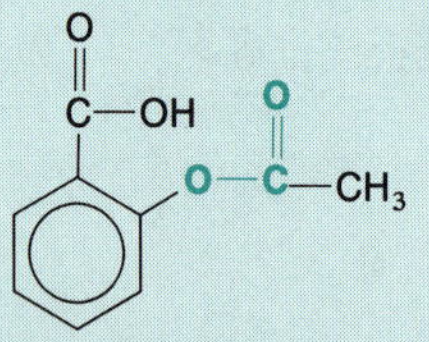

acetylsalicylic acid (aspirin)

Figure 20.3 Two mild analgesics, salicylic acid and acetylsalicylic acid (aspirin), which is made by changing the alcohol group of salicylic acid into an ester group.

Aspirin comes from a traditional herbal medication. In 1763 the English

Figure 20.4 The Bayer Company of Germany introduced the first aspirin preparation in 1899 based on the work of one of its chemists, Felix Hoffman. (Courtesy SSC & B Inc.)

physician and clergyman Edward Stone observed that a mixture of powdered

willow bark (see Figure 19.1) in water given every few hours relieved aches and pains and reduced the fever symptoms of malaria. About seventy-five years later, chemists showed that the active ingredient was salicylic acid (Figure 20.3). By 1870 this compound was widely used as a fever depressant and painkiller, especially for rheumatism. But it is a relatively strong acid (Section 6.1), so it was unpleasant to take orally and it damaged the membranes lining the mouth, esophagus, and stomach.

Felix Hoffman, a young chemist at the Bayer Company in Germany, had a father who suffered from severe rheumatoid arthritis. Because salicylic acid was too irritating to his father's

stomach, he set out to modify salicylic acid to make it more soluble in stomach acids. Hoffman replaced the alcohol functional group in salicylic acid with an ester group (Section 8.6) to produce a compound known as *acetylsalicylic acid* (see Figure 20.3). He tried it on his father with great success.

Hoffman took his laboratory data to the director of pharmacologic research at the Bayer Company, and in 1899 the company began marketing acetylsalicylic acid under the trade name Aspirin (Figure 20.4).

counter pain relievers. These mild **analgesics** relieve minor headache, muscle, and joint pain, usually without interfering with mental alertness or inducing drowsiness. The most widely used analgesic is aspirin, which was discovered in the late 1800s (see the "Chemistry Spotlight"). Today an estimated 20 billion tablets of aspirin or drug combinations containing aspirin are used in the United States each year—an average of eighty tablets per person. We use it to relieve tension headaches and the pain of arthritis, to reduce fever, and in general to keep going in spite of minor aches and pains.

Aspirin is a fairly safe drug if not taken in excessive doses, and its continued use does not lead to addiction or reduce its effectiveness. The most common problem—affecting about 5 percent of the people who take aspirin—is an upset stomach when undissolved aspirin resting on the stomach wall causes small hemorrhages. If this happens frequently, a person could lose enough blood to develop iron-deficiency anemia. Modern aspirin tablets are formulated to dissolve quickly, thus minimizing this problem.

Other concerns are that about 0.5 percent of those taking aspirin, and 3 to 5 percent of asthmatics, suffer from allergic reactions including skin rashes, respiratory difficulty, and shock. In addition, children with viral infections such as flu or chicken pox can develop a toxic allergic response called Reye's syndrome, which can be fatal. All nonprescription aspirin products now are required to have a warning label about this problem.

Aspirin works by blocking the synthesis of prostaglandins (Section 19.4). Certain prostaglandins tighten (constrict) blood vessels, act on heat-regulating centers of the central nervous system, and allow water to pass out of capillaries and into nearby tissues, thus causing swelling and pain. By lowering the concentration of these prostaglandins, aspirin reduces pain, fever, and inflammation. Aspirin also inhibits blood clotting, so small doses (such as one tablet every other day) may reduce the risk of heart attacks.

An aspirin tablet, regardless of brand, is aspirin. All tablets contain a small quantity (usually 325 mg, or 500 mg in "extra-strength" perparations) of acetylsalicylic acid as the active ingredient. They also contain a filler consisting of anything from clay to chalk to glucose. In addition, some preparations contain buffers to reduce the acidity of the aspirin; others coat the tablet with materials that don't dissolve until they reach the intestine. But studies have shown little or no difference in the effectiveness or speed of aspirin-containing pain relievers, despite advertising claims and differences in price. Unless you are allergic to aspirin, the most sensible way to relieve a headache or minor muscular and joint pain is to take the cheapest brand of aspirin available.

People who experience adverse side effects to aspirin and children who want to avoid the risk of Reye's syndrome can take other, more expensive substitutes. One alternative is acetaminophen (Figure 20.5), which is sold

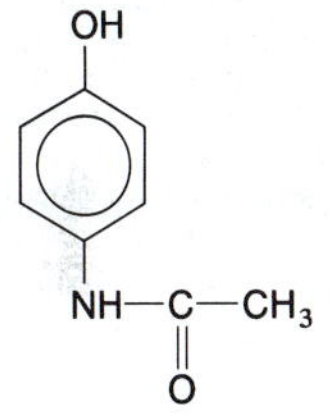

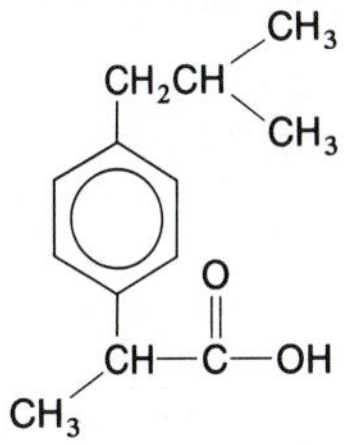

Figure 20.5 Two mild analgesics that are alternatives to aspirin.

by brand names such as Tylenol and Datril. Acetaminophen has the same analgesic and fever-reducing effects as aspirin but is less effective in treating inflammations. Because it is less toxic and has fewer side effects than aspirin, acetaminophen is overtaking aspirin as the drug of choice, particularly in children, to combat fever and minor pain.

Another over-the-counter substitute is ibuprofen (see Figure 20.5). This drug, commonly sold by brand names such as Advil and Nuprin, is similar to aspirin as an analgesic, fever-reducer, and anti-inflammatory agent, but it is a weaker inhibitor of prostaglandin synthesis. Ibuprofen causes less stomach and intestinal bleeding than does aspirin, but people who are allergic to aspirin are usually allergic to ibuprofen.

STRONG ANALGESICS **Narcotics** (from a Greek word meaning "to make numb") are nitrogen-containing drugs commonly derived from the opium poppy. In therapeutic doses, narcotics relieve pain without depressing the central nervous system the way anesthetics do. At higher doses, they act as general depressants and produce stupor or sleep. They work by binding at certain receptor sites in the brain.

Figure 20.6 Opium flower and seed pod (left) found in a field of opium poppies (right). (Courtesy of the U.S. Drug Enforcement Agency)

Figure 20.7 Some strong analgesic narcotics made from opium. Areas in color show the structural differences.

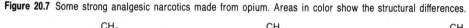

morphine   *codeine*   *heroin*

Opium is a powder with a sharp, bitter taste. It is obtained by squeezing out the white, milky juice in unripened seed pods of opium poppies (Figure 20.6) native to Greece, Turkey, and the Orient. Its euphoric effect was known at least as long ago as 4000 BC.

Morphine (Figure 20.7), which makes up about 10 percent of crude opium, is a very effective pain reliever. Named after Morpheus, the Greek god of dreams, morphine is sometimes used to relieve the intense pain of trauma, heart attacks, cancer, and certain other problems where mild analgesics aren't effective. But physical and psychological addiction readily develops, so it must be used with great caution.

Small structural changes in the morphine molecule produces other narcotics with different levels of activity (see Figure 20.7). Codeine occurs naturally in opium, though it can also be synthesized from morphine. It is much less addictive than morphine but has weaker analgesic activity. It can be taken orally, whereas morphine must be injected. Codeine relieves moderate pain and is used in cough syrup because it depresses the cough center of the brain.

Converting the two hydroxyl groups in morphine to the esters of acetic acid changes morphine into heroin (see Figure 20.7). Heroin is a stronger painkiller than morphine, but it is so much more addictive that it is no longer manufactured in the United States and used as a medical drug. Heroin now has no legal use and is the most widely abused narcotic; an estimated 500,000 people in the United States are addicted. Drug dealers dilute, or cut, heroin with substances such as lactose (milk sugar), so street drugs normally contain 3 to 10 percent heroin (Figure 20.8). About one injection in 100,000 of heroin is fatal, and seriously addicted people may inject heroin three or four times a day. Many addicts share contaminated needles and thus have higher incidences of infectious diseases such as AIDS, hepatitis, and tetanus. At least half of the people who inject heroin intravenously in New York City are now infected with the AIDS virus.

Figure 20.8 Street heroin. (Courtesy of the U.S. Drug Enforcement Agency)

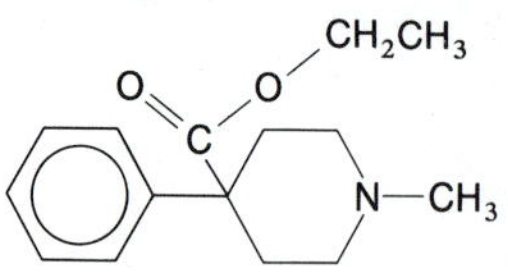

meperidine (Demerol)

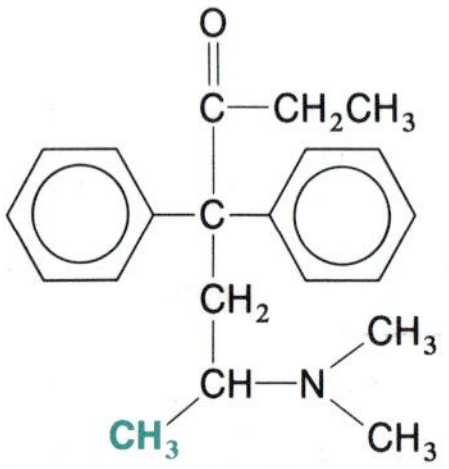

methadone

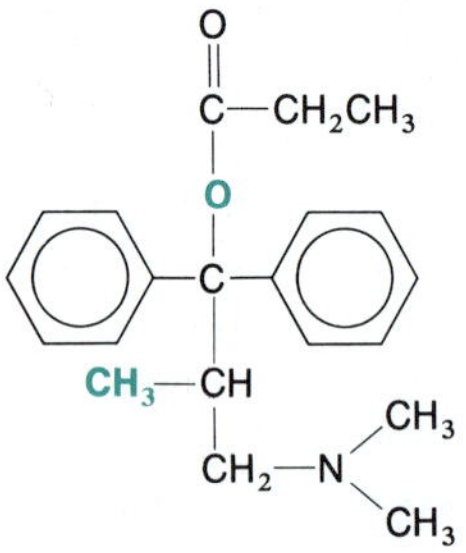

propoxyphene (Darvon)

Figure 20.9 Three synthetic analgesics. Structural differences between methadone and propoxyphene are in color.

For many decades scientists have been trying to synthesize nonaddictive narcotics that are effective pain relievers. Figure 20.9 shows three synthetic compounds. Demerol is between morphine and codeine in its ability to relieve pain and is addictive. At its usual dose, Darvon—despite its popularity and high price—is no more effective than aspirin or acetaminophen in relieving pain.

Methadone, which has a structure similar to that of propoxyphene (Darvon), is an oral drug used to treat heroin addiction. It blocks the heroin high and reduces the craving for heroin. Methadone is longer-acting than heroin, and withdrawal from it is slower and less intense than heroin withdrawal. Some heroin addicts receiving methadone at little or no cost to them (the annual public expense is about $3000 per person) have been able to break their habit. But methadone itself has become an abused drug that is sold illegally on the streets.

Another useful group of drugs are narcotic antagonists. These compounds structurally resemble narcotics and bind to the same receptor cells in the brain; this displaces the narcotic and removes its depressant effects. One such drug, naloxone (Narcan), is used to treat people suffering from overdoses of heroin or other narcotics (Figure 20.10).

Why do our brains have binding sites for narcotics such as morphine? Perhaps those sites normally respond to natural analgesics that we produce. In the early 1970s, scientists discovered that the pituitary gland produces small proteins that bind to the same brain receptors as morphine. One substance, *β-endorphin* (Figure 20.11), contains thirty-one amino acids and was found to be a strong pain reliever. The first five amino acids in β-endorphin make up another analgesic substance, met-enkephalin, that binds to opiate receptors. Other enkephalins also have been discovered.

Our bodies make these natural pain relievers under certain conditions. For example, when we exercise for a long enough time, we produce β-endorphin and enkephalins to help combat fatigue. Some people attribute the "runner's high," a feeling of relaxation and exhilaration after the first few minutes of a long run, to the natural production of these substances. And acupuncture may exert its pain-reducing effects by triggering the release of endorphins.

Figure 20.10 Naloxone, a narcotic antagonist. Structural differences between naloxone and morphine (Figure 20.7) are in color.

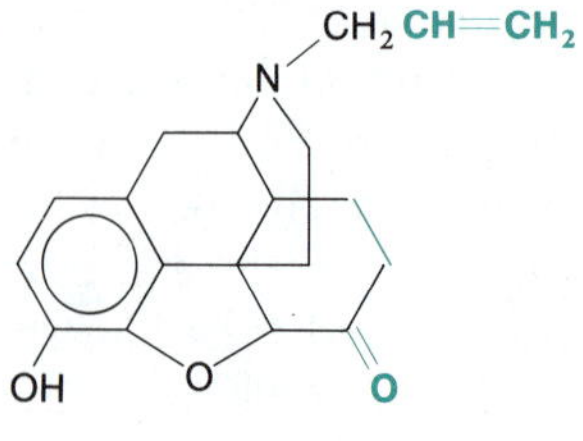

Figure 20.11 β-endorphin (top). The first five amino acids (shown in color) constitute an enkephalin. Abbreviations for the amino acids are in Table 9.3. (Bottom) The tyrosine (tyr) end of endorphins and enkephalins resembles morphine (shown in color) and binds to the same receptor sites in the brain.

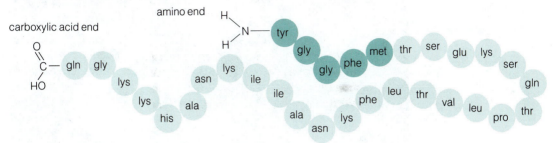

LOCAL ANESTHETICS A **local anesthetic** blocks pain in a specific area when rubbed on or injected just under the skin. Most of you have had a shot of procaine (Novocaine) before dental work (Figure 20.12) or when a physician sutured (sewed up) a cut. These compounds block the flow of Na⁺ ions and K⁺ ions across neuron membranes and thus prevent nerve impulses from reaching the brain.

Local anesthetics usually have the suffix -*caine* and are aromatic amines (Figure 20.13). Lidocaine (Xylocaine), which acts quickly and produces general and local pain relief, has largely replaced Novocaine for dental use. Mepivacaine (Carbocaine) is widely used for dental and medical surgery. Other drugs include tetracaine (Pontocaine) to anesthetize the spinal area and dibucaine (Nupercaine) for local anesthesia during childbirth. Many anesthetic preparations contain a small amount of epinephrine (adrenalin), which tightens the blood vessels and slows blood flow to the area, thus prolonging the anesthetic effect.

The first local anesthetic used successfully in modern medicine was cocaine (see Figure 20.13). It comes from the leaves of the coca plant

Figure 20.13 Some local anesthetics. Notice that they all have an aromatic ring joined to an ester or amide group (in color) and a substituted amine group (in color).

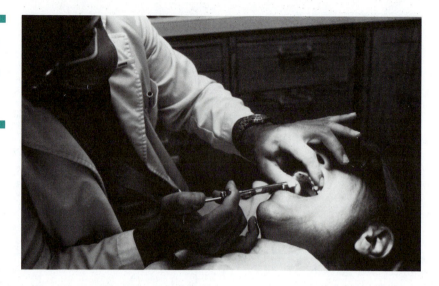

cocaine

procaine (Novocaine)

lidocaine (Xylocaine)

mepivacaine (Carbocaine)

(Figure 20.14), which grows in the Andes Mountains of South America. By the time it reaches the United States (illegally), it is in the form of white crystals.

Besides being an anesthetic, cocaine is a stimulant that triggers the release of certain neurotransmitters (dopamine, serotonin, and norepinephrine) in the brain. Because the euphoric effects last an hour or less, some users inject cocaine as often as twelve times a day. Intense psychological dependence occurs with prolonged use, and physical dependence probably occurs. An overdose depresses the heart and respiration and can be fatal.

Cocaine abuse has risen rapidly in recent years, particularly among the wealthy who could afford the high price. Now the price is dropping,

Figure 20.14 The leaves of the coca plant (left) contain as much as 2 percent cocaine, which can be extracted and sold illegally (right). (Courtesy of the U.S. Drug Enforcement Agency)

particularly for a derivative called "crack," which consists of small pellets of cocaine that can be smoked; this increases the chances for addiction by providing a more rapid and intense euphoria, followed by depression, irritation, and a craving for the drug. With a lower price, more people—including the very young—are abusing this drug. An estimated 6 million people in the United States have recently used cocaine.

GENERAL ANESTHETICS   Substances that produce insensibility to pain and a reversible unconsciousness are classified as **general anesthetics**. Before the introduction of general anesthetics more than 100 years ago, ethyl alcohol (ethanol) and narcotics such as opium were used to reduce the pain of surgery. Usually, however, no drugs at all were used. Patients had to endure unbelievable pain as surgeons tried to operate as quickly as possible, sometimes doing amputations in less than a minute. A physician who amputated his own limb in 1896 said that "the blank whirlwind of emotion, the horror of great darkness [that] swept through my mind and overwhelmed my heart, I can never forget, however gladly I would do so."

The first general anesthetic used in surgery was nitrous oxide (N_2O). Known as *laughing gas* because of the exhilaration and laughter it produces, nitrous oxide was first prepared in 1776. In 1808 Sir Humphrey Davy, an English chemist, discovered the chemical's analgesic properties and suggested that it be used in surgery. In 1844 nitrous oxide was used for the first time to extract a tooth painlessly.

The first well-known use of diethyl ether ($CH_3CH_2-O-CH_2CH_3$), often called simply *ether*, as an anesthetic for an operation was made in 1846 by Dr. William E. Morton at the Massachusetts General Hospital (see Figure 8.16). Diethyl ether is rarely used today because it is slow-acting, highly inflammable, and causes nausea and vomiting in many patients.

In 1847 chloroform ($CHCl_3$) was first used as a general anesthetic in surgery. Although it is not flammable, chloroform is not used now because it can cause liver damage and because only a narrow safety margin separates an anesthetic dose from a fatal dose.

Today, a patient facing major surgery is given a series of drugs. The night before surgery, the patient usually receives a sedative or tranquilizer to help relax the body and ensure a good night's sleep. During surgery, a combination of anesthetics, analgesics, and muscle relaxants is usually given. Table 20.1 shows some of the general anesthetics commonly used

TABLE 20.1 SOME GENERAL ANESTHETICS

Name	Formula	Uses and Characteristics
Inhalation Anesthetics		
Nitrous oxide (laughing gas)	N_2O	Nonflammable and nonirritating; low potency; often used in dentistry and obstetrics
Cyclopropane		Flammable; rapid recovery; used in surgery on babies and small children
Halothane		Nonflammable, nonirritating, and powerful; rapid recovery; usually used with nitrous oxide and muscle relaxants
Enflurane		Nonflammable, nonirritating, and powerful
Methoxyflurane		Nonflammable, nonirritating, and powerful
Injection Anesthetic		
Sodium thiopental (sodium pentothal)		Potent barbiturate hypnotic administered by injection of an aqueous solution; fast acting with few aftereffects; usually used along with an inhalation anesthetic

during surgery. After surgery, strong analgesics such as morphine or Demerol may be given to relieve the pain.

SEDATIVES   **Sedatives** are drugs that cause relaxation; if they put the user to sleep, they are sometimes called **hypnotics**. The best known sedatives and hypnotics are the **barbiturates**, which are made from barbituric acid:

barbituric acid

Barbiturates are widely used for insomnia, seizure control, general anesthetics (sodium pentothal; see Table 20.1), and daytime sedation. Drug companies produce enough to supply each person in the United States with more than fifty pills per year. At least half of these pills end up in the underground drug market as various kinds of "downers."

Table 20.2 shows the structures and actions of some common barbiturates; all are made by attaching hydrocarbon groups to barbituric acid. The drugs range from fast-acting pentobarbital (Nembutal) and secobarbital (Seconal) to slow-acting phenobarbital (Luminal), which is used to treat epilepsy and other seizure disorders. The fast-acting forms are popular with barbiturate addicts but have largely been replaced as prescription sleeping pills by certain tranquilizers (Section 20.3). Barbiturates are similar to (but stronger than) alcohol in removing inhibitions and creating a drowsy feeling of relaxation. A sedated person appears to be drunk, with slurred speech, poor coordination and balance, and difficulty in thinking and controlling emotions.

Physical and psychological dependence develops with these drugs, and the withdrawal effects—cramps, tremors, convulsions, nausea, restlessness, and extreme anxiety—are severe. Regular users develop a tolerance for barbiturates and may need up to fifteen times the usual dose to achieve the desired effects. The lethal dose, however, stays about the same (1500 mg), so addicted users take amounts that progressively approach the lethal dose. Too much sedative depresses the action of the heart and lungs. Each year several thousand people in the United States die from overdoses of barbiturates. Some deaths are suicides; others are accidental, especially when barbiturates are used with alcohol. Another problem is that people under the influence of a sedative may forget how many pills they have already taken.

TABLE 20.2 FOUR WIDELY USED BARBITURATES

Names	Formula	Chief Uses	Effects
Pentobarbital (Nembutal)		Hypnotic	Fast action and short duration
Secobarbital (Seconal)		Hypnotic	Fast action and short duration
Butabarbital (Butisol)		Sedative, hypnotic	Intermediate action and duration
Phenobarbital (Luminal)		Sedative, hypnotic, anticonvulsant	Slow action and long duration

The most widely abused sedative is ethanol. Ethanol depresses the central nervous system and relaxes the user, sometimes relieving inhibitions. Psychological and physical dependence develop, and two types of alcoholism can occur. Type I alcoholism often occurs because of stress but is unrelated to age, sex, or heredity. Type II alcoholism appears mostly before age twenty-five in sons of male alcoholics. These people frequently have aggressive and violent behavior.

Ethanol is metabolized mostly by the liver. Excess ethanol frequently causes the liver to deteriorate, a condition called *cirrhosis*. Ethanol also damages the fetus during pregnancy. Babies born to alcoholic mothers have a range of brain and other neurological problems called *fetal alcohol syndrome*, which is due in part to poor development of nerve cells. In the

Western world, as many as 2 percent of all babies born may be affected in this way.

No one knows exactly how ethanol works. One clue might come from an experimental drug (called Ro15–4513) that blocks certain effects of ethanol. Although it is too toxic for humans to take, this drug may be a useful tool to study how ethanol works.

20.3 Mind Drugs: Treating Mental Diseases and Nerve Disorders

TRANQUILIZERS   Until the middle of this century, there were no drugs to help patients with schizophrenia or other mental disorders. Then Henri Laborit, a French neurosurgeon, tried antihistamines to calm his patients before anesthesia. He found that one of them, chlorpromazine (Thorazine) (Table 20.3), worked especially well and put his patients into a quiet, pleasant mood.

In the early 1950s, a few psychiatrists tried using this drug on patients with mental disturbances. Chlorpromazine calmed manic-depressive patients and treated some of the symptoms of schizophrenia. It worked well enough that hundreds of thousands of patients left mental hospitals to live at home instead of being confined to straitjackets and padded cells. Schizophrenia, however, still afflicts about 1 percent of the world's population and about 300,000 people in the United States.

Chlorpromazine appears to block receptors in the brain for dopamine (see Figure 20.17), one of the neurotransmitters. Because dopamine cannot bind there in the presence of the drug, nerve impulses that depend on dopamine are not transmitted. The metabolic rate drops, body temperature falls, blood pressure drops, and the patient has a calm, tranquil feeling.

Table 20.3 shows the structures and action of chlorpromazine and several other **tranquilizers** that act by depressing the central nervous system, but generally to a milder degree than do barbiturates. Diazepam (Valium) was originally used to tame tigers, but now it is commonly prescribed to tame anxiety, muscle tension, and epilepsy. Meprobamate (Equanil, Miltown), and chlordiazepoxide (Librium) are also used to relieve anxiety, but not to treat psychotic disorders. These drugs are often classified as minor tranquilizers. Flurazepam (Dalmane), temazepam (Restoril), and triazolam (Halcion) are widely prescribed for people who have difficulty falling asleep and may be classified as sedatives or hypnotics. Indeed,

TABLE 20.3 THREE TRANQUILIZERS AND A HYPNOTIC TRANQUILIZER

Names	Structure	Chief Effects	Chief Uses
Chlorpromazine (Thorazine)		Strong tranquilizer that affects the brain as well as the autonomic nervous system	Treatment of psychotic disorders; has largely replaced electroconvulsive shock therapy and psychosurgery
Promazine (Compazine)		Most potent antipsychotic effects; also the most toxic	Similar to chlorpromazine
Diazepam (Valium)		Mild tranquilizer; depresses central nervous system and relaxes skeletal muscles	Relief from anxiety and tension; treatment of muscle tension and epilepsy
Flurazepam (Dalmane)		Hypnotic; depresses central nervous system	Treatment of insomnia

they have largely replaced barbiturates as sleeping pills. All these drugs, however, are potentially addictive.

Drugs used to treat depression are commonly called **antidepressants**. One of the first antidepressants was imipramine (Tofranil), which is also used to treat bed-wetting by children. Its structure (Figure 20.15) is similar to promazine (see Table 20.3), and it works by blocking the uptake of

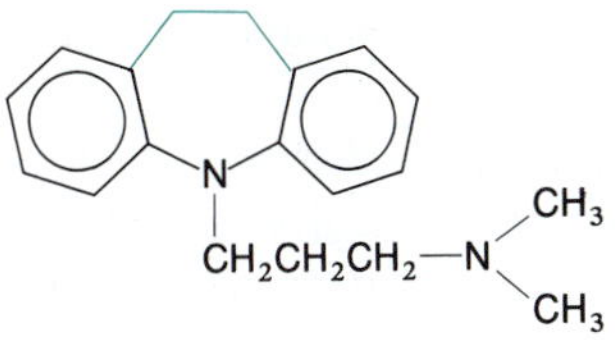

Figure 20.15 Imipramine (Tofranil), an antidepressant. Structural differences from promazine (Compazine; see Table 20.3) are in color.

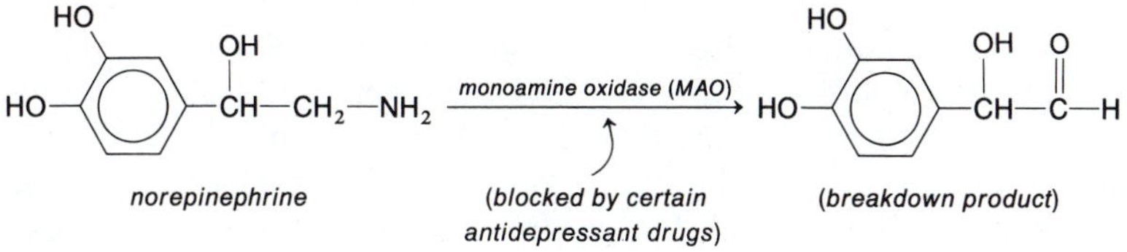

norepinephrine

monoamine oxidase (MAO)

(blocked by certain antidepressant drugs)

(breakdown product)

Figure 20.16 By inhibiting MAO, certain antidepressants increase the amount of norepinephrine and other neurotransmitters in the brain.

norepinephrine (a neurotransmitter) on the dendrities of neurons. Other antidepressants such as doxepin (Sinequan), amitryptiline, and trazodone (Desyrel) interfere with the uptake of norepinephrine or the neurotransmitter serotonin.

Another group of antidepressants, including tranylcypromine (Parnate) and isocarboxazid (Marplan), inhibit the enzyme monoamine oxidase (MAO), which normally breaks down neurotransmitters. By blocking the action of MAO, these drugs increase the brain's level of neurotransmitters such as serotonin and norepinephrine and thus stimulate those nerve pathways (Figure 20.16).

A major drug is the lithium (Li^+) ion, which is supplied as a salt such as lithium carbonate (Li_2CO_3). In clinical tests, lithium salts calm hyperactive children, protect people against depression, and control the outbursts of excitement in schizophrenia and (especially) manic–depression. At high doses, the lithium ion is toxic. Lithium salts are common substances that cannot be patented. Although no one knows exactly how it works, lithium helps prevent brain cells from producing too much or too little neurotransmitters.

CHEMISTRY AND THE BRAIN Some diseases occur because of chemical imbalances in certain parts of the brain. Often the problem is too much or too little of a particular neurotransmitter. Too much ACH, for example, causes nerve impulses to travel continuously, producing convulsions and even death. Too little of a neurotransmitter, on the other hand, prevents the nerve messages from reaching their destination.

One example of the latter condition is **Parkinson's disease**, which is characterized by uncontrolled tremors, stiffness, and difficulty in walking. More than 1 million people in the United States, most of them age sixty or older, have this disease. They have abnormally low concentrations of the neurotransmitter dopamine in certain regions of the brain, which causes them to lose some control over their muscles. Dopamine injections don't help because the chemical cannot enter the brain from the bloodstream.

Now, however, these patients can take a drug called *L-Dopa* (L-*di*hydroxyphenyl*al*anine), which does reach the brain and is converted there into dopamine (Figure 20.17). L-Dopa is taken orally, often in combination with another substance that blocks the conversion of L-Dopa into dopamine until it reaches the brain. This drug has produced dramatic improvement in patients with Parkinson's disease. After several years of use, however, most patients progressively fail to respond to L-Dopa and suffer from hallucinations and depression. More recently, physicians are experimenting with the surgical implantation of dopamine-producing cells (from adrenal glands or fetal brains) into the brains of patients with Parkinson's disease.

Another example is **Alzheimer's disease** (AD), which was first described in 1906 by the German physician Alois Alzheimer. This disease occurs mainly in older people, and its victims typically lose their memory (especially of recent events), get confused, and gradually lose their ability to read, write, calculate, and speak. No one knows what causes Alzheimer's disease, but scientists recently found that AD patients have lost 75 to 90 percent of the neurons in the basal nucleus region of the brain. Nearby regions also have unusually low levels of the enzyme that makes ACH. In addition, large amounts of the protein amyloid accumulates in the brain.

The loss of neurons and failure of nerve pathways depending on ACH may help explain why AD people progressively lose their ability to reason, remember, comprehend, and communicate. One strategy, which is similar to the approach used with Parkinson's disease, is to give AD patients materials that increase the amount of ACH in their brains. The main drugs tried so far are choline, lecithin (both of which can be made into ACH),

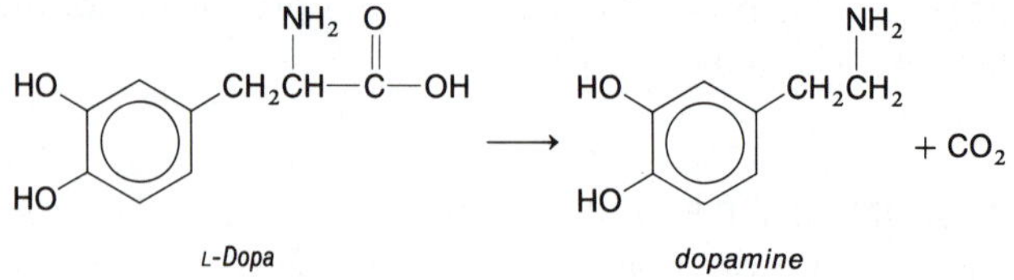

Figure 20.17 L-Dopa is converted in the brain into dopamine, a neurotransmitter that relieves symptoms of Parkinson's disease.

and physostigmine (which increases ACH levels by blocking its breakdown in the brain). The results so far have been disappointing.

Brain chemistry may be related to behavior in many ways. For example, a study of adolescents showed that people with low levels of serotonin are at high risk for committing suicide. Girls with low serotonin levels also are more likely to develop bulimia, an eating disorder. Some studies indicate that autistic children have excessive amounts of serotonin or endorphins and enkephalins in the brain. And schizophrenics have an unusually large number of receptor sites in the brain that bind dopamine. Further research on the chemistry of the brain will likely lead to more effective treatments for these and other disorders.

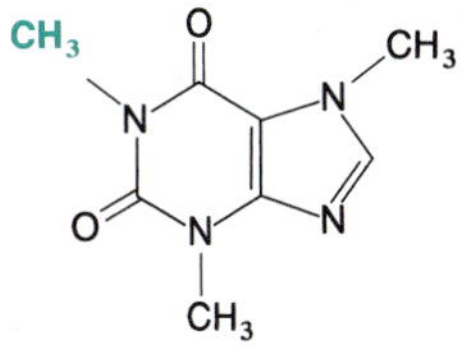

caffeine

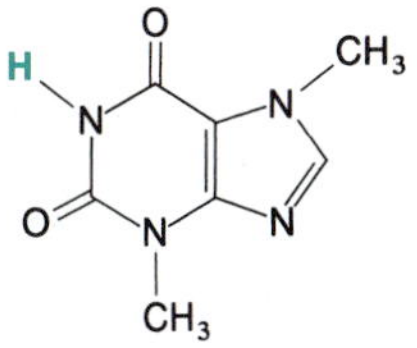

theobromine

Figure 20.18 Two naturally occurring stimulants with similar structures. The difference is shown in color.

20.4 Stimulants and Hallucinogens: Tripping Out

STIMULANTS   **Stimulants** are drugs that increase the activity of the brain and central nervous system. They cause people to feel more alert and wakeful, less tired, and sometimes less hungry. Their effect is the opposite of depressants such as the barbiturates.

Caffeine (Figure 20.18)—which is found in coffee, tea, cola and other soft drinks, certain combination pain relievers (Anacin, Midol, Empirin), and antisleep agents (NoDoz and Vivarin)—is a mild stimulant. If not taken in excess, it is relatively harmless. Some dependence occurs, and heavy users may experience the withdrawal symptoms of headaches and fatigue within eighteen to twenty-four hours of discontinuing use. These symptoms may last a week or so when a person is giving up a coffee habit.

Caffeine stimulates the central nervous system, increases the heart rate, and increases the metabolic rate. Excess caffeine (and other stimulants) is especially dangerous for people with heart trouble. A structurally similar compound, theobromine (see Figure 20.18), occurs naturally in cocoa drinks and is a weaker stimulant than caffeine.

Some of the most widely used and abused stimulants are the **amphetamines** (Figure 20.19). They have a structure similar to epinephrine

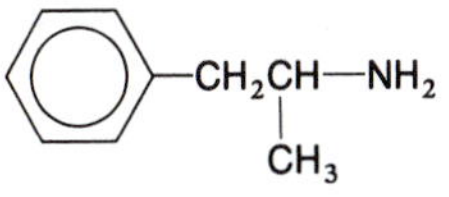

amphetamine
(Benzedrine)

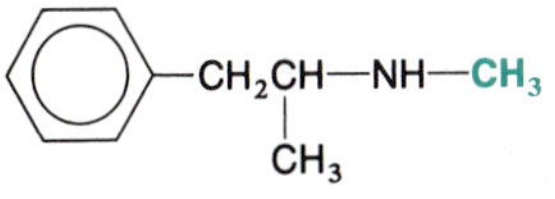

methamphetamine
(Methedrine, "speed")

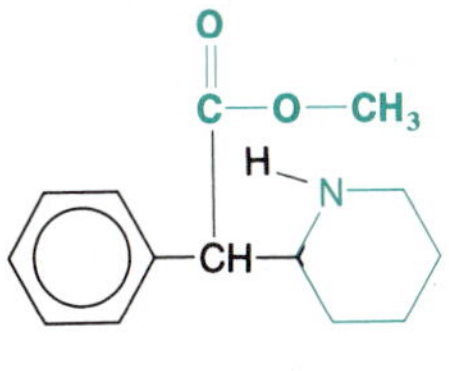

methylphenidate
(Ritalin)

Figure 20.19 Three amphetamines. Colored atoms show differences in structure.

Figure 20.20 The web pictured on the left was spun twelve hours after the spider was given a minute amount of methamphetamine. It took several days before the spider could again build a web as flawless as the one shown on the right. (Dr. Peter N. Witt)

(adrenalin) and norepinephrine (see Figure 20.24), but the stimulant effect lasts longer. Amphetamines help release norepinephrine and other neurotransmitters from storage sites and prolong their effects on the nervous system. People taking methedrine ("speed") may become disoriented and suddenly experience feelings of excitement, energy, and restlessness. Taking the drug intravenously produces intense, orgasmlike "rushes." Figure 20.20 shows the effects of "speed" on the web-building ability of an adult female spider.

At first amphetamines were thought to be safe and nonhabit-forming. This incorrect information led to their widespread use and abuse. Now enough amphetamine tablets are produced in the United States to supply every person there with forty pills per year (Figure 20.21). Typical users are people who are overtired, students cramming for exams, truck drivers trying to keep awake, and athletes seeking a temporary boost in energy.

Figure 20.21 Amphetamine tablets. (Courtesy of the U.S. Drug Enforcement Agency)

Amphetamines have been used to treat obesity (in diet pills), mild depression, and narcolepsy (the tendency to fall asleep at any time). A paradoxical and very controversial use is to treat hyperactive children with amphetamines—especially methylphenidate (Ritalin) (see Figure 20.19)—to calm them and improve their learning. About 4 million American children now take this drug daily.

Dependence and tolerance develop with continued use. The user becomes nervous, irritable, unstable, and unable to sleep. Continued heavy use may produce aggressive behavior, hallucinations, and paranoia. Some people try to alleviate these effects by using other drugs such as barbiturates, marijuana, ethanol, or heroin. Withdrawal from amphetamines can cause fatigue, depression, and lethargy.

HALLUCINOGENS   Known also as psychedelic drugs, **hallucinogens** are mind-affecting chemicals that cause vivid illusions, fantasies, and hallucinations.

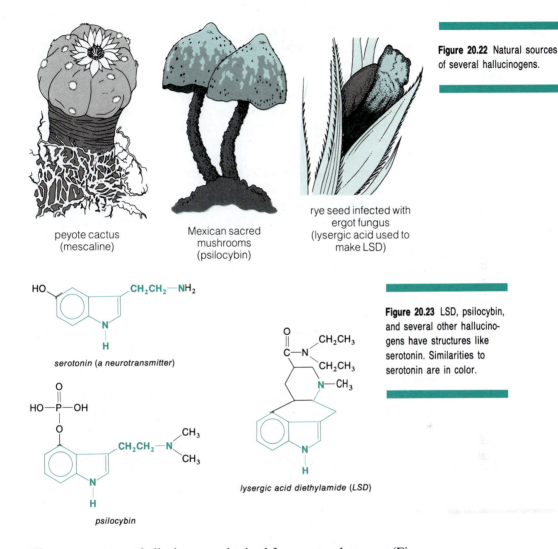

Figure 20.22 Natural sources of several hallucinogens.

peyote cactus
(mescaline)

Mexican sacred
mushrooms
(psilocybin)

rye seed infected with
ergot fungus
(lysergic acid used to
make LSD)

serotonin (a neurotransmitter)

Figure 20.23 LSD, psilocybin, and several other hallucinogens have structures like serotonin. Similarities to serotonin are in color.

lysergic acid diethylamide (LSD)

psilocybin

The most common hallucinogens obtained from natural sources (Figure 20.22) are (1) mescaline, which comes from the fruit of the peyote cactus that grows in Mexico and the southwestern United States; (2) psilocybin and psilocin, which come from certain mushrooms; and (3) LSD (lysergic acid diethylamide), which is made from lysergic acid derived from the morning glory and from ergot, a fungus that grows on wheat, rye, and other grasses. Besides LSD, a whole alphabet soup of other synthetic street drugs—MDA, DOM (STP), PCP, and DMT—are hallucinogens.

LSD, psilocybin, psilocin, and DMT are similar in structure to the neurotransmitter serotonin (Figure 20.23). They act as hallucinogens by altering the serotonin nerve pathways.

As little as 0.05 mg (1/500,000 oz) of LSD produces noticeable effects in most people. Their blood pressure and heart beat increase, their pupils dilate, their faces become flushed, and they may feel euphoric with sounds and colors taking on different dimensions. They also may become anxious, depressed, nauseated, and psychotic. The hallucinogenic experience is highly personal. It varies with the dose, the expectations of the person, and the setting in which the experience takes place. Accidental high doses are common because of the extremely small amounts involved and because LSD purchased illegally varies widely in purity and strength. Street drugs—particularly those sold as LSD—often contain other dangerous drugs, one of which is strychnine (Section 21.3).

In laboratory experiments, LSD causes chromosome damage in white blood cells and skin cells, though there is no evidence of genetic damage in LSD users. LSD taken during pregnancy may harm the fetus. People who take LSD regularly may develop tolerance and psychological, but not physical, addiction. Users sometimes have "flashbacks," in which they suddenly experience hallucinations they previously had under the influence of the drug.

Psilocybin and psilocin have effects similar to LSD but are less potent. Psilocybin is metabolized in the body into psilocin, which is the active hallucinogen. As long as 3000 years ago, Indian cultures of Mexico and South America used mushrooms containing these drugs as part of their religious rituals.

Other hallucinogens, including mescaline, DOM (STP), and MDA, are similar in structure to amphetamine and norepinephrine (Figure 20.24). Indians in Mexico ingested mescaline by chewing the buttons from the peyote cactus as part of their religious ceremonies, apparently with little harm. Though less potent than LSD, mescaline can produce a hallucinatory state lasting up to ten hours. MDA is about twice as active as

Figure 20.24 Mescaline structurally resembles norepinephrine and amphetamine. Similarities are in color.

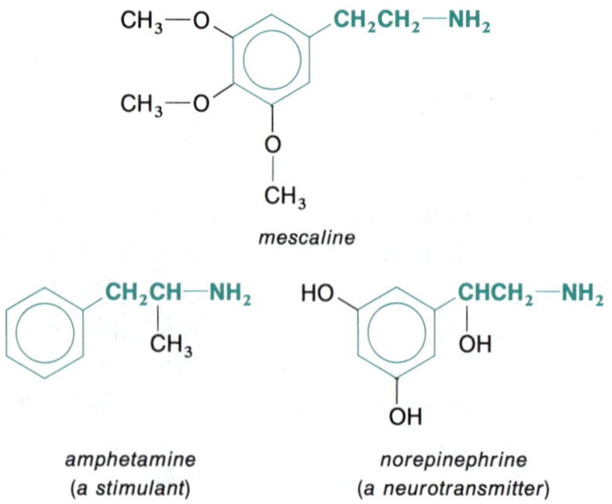

mescaline

amphetamine
(a stimulant)

norepinephrine
(a neurotransmitter)

mescaline, whereas STP (DOM) is 50 to 100 times more active. STP is often sold as mescaline but produces a higher incidence of "bad trips" and feelings of panic.

Only veterinarians use phenylcyclidine (PCP) legally—as a tranquilizer for animals. Users take PCP ("angel dust") orally or inhale it when it is sprinkled on other smokable drugs such as marijuana. They quickly go into a trance, often remaining motionless with their eyes open. PCP produces unpleasant and frightening hallucinations, and users may experience sweating, dizziness, nausea, and vomiting.

Marijuana is a mild hallucinogen and sedative made from the flowering tops, seeds, leaves, and stems of the female hemp plant *Cannabis sativa* (Figure 20.25). This plant grows like a weed in almost any climate or soil condition, but plants grown in tropical areas (especially in Africa, India, the Middle East, and Mexico) produce the most potent drugs. Hashish, made from the resin of the plant, is two to five times as strong as marijuana. In the United States, marijuana is usually rolled into cigarettes (Figure 20.25), but in other countries it is mixed with liquids ("tea") or baked in sweet foods. Smoking brings special dangers. In a recent study, people smoking marijuana were found to have five times as much carbon monoxide and three times as much tar inhaled into their lungs as when smoking tobacco cigarettes.

The active ingredient in marijuana is tetrahydrocannabinol (THC) (Figure 20.26), which produces hallucinations only at high doses. Common effects include a change in mood, dryness of the mouth, sleepiness, time distortion, reduced ability to concentrate, decreased reaction time, and increased hunger. Marijuana is not physically addicting although heavy users occasionally experience mild withdrawal symptoms such as restlessness, irritability, and insomnia. Tolerance and psychological dependence occur with extended use. An estimated 62 million people in the United States have tried marijuana, and 18 million are recent users.

The federal government has been making marijuana available to research scientists to study potentially useful effects of THC. It has shown

Figure 20.25 Female hemp plant (top) used to obtain ground marijuana leaves, seeds, and cigarettes. (Courtesy of the U.S. Drug Enforcement Agency)

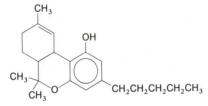

Figure 20.26 Tetrahydrocannabinol (THC) is the active ingredient in marijuana and hashish. Actually, THC is a mixture of two isomers having the same chemical formula but slightly different geometric structures.

promise in relieving the eye pressure from glaucoma, and in 1985 it was approved to help reduce the side effects, such as nausea and vomiting, experienced by cancer patients taking chemotherapy. But the other effects of marijuana and its widespread abuse make it unlikely to become widely used for medical treatments.

20.5 Drug Abuse: Getting Hooked and Unhooked

DRUG ABUSE During this century, drugs have saved millions of lives and relieved the suffering of many millions of people. But many drugs are misused. Drug abuse is the taking of a drug in a way that is not medically or legally approved in a culture. The causes are a complex mix of physiological, psychological, and social factors that vary from drug to drug and from user to user.

The two most widely abused drugs in the United States and throughout much of the world are ethyl alcohol and nicotine (found in tobacco products). Although both drugs are socially acceptable in most countries, they cause more illness, death, loss of time from work, family disruption, and economic loss than any of the more publicized, illegal drugs.

When certain drugs are taken over long periods of time they may produce a type of dependence known as **addiction**. The user has a strong need for the drug and undergoes pronounced physical or emotional reactions when deprived of a supply. Table 20.4 lists some commonly abused drugs and their potential for addiction.

Psychological dependence (habituation) occurs if the user feels uneasy, anxious, nervous, or distressed when not using a particular drug such as marijuana, cocaine, or LSD. These reactions during withdrawal can also produce physical symptoms. Psychological dependence can occur with any drug, and it can be as compelling as physical dependence in turning a person into a compulsive drug abuser.

Some drugs, however, also produce **physical dependence**. As Table 20.4 shows, ethanol, barbiturates, and narcotics have this effect. After long and heavy usage, the user suffers not only an emotional disturbance but also a disruption in body functions when the drug is withdrawn. Withdrawal symptoms vary from drug to drug and can include severe muscle pain, cramps, vomiting, diarrhea, convulsions, tremor, and even death.

No one knows for sure what causes physical addiction. In narcotic addicts, brain receptor sites for β-endorphin and enkephalins may be occupied so much of the time by the drug (such as heroin) that those receptors lose their natural function. Then when the heroin is withdrawn,

TABLE 20.4 DRUGS WITH POTENTIAL FOR ADDICTIONS AND TOLERANCE

Drug	Potential for Addiction and Tolerance
Narcotics	
Opium, morphine, meperidine (Demerol), heroin, codeine, methadone	Tolerance, high psychological dependence, physical dependence
Depressants	
Ethanol, barbiturates	Tolerance, psychological dependence, physical dependence
Tranquilizers and Antidepressants	
Thorazine, Valium, Tofranil	Tolerance, psychological dependence, physical dependence
Stimulants	
Amphetamines	Tolerance, high psychological dependence, probably no physical dependence
Cocaine, nicotine	Tolerance, high psychological dependence, probably physical dependence
Caffeine	Possible tolerance, moderate psychological dependence, mild physical dependence
Hallucinogens	
LSD, mescaline, psilocybin	Tolerance, moderate psychological dependence, probably no physical dependence
Marijuana	Mild tolerance, psychological dependence, probably no physical dependence

the receptors may fail to operate normally, causing the addicted person to suffer unpleasant symptoms.

Drugs can help treat addiction in some cases. We discussed earlier (Section 20.2) how a synthetic drug such as methadone blocks the effects of morphine compounds and can make the withdrawal process a little easier. Another strategy, called aversion therapy, has been used to treat ethanol addiction. The most widely used drug is disulfiram (Antabuse), which blocks the metabolism of ethanol and causes acetaldehyde to accumulate (Figure 20.27); this produces very unpleasant effects. Antabuse is a deterrent to drinking because people taking this drug know that if they

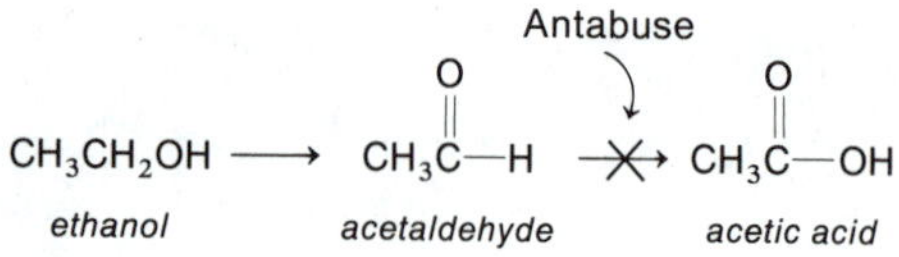

Antabuse

Figure 20.27 Antabuse (disulfiram) blocks the metabolism of ethanol and causes acetaldehyde to accumulate.

$$CH_3CH_2OH \longrightarrow CH_3\overset{\overset{\displaystyle O}{\|}}{C}-H \;\;\overset{}{\times\!\!\!\!\!\!\longrightarrow}\; CH_3\overset{\overset{\displaystyle O}{\|}}{C}-OH$$

ethanol   *acetaldehyde*   *acetic acid*

then drink ethanol, they will experience a pounding headache, nausea, vomiting, palpitations, and other very unpleasant effects.

Another characteristic of many drugs is **tolerance**, which means that a user must take progressively larger doses to achieve the same effects. Tolerance usually accompanies physical addiction, but it also can occur with habituating drugs such as amphetamines. Table 20.4 shows the tendency for tolerance to develop with certain drugs.

Tolerance is better understood, chemically, than addiction. It often occurs because regular users become able to metabolize and detoxify the drug faster. For example, people who frequently use barbiturates produce a greater amount of the liver enzymes that metabolize (oxidize) those drugs; because they break down barbiturates faster, they have to take more of the drug to get the desired effect. Tolerance to narcotics, however, may develop because brain neurons adapt to those drugs by producing and storing different amounts of endorphins and enkephalins.

Related drugs often are metabolized by the same enzymes in the body, so developing a tolerance to one drug may automatically produce a tolerance to other related drugs. This is called **cross-tolerance**. For example, a person who has built up a tolerance for LSD also will have a tolerance for most other hallucinogens (except marijuana and PCP, which are in different chemical classes) without ever having taken them.

Taking drugs in combination may produce unexpected effects. Ethanol and barbiturates are particularly dangerous combination because they accentuate each other's effects. Sublethal doses of these drugs can be fatal in combination, and this is a fairly common way to commit suicide or to die accidentally. This combination occurs when people use barbiturates to shorten the time for getting intoxicated with alcohol or when alcoholics take barbiturates to ease the pain of a hangover.

Chloral hydrate and ethanol also increase the potency of each other's effects. Chloral hydrate, the first synthetic hypnotic, is metabolized by the same enzyme that metabolizes ethanol (Figure 20.28). When both drugs are present, they compete for access to that enzyme and are both metabolized more slowly than normal; this prolongs and enhances their depressant effects. Popularly known as "knockout drops" or a "Mickey Finn," this drug combination rapidly puts the user to sleep.

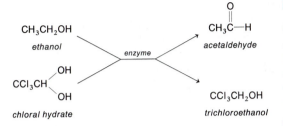

Figure 20.28 Ethanol and chloral hydrate are metabolized by the same enzyme, so in combination neither drug can be metabolized as rapidly as normal.

Although drugs are very valuable in medicine, their abuse is itself a major medical problem. Depressants, antidepressants, stimulants, narcotics, hallucinogens, and the like are all grossly overused. The answer to our personal, nonmedical problems are not found inside bottles.

SHERMAN M.
MELLINKOFF

The body chemistry is exceedingly complex and usually does not call for intervention. If it does require intervention, the less the better.

Summary

Neurotransmitters released from axons of neurons bind to receptors on receiving neurons, thus enabling nerve impulses to travel from one neuron to another. Many drugs alter the action of neurotransmitters.

Mild pain relievers (analgesics), including aspirin, acetaminophen, and ibuprofen, inhibit the synthesis of prostaglandins. Narcotics such as morphine are strong analgesics that bind to receptor sites in the brain that also bind β-endorphin and enkephalins. Local and general anesthetics prevent nerve impulses from reaching the brain. Sedatives and hypnotics such as barbiturates depress the central nervous system.

Many drugs are used to treat mental diseases or nerve disorders. Tranquilizers treat schizophrenia and anxiety; these drugs depress the central nervous system, and some block nerve pathways using dopamine. Lithium salts help treat manic–depression. Some antidepressants increase the concentration of neurotransmitters in the brain by inhibiting their breakdown by monoamine oxidase (MAO). L-Dopa is used to treat patients with Parkinson's disease.

Stimulants such as amphetamines increase the action of neurotransmitters, especially norepinephrine. Hallucinogens also alter the actions of neurotransmitters such as norepinephrine.

Drugs that affect the nervous system are widely abused. Psychological and physical dependence develop with many drugs, and tolerance (and cross-tolerance) also may occur with regular use.

Terms for Review

After completing this chapter, you should know and understand the meaning of the following terms:

addiction (p. 568)

Alzheimer's disease (p. 562)

amphetamine (p. 563)

analgesic (p. 549)

antidepressant (p. 560)

axon (p. 545)

barbiturate (p. 557)

cross-tolerance (p. 570)

dendrite (p. 545)

general anesthetic (p. 555)

hallucinogen (p. 564)

hypnotic (p. 557)

local anesthetic (p. 553)

narcotic (p. 550)

neuron (p. 545)

neurotransmitter (p. 546)

Parkinson's disease (p. 562)

physical dependence (p. 568)

psychological dependence (p. 568)

sedative (p. 557)

stimulant (p. 563)

synapse (p. 546)

tolerance (p. 570)

tranquilizer (p. 559)

Questions

Odd-numbered questions are answered at the back of this book.

1. How does acetylcholine (ACH) work as a neurotransmitter?

2. Which positive ions cross the neuron membrane when a neurotransmitter binds to the cell?

3. Are any brand-name aspirins 100 percent acetylsalicylic acid? Why? What is the difference in chemical content and effectiveness among major brands of aspirin?

4. Compare aspirin with acetaminophen with regard to (a) price, (b) safety, and (c) effectiveness.

5. What is the most abundant narcotic in opium?

6. Which narcotic is commonly used in cough medicine?

7. Explain how naloxone works in treating a person with a heroin overdose.

8. How does a local anesthetic differ from a general anesthetic?

9. Name the most widely abused local anesthetic.

10. Which barbiturate is widely used to treat epilepsy and other seizure disorders?

11. Why are accidental deaths from drug overdoses more common with barbiturates than with most other nonnarcotic drugs?

12. Barbiturates have largely been replaced by what type of drugs as prescription sleeping pills?

13. Besides alcoholism, excess alcohol consumption often produces what other health disorders?

14. Some antidepressants block the action of monoamine oxidase (MAO). How does this alleviate depression?

15. Explain how L-Dopa works in treating patients with Parkinson's disease.

16. People with Alzheimer's disease typically have low concentrations of which neurotransmitter in a certain region of the brain?

17. Identify mood-altering drugs that are used for each of the following: (a) diet pills, (b) treating side effects of anticancer drugs, (c) calming hyperkinetic children, and (d) treating manic–depression.

18. What neurotransmitters do hallucinogens and amphetamines chemically resemble?

19. Many hallucinogens produce cross-tolerance to other hallucinogens. What does this mean?

20. Classify each of the following as a mild analgesic, tranquilizer, sedative or hypnotic, strong analgesic, local anesthetic, general anesthetic, hallucinogen, or stimulant: (a) aspirin, (b) phenobarbital, (c) Novocaine, (d) caffeine, (e) ibuprofen, (f) Seconal, (g) mescaline, (h) codeine, (i) methadone, (j) Valium, (k) cocaine, (l) methamphetamine, (m) nitrous oxide, (n) ethanol, and (o) flurazepam (Dalmane).

21. Distinguish among psychological dependence, physical dependence, and tolerance. Identify which of these are characteristics of the following drugs: (a) morphine, (b) caffeine, (c) barbiturates, (d) LSD, (e) marijuana, (f) amphetamines, (g) codeine, (h) cocaine, (i) heroin, and (j) meprobamate.

Topics for Discussion

1. Should marijuana become a legal drug for recreational use? Why?

2. As a matter of public policy, heroin addicts in England can get heroin at little or no expense. What are the advantages and disadvantages of this approach? Do you favor this idea? Why?

3. Making a slight chemical modification in an illegal drug produces a new drug that is not illegal; these new drugs are called "designer drugs." Should each designer drug be legal until it is evaluated and possibly made illegal? Is there a practical way to decide how much of a chemical change is necessary to consider the substance a new and different drug?

4. Drug addicts sharing needles to inject drugs transmit infectious diseases such as AIDS. Should the government provide sterile needles free to drug addicts? Will this be an effective way to reduce the spread of AIDS?

5. What do you consider to be the most serious drug problem in the United States today? Why? What can be done to reduce the problem?

CHAPTER 21

Toxicology: Dealing with Poisons, Mutagens, Carcinogens, and Teratogens

General Objectives

1. What is a poison, and what constitutes a normally lethal dose for a chemical?

2. What types of poisons damage skin, interfere with enzyme action, or deplete the body's oxygen supply? How do they work?

3. What types of poisons disrupt the transmission of nerve signals, and what are the antidotes for such poisons?

4. How do certain metal substances act as poisons, and how can metal poisoning be treated?

5. What types of compounds can cause genetic changes, cancers, and birth defects?

Many centuries ago, people lived in a simpler chemical world. Most substances that our ancestors encountered occurred naturally on earth. Their organic materials came from living things, and their minerals came from rocks, soil, and water.

Some of those natural materials were toxic. The philosopher Socrates died by drinking a brew from the poison hemlock plant. South American natives used curare (a black, thick syrup from plants) as an arrow poison to paralyze their prey. Another plant poison, strychnine, sent its victims into convulsions before they died. And several forms of arsenic were well known for their ability to kill.

Today these natural poisons are joined by a wide variety of toxic synthetic materials. Some poisons inflame the tissues locally or upset metabolism by interfering with the actions of enzymes. Others block the nervous system. Still others cause genetic changes, cancers, or birth defects. We now examine how some of these toxic substances work.

21.1 Poisons and Toxicity Ratings: How Much Is Too Much?

More than 1 million people in the United States are poisoned each year. Most cases are accidents and involve compounds such as ethanol, carbon monoxide, aspirin, and barbiturates. Other household and farm products poison thousands of people each year, with children younger than age five accounting for about one-third of the accidental poisonings.

TABLE 21.1 TOXICITY RATINGS AND AVERAGE LETHAL DOSES FOR HUMANS

Toxicity Rating	LD_{50} (mg/kg Body Weight)	Average Lethal Dose* for a 70.4-kg (155-lb) Human	Examples
Supertoxic	Less than 0.01	Less than 1 drop	Nerve gases, botulism toxin, mushroom toxins
Extremely toxic	Less than 5	Less than 7 drops	Potassium cyanide, heroin, atropine, arsenic trioxide, parathion
Very toxic	5–50	7 drops to 1 teaspoon	Mercury salts, morphine, codeine
Toxic	50–500	1 teaspoon to 1 ounce	Lead salts, DDT, sodium hydroxide, fluoride, sulfuric acid, caffeine, carbon tetrachloride
Moderately toxic	500–5000	1 ounce to 1 pint	Methyl (wood) alcohol, ether, phenobarbital, amphetamine, kerosene, nicotine, aspirin
Slightly toxic	5000–15,000	1 pint to 1 quart	Ethyl alcohol, lysol, soaps
Essentially nontoxic	15,000 or greater	More than 1 quart	Water, glycerin

*For substances that are liquids at room temperature.

Practically any chemical, even water, can be harmful if you ingest enough of it. For this reason, a **poison** (or toxic substance) is defined as any chemical that causes illness or death in an organism when used in a relatively small amount.

As Table 21.1 shows, the lethal dose of poisons varies widely. The term **LD_{50}** refers to the dose of a given substance (in milligrams of the substance per kilogram of body weight) that kills 50 percent of the subjects exposed to that dose. For example, the LD_{50} of potassium cyanide (KCN) for humans is about 1 mg/kg body weight. This means that about 70 mg (0.0025 oz) of cyanide has a 50 percent chance of killing an average 70-kg (154-lb) person.

Lethal doses are not determined with humans; besides, there is an acute shortage of volunteers for such studies. Instead, scientists administer various doses of a substance to laboratory animals, often rats or mice. If it takes 2 mg/kg body weight of a particular chemical to kill 50 percent of the rats tested, the LD_{50} for that poison in rats would be 2 mg/kg. The LD_{50} for a chemical varies with different animals, so the data for rats, rabbits, or monkeys don't automatically apply to humans. But any substance with a low LD_{50} for several animal species will probably be very toxic to humans.

Some poisons are immediately life-threatening; they may destroy tissue on contact or block a critical chemical process in the body. Other poisons act over a longer period of time. For example, they may cause genetic changes, cancer, or birth defects. In the following sections, we examine several examples from each of these categories to see how they work and what antidotes can combat them.

The single most valuable emergency treatment for ingested poisons, however, is activated charcoal (carbon). You should administer 1 to 2 tablespoons (50 g for adults, 20 g for children) of activated charcoal powder, mixed with water or fruit juice, to the victim. The powdered charcoal rapidly binds poisons and carries them into the intestine for excretion. Charcoal doesn't dissolve in the body or enter the blood, so it is quite safe to use.

Another useful treatment to have the victim swallow syrup of ipecac to induce vomiting. This removes the poison from the body within a few minutes. Vomiting should not be induced, however, to expel corrosive poisons (Section 21.2).

21.2 Corrosive and Metabolic Poisons: Destroying Skin, Enzymes, and the Blood's Oxygen Supply

CORROSIVE POISONS   Strong acids, bases (Section 6.1), and oxidizing agents (Section 6.3) are **corrosive poisons** that can destroy tissues on contact. The amount of damage depends on the amount of the chemical and how long it is in contact with the tissue. This is why it is important to flush the skin immediately with water to remove (or at least dilute) a corrosive poison. People who swallow the poison should drink water or milk right away to reduce the poison's concentration. They should not try to vomit, however, because this again exposes their throat and nose tissues to the poison.

Table 21.2 shows some of the most common corrosive poisons. Strong acids damage tissues in several ways. They often dehydrate tissues by reacting with water. Sulfuric acid, for example, reacts with water in body tissues to form the hydronium (H_3O^+) ion and hydrogen sulfate (HSO_4^-) ion:

$$H_2SO_4 + H_2O \longrightarrow H_3O^+ + HSO_4^-$$

TABLE 21.2 SOME COMMON CORROSIVE POISONS

Toxic Substance	Formula	Source
Acids		
Hydrochloric (or muriatic) acid	HCl	Floor, toilet-bowl, brick, and metal cleaners
Oxalic acid	$H_2C_2O_4$	Bleaches, metal polishes, ink remover, rhubarb, spinach
Sulfuric acid	H_2SO_4	Automobile batteries
Bases		
Ammonia	NH_3	Household cleaning solutions
Sodium carbonate	Na_2CO_3	Washing soda, detergents
Sodium hydroxide	NaOH	Lye, caustic soda, drain and oven cleaners
Sodium perborate	$Na_2[B_2O_4(OH)_4]$	Detergents
Sodium silicate	$NaSiO_3$	Detergents
Sodium phosphate	Na_3PO_4	Detergents
Oxidizing Agents		
Hydrogen peroxide	H_2O_2	Bleaches, antiseptics
Ozone	O_3	Air
Sodium hypochlorite	NaOCl	Bleaches

Phosgene ($COCl_2$), a deadly gas used in World War I, reacts with moisture in the lungs to form hydrochloric acid (HCl), another strong acid:

$$Cl-\overset{\overset{\textstyle O}{\|}}{C}-Cl + H_2O \longrightarrow 2HCl + CO_2$$

phosgene

The HCl then attracts water into the lungs, causing the victim to suffocate.

Strong acids and bases destroy the proteins in tissues by breaking down the amide (peptide) bonds:

$$R-\overset{\overset{\displaystyle O}{\|}}{C}-\underset{\underset{\displaystyle H}{|}}{N}-R' + H_2O \xrightarrow{acid\ (H^+)} R-\overset{\overset{\displaystyle O}{\|}}{C}-O-H + H-\overset{\overset{\displaystyle H}{|}}{\underset{\underset{\displaystyle H}{|}}{N}}{}^{\pm}-R'$$

peptide linkage
in protein

$$R-\overset{\overset{\displaystyle O}{\|}}{C}-\underset{\underset{\displaystyle H}{|}}{N}-R' + NaOH \longrightarrow R-\overset{\overset{\displaystyle O}{\|}}{C}-O^-Na^+ + H-\underset{\underset{\displaystyle H}{|}}{N}-R'$$

peptide linkage   base
in protein

Because amide bonds join amino acids to make proteins (Section 9.4), this action by acids and bases causes structural proteins and enzymes in cells to disintegrate.

Oxidizing agents attack many substances in the cell. Because oxidizing agents seek electrons (Section 6.3), they readily react with unsaturated fatty acids, which have electron-rich double bonds between carbon atoms (Section 9.3). This action especially damages cell membranes, which contain large amounts of unsaturated fatty acids.

Another common target is proteins that have —SH groups from the amino acid cysteine (see Table 9.2). Oxidizing agents cause two nearby —SH groups to join together into one —S—S— group (Figure 21.1); this changes the shapes of proteins and often makes them less able to function.

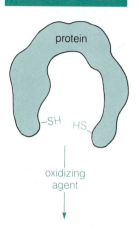

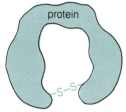

Figure 21.1 Oxidizing agents change —SH groups into —S—S— bonds, thus changing the shape of a protein.

METABOLIC POISONING FROM CYANIDE   **Metabolic poisons** are toxic because they interfere with normal metabolic processes. They may block the action of an important enzyme (Section 9.4), disrupt oxygen metabolism, disable a crucial substance such as DNA, or interfere in some other way. Common examples of metabolic poisons are the cyanide (CN^-) ion, hydrogen cyanide (HCN), carbon monoxide (CO), the fluoride (F^-) ion, and alcohols such as methanol and ethylene glycol (Section 8.6).

Cyanide poisoning occurs from inhaling gaseous hydrogen cyanide (HCN), which smells like bitter almonds. In fact, the chemist who discovered HCN, Karl Wilhelm Scheele, died from its vapors. Solutions of soluble cyanide salts, such as potassium cyanide (KCN) or sodium cyanide (NaCN), are also dangerous. In a slightly acid solution, cyanide salts form hydrogen cyanide gas, which can easily be inhaled:

$$\text{NaCN} + \text{HCl} \longrightarrow \text{HCN} + \text{NaCl}$$

(poisonous   (poisonous
solid)   gas)

This reaction is why cyanide salts should not be flushed down drains or into acidic bodies of water.

Cyanide works as a poison by rapidly attacking cytochrome oxidase, an enzyme that helps metabolize oxygen. This blocks cells from using oxygen, so the victim dies a painful death, usually within about fifteen minutes. If the poisoning is discovered in time (which is rare), the victim should receive an injection of sodium nitrite ($NaNO_2$) and sodium thiosulfate ($Na_2S_2O_3$). Nitrite oxidizes hemoglobin (Section 9.4) to the form methemoglobin, which cannot transport oxygen but does attract cyanide away from cytochrome oxidase; this allows oxygen metabolism to resume. The thiosulfate ion ($S_2O_3^{2-}$) then removes the cyanide by changing it into thiocyanate (SCN^-), which is rapidly excreted in the urine:

$$CN^- + S_2O_3^{2-} \longrightarrow SCN^- + SO_3^{2-}$$

cyanide　　thiosulfate　　　thiocyanate　　sulfite
 ion　　　　 ion　　　　　　　 ion　　　　 ion

Treating cyanide poisoning is tricky because too much nitrite oxidizes so much hemoglobin to methemoglobin that the patient could die of oxygen starvation. Indeed, it is often the case that the antidote to a poison is a poison itself.

METABOLIC POISONING FROM CARBON MONOXIDE   Carbon monoxide (CO) is a colorless, odorless, poisonous gas produced by the incomplete combustion of carbon and carbon compounds such as coal, gasoline, and cigarettes:

$$2C + O_2 \longrightarrow 2CO$$
(coal)

$$C_8H_{18} + 11O_2 \longrightarrow 5CO_2 + 3CO + 9H_2O$$
(octane
in gasoline)

People who direct traffic, work in enclosed areas such as tunnels or garages, smoke tobacco in any form, or work near kilns or blast furnaces in steel mills may be exposed to high CO concentrations. Indoors, the dangers come from smoking, using charcoal grills or hibachis, and burning fuels such as kerosene in heaters without proper ventilation.

CO interferes with the action of hemoglobin (*Hb*), which normally binds O_2 as the blood circulates through the lungs and then releases O_2 at the tissues. The normal reaction forms an equilibrium that illustrates Le Châtelier's principle (Section 7.4):

$$Hb \quad + O_2 \; \rightleftharpoons \quad O_2Hb$$

hemoglobin   *oxyhemoglobin*

In the lungs, where oxygen concentration is relatively high, equilibrium shifts to the right. When the hemoglobin reaches the tissues, where there is much less oxygen, the equilibrium shifts to the left and releases the needed oxygen. But when the hemoglobin encounters CO in the lungs, it binds CO about 200 times more strongly than it binds O_2:

$$Hb \quad + CO \; \longrightarrow \quad COHb$$

hemoglobin   *carboxyhemoglobin*

So when CO levels rise in the air and that air is breathed, an increasing amount of hemoglobin forms carboxyhemoglobin and cannot carry oxygen to the tissues.

Table 21.3 shows the effects of exposure to various levels of CO. Cigarette smoke contains 200 to 400 ppm of CO and can tie up about 20 percent of the smoker's hemoglobin. Automobile exhaust contains 1000 to 7000 ppm of CO. Typical concentrations inside automobiles (for non-smokers) range from 25 to 115 ppm in downtown traffic and traffic jams,

TABLE 21.3 EFFECTS OF CARBON MONOXIDE

CO Levels in the Air (ppm)	Exposure	% of Hb as COHb	Symptoms
Under 100	Indefinite	0–10	None
100–200	Indefinite	10–20	Slight headache, tightness across forehead
200–300	5–6 hrs	20–30	Throbbing headache
400–600	4–5 hrs	30–40	Severe headache, dizziness, nausea, weakness, impaired vision, collapse
700–1000	3–4 hrs	40–50	Increased pulse and breathing, increased tendency to collapse
1000–3000	1–3 hrs	50–70	Coma with intermittent convulsions, depressed heart action, and possible death at 1600 ppm and higher

SOURCE: Jay M. Arena and Richard H. Drew (Eds.), Poisoning: Toxicology, Symptoms, Treatments, 5th ed., Springfield, IL: Thomas, 1986.

from 10 to 75 ppm on freeways during the rush hour, and from 5 to 20 ppm in urban residential areas. So in traffic jams and on crowded freeways, there may be enough CO to impair a person's judgment and reflexes.

The treatment for CO poisoning makes use of Le Châtelier's principle. The reaction by which CO displaces oxygen from hemoglobin is

$$O_2Hb \quad + CO \rightleftharpoons \quad COHb \quad + O_2$$

oxyhemoglobin   carboxyhemoglobin

So by lowering CO concentration and increasing O_2 concentration, we can shift the equilibrium to the left, with Hb binding O_2 and releasing CO. This is why victims are removed from the CO-contaminated air and treated with pure oxygen, if possible, or at least fresh air.

FLUORIDE POISONING   In Section 16.1, we discussed how fluoride (F^-) ions help prevent tooth decay by replacing some of the hydroxide (OH^-) ions in tooth enamel. Fluoride levels in the water vary considerably. About half of U.S. residents live in communities that add fluoride to their water supplies to a concentration of 1 ppm in order to prevent tooth decay. Children drinking water with higher levels of fluoride (2 to 3 ppm or more) develop mottled, or discolored, teeth (see Figure 16.3). Drinking water with 10 ppm of fluoride causes dark stains and pitting of the teeth as parts of the enamel become more fragile (Figure 21.2), a condition called *fluorosis*.

Fluoride compounds are used extensively in industries making phosphate fertilizer, aluminum, and tile; geysers and volcanoes are important natural sources. People exposed to 10 to 15 mg a day develop skeletal fluorosis as excess fluoride deposits make their bones abnormal and enlarged. In addition, fluoride ions in the blood react with calcium ions to form insoluble calcium fluoride (CaF_2):

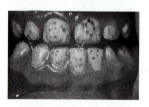

Figure 21.2 People who regularly drink water with 10 ppm of fluoride ion have teeth with severe mottling, or fluorosis. (Courtesy of National Institutes of Health)

$$Ca^{2+} + 2F^- \longrightarrow \quad CaF_2$$

(*insoluble*)

This removes Ca^{2+} from the blood and causes painful muscle spasms, depressed breathing, and irregular heartbeats.

The lethal dose of sodium fluoride (NaF) for an adult is about 5 g. The most common cause of death from fluoride poisoning is by accidentally ingesting an insecticide such as roach powder, which is 30 to 90 percent NaF. The treatment is to have the victim drink milk, limewater—0.15 percent $Ca(OH)_2$—or a dilute $CaCl_2$ solution, plus slow intravenous injections of calcium salts. As the equation above shows, calcium ions com-

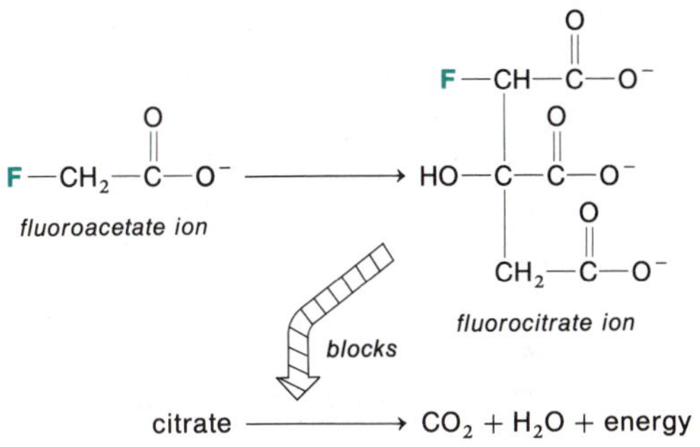

Figure 21.3 Fluoroacetate ion in sodium fluoroacetate (Compound 1080) is metabolized into fluorocitrate ion, which blocks the normal metabolism of citrate to produce energy.

bine with toxic fluoride ions to form insoluble, nontoxic calcium fluoride, which eventually is excreted.

A type of plant in South Africa converts excess fluoride into fluoroacetic acid, which is very toxic to cattle eating the plant. A salt of this acid, sodium fluoroacetate (Figure 21.3), is a commercial poison (called Compound 1080) used on rats and livestock predators; it is also toxic to humans and other animals. Animals metabolize the fluoroacetate ion to fluorocitrate ion, which then inhibits the normal metabolism of citrate to produce energy (Figure 21.3). By blocking the major energy-producing process in the cell, fluoroacetate causes seizures, coma, and death, usually from heart failure.

POISONING FROM ALCOHOLS   Although methyl alcohol and ethylene glycol are more toxic than ethyl alcohol (ethanol), alcoholics occasionally drink these alcohols—and isopropyl alcohol (rubbing alcohol)—as cheap substitutes for ethanol. Children accidentally drink these alcohols too—some being attracted to ethylene glycol because of its sweet taste.

Methanol is a major ingredient in duplicator fluid, windshield deicers, and gas-line antifreeze, whereas ethylene glycol is the major ingredient in antifreeze. Both of these alcohols, like ethanol, depress the central nervous system and produce symptoms similar to ethanol intoxication. In addition, methanol causes blurred vision and eventually blindness.

As Figure 21.4 shows, these alcohols are oxidized in the body to form aldehydes and carboxylic acids. Those products, such as formaldehyde and oxalic acid, account for much of the toxicity of methanol and ethylene glycol.

The enzyme—alcohol dehydrogenase—that oxidizes methanol and ethylene glycol is the same enzyme that metabolizes ethanol, and this enables ethanol to work as an antidote for both methanol and ethylene glycol poisoning. Once the victim receives ethanol, usually by intravenous

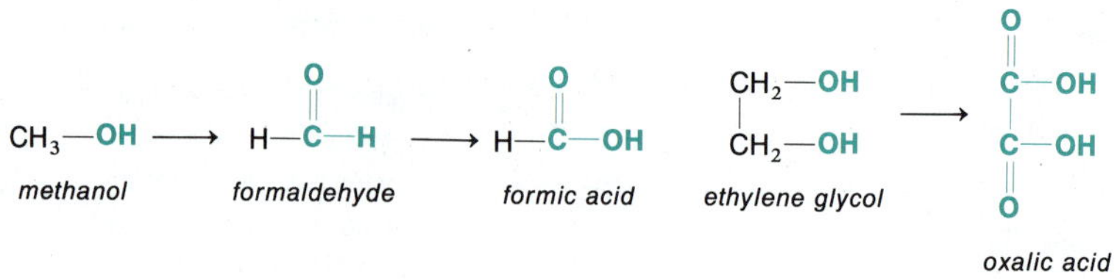

Figure 21.4 Oxidation of methanol and ethylene glycol to form toxic products.

$$CH_3-OH \longrightarrow$$ methanol

formaldehyde

formic acid

ethylene glycol

oxalic acid

injection, the enzyme begins to metabolize ethanol instead of the methanol or ethylene glycol. This slows the production of toxic products and gives the body time to excrete the toxic alcohol. In effect, the antidote substitutes a less toxic alcohol (ethanol) for a more toxic one.

21.3 Nerve Poisons: Causing the Ultimate Breakdown

A number of deadly poisons disrupt the transmission of nerve signals in the nerve cells. These **neurotoxins** include nerve gases, various organophosphate and carbamate insecticides (Section 17.2), strychnine, curare, nicotine, red pigments secreted from dinoflagellates (microscopic marine organisms in what are known as *red tides*), and the toxin from the bacterium *Clostridium botulinum* that causes botulism (a form of food poisoning). The toxin that causes botulism is one of the most potent human poisons known (see Table 21.1); a dose of less than one/ten-millionth of a gram (10^{-7} g) can kill a person.

We discussed in Section 20.1 how neurotransmitters such as acetylcholine (ACH) help nerve impulses travel from cell to cell. Figure 21.5 shows the role of ACH and how neurotoxins can alter the transmission of nerve signals.

Some compounds, such as dinoflagellate and botulin toxins, slow or block the synthesis of ACH, preventing the normal transmission of nerve signals. This paralyzes key organs and causes death by respiratory failure.

Other neurotoxins chemically resemble ACH (Figure 21.6) and thus bind tightly to acetylcholinesterase, the enzyme that normally breaks down ACH after the signal has been sent. When this happens, excess levels of ACH accumulate and overstimulate the neurons, causing convulsions, paralysis, and death. Neurotoxins that do this include nerve gases, some mushroom toxins, and organophosphate and carbamate insecticides.

Still other toxins keep ACH from binding to the dendrite of the receiving nerve cell; this prevents the nerve signal from being received and

Figure 21.5 Effects of three different kinds of neurotoxins on the transmission of nerve signals between two neurons.

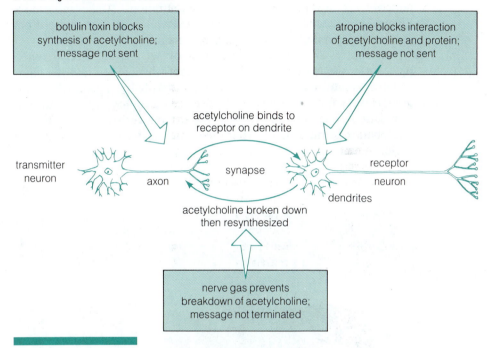

Figure 21.6 Acetylcholine (ACH) and toxic substances that block its action as a neurotransmitter. All have a carbon- or phosphorus-type ester structure (shown in color).

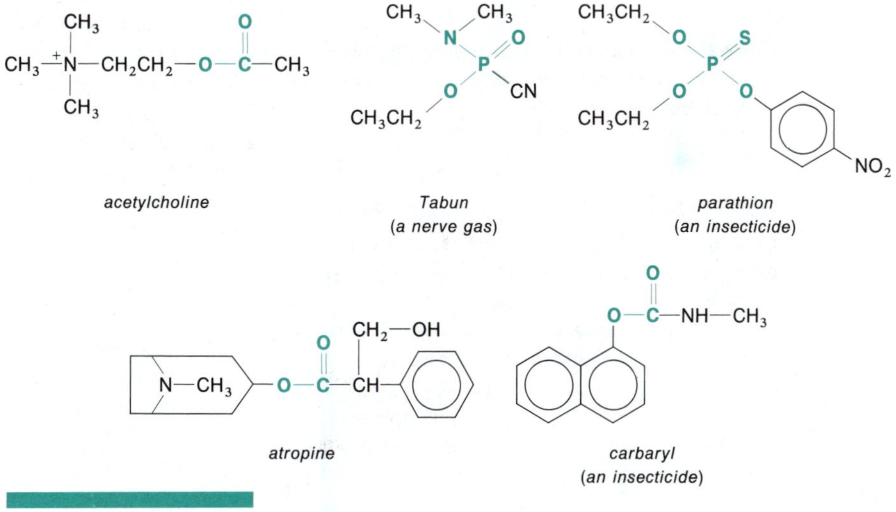

passed on. Cobra venom works in this way, as do the local anesthetics nupercaine, procaine, cocaine, and tetracaine. Curare, a plant extract used in the poisoned arrows of Amazon Indians to paralyze their prey, is a muscle relaxant when used at low concentrations. And atropine (see Figure 21.6), another poison, is used at low concentrations in eye drops to dilate the pupil of the eye during examination.

Atropine is an antidote for poisoning by nerve gases or organophosphate and carbamate insecticides, but it must be injected into the blood within a minute or so after exposure to the poison. By blocking the action of ACH on receptor neurons (see Figure 21.5), atropine prevents the over-stimulation of neurons caused by nerve gas and carbamate and organophosphate pesticides. Once again, the antidote to a poison is another poison.

C H E M I S T R Y   S P O T L I G H T   PUFFER FISH AND ZOMBIES

Figure 21.7 The puffer fish (fugu) contains the poison tetradotoxin. (Courtesy of B. W. Halstead, World Life Research Institute)

Puffer fish (Figure 21.7) contain a deadly neurotoxin called tetrodotoxin, which blocks the transmission of nerve impulses by preventing sodium (Na^+) ions from entering neurons. In Japan only chefs with special training are licensed to prepare and serve this delicacy, which kills about 100 of its customers each year. About six hours after ingesting the poison, a victim either dies of respiratory failure or recovers completely. A few people have been pronounced dead, placed in coffins, and then revived.

According to Haitian folklore, a sorcerer can turn a person into a zombie, someone who exists in a twilight state between life and death. One scientist recently reported that victims are injected with a "zombie powder" containing tetrodotoxin. According to folk belief, the person is then taken out of the ground after a brief burial, fed a paste containing drugs that induce delirium, and taken away to become a slave. Other scientists, however, contend that the zombie powder contains too little tetrodotoxin to be effective.

21.4 Toxic Metals: Why the Mad Hatter Went Mad

Certain metals and metal compounds are well-known poisons. *Arsenic and Old Lace* is a play that features one such chemical killer. The Mad Hatter in *Alice in Wonderland* is based on the strange behavior of people working in felt-hat factories. Mercury nitrate, $Hg(NO_3)_2$, was used to tan leather and felt to prevent fungi from growing. The Mad Hatter chewed the felt to soften it and thus ingested the mercury. Analysis of hair samples of the famous scientist Sir Isaac Newton indicates that he also suffered from exposure to mercury compounds and other toxic metals. And there are suspicions that more than one painter (including van Gogh) suffered neural damage from the lead salts in their paints.

Table 21.4 lists some widely used metals and their toxic effects. Although arsenic is a metalloid (Section 2.6), it is included here because it

TABLE 21.4 SOME WIDELY USED TOXIC METALS

Element	Sources	Health Effects
Serious Threats Now		
Cadmium	Burning coal; mining zinc; water pipes; tobacco smoke; rubber tires; plastics; superphosphate fertilizers; electroplating	Heart and artery disease; high blood pressure; brittle bones; kidney disease; fibrosis of lungs; possibly cancer
Lead	Automobile exhaust (leaded gasoline); paints (made before 1940); lead batteries; shot; solder; pottery glaze	Brain damage; convulsions, behavioral disorders; death
Mercury	Burning coal; electrical batteries; many industrial uses; dental fillings	Nerve damage; death
Nickel	Diesel oil; burning coal; tobacco smoke; steel; gasoline additives	Lung cancer
Beryllium	Burning coal; industrial uses including nuclear power industry and rocket fuel	Acute and chronic respiratory diseases; lung cancer
Potential Hazards if Levels Increase		
Antimony	Industry; typesetting; enamelware	Heart disease; skin disorders
Arsenic	Burning coal and oil; pesticides; mine tailings	Cumulative poison at high levels; possibly cancer
Manganese	Metal alloys; smoke suppressant in power plants; possible future gasoline additive	Nerve damage

SOURCE: G. T. Miller, Jr., *Living in the Environment*, 6th ed., Belmont, CA: Wadsworth, 1990.

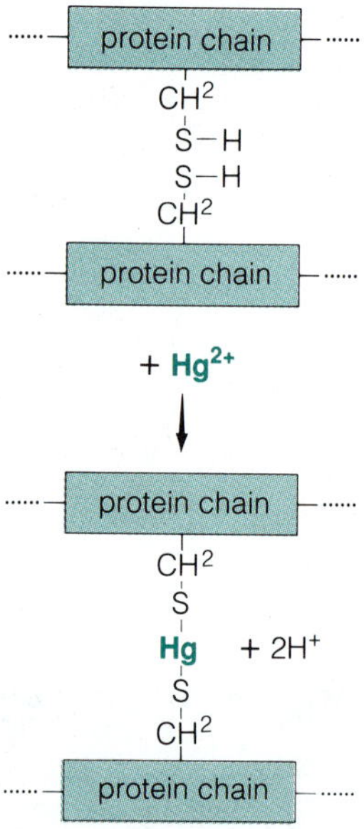

Figure 21.8 How mercury inhibits key enzyme proteins. Other heavy-metal compounds act in a similar way.

often reacts like a metal. Five of the substances listed are known public health hazards, whereas the other three could become such. We have already examined how three of these toxic metals—lead (Section 12.3), cadmium (Section 13.5), and mercury (Section 13.5)—act as pollutants. Now we see how metals work as metabolic poisons.

The difference between a wholesome metal such as zinc and a metabolic poison such as cadmium is often subtle. Several important enzymes in your body contain zinc (Zn^{2+}) ions, which are necessary for certain enzymes to function. Cadmium, which is in the same group on the periodic table as zinc, is enough like zinc to replace it in those enzymes:

$$\text{enzyme–}Zn^{2+} + Cd^{2+} \rightleftharpoons \text{enzyme–}Cd^{2+} + Zn^{2+}$$

But cadmium is also different enough to make the enzymes less effective catalysts; this interferes with the metabolic processes that need those enzymes. Consuming more zinc in the diet (Section 18.2) shifts the reaction above to the left and reduces some of the toxic effects of cadmium.

Other toxic metals—especially arsenic, mercury, and, to a lesser extent, lead—inactivate enzymes and other proteins by binding to —SH groups in the amino acid cysteine (Table 9.3). Figure 21.8 shows how mercury does this. Acting similarly, lead blocks the action of a particular enzyme necessary for making hemoglobin; this is why lead poisoning causes anemia, a shortage of red blood cells.

Attracting metals to —SH groups also is a way to counteract poisoning by metals. During World War I, British scientists searched for an antidote to Lewisite (Figure 21.9), a poisonous arsenic-containing gas that could be used for gas warfare. The most effective compound they discovered contained two —SH groups and was given the name *British Anti-Lewisite* (*BAL*). BAL is a **chelating agent**, which means that it has a clawlike structure that binds metals effectively. The —SH groups in BAL tie up the metal in a BAL–metal complex (Figure 21.9) and thus keep the metal ion from doing damage elsewhere. Now BAL is used in hospitals as an antidote for poisoning by arsenic and by such metals as mercury, lead, nickel, antimony, and manganese.

Figure 21.9 Lewisite, BAL, and a BAL–lead complex.

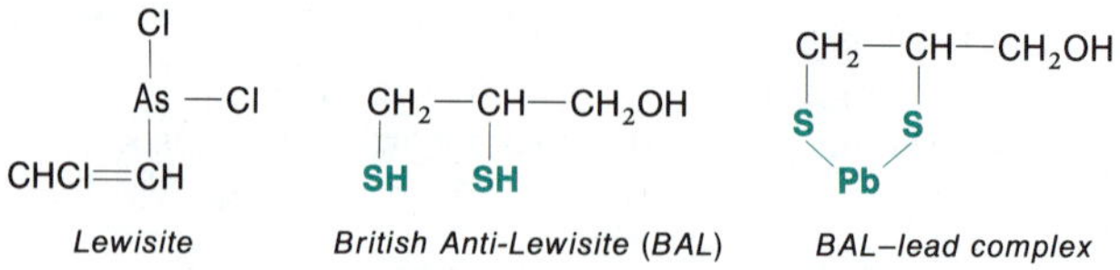

Another chelating agent is *EDTA* (ethylenediaminetetra-acetic acid). The calcium salt of this acid (Figure 21.10) is an antidote for poisoning from lead, beryllium, cadmium, or iron. EDTA is more effective than BAL in binding to these metals. The metal ions displace calcium and form metal–EDTA complexes as follows:

$$Ca^{2+}\text{–EDTA} + Cd^{2+} \longrightarrow Cd^{2+}\text{–EDTA} + Ca^{2+}$$

BAL is often used in combination with EDTA to treat lead poisoning because both chelating agents bind lead effectively.

In an emergency, drinking egg whites and milk, followed by vomiting, helps treat mercury, lead, and other heavy-metal poisons. Egg whites and milk work by providing protein —SH groups to bind the metal ion and prevent it from doing further damage. But most cases of metal poisoning occur because of a lengthy exposure to the toxic material. When antidotes are given after much of the damage already has been done, they have little effect.

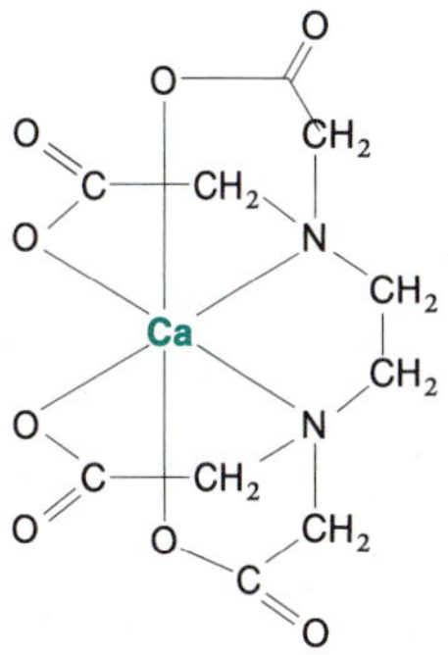

Figure 21.10 The calcium salt of EDTA.

C H E M I S T R Y S P O T L I G H T   **PLATINUM: A USEFUL TOXIC METAL**

In 1961 Barnett Rosenberg, a biophysicist at Michigan State University, placed platinum electrodes in a colony of bacteria to study the effects of an electrical field on cell division. He noticed that after the electrical field was on for awhile, the bacteria developed a long, filamentous shape. They were growing, but they weren't dividing.

What was blocking cell division? It turned out that the culprit wasn't the electrical field. Instead, it was the compound $Pt(NH_3)_2\,Cl_2$ that had formed from the platinum electrodes. The active isomer is now called cisplatin.

Further research showed that cisplatin blocks cell division by cross-linking DNA. Because certain cross-linking agents are useful in treating cancer (Section 19.3), cisplatin was tested as an anticancer drug. It worked. Now it is one of the most effective drugs used for treating ovarian, head and neck, bladder, testicular, prostate, and several other types of cancer.

Many people are alive today because scientists followed the clue left by an unintentional experiment with platinum.

TABLE 21.5 SOME KNOWN OR SUSPECTED CHEMICAL MUTAGENS

Substance	Examples	Common Occurrence
Some epoxides	ethylene oxide propylene oxide	Industrial and automobile emissions into air
Some molds	aflatoxin	Mold on peanuts and other plants
Some mercury compounds	$[CH_3—Hg]^+$ methyl mercury	Industrial (such as chloralkali plants) emissions into water
Some solvents	benzene toluene acetone	Glue, gasoline vapors
Some polycyclic aromatic hydrocarbons	3,4-benzopyrene chrysene	Tobacco and coal smoke
Acridines	quinacrine	Dyes, treating tapeworm
Nitrous acid, nitrites, some nitrosamine compounds	HNO_2, $NaNO_2$, dimethyl nitrosoamine	Preservative in meats, cooked bacon
Some aldehydes	formaldehyde	Mobile homes, clothing industry, glues in plywood and particle board

21.5 Mutagens, Carcinogens, and Teratogens: Causing Genetic Errors, Cancers, and Birth Defects

MUTATIONS Chemical changes in the nitrogen bases in an organism's DNA are **mutations**. Some mutations have little or no effect on an organism. A few turn out to be beneficial, enabling living things to adapt to their environment. Many mutations, however, are detrimental and cause the cell to synthesize a faulty protein. For example, an abnormal protein in hemoglobin causes sickle-cell anemia (Section 9.4). An abnormal, nonfunctional enzyme in the synthesis of the skin pigment melanin is responsible for albinism.

Mutations occasionally occur when a dividing cell makes new DNA. They also arise when chromosomes relocate portions of DNA onto the same or another chromosome. In addition, exposure to certain chemicals or to radiation (including natural background radiation) can cause mutations.

Mutations can occur anywhere in the body. In body cells, they produce little damage if the affected cells die and are replaced by healthy cells. But if the altered DNA passes on to other cells when the affected cell divides, a tumor or other extensive damage may result. Mutations in an egg or sperm cell may cause genetic disorders in the next generation. Sickle-cell anemia, albinism, and many other genetic disorders stem from some original mutation that has been passed on from generation to generation.

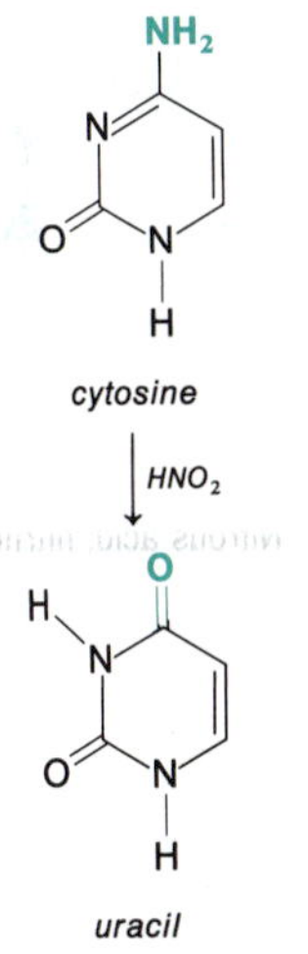

MUTAGENS Chemicals that cause mutations are **mutagens**. Different mutagens (Table 21.5) work in different ways. For example, nitrous acid (HNO_2) and formaldehyde react with adenine, cytosine, and guanine (normal nitrogen bases in DNA; Section 9.5) to change their amino groups into ketone groups. For example, Figure 21.11 shows the conversion of cytosine into a different base, uracil.

One concern with using sodium nitrite ($NaNO_2$) as a preservative and color enhancer in meat (Section 18.4) is that it may react with hydrochloric acid (HCl) in the stomach to form nitrous acid:

$$NaNO_2 + HCl \longrightarrow HNO_2 + NaCl$$

Nitrites also can be metabolized to nitrosamines, another type of mutagen (see Table 21.5). Whether these reactions occur to a significant extent in humans continues to be debated.

Figure 21.11 Nitrous acid changes cytosine into a different base, uracil.

Other mutagens act in other ways. Acridines (see Table 21.5), for example, are flat, aromatic molecules that slip in between adjacent bases in DNA, causing normal bases to be inserted in or removed from the DNA chain. Certain epoxides are mutagens that form when oxygen or ozone (O_3) reacts with unsaturated hydrocarbons:

$$R-CH{=}CH_2 + O_2 \longrightarrow R-CH-CH_2 + O$$
$$\underset{O}{\diagdown\diagup}$$

(*an unsaturated*   (*an epoxide*)   *oxygen*
  hydrocarbon)   *atom*

Epoxides occur in automobile and industrial emissions and in certain pesticides and their decomposition products.

Many chemicals may cause mutations in humans, but the evidence for this is often indirect. Mutagens in other organisms, for example, are prime candidates as mutagens in humans. The most widely used test, originated by Bruce Ames at the University of California at Berkeley, detects chemicals that cause mutations in the bacterium *Salmonella typhimurium* (Figure 21.12).

Some chemicals are suspected mutagens because of statistical data linking genetic disorders with exposure to the chemical. But this analysis is extremely difficult. Specific genetic disorders may occur only once in 10,000 or 100,000 births, so getting meaningful statistical data requires a controlled study of a large population. Another problem is in tracing a disorder to a specific chemical from among the tens of thousands that any human encounters.

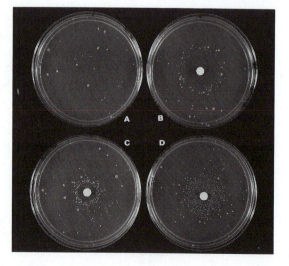

Figure 21.12 Plates containing bacteria in the absence (A) or presence (B–D) of a mutagen. In the Ames test, mutagens reverse a genetic error in the bacteria and thus enable them to grow around the disc (in the center of plates B–D) that contains a mutagen. (Dr. Bruce N. Ames, from *Mutation Research* 31: 347–364 (1975) by Ames, B. N., McCann, J., and Yamasaki, E.)

CARCINOGENS   Cancer is a collection of more than 100 diseases, all characterized by abnormal, uncontrolled cell growth (Section 19.3). The malignancy can spread to nearby tissues and disperse to other parts of the body through the blood or the lymph system.

Various forms of cancer have different causes. Skin cancer can result from repeated exposure to the sun's ultraviolet rays or to gamma radiation. Some types of cancer may be related to the action of viruses or of specific genes. Cancers also can be caused by chemicals known as **carcinogens**. Table 21.6 shows a few known or suspected carcinogens. According to several estimates, 60 to 90 percent of all cancers are caused by environmental factors such as carcinogens (Figure 21.13).

Scientists know little about how to determine which chemicals cause cancer in humans. The most common screening method, the Ames test, identifies whether chemicals are mutagens in bacteria; about 80 percent of the mutagens identified by this test are carcinogens. But our bodies convert some substances into carcinogens. Bacteria don't carry out this metabolism, so the Ames test may not identify those hazardous substances. Suspected carcinogens also are tested by experiments with lower animals, especially mice and rats, but those findings don't always apply to humans.

Chemicals are often chosen for testing because of statistical data linking them with certain types of cancer. Such a correlation doesn't prove that the chemical causes the cancer, but it does suggest some kind of connection. Statistical data, for example, established the relationship between smoking and lung cancer. We know now that cigarette smoke contains cancer-causing compounds such as 3,4-benzopyrene (see Table 21.6), although benzopyrene itself is not the active agent; it is oxidized in the liver to form other products that are carcinogens.

Some compounds that are nontoxic in a large, single dose may cause cancer with repeated exposure to lower concentrations over many years. These chemicals are not immediately recognized as carcinogens, so people continue to be exposed to them. For example, it took thirty years for bladder cancer to appear in people who worked in the dye industry; finally, it was discovered that 2-naphthylamine (see Table 21.6) and certain other dye intermediates are carcinogens. Similarly, factory workers were exposed to vinyl chloride, which is used to make polyvinyl chloride (PVC) (Section 14.2), for about ten years before some of them developed a rare and fatal form of liver cancer.

Cancer develops in three stages: initiation, promotion, and progression. *Initiators*—ionizing radiation, tumor viruses, and chemicals such as cyclophosphamide—require only a brief exposure and apparently work by causing mutations. If this is followed by repeated exposure to a second chemical known as a *promoter*, cancer can develop. Promoter agents include asbestos, cigarette smoke, and saccharin. Exposure to an initiator

Figure 21.13 Skin cancer developed on this mouse after its skin was painted with pollutants extracted from urban air. (Environmental Protection Agency)

TABLE 21.6 SOME KNOWN OR SUSPECTED CARCINOGENS

Substance	Use or Source	Formula
Aflatoxin	Mold on peanuts and other plants	
Asbestos	Brake linings, pipes, insulation, fireproofing	$3MgO \cdot 2SiO_2 \cdot 2H_2O$ *(one type)*
Benzidine	Manufacture of dyes, synthetic rubber, and some polymers	NH_2—⬡—⬡—NH_2
3,4-Benzopyrene	Cigarette and coal smoke	
Carbon tetrachloride	Dry-cleaning agent, solvent	CCl_4
Dioxane	Solvent in chemical industry, cosmetics, glues, deodorants	
Mustard gas	War gas	CH_2Cl—CH_2—S—CH_2—CH_2Cl
2-Naphthylamine	Optical brightener in bleach, dye intermediate	
Sodium saccharin	Diet drinks	
Urethane	Beer, sake	CH_3CH_2—O—$C(=O)$—NH_2
Vinyl chloride	Monomer for polyvinyl chloride (PVC)	$CH_2{=}CHCl$

alone or a promoter alone would not lead to the *progression* stage, in which the cancer develops and spreads. This may explain why some people who are exposed to low doses of cigarette smoke for many years don't develop cancer.

Scientists still don't know exactly how cancer develops. But recently they have identified more than fifty different cancer-causing genes, called **oncogenes**. Some oncogenes appear to be mutated forms of genes that normally code for cell growth and development; the mutated genes, however, cause uncontrolled cell growth. Other oncogenes normally code for growth-control substances; once mutated, they allow uncontrolled growth. Despite these advances, however, it may not prove possible for a single, comprehensive theory to explain how all cancers develop.

TERATOGENS   Chemicals that cause birth defects are called **teratogens**. Such defects, unlike genetic disorders, are not inherited by later generations. Teratogens interfere with normal embryonic and fetal development, causing organs or other body parts to develop abnormally. The most sensitive time of exposure seems to be the first two months of pregnancy, when organ systems are forming.

Figure 21.14 shows some chemicals that are known or suspected teratogens. The best known is *thalidomide*, which in 1961 was used widely (especially in West Germany and Great Britain) as a tranquilizer and to treat nausea during pregnancy. More than 5000 women in those countries who used the drug during pregnancy gave birth to babies with underdeveloped or absent arms and legs, plus various other defects (Figure 21.15). Thalidomide also caused the death of many other fetuses before they were born. Animal tests in Europe had shown thalidomide to be safe, but the FDA did not approve the use of the drug in the United States. As a result, the birth defects in the United States were limited to about twenty babies whose mothers had used thalidomide in Europe.

A dioxin called TCDD (see Figure 21.14) is a contaminant in Agent Orange, an herbicide used as a defoliant by U.S. military forces in Vietnam. Tests with laboratory animals indicate that TCDD is a carcinogen and a teratogen. Veterans from the Vietnam War attribute these

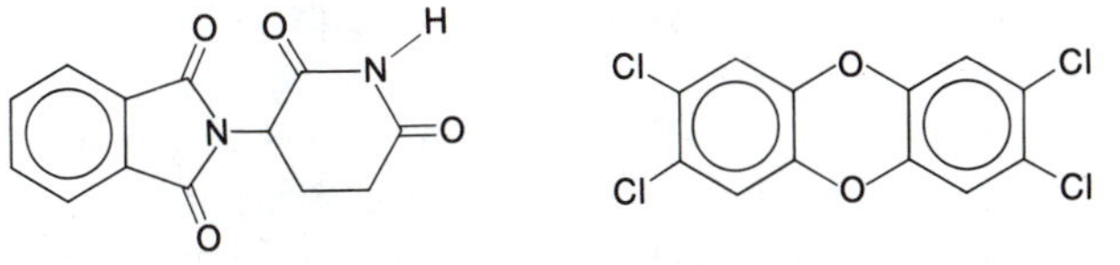

thalidomide

2,3,7,8-tetrachlorodibenzo-paradioxin (TCDD)

CH_3CH_2OH $CH_2{=}CHCl$ $CHCl{=}CCl_2$
ethanol   vinyl chloride   trichloroethylene

Figure 21.14 Some known or suspected teratogens.

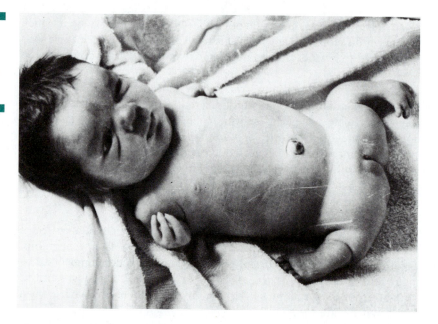

Figure 21.15 A baby whose mother took thalidomide during pregnancy. (Courtesy Photo Researchers)

and other health problems to their exposure to Agent Orange. But TCDD hasn't yet been established as a carcinogen or teratogen in humans, and more studies are in progress.

In the last decade, ethyl alcohol has become recognized as a teratogen. Women who drink large amounts of alcoholic beverages during pregnancy—especially in the first few months—increase their risk of having babies with mild facial and other structural abnormalities and below-normal mental abilities. This condition is called *fetal alcohol syndrome.*

Polychlorinated biphenyls (PCBs; see Figure 14.14) and their breakdown products, polychlorinated dibenzofurans (PCDFs), are also teratogens. Taiwanese mothers who ate food cooked in oil contaminated with these compounds gave birth to children with low weight and slow growth and mental development. Being nonpolar, PCBs and PCDFs remain stored in fatty tissues for long periods. Indeed, some of the affected children were born several years after their mothers had eaten the contaminated food.

Certain industrial metals may be teratogens, but most of the data come from experiments with animals. The Minimata Bay disaster (Section 13.5) demonstrated that methyl mercury is a teratogen in humans that damages the central nervous system. Lead compounds can do similar damage, at least in animals. Certain cadmium, arsenic, nickel, and chromium compounds are also potential teratogens.

Many other teratogens may as yet remain undiscovered. So it is important for pregnant women to avoid, as much as possible, drugs and other substances that could possibly harm the young lives they are carrying.

DAVID P. RALL

There's no question, we'll have to accept some degree of chemical risk. But let's do it in the most knowledgeable way possible—not for chemicals that have no particular utility. Let's save the risk for a critical food preservative, a drug, or effluent from a critical power source.

Summary

The toxicity of a substance is often measured by its LD_{50} value, the dose that kills 50 percent of the test population. Corrosive poisons are acids, bases, or oxidizing agents that damage tissue on contact. Metabolic poisons interfere with a normal chemical process in the body, usually by disrupting the action of an enzyme or other protein.

Neurotoxins disrupt the transmission of nerve impulses, often by interfering with the action of neurotransmitters. Toxic metals typically disable proteins by binding to their —SH groups.

Chemicals that alter the structure of DNA are mutagens, and many mutagens are carcinogens. Cancers are thought to develop in three stages—initiation, promotion, and progression. Substances that cause nongenetic birth defects are teratogens.

Terms for Review

After completing this chapter, you should know and understand the meaning of the following terms:

carcinogen (p. 593)

chelating agent (p. 588)

corrosive poison (p. 577)

LD_{50} (p. 576)

metabolic poison (p. 579)

mutagen (p. 591)

mutation (p. 591)

neurotoxin (p. 584)

oncogene (p. 595)

poison (p. 576)

teratogen (p. 595)

Questions

Odd-numbered questions are answered at the back of this book.

1. What is LD_{50}, and how is it determined?

2. The LD_{50} values (in mg/kg body weight) for rats for DDT, dieldrin, and parathion are 120, 50, and 5, respectively. On this basis, which insecticide is the most toxic to rats?

3. How does activated charcoal (carbon) work as an antidote to many poisons?

4. For which of the following materials, if swallowed, should the victim *not* take syrup of ipecac? (a) Lead compounds, (b) hydrochloric acid (HCl), (c) an organophosphorus insecticide, (d) antifreeze, (e) bleach, (f) benzene, (g) lye (NaOH), and (h) arsenic compounds.

5. List some common household products that are corrosive poisons.

6. Explain the mechanism of poisoning for each of the following: (a) potassium cyanide, (b) toxic mushrooms, (c) sodium hydroxide, (d) organophosphorus pesticides, (e) methyl mercury, and (f) carbon monoxide.

7. Even in highly polluted air, there is much more oxygen than carbon monoxide. How is it possible for 50 percent or more of a person's hemoglobin to be bound to CO instead of O_2?

8. Breathing oxygen-rich air is an effective way to treat carbon monoxide poisoning. Why is this *not* an effective treatment for cyanide poisoning?

9. Victims of fluoride poisoning are treated by drinking solutions containing calcium ions. Considering how calcium ions work in this situation, suggest a reason why drinking a solution of table salt (NaCl) would not be effective.

10. How does ethanol work as an antidote for methanol poisoning? Would methanol be a suitable antidote for ethanol toxicity? Why?

11. Would a chelating agent such as BAL or EDTA work as an antidote to nitrite poisoning? Explain.

12. A nerve gas called Agent VX has the structure below. Compare its structure with those shown in Figure 21.6 and suggest how it works as a nerve gas.

$$CH_3 \quad CH_3$$
$$CH$$
$$CH_3$$
$$CH-N-CH_2CH_2-S-P-O-CH_2CH_3$$
$$CH_3 \qquad\qquad\qquad O$$
$$CH_3$$

13. Identify the major sources of poisoning by the following metal or metalloid compounds: (a) cadmium, (b) lead, and (c) arsenic.

14. Why is cesium (Cs) less likely than cadmium (Cd) to disrupt the action of enzymes containing zinc (Zn)? [HINT: Look at the periodic table.]

15. Which one of the following would be the most likely to bind metal ions and function as a chelating agent?

a.
$$
\begin{array}{c}
O \\
\parallel \\
C-O^- \\
\mid \\
C-O^- \\
\parallel \\
O
\end{array}
$$

oxalate

b.

benzene

c. N_2

nitrogen

d.
$$
CH_3-\overset{\overset{\displaystyle O}{\parallel}}{C}-O-CH_2CH_2-\overset{\overset{\displaystyle CH_3}{\mid}}{\underset{\underset{\displaystyle CH_3}{\mid}}{N^+}}-CH_3
$$

acetylcholine

16. What is the Ames test? What is the relationship between a mutagen and a carcinogen?

17. Describe the three stages of cancer development.

18. What carcinogen listed in Table 21.6 is used (legally) as a food additive?

19. Do mutations necessarily cause birth defects? Why?

20. Both mutagens and teratogens can produce birth defects, but they do so in different ways. Explain.

Topics for Discussion

1. Are mutations always harmful? Are they ever neutral or helpful? Explain.

2. The tobacco industry maintains that it has not been proved that smoking causes cancer. What are the arguments for and against this statement? Should smoking be banned by law in all enclosed public places? Why?

3. Does your community fluoridate its water supplies? Do you favor or oppose this practice? Why?

4. Suppose a drug that relieves pain has been shown to be a carcinogen in rats. Would you favor banning it, restricting its use, or allowing free usage but with a warning label? Why?

5. Suppose the drug described in the previous question were a teratogen instead of a carcinogen. Would your answer regarding usage be any different? Why?

CHAPTER 22

Better Bodies Through Chemistry: Changing Genes, Body Parts, and Athletic Performance

General Objectives

1. How can we change the genetic instructions in DNA so that cells produce useful proteins such as insulin, growth hormone, and vaccines?

2. How can we modify genes to cure people with genetic diseases?

3. How can we use chemical knowledge to provide artificial parts for the body?

4. How can we use chemical knowledge to improve athletic performance?

Living things are special. No scientist yet has mixed chemicals together and produced something alive. But scientists who believe it might become possible to do this someday point to one key reason: Living things follow the same laws of chemistry and physics as everything else. They are held together by the same types of chemical bonds as simpler materials, and their actions are governed by the same laws of energy that apply to automobiles and hydroelectric power plants.

We know that a chemical such as insulin has the same structure and properties whether it is made in a cell or in a laboratory. In fact, we will see that it's even possible (though difficult) to chemically synthesize something as complex as a gene and have it work normally in a cell. We have also learned how to make artifical body parts from materials such as metals and plastics. And we can use our knowledge of biochemistry and physiology to improve athletic performance.

Indeed, we can understand a great deal about living things by their chemistry. And the more we understand these things, the more we can chemically change the way organisms—including people—look and act.

22.1 Genetic Engineering: Redesigning Organisms

SPLICING IN NEW GENES Our genes specify what proteins we can make, and proteins provide all our hereditary features such as our height, the color of our eyes and skin, and whether we have a genetic disorder such as albinism or sickle-cell anemia. A **gene** is a section of DNA in a chromosome that carries the information for making one protein. **Chromosomes** are huge molecules of DNA (with attached proteins) in the nuclei of our cells. We have forty-six chromosomes (twenty-three pairs) in our cells and

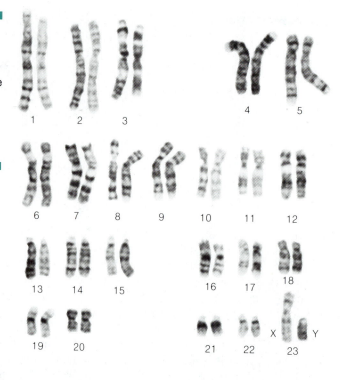

Figure 22.1 Normal chromosome pattern for a male, indicated by one X and one Y chromosome. Females have two X (and no Y) chromosomes. (A. D. Stock, City of Hope National Medical Center)

Figure 22.2 This tobacco plant, redesigned to contain a firefly gene, glows in the dark. (Courtesy of Keith V. Wood, Ph.D., *Science Vol. 234*, pages 856–859, 14 November 1986, by Dr. David W. Ow. Copyright 1986 by the AARS.)

50,000 to 100,000 different genes, so each chromosome contains many genes (Figure 22.1).

Genetic engineering is a way to change the genetic instructions in a cell so it will make different proteins. Scientists have discovered ways to insert genes into cells and produce new combinations that don't occur in nature. Any combination is possible: animal genes in plants (Figure 22.2), plant genes in bacteria, and animal (including human) genes in bacteria, plants, or other animals. DNA containing material from two or more different sources is called **recombinant DNA**.

Gene splicing became possible in the early 1970s because of two developments. First came the discovery of a material (or *vector*, as it is now called) to carry new genes into the cells and take up permanent residence there. The usual vectors for bacterial cells are **plasmids**, which are small, freely floating ringlets of DNA inside most cells that often carry genetic information such as resistance to antibiotics (Figure 22.3). Scientists discovered simple treatments to isolate plasmids from cells and to insert them (carrying new genes in the form of recombinant DNA) back into cells. The other vector, especially for animal cells, was chemically modified viruses that insert DNA into cells when they infect them.

The second development was finding a way to splice new genes into the vectors. Scientists discovered **restriction enzymes** that cut double-

Figure 22.3 An *E. coli* cell that has ruptured, releasing a portion of its main DNA and several plasmids (arrows). (Courtesy Dr. Huntington Parker and Dr. David Dressler, Harvard Medical School)

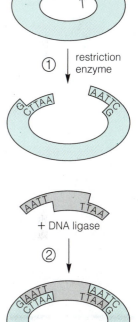

stranded DNA at very specific places and leave the DNA with frayed ends containing a short segment of just one of the strands (Figure 22.4); these are sometimes called "sticky ends." Restriction enzymes typically cut DNA at places where there is a specific palindrome base sequence. A *palindrome* is an arrangement of letters that reads the same forward or backward, such as OTTO and ROTOR. Notice in Figure 22.4 the palindrome sequence of bases on the two strands where DNA is cut.

The idea then is to cut the plasmid (or other vector) with a restriction enzyme and mix it with new genes containing the same kind of sticky ends. The ends naturally join together because of the natural pairing between adenine (A) and thymine (T) and between cytosine (C) and guanine (G) (Section 9.5). After another enzyme (DNA ligase) seals together the DNA pieces, the hybrid plasmid (see Figure 22.4) can carry its new gene(s) into a cell and direct the synthesis of new proteins. And the original source of the genes doesn't matter. Bacteria, for example, accept plant, animal, or other bacterial genes and make the proteins they prescribe.

Obtaining genes to splice into vectors, however, can be a problem. One solution is to chemically synthesize the desired gene. Scientists have synthesized a number of animal (including human), plant, and bacterial genes; spliced them into plasmids; put them into bacteria; and found that the genes worked in their new homes.

It isn't even necessary to know the structure of the natural gene in order to synthesize one that works. Scientists know the entire genetic code, the pattern by which a sequence of nucleotides in DNA specifies a particular sequence of amino acids in a protein (Section 9.5). So if they know the amino acid sequence of the desired protein, they can use the genetic code to devise a nucleotide sequence (gene) that codes for that protein.

Figure 22.4 (1) Restriction enzyme cuts plasmid DNA at a specific palindrome region to leave sticky ends. (2) When genes with similar ends and DNA ligase are added, a hybrid plasmid forms containing the new genes.

USING RECOMBINANT DNA TO MAKE INSULIN   Genetic engineering is a way to make cells into miniature factories to manufacture proteins. The U.S. Supreme Court has ruled that new forms of life made by genetic engineering can be patented. So gene splicing is big business, and now hundreds of companies are genetically redesigning bacteria, yeasts, and other cells to make products they can sell.

The first such product was human insulin. Insulin is a small protein consisting of two chains of amino acids joined together by disulfide bonds. Human, beef, and pork insulin differ slightly in their amino acid sequences (Figure 22.5). Until scientists discovered how to use gene splicing to make human insulin, diabetics had to use beef or pork insulin prepared from the pancreases of slaughtered animals. It takes about 180 kg (400 lb) of beef or pork pancreases to produce just 30 g (1 oz) of insulin.

Now diabetics can take human insulin made by genetically altered bacteria. Scientists chemically synthesized the genes for both chains of human insulin, spliced them into plasmids, and put the plasmids into bacteria. Then they isolated the individual protein chains from the bacteria and joined (oxidized) them together by disulfide bonds (—S—S—) to produce the complete insulin molecule (Figure 22.6). Diabetics taking human insulin may have fewer allergic reactions than with beef or pork insulin.

Figure 22.5 The structure of human insulin. The A chain, with twenty-one amino acids (abbreviated by three-letter symbols), and the B chain, with thirty amino acids, are joined together by two disulfide (—S—S—) bonds (in color). Pork insulin is the same except for having a different amino acid at the end of the B chain (shaded). Beef insulin has the same difference plus different amino acids at two other positions in the A chain (shaded).

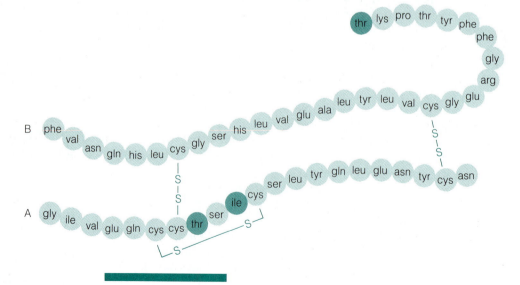

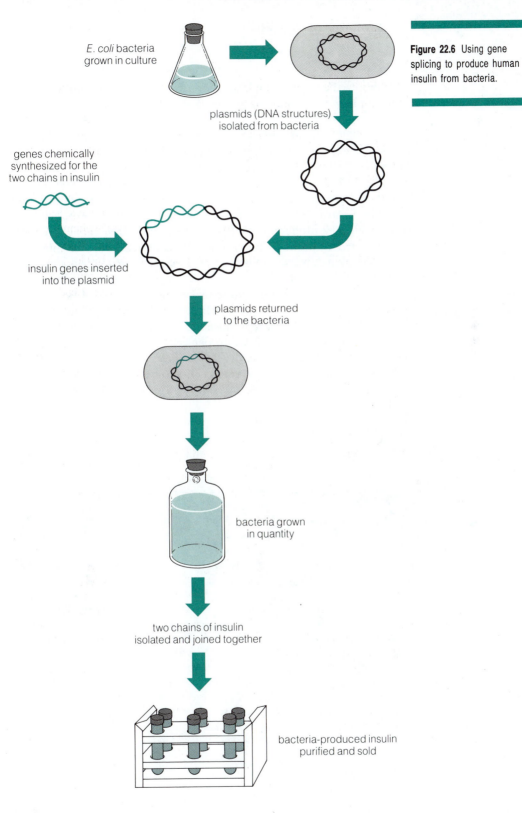

E. coli bacteria
grown in culture

plasmids (DNA structures)
isolated from bacteria

Figure 22.6 Using gene splicing to produce human insulin from bacteria.

genes chemically
synthesized for the
two chains in insulin

insulin genes inserted
into the plasmid

plasmids returned
to the bacteria

bacteria grown
in quantity

two chains of insulin
isolated and joined together

bacteria-produced insulin
purified and sold

In addition, the supply should be more reliable because it no longer depends on how many animals are being slaughtered.

OTHER PRODUCTS OF RECOMBINANT DNA TECHNOLOGY   Many other products are on the way. Genetically engineered bacteria now make human growth hormone (HGH) for treating children who are dwarfed because they produce too little HGH. Previously, HGH was very expensive and difficult to obtain from the brains of human cadavers. Bovine growth hormone will soon be available, and injections into dairy cattle is expected to increase their milk production by 10 to 30 percent. Recombinant DNA also is being used to produce larger amounts of proteins such as interferon and interleukin-2, which are being tested for the treatment of several diseases including cancer.

Another product, human tissue plasminogen activator (tPA), is valuable in treating patients during hearts attacks because it dissolves blood clots. Re-engineered bacteria produce this protein. But scientists have also incorporated the gene for tPA into one-cell mouse embryos; those cells, implanted in a female mouse, developed into mice that produce tPA in their milk. If this can be done with cows or goats, the milk from a herd of several dozen animals may be enough to provide the world's supply of tPA.

Vaccines are another important product. When disease-causing bacteria or viruses infect us, our immune system recognizes certain proteins on the surface of the infectious agent and produces antibodies to destroy

---

C H E M I S T R Y   A NEW WAY TO
S P O T L I G H T   TELL WHODUNIT

Restriction enzymes cut DNA only where specific base sequences occur. If several different restriction enzymes cut a person's DNA, they produce DNA fragments that can be separated and visualized. The fragment pattern resembles a bar code on supermarket products. Like a fingerprint, the pattern is unique for each person.

This new method of chemical sleuthing works on DNA from such materials as blood, hair roots, and semen. Tiny samples, as little as a single hair, are enough to do the analysis. DNA "fingerprinting" already has been used to convict a person of murder and rape; in another case, it showed that a person accused of murder was

innocent. And it settled a dispute over who a boy's parents were, enabling him to rejoin his true family in another country. Many more applications are on the way.

it. But now bacteria given the right genes can manufacture those surface proteins that the immune system recognizes; this protein then works as the vaccine. Using proteins should be safer than the current practice of exposing the immune system to a weakened form of the infectious agent. Already in use are protein vaccines for one type of hepatitis and for hoof-and-mouth disease, a viral disease that afflicts livestock.

Genetic engineering also is a way to redesign crops and other plants to make them resistant to pests, viruses, and herbicides. For example, plants given genes to make them resistant to an herbicide can grow in fields where weeds are controlled by spraying that herbicide.

Some people are concerned that redesigned microbes, once released into the environment, could be harmful; they might carry diseases or in some other way upset the normal balance of nature. But the experience of the last decade, plus safety measures that have been implemented, have reduced these dangers. A greater danger might stem from the intentional, rather than accidental, use of these techniques. For example, genetic engineering could be used to produce highly dangerous and lethal bacteria for biological warfare.

22.2 Genetic Therapy: Treating Genetic Diseases

REDESIGNING HUMANS: CAN WE DO IT?   If we can redesign yeasts and bacteria, we also should be able to redesign plants and animals, including humans. Perhaps we could provide the genes to modify a person's height, hair and eye color, and intelligence. We might even imagine giving people new genes to make them superathletes or geniuses.

But there are practical and ethical limits to what we could and should do. Reputable scientists regard most of these prospects as fantasies. It is one thing to understand the chemical blueprint in human genes; it is quite another to translate this into full-blown individual characteristics.

We couldn't, for example, re-engineer many of our personal characteristics, which depend on our environment as well as our genes. And qualities that do have a genetic component—such as height and intelligence—depend both on the action of genes and on environmental factors. So we couldn't provide these qualities just by supplying the right genes. Using genetic engineering to correct certain genetic disorders, however, is a more likely possibility.

TREATING GENETIC DISORDERS BY GENETIC ENGINEERING   More than 2000 disorders occur, at least in part, because of faulty genes (Table 22.1). The most likely candidates for genetic therapy will be people who have disorders caused by the lack of one normal gene. Called recessive disorders,

TABLE 22.1 TYPES OF GENETIC DISORDERS AND SOME EXAMPLES	
Type of Disorder	**Examples**
Chromosome disorders (abnormal amounts of chromosome material)	Down's syndrome Klinefelters syndrome Turner's syndrome
Recessive disorders*	Cystic fibrosis Albinism Sickle-cell anemia Phenylketonuria (PKU) Tay–Sachs disease β-Thalassemia Galactosemia
Recessive disorders carried on the X chromosome*	Hemophilia Muscular dystrophy Color blindness
Dominant disorders	Huntington's chorea Marfan syndrome
Polygenic disorders	Club feet Cleft palate Spina bifida

* Recessive disorders are more likely to respond to gene therapy.

Figure 22.7 Although his mother has dark skin, the boy (right) is an albino because he lacks a gene needed for the production of pigment. (Courtesy Lynn McLaren/Photo Researchers)

they include phenylketonuria (PKU), albinism (Figure 22.7), sickle-cell anemia, cystic fibrosis, hemophilia, muscular dystrophy, and many other rare diseases.

In these diseases, the gene coding for a critical protein has an error in its nucleotide sequence. This change, called a **mutation**, keeps the cell from making the normal protein. People with sickle-cell anemia, for example, make an abnormal version of a protein that is part of the red blood cell; this gives the red blood cell its unusual sickle shape (Figure 9.13) and its tendency to block small blood vessels. And people with albinism genetically cannot make an enzyme that helps produce melanin, the normal pigment in skin.

One way to cure these genetic errors would be to correct the nucleotide sequence in the faulty genes. But this is impractical; there is no reasonable way to isolate, in each cell, the 1 gene out of 100,000 or so that needs to be fixed and then fix it. A more promising idea is to use recombinant

DNA to supply a normal version of the gene to some of the cells in the body that need it most. Diabetics, for example, would need pancreas cells with insulin genes, and people with sickle-cell anemia would need re-engineered bone marrow cells, which make hemoglobin.

But supplying normal genes to the cells isn't enough. Those genes have to work normally in the cell if they are to do their new owners much good; they need to produce just the right amount of protein product, neither too little nor too much. For example, if diabetics receiving insulin genes produced too much insulin, they would have very low blood sugar levels and would go into a coma (insulin shock) because their brains didn't have enough glucose to work properly.

The first attempt to provide genetic therapy was in 1980. Scientists attempted to treat two young women who had β-thalassemia, a disease in which the bone marrow cells don't make enough normal hemoglobin. The needed gene was transferred into bone marrow cells, and the cells were implanted in the young women. But the experiment failed; the new genes didn't work, and the young women did not improve. Nevertheless, animal studies now are in progress for several genetic diseases. Recently, for example, scientists were able to correct β-thalassemia in mice by gene therapy. The next attempt to correct a human genetic disorder—one causing a deficient immune system—began in 1990.

This type of genetic therapy, however, has an important drawback: People receiving the genes in certain body cells would still have reproductive cells (sperm or ova) that could pass the genetic disorder on to their children, who in turn would need genetic therapy. Perhaps it would be better to re-engineer all cells, including reproductive cells, with the needed genes. But this would be difficult unless the genetic engineering were done very early in life, perhaps even during embryonic development.

Would that be possible to do? It already has been done—in mice. Scientists have spliced a rat gene for growth hormone into mouse DNA, and they injected the DNA into fertilized mouse eggs. A few re-engineered embryos developed into baby mice and produced the rat growth hormone. This extra hormone made them an average of 50 percent larger than normal (Figure 22.8).

Figure 22.8 The mouse on the left has a gene for rat growth hormone and is nearly twice as large as its normal-size littermate, who doesn't have the rat gene. Both mice are ten weeks old. (Courtesy Dr. R. L. Brinster)

22.3 New Body Parts: Trading In Before It's Too Late

THE PARTIALLY REPLACEABLE PERSON Chemists and physicists have developed many artificial body parts (Figure 22.9). These new devices are less vulnerable to rejection, infection, and degenerative diseases, but they also

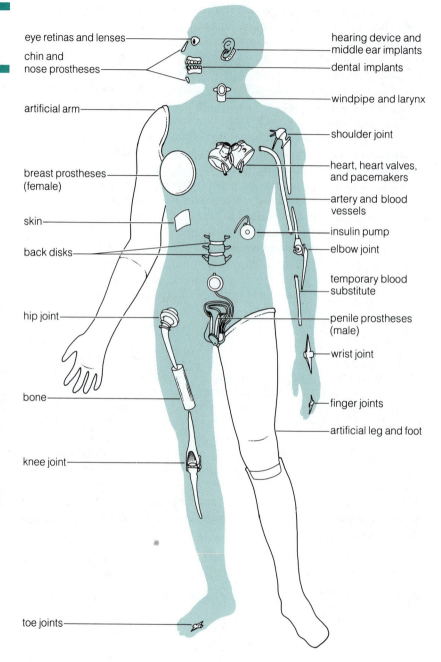

Figure 22.9 Some artificial parts used in humans.

eye retinas and lenses

chin and nose prostheses

artificial arm

breast prostheses (female)

skin

back disks

hip joint

bone

knee joint

toe joints

hearing device and middle ear implants

dental implants

windpipe and larynx

shoulder joint

heart, heart valves, and pacemakers

artery and blood vessels

insulin pump

elbow joint

temporary blood substitute

penile prostheses (male)

wrist joint

finger joints

artificial leg and foot

have drawbacks. They are usually less versatile than the original part, less adaptable to growth and stress, incapable of self-repair, and cannot use the body's own chemical energy system.

A major problem is finding materials that work as well as the original materials. Bone, for example, has remarkable strength compared to its mass; few materials can match this combination. But artificial thigh and other bones now are being made from lightweight alloys containing titanium and other metals. Artificial parts also can cause blood clots or calcium deposits to form, or they may not connect properly to the natural body tissues. One answer is to coat porous, high-density polyethylene (Section 14.2) onto joint prostheses because it allows natural bone and soft tissues to attach and grow.

ARTIFICIAL JOINTS AND EYE, EAR, AND LARYNX IMPLANTS Replacing worn-out joints is now a reality. Artificial hips, for example, are a ball-and-socket joint (Figure 22.10). The artificial unit is a cup made of a special kind of polyethylene reinforced with carbon fibers; the polished ball fitting into the cup is made of stainless steel or a titanium or chromium–cobalt alloy. Once sealed into place with an acrylic bone cement, the joint performs many of the motions of the natural hip. In the United States, more than 100,000 people a year—especially older people—have an artificial hip installed. Replacement joints also are coming of age for the wrists, fingers, elbow, shoulder, and knee.

Artificial parts also can help restore our ability to see, hear, and speak. Polymethyl methacrylate (Plexiglas, Lucite; Section 14.2) is very clear and is often implanted as an artificial lens in the eye following cataract surgery. Middle-ear implants made of Teflon, stainless steel, or platinum can restore hearing to people whose natural bone structure there has become too spongy. And people whose larynx was surgically removed, usually for cancer, can speak again with the help of a prosthetic valve implanted in the throat.

PLASTIC SURGERY IMPLANTS, BLOOD VESSELS, AND ARTIFICIAL BLOOD Silicone rubber, which has the structure shown below,

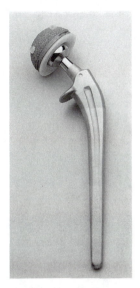

Figure 22.10 An artificial hip joint. The ball-and-cup surfaces are polished to a mirror finish to minimize friction during movement. (Courtesy Zimmer Corporation)

can be made in varying degrees of softness and is easily molded. Plastic surgeons frequently implant this material to improve the appearance of the ears, chin, nose, cranium, breasts, or other body parts.

More than 100,000 artery segments of texturized Dacron or Teflon have been put in people with damaged blood vessels. Thousands of people have received silicone rubber tubes that pass from the brain cavity, through the skull, and into the heart or abdominal area; these tubes drain excess fluid into the blood to relieve the pressure of hydrocephalus ("water on the brain"). And tens of thousands of people have synthetic heart valves containing such materials as titanium, Dacron, Teflon, or a special form of carbon.

One kind of artificial blood is a mixture containing the clear fluorocarbon liquid perfluorodecalin (trade name Fluosol), which has the structure

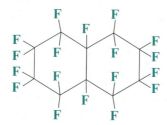

perfluorodecalin (Fluosol)

This material dissolves very large amounts of oxygen (Figure 22.11) and is not metabolized, though the body slowly excretes it. It is used as artificial blood in emergencies when the needed blood or blood type isn't available. Another advantage is that it avoids the risk of transmitting infectious diseases such as AIDS. Hundreds of people have survived because of this temporary blood substitute.

ARTIFICIAL KIDNEYS AND LIVERS   People with failing kidneys can use a kidney dialysis machine to cleanse their blood while they await a kidney transplant. Their blood circulates through cellophanelike membranes that remove toxic waste materials. Now portable, 9-lb units can be strapped to the chest and used for short periods (Figure 22.12).

People with liver failure may have toxic materials in their blood that dialysis does not remove. One solution is to use charcoal, whose ability to absorb impurities onto its surface has found many uses including cigarette filters and gas masks. Physicians have detoxified patients with severe liver disease by circulating their blood through a chamber containing granules of plastic-coated charcoal. This buys time for the patients, giving their own livers a chance to recover or providing more time for a possible liver transplant.

Figure 22.11 This rat suspended in a container of perfluorodecalin continues to breathe the oxygen dissolved in that liquid. (Courtesy of Leland C. Clark, Ph.D., Professor of Pediatrics, Children's Hospital Research Foundation)

Figure 22.12 A 4-kg (9-lb) battery-powered kidney dialysis unit that can be strapped to the front of the body. It must periodically be filled with dialysis fluid from a portable 20-L (20-qt) tank. (Courtesy University of Utah)

OXYGENATORS AND INSULIN PUMPS   An implantable artificial lung is still many years away, largely because few materials can match the natural lung in letting gases pass in and out of the body. But temporary assistance is quite a different matter. The *respirator* is a machine that in essence breathes air in and out of a patient's lungs. Another device, the *membrane oxygenator*, bypasses the lungs entirely. It circulates blood out of the body and passes it through membranes that remove carbon dioxide; then oxygen is added before returning the blood to the patient. Although membrane oxygenators can be used for only a few hours or days at a time, respirators can keep people alive for years.

The pancreas secretes insulin into the blood when our glucose levels rise; this causes glucose to leave the blood and enter our cells, enabling us to metabolize glucose. But now insulin-dependent diabetics can use artificial *insulin pumps* to do this. Scientists are developing devices with a sensor to monitor glucose concentrations and a tiny computer to process the information and activate a pump that injects insulin into the blood when necessary. This moment-to-moment action is more effective than taking one or two large doses of insulin during the day. Most insulin pumps are powered by batteries and worn on a belt, but implantable units also are being developed.

ARTIFICIAL HEARTS   One of the most dramatic developments is the *artificial heart*. External pumps are increasingly used to help patients during and after open-heart surgery. But in 1982, Barney Clark, a retired dentist, became the first person to have a permanent artificial heart implanted in his body. He lived 112 days with his artificial heart. The longest survivor with an artificial heart lived 620 days.

The artificial heart implanted in Clark (Figure 22.13) was made of aluminum and polyurethane. Its valves, which did not work satisfactorily, have now been replaced by solid titanium ones. The heart weighs about

Figure 22.13 An artificial heart that includes artificial right and left ventricles and one-way valves to regulate blood flow. It connects to an air compressor, which inflates a diaphragm that pumps the blood. (Courtesy Symbion, Inc.)

12 oz and is connected to an external compressor through hoses that pass through the skin. But now patients can make brief excursions while using an 11-lb portable power pack that works for up to 3 hours. Eventually, artificial hearts may have miniature drive systems implanted, so the recipient will only need to wear a battery pack around the waist.

For now, artificial hearts are used mostly to sustain patients until they can receive a heart transplant. Because of improvements in heart transplants and practical difficulties with artificial hearts, the future of a permanent, implantable, artificial heart is uncertain.

22.4 Better Athletic Performance: Increasing Oxygen Supply, Muscle, and Stamina

MORE OXYGEN AND LESS MUSCLE SORENESS   As we understand how the body works chemically, we can make chemical changes to help our bodies work better. One use of this knowledge is to improve athletic performance through artificial and natural substances.

Oxygen reacts with our carbohydrate, lipid, and protein fuels to produce CO_2 and H_2O, generating energy in the process. When our muscles work intensely, as in sprinting or lifting heavy weights, they have a temporary shortage of oxygen, called **oxygen debt**. When this happens, muscle cells convert carbohydrates to lactic acid, which requires less oxygen but yields only about 5 percent as much energy as when we oxidize carbohydrates completely to CO_2 and H_2O (Figure 22.14). Lactic acid, in turn, may impair muscle function and help make us feel stiff and sore the next day. So now an increasing number of athletes monitor the appearance of lactic acid in their blood during workouts. Then they train at the maximum level at which they don't produce lactic acid.

Athletes (and everyone else) can also oxidize carbohydrates, fats, and proteins more quickly by having more oxygen-carrying red blood cells.

Figure 22.14 Metabolizing glucose during oxygen debt produces only about 5 percent as much energy as when oxygen is used.

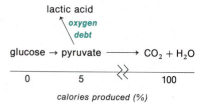

People living at higher elevations naturally produce more red blood cells to compensate for the lower oxygen concentrations in the air. So some athletes, such as distance runners, live and train at high elevations.

Another way to do this is called **blood doping**. Although the details vary, athletes typically have a liter of their own blood removed and the red blood cells stored frozen for eight to twelve weeks while their bodies replenish their own supply. Then about a day before competition, they are injected with their stored red blood cells. This extra dose of the person's own cells instantly boosts the amount of oxygen their blood carries and is an advantage in events such as the marathon or long bicycle races. Indeed, several medalists on the U.S. cycling team admitted receiving such injections just before the 1984 Olympic Games.

Sports federations, such as the U.S. Olympic Committee, are trying to ban the practice. But enforcement is difficult because there is no practical way to detect blood doping; no abnormal substance is present in the body, and individuals vary considerably in their natural levels of red blood cells. One danger with blood doping is that the high concentration of red blood cells makes the blood thicker, and this may increase the risk of blocking small blood vessels. In addition, a few athletes have reported that the sudden infusion of red blood cells made them sick, with flulike symptoms.

Now on the horizon is a compound that may make blood doping obsolete. A natural kidney hormone, erythropoietin, is being produced by recombinant DNA technology. Erythropoietin can increase the number of red blood cells by as much as 10 percent and is intended for treating people with anemia. But the customers for this new product will surely include some athletes.

IMPROVING BODY FUEL SUPPLIES   Athletes use several tactics to alter their bodies' supply of fuel. Body builders go on very low-calorie diets just before competition to reduce their amount of fat. This makes their muscles (which are mostly protein) stand out in greater detail.

Endurance athletes, on the other hand, want to increase their supply of fuel, especially carbohydrates. Through a process called **carbohydrate** (or glycogen) **loading**, they can store two to three times their normal amount of glycogen. First, they deplete their glycogen for about three days by going on a low-carbohydrate diet or, more commonly, by extensive exercise. Second, they pack in glycogen by changing to a carbohydrate-rich diet three or four days before competition.

During long events, athletes have to replenish their supply of carbohydrates, minerals, and water. They often do this by drinking a solution such as Gatorade that contains all these ingredients. The drink usually contains carbohydrates and small amounts of sodium and potassium, which are needed for normal muscle function and are lost in sweat. Water is a critical ingredient, for dehydration sharply reduces athletic performance and can cause the body temperature to rise dangerously. The effect is like a car that overheats because it doesn't have enough water in its radiator. A 70-kg (150-lb) runner typically loses about 1 L of sweat per hour in hot weather.

After 60 to 90 minutes of exercise, blood glucose levels begin to decrease. Some athletic drinks (Table 22.2) provide carbohydrates in the form of glucose or sucrose (which is digested to produce glucose and fructose). But too much glucose entering the blood triggers the release of insulin, which often overcorrects the problem and temporarily produces low blood sugar levels (hypoglycemia). This is why athletes should not consume food high in simple sugars an hour or so before competition. Moderate amounts of glucose taken during exercise, however, are unlikely to cause hypoglycemia.

The insulin problem can be avoided by supplying carbohydrates in other forms. One solution is to ingest fructose, which doesn't trigger the release of insulin. Another way is to take a complex carbohydrate that is gradually digested to provide a slow but steady supply of glucose into the blood. Several commercial products contain either maltodextrin or glucose polymer, which has about five to twelve glucose units per molecule of polymer.

Too much sugar or salt is unpalatable and can cause nausea. In addition, the higher concentration of dissolved materials in the stomach than

TABLE 22.2 CONTENTS OF SPORTS DRINKS PER 227 g (8 oz)

Drink	Recommended Concentration (%)	Form of Carbohydrate	Major Ions	Calories
Carbo Plus	16	Maltodextrin	Na^+ (5 mg) K^+ (100 mg) Mg^{2+} (100 mg)	170
E.R.G.	5.7	Glucose	Na^+ (70 mg) K^+ (100 mg)	45
Exceed	7.2	Glucose polymer	Na^+ (66 mg) K^+ (56 mg) Mg^{2+} (6 mg)	68
Gatorade	6	Glucose, sucrose	Na^+ (110 mg) K^+ (25 mg)	50
Max	7.5	Glucose polymer	Na^+ (15 mg)	70

in surrounding fluids and tissues causes water to flow into the stomach; this dehydrates other parts of the body and impairs athletic performance. To reduce this risk, an athlete should avoid concentrations higher than an 8 percent solution of glucose or sucrose or a 5 to 15 percent solution of glucose polymer. The problem is less common with complex carbohydrates because they supply an equivalent amount of glucose in a much smaller number of dissolved molecules.

USING DRUGS TO IMPROVE PERFORMANCE AND TREAT INJURIES Athletes sometimes use drugs to gain a competitive edge. Five classes of drugs now are illegal to use in competition and are tested for: narcotics and analgesics, diuretics, amphetamines and other stimulants, beta blockers, and anabolic steroids. Narcotics and analgesics are used to relieve pain. Diuretics are banned because they increase the volume of urine; this dilutes drugs that may be present in the urine and may make their concentrations too low to be detected during testing.

Amphetamines are used to reduce fatigue and to increase aggressiveness, attentiveness, and a perception of energy. Another stimulant, caffeine, enhances performance in endurance events. In one experiment, for example, bicycle riders were able to exercise 20 percent longer in the laboratory when they first received caffeine. This drug helps break down glycogen fuel in the muscles and releases stored fat into the blood so it can be used as fuel by muscle and other cells. For this reason, coffee is a popular drink an hour or so before a marathon begins. In athletic competition, the allowable limit for caffeine is equivalent to drinking six cups of coffee. The allowable limit for amphetamines is zero.

Drugs such as propranolol (Inderal) are beta blockers; they slow the heart rate and are used to treat high blood pressure. In sports such as archery and shooting, a slower pulse is an advantage because the competitor can more easily shoot in between heart beats; this gives a steadier aim and greater accuracy. Thus, in competition shooters are tested for beta blockers, ethanol, and diazepam (Valium).

The most abused drugs for athletic competition are the **anabolic steroids**, which include testosterone and many synthetic derivatives (Figure 22.15). These hormones, which have steroid structures (Figure 9.11), stimulate protein synthesis and are used to increase muscle mass. Some weightlifters, sprinters, body builders, football players, and people who compete in strength events such as the shot put and discus use these drugs. The drugs enable users to build muscle quickly and to do more rigorous training.

Because anabolic steroids are banned in amateur athletic competition, some athletes take the drugs for training and then discontinue their use early enough before competition so the drugs cannot be detected. But these compounds are nonpolar and are stored in fatty tissues for

Figure 22.15 Testosterone and two synthetic anabolic steroids. The long hydrocarbon chain (shown in color) in Deca-Durabolin makes it especially nonpolar.

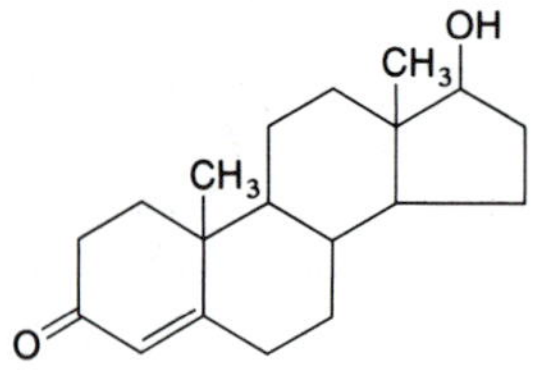

testosterone

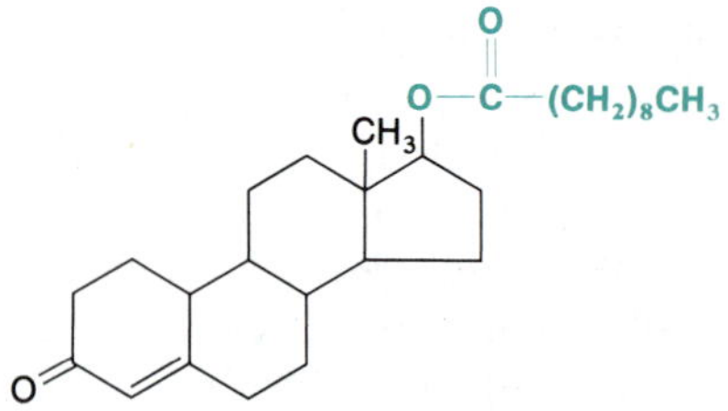

handrolone decanoate
(Deca-Durabolin)

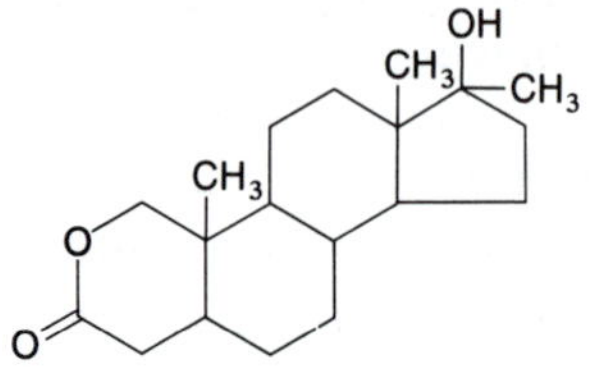

oxandrolone
(Anavar)

CHEMISTRY SPOTLIGHT   AN ATHLETIC DRUG BUST

The International Olympic Committee has banned more than 300 drugs for athletic competition, but until recently the chemical tests for these substances were not very sophisticated. Now that has changed.

The new tests were first used at the Pan American Games in 1983. Nine medalists and five other athletes from the United States, Canada, and Cuba were disqualified because of stimulants or anabolic steroids found in their urine. Twelve other athletes decided to withdraw and not be tested for drugs. The disqualified athletes were mostly weightlifters.

But weightlifters aren't the only athletes found to use anabolic steroids. A Finnish runner lost his silver medal in the 10,000-m run at the 1984 Olympic Games when a banned steroid was detected in his urine. And a Polish hockey player was disqualified in the 1988 Winter Olympic Games.

Athletes are caught in a bind. One authority recently explained, "My position is the same as that of most sports medicine professionals: Anything that probably gives you an edge and is not expressly forbidden by the rules should be compulsory."

The methods for detecting drugs will get even better and will further discourage drug use. The main beneficiaries will be the athletes, who no longer will feel forced to try dangerous drugs in order to keep up with their competitors.

months. The most nonpolar steroids, such as nandrolone decanoate (Deca-Durabolin; see Figure 22.15), are taken by injection and can be detected in the urine for six to twelve months. Oxandrolone (Anavar) is slightly more polar and is taken orally; it can be detected up to ten weeks after use. Testosterone injections are detected by its elevated levels in urine compared to its metabolic product epitestosterone (Figure 20.16).

Anabolic steroids have serious side effects. They alter the normal hormonal balance in the body and in men can cause aggressiveness, the development of female breasts, reduced fertility, and underdeveloped testicles. In women they cause menstrual irregularities and masculizing effects such as a deeper voice and excessive hair growth. These drugs also increase the risk of liver cancer and heart disease.

Now some athletes are using human growth hormone (HGH), produced by genetic engineering, to increase their size and muscle mass. HGH is expensive, and large amounts cause bones of the skull, face, hands, and feet to thicken, while the skin becomes coarse. HGH also increases the risk of diabetes and high blood pressure.

Many chemicals are used to treat athletic injuries. As we discussed earlier (Section 8.5), trainers often spray ethyl chloride on bruised areas

Figure 22.16 Graph of testosterone levels relative to epitestosterone, a metabolic product, in a normal subject and in a person taking testosterone injections (color). A testosterone–epitestosterone ratio of 6.0 or more indicates the use of testosterone.

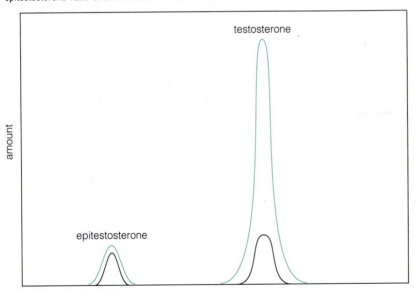

to "freeze" the skin and temporarily relieve the discomfort. A variety of substances also serve as liniments, pain relievers, and muscle relaxants. Cortisone, for example, is often used to reduce the inflammation in damaged joints.

Some athletes also use dimethylsulfoxide (DMSO) to relieve muscle and joint pain. DMSO is a clear liquid that rapidly penetrates into the underlying tissue when it is rubbed onto the skin. It typically gives users a skin rash and garliclike taste and breath. DMSO is widely used in veterinary medicine, especially to treat leg injuries in horses, but it is not approved for use in humans because it may increase the risk of eye and liver damage.

MARC LAPPÉ   Recombinant DNA really is the millennium in biology, just as nuclear fission was the millennium in physics. The power to do good or ill from both techniques is virtually limitless. . . . We must hold accountable what industries or governments will choose to create with this new found power.

Summary

Genetic engineering is done by splicing genes into plasmids or viruses, which carry the genes into cells. The cells then synthesize the proteins specified by the genes. Redesigned bacteria and yeasts now make protein products for medical and other purposes.

Providing genes to people with genetic disorders hasn't yet cured them. Disorders due to the lack of one gene are the most likely candidates for genetic therapy.

Many artificial body parts are available, including bones, joints, blood vessels, heart valves, and cosmetic implants. Artificial organs can perform some, but not all, functions of the kidney, lung, pancreas, and heart.

Athletes use a wide variety of chemicals to improve performance. Some—including stimulants, narcotics and analgesics, diuretics, beta blockers, and anabolic steroids—are illegal in competition and dangerous to use. Carbohydrate loading increases the glycogen supply for endurance events, and blood doping increases the number of red blood cells. Sports drinks replenish water, carbohydrates, and minerals.

Terms for Review

After completing this chapter, you should know and understand the meaning of the following terms:

anabolic steroid (p. 617)

blood doping (p. 615)

carbohydrate loading (p. 615)

chromosome (p. 601)

gene (p. 601)

genetic engineering (p. 602)

mutation (p. 608)

oxygen debt (p. 614)

plasmid (p. 602)

recombinant DNA (p. 602)

restriction enzyme (p. 602)

Questions

Odd-numbered questions are answered at the back of this book.

1. Distinguish between a gene and a chromosome.

2. What are the main vectors used to carry new genes into cells?

3. When "sticky ends" of DNA join together, which base pairs with each of the following? (a) Adenine (A), (b) cytosine (C), (c) guanine (G), and (d) thymine (T). [Review Section 9.5 if necessary.]

4. Explain how genes can be spliced into plasmids to produce recombinant DNA.

5. How do vaccines work? Why might vaccines produced by genetic engineering be safer to use?

6. What are the advantages for insulin-dependent diabetics to use insulin produced by genetically engineered bacteria instead of beef or pork insulin?

7. What kinds of genetic diseases are the most likely to be treatable by genetic therapy? Why?

8. People with Down's syndrome have an extra chromosome, number 21. Is this disorder likely to become treatable with recombinant DNA? Why?

9. To treat a genetic disorder, would it be necessary to genetically alter all cells in a person's body? Why?

10. In general, what are the advantages and disadvantages of artificial body parts compared to the natural parts? Can you think of examples where you would prefer to have the artificial part?

11. Explain how an "artificial kidney" works. Is it a permanent solution for a person with failing kidneys? Why?

12. In what situations might artificial blood be preferable to receiving a transfusion of real blood?

13. Both respirators and membrane oxygenators help keep patients' blood oxygenated. How do these devices differ in the way they do this?

14. What is oxygen debt, and how does it impair athletic performance?

15. Why might erythropoietin reduce the incidence of blood doping?

16. During long-distance running events, some athletes drink defizzed cola. What advantages might this have compared with drinking water?

17. Explain how carbohydrate loading works. What kinds of athletes are the least likely to benefit from this practice?

18. Which sports drinks listed in Table 22.2 are the most likely to trigger excess insulin secretion if large amounts are consumed before competition?

19. List the main reasons for athletes taking the following just before competition: (a) caffeine, (b) beta blockers, and (c) amphetamines.

20. Why are anabolic steroids used by some athletes? What are the disadvantages of using these drugs?

Topics for Discussion

1. According to U.S. law, companies can patent new genetic organisms—bacteria, yeasts, plants, and nonhuman animals—that they produce. In 1988, for example, Harvard University was granted the first U.S. patent on a genetically engineered animal—a tumor-prone mouse to be used in cancer research. What limits, if any, would you put on patenting forms of life? Why?

2. Under what conditions is genetic engineering appropriate in people? What limits to this technology would you set? Why? How would you implement those restrictions?

3. A patient underwent surgery, which saved his life. The removed tissue turned out to be useful to a company in producing a commercially valuable substance. The patient sued to receive a share of the commercial profits. How would you judge this case? Explain your reasoning.

4. DNA fingerprinting (see the "Chemistry Spotlight") is a very effective way to identify people. If the DNA fingerprint of an unknown murderer were obtained, would you favor doing DNA fingerprints on a large population to identify the criminal? Why?

5. What should be the criteria for implanting an artificial heart in a person? Is this form of "human experimentation" appropriate now? By the year 2010, do you expect heart transplants or artificial hearts to be the main way for a person to receive a new heart? Why?

6. Some athletes argue that blood doping should be legal in athletic competition because the athlete is using his or her own cells, not foreign substances. Do you agree or disagree? Why? Why is blood doping difficult to detect?

7. Should the same restrictions on drug use by amateur athletes be applied to professional athletes? Why?

Exponential Numbers: Powers of Ten

EXPONENTIAL NUMBERS   Any decimal number can be written as the product of two factors: a number multiplied by 10 raised to some power or exponent. Examples are 1×10^2, 4.2×10^7, 1×10^{-4}, and 8.3×10^{-9}. Numbers written in this form are called *exponential numbers*.

A positive power of 10 or exponent indicates how many times a number is multiplied by 10. For example, 1×10^2 is equal to $1 \times 10 \times 10$, or 100. Often 1×10^2 is written as 10^2, with the implicit understanding that it is multiplied by 1. Some other examples are

1×10^1 or $10^1 = 1 \times 10 = \mathbf{10}$

1×10^3 or $10^3 = 1 \times 10 \times 10 \times 10 = \mathbf{1000}$

$1 \times 10^6 = 1 \times 10 \times 10 \times 10 \times 10 \times 10 \times 10 = \mathbf{1,000,000}$

$4.2 \times 10^7 = 4.2 \times 10 \times 10 \times 10 \times 10 \times 10 \times 10 \times 10 = \mathbf{42,000,000}$

A negative power of 10 or exponent indicates how many times a number is divided by 10. For example,

$$1 \times 10^{-2} \text{ or } 10^{-2} = \frac{1}{10^2} = \frac{1}{10 \times 10} = \frac{1}{100} = \mathbf{0.01}$$

$$1 \times 10^{-4} \text{ or } 10^{-4} = \frac{1}{10^4} = \frac{1}{10 \times 10 \times 10 \times 10} = \frac{1}{10,000} = \mathbf{0.0001}$$

$$8.3 \times 10^{-9} = \frac{8.3}{10^9} = \frac{8.3}{10 \times 10 \times 10 \times 10 \times 10 \times 10 \times 10 \times 10 \times 10}$$

$$= \frac{8.3}{1,000,000,000} = \mathbf{0.0000000083}$$

Note that the sign of the exponent is reversed when a power of 10 is moved from the numerator (top) to the denominator (bottom) and vice versa.

$$6 \times 10^{-4} = \frac{6}{10^4} \quad \text{and} \quad \frac{1}{10^{12}} = 1 \times 10^{-12}$$

WRITING EXPONENTIAL NUMBERS IN SCIENTIFIC NOTATION   In scientific measurements and calculations, we often have to deal with very small or very large numbers. For example 12.0 g of carbon contains 602,000,000,000,000,000,000,000 atoms of carbon and the mass of a single hydrogen atom is 0.00000000000000000000000167 g. Writing large and small numbers in this manner is cumbersome and can easily lead to careless errors in getting the correct number of zeros. To avoid these problems, scientists express such numbers in scientific notation as powers of 10.

In scientific notation, a number is written in exponential form. The first factor is a decimal number equal to or greater than 1 and less than 10 (such as 1, 6.02, and 1.67, but not 602 or .167). The second factor is 10 raised to some positive or negative power or exponent (such as 10^6, 10^{23}, 10^{-6}, or 10^{-24}). Positive exponents or powers of 10 are used to specify numbers 10 or greater (such as 100, 1,000,000, or 13,600), and negative exponents are used to express numbers less than 1 (such as 0.001, 0.000001, or 0.0032).

The two very long numbers given in the first paragraph of this section can be expressed in scientific notation as follows:

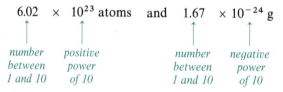

The following simple rules can be used to write these or any numbers in scientific notation:

1. Move the decimal point of the number to the right or left to form a number between 1 and 10.

    6 0 2 0 0 0 0 0 0 0 0 0 0 0 0 0 0 0 0 0 0 0 0 atoms
    Move the decimal point 23 places to the left.

    0.0 0 0 0 0 0 0 0 0 0 0 0 0 0 0 0 0 0 0 0 0 0 0 1 67 g
    Move the decimal point 24 places to the right.

2. *For Numbers Equal To or Greater Than* 1: For each decimal place moved to the *left*, multiply the part of the number between 1 and 10 by a positive power of 10. The power of 10 gives the number of

places the decimal point in the number must be moved back to yield the original number in nonexponential form.

6.02×10^{23} atoms   (Decimal place was moved 23 places to the left, so the power for 10 is $+23$.)

3. *For numbers Between* 0 *and* 1: For each decimal point moved to the *right*, multiply the part of the number between 1 and 10 by a negative power of 10.

1.67×10^{-24} (Decimal point was moved 24 places to the right, so the power for 10 is -24.)

PRACTICE EXERCISE

Write the following numbers in scientific notation: (a) 1398 g, (b) 850,000 m, (c) 0.000056 g, and (d) 0.265 m.

SOLUTION

a. $1398 \text{ g} = 1.398 \times 10^3 \text{ g}$
To obtain a number between 1 and 10, the decimal point must be moved 3 places to the *left*, so the power of 10 is $+3$.

b. $850,000 \text{ m} = 8.5 \times 10^5 \text{ m}$
To obtain a number between 1 and 10, the decimal point must be moved 5 places to the *left*, so the power of 10 is $+5$.

c. $0.000056 \text{ g} = 5.6 \times 10^{-5} \text{ g}$
To obtain a number between 1 and 10, the decimal point must be moved 5 places to the *right*, so the power of 10 must be -5.

d. $0.265 \text{ m} = 2.65 \times 10^{-1} \text{ m}$
To obtain a number between 1 and 10, the decimal point must be moved 1 place to the *right*, so the power of 10 is -1.

MULTIPLYING AND DIVIDING EXPONENTIAL NUMBERS   Exponential numbers can be multiplied or divided using simple rules based on adding the exponents when exponential numbers are multiplied and subtracting the exponents when they are divided. The following rules illustrate this procedure:

1. **Multiplication** To multiply two or more numbers, (a) convert all numbers to scientific notation if they are not in that form, (b) multiply the decimal numbers in the usual way, (c) add the powers of 10, and (d) adjust the answer to scientific notation if necessary.

Examples:

$1\underset{\smile}{00} \times 1\underset{\smile\smile}{000} = (1 \times 10^2) \times (1 \times 10^3)$ (*Convert to scientific notation.*)

$\qquad\qquad\qquad = (1 \times 1) \times (10^{2+3})$ (*Multiply decimal numbers and add powers of 10.*)

$\qquad\qquad\qquad = \mathbf{1 \times 10^5}$ (*Answer is already in scientific notation.*)

$(0.\underset{\smile\smile\smile}{000}9)(0.\underset{\smile\smile\smile\smile\smile\smile\smile}{0000000}2)$

$\qquad = (9 \times 10^{-4})(2 \times 10^{-8})$ (*Convert to scientific notation.*)

$\qquad = (9 \times 2)(10^{-4-8})$ (*Multiply decimal numbers and add powers of 10.*)

$\qquad = 18 \times 10^{-12}$ (*Answer is not in scientific notation.*)

$\qquad = 1.\underset{\smile}{8} \times 10^1 \times 10^{-12} = 1.8 \times 10^{1-12}$ (*Convert to scientific notation.*)

$\qquad = \mathbf{1.8 \times 10^{-11}}$

2. **Division** To divide two or more numbers, (a) convert all numbers to scientific notation, (b) divide the decimal numbers in the usual way, (c) subtract the exponents (numerator minus denominator), and adjust the answer to scientific notation if necessary.

Examples:

$\dfrac{10^6}{10^2} = \dfrac{1 \times 10^6}{1 \times 10^2} = \dfrac{1}{1} \times 10^{6-2}$ (*Divide decimal numbers and subtract denominator power from numerator power of 10.*)

$\qquad\qquad\qquad = \mathbf{1 \times 10^4 \text{ or } 10^4}$

$\dfrac{3\overset{\frown\frown\frown\frown\frown\frown\frown\frown}{00000000}}{0.000\underset{\smile\smile\smile\smile\smile}{06}} = \dfrac{3 \times 10^8}{6 \times 10^{-5}}$ (*Convert to scientific notation.*)

$\qquad\qquad = \dfrac{3}{6} \times 10^{8-(-5)}$ (*Divide decimal numbers and subtract denominator power from numerator power.*)

$\qquad\qquad = 0.\underset{\curvearrowright}{5} \times 10^{13}$ (*Answer is not in scientific notation.*)

$\qquad\qquad = 5 \times 10^{-1} \times 10^{13}$ (*Convert to scientific notation.*)

$\qquad\qquad = 5 \times 10^{-1+13}$ (*Add exponents.*)

$\qquad\qquad = \mathbf{5 \times 10^{12}}$

PRACTICE EXERCISE

Solve the following by using exponential numbers and express each answer in scientific notation.

a. $\dfrac{3.8 \times 10^{-7}}{9.5 \times 10^{-12}}$

b. $(6.9 \times 10^6)(3.2 \times 10^4)$

c. $(0.0000006)(80,000)$

d. $\dfrac{15,000}{3,000,000}$

SOLUTION

a. $\dfrac{3.8 \times 10^{-7}}{9.5 \times 10^{-12}} = \dfrac{3.8 \times 10^{-7+12}}{9.5} = 0.40 \times 10^5$
$= 4.0 \times 10^{-1} \times 10^5 = 4.0 \times 10^{-1+5} = 4.0 \times 10^4$

b. $(6.9 \times 10^6)(3.2 \times 10^4) = (6.9 \times 3.2)(10^6 \times 10^4)$
$= (22.1)(10^{6+4}) = 22.1 \times 10^{10}$
$= 2.21 \times 10^1 \times 10^{10} = 2.21 \times 10^{11}$

c. $(0.0000006)(80000) = (6 \times 10^{-7})(8 \times 10^4)$
$= (6 \times 8) \times 10^{-7+4} = 48 \times 10^{-3}$
$= 4.8 \times 10^1 \times 10^{-3} = 4.8 \times 10^{1-3}$
$= 4.8 \times 10^{-2}$

d. $\dfrac{15,000}{3,000,000} = \dfrac{1.5 \times 10^4}{3.0 \times 10^6} = \dfrac{1.5}{3.0} \times 10^{4-6} = 0.5 \times 10^{-2}$
$= 5 \times 10^{-1} \times 10^{-2} = 5 \times 10^{-1-2} = 5 \times 10^{-3}$

Questions

Questions numbered and lettered in color are answered at the back of this book.

1. Express the following numbers in scientific notation.
   a. The 400 billion kg of water used each day by industry in the United States
   b. 306,982
   c. 0.0000000000000000000000000000000000000000000000000392

d. 0.000000000000001
e. 0.00000003892
f. 0.000000009 g of mercury
g. 4,389,256
h. 13,200,000,000,000,000,000,000,000,000,000
i. 680,320,000,000,000
j. 0.00003060

2. Multiply the following exponential numbers and express the answers in scientific notation:
   a. $(1 \times 10^{23})(6 \times 10^{-39})$
   b. $(8.40 \times 10^{-6})(2.50 \times 10^{-16})$
   c. $(0.0000083)(1,600,000)$
   d. $(0.0000000000014)(0.00000030)$
   e. $(800 \times 10^{16})(0.0012 \times 10^{-10})$
   f. $(6.2 \times 10^{9})(1,200,000,000)$

3. Divide the following numbers and express the answers in scientific notation:
   a. $6.88 \times 10^{13}/2.00 \times 10^{-16}$
   b. $(0.000004800)/(3200)$
   c. $(14 \times 10^{-9})/(29 \times 10^{-7})$
   d. $(4,500,000,000,000)/6500$
   e. $(3.00 \times 10^{12})/(9.00 \times 10^{-13})$
   f. $(33.0 \times 10^{60})/(0.0004)$

Units and Unit Conversions

THE METRIC AND SI UNITS OF MEASUREMENT   When we measure the length, volume, mass, or temperature of an object, the resulting number must be accompanied by the appropriate unit (Section 1.4). For example, if your height or length is measured with a meter stick marked off in fractions of a meter, then your height is reported as so many meters (m).

As discussed in Section 1.4, most scientists use an updated version of the metric system of measurement that is based on establishing certain fundamental units of measurement for quantities such as length, volume, mass, temperature, and energy. Commonly used fundamental units in chemistry are the meter (m) for length, the liter (L) for volume, the gram (g) for mass, the joule (J) for energy, and degrees Celsius (°C) for temperature (Section 1.4).

Smaller and larger units of these units are obtained by multiplying the fundamental unit by certain positive or negative powers of 10 (Appendix 1). Each of these larger or smaller units of the fundamental unit is identified by attaching a prefix to the name of the fundamental unit. Each prefix corresponds to multiplying the fundamental unit by some positive power (for example, 10^3) or some negative power (such as 10^{-3}) of 10.

The six most commonly used prefixes and their multiplier values were given in Table 1.1. For example.

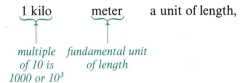

mean 1000, or 10^3 meters. Because the abbreviation for kilo- is k and that for meter is m, the abbreviation for kilometer is km.

1 kilometer, or 1 km = 1000 meters, or 10^3 m

Similarly,

1 milli   gram, a unit of mass,

*multiple   unit
of 10 is 0.001 of mass
or 10^{-3}*

is abbreviated mg and is equal to 0.001 g, or 10^{-3} g.

1 milligram, or 1 mg = 0.001 g, or 10^{-3} g

UNIT CONVERSION FACTORS Each relationship between two different units can be expressed as a unit conversion factor. These factors can be used to convert from one of the units in the factor to the other unit in the factor. We have just seen, for example, that 1 km = 1000 m, or 10^3 m. To change this into a unit conversion factor relating km and m, both sides of the equation are divided by 1 km:

$$\frac{1 \text{ km}}{1 \text{ km}} = 1 = \frac{1000 \text{ m}}{1 \text{ km}} \text{ or } \frac{10^3 \text{ m}}{1 \text{ km}}$$

It can be read as 1000 or 10^3 meters per kilometer, where *per* means *divided by*. Because this numerical ratio is equal to 1, it or any other unit conversion factor can also be written in inverted form:

$$1 = \frac{1 \text{ km}}{1000 \text{ m}} \text{ or } \frac{1 \text{ km}}{10^3 \text{ m}}$$

Table A2.1 lists the relationships between various units within the metric system and between the metric and English systems of units, along with their unit conversion factors.

THE UNIT CONVERSION METHOD Often it is necessary to convert from one unit of measurement to another. Such conversions can be made using a procedure known as the *unit conversion method* or *factor label method*. This method involves multiplying a given number and its unit by one or more

TABLE A2.1 SOME COMMON UNIT CONVERSION FACTORS

Definition	Unit Conversion Factors

Length

Metric–Metric

1 km = 1000 m, or 10^3 m   1000 m/km = **1** = 1 km/1000 m

1 cm = 0.01 m, or 10^{-2} m   0.01 m/1 cm = **1** = 1 cm/0.01 m

1 mm = 0.001 m, or 10^{-3} m   0.001 m/1 mm = **1** = 1 mm/0.001 m

Metric–English

1 mi = 1.61 km   1.61 km/1 mi = **1** = 1 mi/1.61 km

1 m = 39.4 in.   39.4 in./1 m = **1** = 1 m/39.4 in.

1 in. = 2.54 cm   2.54 cm/1 in. = **1** = 1 in./2.54 cm

Mass

Metric–Metric

1 kg = 1000 g, or 10^3 g   1000 g/1 kg = **1** = 1 kg/1000 g

1 mg = 0.001 g, or 10^{-3} g   0.001 g/1 mg = **1** = 1 mg/0.001 g

Metric–English

1 kg = 2.20 lb   2.20 lb/1 kg = **1** = 1 kg/2.20 lb

1 lb = 454 g   454 g/1 lb = **1** = 1 lb/454 g

1 oz = 28.4 g   28.4 g/1 oz = **1** = 1 oz/28.4 g

Volume

Metric–Metric

1 L = 1000 mL, or 10^3 mL   1000 mL/1 L = **1** = 1 L/1000 mL

1 mL = 0.001 L, or 10^{-3} L   0.001 L/1 mL = **1** = 1 mL/0.001 L

Metric–English

1 L = 1.06 qt   1.06 qt/1 L = **1** = 1 L/1.06 qt

1 L = 0.265 gal   0.265 gal/1 L = **1** = 1 L/0.265 gal

Energy

Metric–Metric

1 cal = 4.18 J   4.18 J/1 cal = **1** = 1 cal/4.18 J

1 kJ = 1000 J, or 10^3 J   1000 J/1 kJ = **1** = 1 kJ/1000 J

1 kcal = 1000 cal, or 10^3 cal   1000 cal/1 kcal = **1** = 1 kcal/1000 cal

1 kcal = 4180 J   4180 J/1 kcal = **1** = 1 kcal/4180 J

unit conversion factors (which usually can be obtained from Table A2.1), to convert it to the desired number and unit (the unknown).

$$\begin{matrix} \text{given quantity} \\ \text{and unit} \end{matrix} \times \text{unit conversion factor(s)} = \begin{matrix} \textbf{desired quantity} \\ \textbf{and unit} \end{matrix}$$

The first step is to read the problem and identify (1) the given quantity and its unit and (2) the unknown quantity whose numerical value is to be calculated and its desired unit. Then use your memory or Table A2.1 to find one or more unit conversion factors that relate the given unit to the desired unit. Sometimes no factor directly relates the given unit and the desired unit. In this case, try to find a unit conversion factor that converts the given unit to a new unit and then a second unit conversion factor that relates the new unit to the desired unit.

Use this information to plan a solution for converting the given unit to the desired unit. Set up the plan so that all the units you don't want are canceled and only the desired unit is left:

$$\text{given unit} \times \frac{\text{desired unit}}{\text{given unit}} = \textbf{desired unit}$$

or

$$\text{given unit} \times \frac{\text{new unit}}{\text{given unit}} \times \frac{\text{desired unit}}{\text{new unit}} = \textbf{desired unit}$$

Then insert the known numbers that go with the given unit and each unit conversion factor and calculate the desired quantity and its unit.

Such conversions can be made between units within the same system of measurement (for example, within the metric system or within the English system) or between units within different systems of measurement (for example, between the metric and English systems or vice versa), as illustrated by the worked-out "Practice Exercises" that follow.

CONVERSIONS WITHIN THE METRIC SYSTEM   The examples below illustrate conversions from one metric unit to another.

PRACTICE EXERCISE

A package has a mass of 1.50 kg. What is its mass in grams?

SOLUTION

$$\text{given quantity} \times \text{unit conversion factor(s)} = \begin{array}{c}\text{desired quantity}\\ \text{and unit}\end{array}$$

$$1.50 \text{ kg} \times \text{unit conversion factor(s)} = \text{g}$$

From Table A2.1, we see that the unit conversion factors, 1000 g/1 kg and 1 kg/1000 g relate the given unit (kg) and the desired unit (g):

$$\text{given unit} \times \frac{\text{desired unit}}{\text{given unit}} = \text{desired unit}$$

$$\text{kg} \times \frac{\text{g}}{\text{kg}} = \text{g}$$

$$1.50 \text{ kg} \times \frac{1000 \text{ g}}{1 \text{ kg}} = 1500 \text{ g, or } 1.50 \times 10^3 \text{ g}$$

PRACTICE EXERCISE

The distance from the chemistry building to a student's dormitory is 0.520 km. What is this distance in centimeters?

SOLUTION

$$\text{given quantity} \times \text{unit conversion factor(s)} = \begin{array}{c}\text{desired quantity}\\ \text{and unit}\end{array}$$

$$0.520 \text{ km} \times \text{unit conversion factor(s)} = \text{cm}$$

From Table A2.1 we see that there is no unit conversion factor directly relating the given unit (km) and the desired unit (cm). However, we can convert kilometers to meters using 1000 m/1 km and then convert meters to centimeters using 1 cm/0.01 m.

$$\text{given unit} \times \frac{\text{new unit}}{\text{given unit}} \times \frac{\text{desired unit}}{\text{new unit}} = \text{desired unit}$$

$$0.520 \text{ km} \times \frac{1000 \text{ m}}{1 \text{ km}} \times \frac{1 \text{ cm}}{0.01 \text{ m}} = 52{,}000 \text{ cm, or } 5.2 \times 10^4 \text{ cm}$$

PRACTICE EXERCISE

What is the weight in kilograms of a person whose weight is 154 lb?

SOLUTION

$$154 \text{ lb} \times \text{unit conversion factor(s)} = \text{kg}$$

From Table A2.1, we see that the unit conversion factor, kg/2.20 lb, relates the given unit (lb) to the desired unit (kg).

$$154 \text{ lb} \times \frac{1 \text{ kg}}{2.20 \text{ lb}} = 70 \text{ kg}$$

PRACTICE EXERCISE

The *density* of a sample of matter is defined as its mass per unit volume.

$$\text{density} = \frac{\text{mass}}{\text{volume}}$$

If the density of a urine sample is 1.020 g/mL, what is the mass in grams of 1.25 L of this fluid?

SOLUTION

$$\text{given quantity} \times \text{unit conversion factor(s)} = \frac{\text{desired quantity}}{\text{and unit}}$$

$$1.250 \text{ L urine} \times \text{unit conversion factor(s)} = \text{g urine}$$

The density of 1.020 g urine/mL urine can be used as a unit conversion factor to convert mL urine to g urine. However, to use this factor we must first convert L urine to mL urine using the unit conversion factor, 1000 mL/L, that relates milliliters and liters.

$$\frac{\text{given unit}}{} \times \frac{\text{new unit}}{\text{given unit}} \times \frac{\text{desired unit}}{\text{new unit}} = \text{desired unit}$$

$$1.250 \text{ L urine} \times \frac{1000 \text{ mL urine}}{1 \text{ L urine}} \times \frac{1.020 \text{ g urine}}{1 \text{ mL urine}} = 1275 \text{ g urine, or}$$

$$1.275 \times 10^3 \text{ g urine}$$

CONVERSIONS BETWEEN SYSTEMS   The examples below illustrate conversions between the metric and English systems of measurement.

PRACTICE EXERCISE

What is the speed in kilometers per hour of a car traveling at 50 miles per hour?

SOLUTION

$$50 \frac{mi}{hr} \times \text{unit conversion factor(s)} = \frac{km}{hr}$$

From Table A2.1, we see that there are 1.61 km/mi:

$$50 \frac{mi}{hr} \times \frac{1.61 \text{ km}}{1 \text{ mi}} = 80.5 \text{ km/hr}$$

PRACTICE EXERCISE

If a car has an average fuel consumption of 10.7 km/L of gasoline, what is its average fuel consumption in miles per gallon?

SOLUTION

$$10.7 \frac{km}{L} \times \text{unit conversion factor(s)} = \frac{mi}{gal}$$

Think of this as a two-step problem. First, we use the unit conversion factor mi/1.61 km (Table A2.1) to convert the numerator unit of kilometer to mile. Then we use the unit conversion factor L/0.265 gal to convert the denominator unit of liter to gallon.

$$10.7 \frac{km}{L} \times \frac{1 \text{ mi}}{1.61 \text{ km}} \times \frac{0.265 \text{ gal}}{1 L} = 25.1 \frac{mi}{gal}$$

CONVERSIONS BETWEEN THE CELSIUS AND FAHRENHEIT TEMPERATURE SCALES   The relationships between the Celsius scale used to make scientific measurements of temperature and the Fahrenheit scale used to measure many everyday values of temperature in the United States are shown in Figure 1.8. We can use the following formulas to convert temperature values between these two scales:

°F to °C	°C to °F

$$°C = \frac{(°F - 32.0)}{1.80} \qquad °F = (°C × 1.80) + 32.0$$

The examples below illustrate the use of these formulas.

PRACTICE EXERCISE

Many thermostats are set to 68°F in cold weather. What is this temperature on the Celsius scale?

$$°C = \frac{(°F - 32.0)}{1.80} = \frac{68°F - 32.0}{1.80} = \frac{36.0}{1.80} = 20°C$$

SOLUTION

PRACTICE EXERCISE

A patient has a body temperature of 40.0°C. What is this temperature on the Fahrenheit scale?

$$°F = (°C × 1.80) + 32.0 = (40.0°C × 1.80) + 32.0$$
$$= (72.0) + 32.0 = 104°F$$

SOLUTION

Questions

Questions numbered or lettered in color are answered in the back of this book.

1. Make the following unit conversions: (a) 25 mm to m; (b) the length in meters of a 100-yard football field; (c) the speed in miles per hour of a car traveling at 125 km per hour; (d) the mass in milligrams of a package with a mass of 2.50 kg; (e) mass

in kilograms of a 3.60-ton car; (f) the number of cigarettes a person must smoke to inhale 1.0 lb of tar if each cigarette contains 25 mg of tar; (g) the number of milliliters of liquid in a 3.0-L bottle; (h) the number of liters in a 25-gal. tank; (i) the cost per pound for a medicine that sells for $22.00/g; (j) volume in milliliters of a 225-g sample of ethyl alcohol with a density of 0.789 g/mL; (k) the number of kilograms of carbon tetrachloride with a density of 1.59 g/mL in 6.80 L of carbon tetrachloride; (l) the number of joules of energy in a piece of pie containing 675 kcal of energy.

2. The boiling point of ethyl alcohol at normal atmospheric pressure is 79.0°C. What is its boiling point in degrees Fahrenheit?

3. Dermatologists sometimes treat acne by wiping a patient's face with a sponge that has been cooled to −321°F by dipping it in liquid nitrogen. What is the temperature of the sponge on the Celsius scale?

Answers to Odd-Numbered Questions

Chapter 1

1. (b), (d), and (g) are laws; (a) and (e) are theories; (c), (f), and (h) are hypotheses.

3. With further testing, hypotheses may become more widely accepted; when this happens, they become theories. But hypotheses and theories do not become laws. Instead, they help *explain* scientific laws.

5. Scientists learn about nature in many ways, not just one.

7. The elements are (e) pure phosphorus and (l) pure mercury. The compounds are (f) pure table salt, (h) pure water, and (k) pure ice. The mixtures are (a) wood, (b) cup of coffee, (c) toothpaste, (d) paint, (g) beer, (i) muddy water, (j) iced tea, and (m) rubber.

9. (b), (c), (d), (e), (g), (h), and (i) are physical changes because no new pure substance forms. (a), (f), and (j) are chemical changes because a new pure substance forms.

11. (a) 78,000,000 mg; (b) 12,000 mL; (c) 0.725 kcal; (d) 0.041 km

13. 357°C

15. 1100 millidollars

17. (c)

19. (b)

Chapter 2

1. The "law" of conservation of mass summarized how nature was consistently observed to behave.

3. (a) A compound consists of atoms of elements in a specific, simple, whole-number ratio. (b) Atoms cannot be created or destroyed in a chemical reaction.

5. The number of positively charged particles (protons) equals the number of negatively charged particles (electrons).

7. (a) $^{32}_{16}X$ and (c) $^{30}_{16}X$ are isotopes of the same element, and (b) $^{16}_{8}X$ and (d) $^{15}_{8}X$ are isotopes of another element.

9. (a) 48p, 48e, 64n; (b) 84p, 84e, 126n; (c) 80p, 80e, 121n; (d) 56p, 56e, 82n

11. $(0.75 \times 157 \text{ amu}) + (0.25 \times 155 \text{ amu}) = 156.5$ amu

13. (a) An electron is in its lowest energy level. (b) An electron is in a higher energy level than the lowest one.

15. The Bohr and quantum mechanical models both account for the line spectrum of hydrogen in terms of excited electrons emitting energy as they return to lower energy levels. The Bohr model does not work well for atoms more complex than hydrogen.

17. (a) Hg (mercury); (b) P (phosphorus); (c) At (astatine); (d) Ni (nickel); (e) Tc (technetium); (f) Br (bromine); (g) Zn (zinc) and Hg (mercury); (h) N (nitrogen), As (arsenic), Sb (antimony), and Bi (bismuth); (i) Be (beryllium), Mg (magnesium), Ca (calcium), Ba (barium), and Ra (radium).

19. (a) Group 1 or IA, Period 7; (b) Group 16 or VIA, Period 5; (c) Group 15 or VA, Period 4; (d) Group 6 or VIB, Period 5; (e) Group 2 or IIA, Period 7.

21. (a) helium, neon, argon, krypton, xenon, and radon; (b) lithium, sodium, potassium, rubidium, cesium, and francium; (c) fluorine, chlorine, bromine, iodine, and astatine

23. (a) $\cdot \overset{\cdot}{\underset{\cdot}{C}} \cdot$, (b) Sr:, (c) $\cdot \overset{\cdot}{As} \cdot$, (d) $: \overset{\cdot \cdot}{\underset{\cdot \cdot}{Kr}} :$

Chapter 3

1. Although they travel a shorter distance than gamma rays, alpha particles do more damage inside the body because of their size and electrical charge.

3. a. $^{40}_{19}K \longrightarrow {}^{40}_{20}Ca + {}^{0}_{-1}e$
   b. $^{27}_{13}Al + {}^{2}_{1}H \longrightarrow {}^{28}_{13}Al + {}^{1}_{1}H$
   c. $^{239}_{92}U \longrightarrow {}^{239}_{93}Np + {}^{0}_{-1}e$
   d. $^{233}_{92}U \longrightarrow {}^{4}_{2}He + {}^{229}_{90}Th$
   e. $^{85}_{36}Kr \longrightarrow {}^{85}_{37}Rb + {}^{0}_{-1}e$
   f. $^{226}_{88}Ra \longrightarrow {}^{4}_{2}He + {}^{222}_{86}Rn$
   g. $^{14}_{7}N + {}^{4}_{2}He \longrightarrow {}^{17}_{8}O + {}^{1}_{1}H$
   h. $^{2}_{1}H + {}^{2}_{1}H \longrightarrow {}^{4}_{2}He$

5. Sulfur (S), which has an atomic number of 16.

7. (a) 0.6–76 mrem, (b) 910 mrem, (c) 40 + 50 = 90 mrem

9. Radioactivity can kill cells, particularly those in the act of dividing. Cancer cells are affected more than most normal cells because cancer cells divide rapidly.

11. Alpha, short

13. ^{14}C has a half-life of 5730 years. After three half-lives, beginning in 2050, the year would be $(3 \times 5730) + 2050 = 19,240$.

15. (a)

17. In both nuclear fission and fusion, nuclei are changing to a more stable (lower energy) state; thus, energy is released.

19. Nuclear fusion is the joining together of very small nuclei, usually isotopes of hydrogen, into a larger nucleus.

Chapter 4

1. Chemical bonds form in order for atoms to achieve a state of lower potential energy.

3. (a) Na^+, (b) H_2 molecule

5. (a) 1, (b) 2, (c) 0, (d) 2, (e) 1, (f) 3

7. The nonmetal atoms except Ne could form either ionic compounds (if they reacted with an appropriate metal atom) or covalent compounds (if they reacted with an appropriate nonmetal atom).

9. (a) ionic, (b) covalent, (c) covalent, (d) ionic, (e) covalent, (f) metallic

11. (a) Ra^{2+}, (b) no ion, (c) Sb^{3-}, (d) Al^{3+}, (e) Sr^{2+}, (f) I^-, (g) As^{3-}

13. (a) 36, (b) 54, (c) 54, (d) 10, (e) 10, (f) 46

15. (a) CsF, (b) $GaCl_3$, (c) SrI_2, (d) SrO, (e) K_2O, (f) AlI_3, (g) Ca_3P_2

17. (a) cesium fluoride, (b) gallium chloride, (c) strontium iodide, (d) strontium oxide, (e) potassium oxide, (f) aluminum iodide, (g) calcium phosphide

19. (a) Na_3PO_4, (b) $Ca(OH)_2$, (c) $(NH_4)_2O$, (d) $NaHCO_3$

21. (a) Cl—P—Cl, (b) H—Br,
    |
    Cl

(c) $F-\overset{\displaystyle F}{\underset{\displaystyle F}{C}}-F$, (d) $\overset{\displaystyle }{\underset{\displaystyle H}{S}}-H$

23. (a) PCl_3, (b) HBr, (c) CF_4, (d) H_2S

25. (c) because it is symmetric

27. The electrons originally were valence electrons in the metal atoms.

Chapter 5

1. (b) and (e) are physical changes; (a), (c), (d), (f), and (g) are chemical reactions.

3. See Table 5.1.

5. (a) increase temperature or decrease volume, (b) increase pressure or decrease temperature, (c) increase temperature

7. Butyl alcohol

9. (b), (e), and (f)

11. No (Nonpolar substances don't dissolve in polar solvents.)

13. A chemical reaction occurs when a new substance is formed. Signs of a chemical reaction include bubbles of gas forming, odors or colors changing, formation of a precipitate, burning, or giving off large amounts of heat or light or both.

15. In a balanced equation, the products must have the same number of atoms of each element as the reactants in order to comply with the law of conservation of mass.

17. a. $2Na + Cl_2 \longrightarrow 2NaCl$
    b. $2KNO_3 \longrightarrow 2KNO_2 + O_2$
    c. $2NaHCO_3 \longrightarrow Na_2CO_3 + H_2O + CO_2$
    d. $2K + 2H_2O \longrightarrow 2KOH + H_2$
    e. $2NaCl + Pb(NO_3)_2 \longrightarrow PbCl_2 + 2NaNO_3$

19. (a) 58.5 g, (b) 106.0 g, (c) 114.0 g, (d) 39.9 g, (e) 28.0 g

21. *Reaction 1*: $C + O_2 \longrightarrow CO_2$
    Reaction 2: $2C_8H_{18} + 25O_2 \longrightarrow$
    $$16CO_2 + 18H_2O$$
    Reaction 3: $N_2 + O_2 \longrightarrow 2NO$
    Reaction 4: $2NO + O_2 \longrightarrow 2NO_2$
    Reaction 5: $S + O_2 \longrightarrow SO_2$
    Reaction 6: $2SO_2 + O_2 \longrightarrow 2SO_3$
    Reaction 7: $SO_3 + H_2O \longrightarrow H_2SO_4$
    Reaction 8: $N_2 + 3H_2 \longrightarrow 2NH_3$
    Reaction 9: $6CO_2 + 6H_2O \longrightarrow$
    $$C_6H_{12}O_6 + 6O_2$$

23. 300 mol N_2

25. 2400 g O_2

27. 119 g SO_3

29. 85 g glucose, 91 g O_2

Chapter 6

1. An acid is a substance that in water donates hydrogen ions and thus produces hydronium ions. A base produces hydroxide ions in water or accepts protons, or both.

3. (c), (e), and (f) are acids; (a), (b), (d), and (g) are bases.

5. Strong acids react completely (or nearly completely) with water to produce hydronium ions. Weak acids do not react as completely and thus produce fewer hydronium ions.

7. $2HCl + Ca(OH)_2 \longrightarrow CaCl_2 + 2H_2O$

9. a. $NaOH + HCl \longrightarrow NaCl + \mathbf{H_2O}$
   b. $Mg(OH)_2 + 2HCl \longrightarrow MgCl_2 + \mathbf{2H_2O}$
   c. $NaOH + HCN \longrightarrow \mathbf{NaCN} + H_2O$
   d. $2HNO_3 + Ca(OH)_2 \longrightarrow$
   $$\mathbf{Ca(NO_3)_2} + \mathbf{2H_2O}$$
   e. $H_2SO_4 + 2KOH \longrightarrow \mathbf{K_2SO_4} + \mathbf{2H_2O}$

11. (a) acidic, (b) basic, (c) neutral, (d) basic

13. 3

15. (b) and (c)

17. An oxidizing agent is a substance that gains electrons and thus causes another substance to be oxidized. A reducing agent is a substance that loses electrons and thus causes another substance to be reduced.

19. a. Oxidizing agent is N_2; reducing agent is H_2.
    b. Oxidizing agent is WO_3; reducing agent is H_2.
    c. Oxidizing agent is Fe_2O_3; reducing agent is Al.
    d. Oxidizing agent is PbO_2; reducing agent is Pb.
    e. Oxidizing agent is CO_2; reducing agent is H_2O.
    f. Oxidizing agent is HCl; reducing agent is Al.

21. a. Mg is oxidized; SiO_2 is reduced.
    b. H_2 is oxidized; BCl_3 is reduced.
    c. Fe is oxidized; PbS is reduced.
    d. Cr is oxidized; HgO is reduced.
    e. CH_2O is oxidized; $Cu(OH)_2$ is reduced.
    f. Al is oxidized; Cu^{2+} is reduced.
    g. $C_{55}H_{100}O_6$ is oxidized; O_2 is reduced.

23. $2H_2 + O_2 \longrightarrow 2H_2O$ (H_2 is oxidized; O_2 is reduced.)

25. O_2

Chapter 7

1. (a) In all chemical and physical changes, energy is neither created nor destroyed but merely transformed from one form to another.

The total energy of the observable universe remains constant. You can't get something (energy) for nothing. (b) In any conversion of high-quality energy to useful work, some energy is always degraded to a more dispersed and less useful form. Heat energy flows spontaneously from hot to cold. Any process increases the entropy of the universe.

3. (a) An exothermic reaction gives off heat to the surroundings; (b) an endothermic reaction absorbs energy from the surroundings.

5. (a) and (c) are due to decreased potential energy; (b) is due to increased entropy.

7. Some processes are spontaneous despite a decrease in entropy in the system because of a sufficient decrease in potential energy of the system. Any spontaneous process, however, results in an increase in the entropy of the system plus its surroundings.

9. By falling off the wall and breaking, Humpty Dumpty went to a state of increased entropy. That process cannot be reversed without a great expenditure of energy and a further increase in entropy of the surroundings.

11. According to the second law, lower-quality forms of energy (often heat) are produced in energy conversions.

13. Collision frequency, orientation, and energy

15. A catalyst

17. Decomposing sugar in your hand (left) has a higher activation energy than in your body (right), which uses enzyme catalysts.

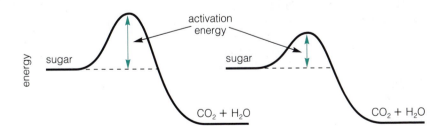

19. (b), (c), and (d) are true; (a) is false.

21. Le Châtelier's principle states that if a system at equilibrium is placed under stress, the system will adjust to relieve the stress. See Section 7.4 for examples.

23. (a), (d), and (e) go to the left; (b), (c), and (f) go to the right.

25. Removing NO shifts the equilibrium (of the reaction in question 24) to the right.

27. Shift to the right

Chapter 8

1. Carbon can form stable chains and rings and single, double, or triple covalent bonds with carbon and other atoms; bond with a variety of other nonmetals; and form many different isomers. Carbon forms covalent bonds; thus, it bonds to nonmetals.

3. An alkene contains one or more carbon-to-carbon double bonds. An aromatic compound contains a benzene or benzenelike ring.

5. $CH_3CH=CH_2$; $\overset{CH_2}{\underset{CH_2-CH_2}{\diagup\diagdown}}$ or $\triangle$

7. $CH_3CH=CHCH_3$, $CH_2=CHCH_2CH_3$,
$\underset{CH_3}{\overset{|}{CH_3C=CH_2}}$, $\underset{CH_2-CH_2}{\overset{|}{CH_2-CH_2}}$, $\underset{CH_2-CH_2}{\overset{CH-CH_3}{\diagup}}$

9. Gasoline. Being a nonpolar hydrocarbon, gasoline lacks the strong hydrogen bonding present in water; thus, at warm temperatures, gasoline molecules separate and vaporize more readily than does water.

11. The gasoline obtained directly from fractional distillation doesn't work well in modern internal combustion engines; its octane number is too low. This gasoline can be upgraded by cracking, polymerization, and catalytic reforming as described in Table 8.8.

13. The octane number is based on the performance of a fuel in a standard engine compared with pure iso-octane (octane number = 100) and pure n-heptane (octane number = 0). An octane number of 87, for example, means the performance is equivalent to that of a mixture of 87 percent iso-octane and 13 percent n-heptane.

15. (a) n-butane, (b) propene, (c) ethanol (ethyl alcohol), (d) benzene, (e) ethyl chloride, and (f) aminoethane (ethyl amine)

17. A functional group is an arrangement of atoms in a molecule that accounts for most of the chemical behavior of the substance.

19. Oxygen; nitrogen

21. Ethanol. Both ethanol and octanol have a polar —OH group, but octanol contains much more nonpolar, hydrocarbon material than does ethanol; this nonpolar material decreases the solubility in water.

Chapter 9

1. Both dogs and people are composed of carbohydrates, lipids, proteins, nucleic acids, vitamins, minerals, and water; the types of these substances, particularly nucleic acids and proteins, are different in different living things. Inanimate objects such as rocks are composed primarily of minerals.

3. (a) fructose and (c) glucose

5. Celery contains cellulose, a polymer of glucose that we cannot digest and thus furnishes no calories. Potato chips contain starch, a polymer of glucose we can digest that provides calories.

7. All these lipids consist mostly of nonpolar hydrocarbon material.

9. CH_2—O—$\overset{\overset{\textstyle O}{\|}}{C}$—$(CH_2)_{16}CH_3$ (stearic acid unit)

  CH—O—$\overset{\overset{\textstyle O}{\|}}{C}$—$(CH_2)_7(CH{=}CHCH_2)_3CH_3$ (linolenic acid unit)

  CH_2—O—$\overset{\overset{\textstyle O}{\|}}{C}$—$(CH_2)_7CH{=}CH(CH_2)_7CH_3$ (oleic acid unit)

  This triglyceride would be more likely to occur in an oil because of its unsaturation.

11. Carbohydrates

13. Amine group and carboxylic acid group; amino acids differ in their R groups.

15. H_2N—$\overset{\underset{\textstyle CH_2}{|}}{CH}$—$\overset{\overset{\textstyle O}{\|}}{C}$—$\overset{\underset{\textstyle H}{|}}{N}$—$\overset{\underset{\textstyle CH_2}{|}}{CH}$—$\overset{\overset{\textstyle O}{\|}}{C}$—OH

17. Sickle-cell anemia is caused by substituting one amino acid in a protein chain in hemoglobin. This change makes red blood cells sickle-shaped, more fragile, less able to bind oxygen, and more likely to clog small blood vessels.

19. Base, pentose, and phosphate

21. hydrogen bonds; weak (but numerous)

Chapter 10

1. (a) biosphere, (b) lithosphere, (c) biosphere, (d) hydrosphere, (e) atmosphere, (f) biosphere, (g) lithosphere, and (h) atmosphere and hydrosphere

3. (a) nitrogen oxides, NO_3^-, NO_2^-; (b) proteins, NH_3, NH_4^+

5. Redwood forests and diamonds are nonrenewable unless they are used much more slowly. Slag is not currently a resource. Broken glass could become more of a resource and could be renewable.

7. Human ingenuity makes a substance useful (and thus a resource) if it is affordable.

9. Wind and water

11. Sand and quartz (silica)

13. Semiconductor

15. a. $Fe_2O_3 + 3C \xrightarrow[\text{temperature}]{\text{high}}$
    $3CO + 2Fe$ (pig iron)

    b. $2C + O_2 \xrightarrow[\text{temperature}]{\text{high}} 2CO$ (purifies Fe)

17. $ZnO + C \longrightarrow Zn + CO$

19. Because of the energy needed for electrolytic reduction, it takes about five times more energy to produce a ton of Al than a ton of steel.

21. Recycling is collecting, reprocessing, and refabricating products. Reuse means to use the same product repeatedly.

23. See Table 10.3.

Chapter 11

1. Most oil in the Mideast is more accessible and takes less energy to obtain from drilling than oil in the United States.

3. According to the second law of thermodynamics, some useful energy is lost in the process of making hydrogen gas from water. Not all of the original useful energy can be regained in burning the hydrogen gas back to water, so there is a negative net energy.

5. The energy of collecting (cutting down) and transporting the wood and the method of burning all reduce the net energy.

7. (a) Most of the best dam sites are already in use. This limits the potential for producing much more hydroelectric power. (b) Generating the electricity (from fossil fuels or other resources) may be less efficient than burning gasoline in cars directly. (c) All-electric homes and buildings are less efficient because several steps are needed to produce and deliver energy in the form of electricity. At each step, there is some loss in useful energy.

9. Both in coal gasification and in coal liquefaction, some of the useful energy in the coal is lost during the conversion.

11. (a) supply fission fuel, (b) slow neutrons and cool the reactor, and (c) absorb neutrons to slow or stop fission

13. The neutrons produced can make nearby materials radioactive.

15. Active solar uses solar collectors to collect the energy and transfer it to another system. Passive solar uses a building's design and composition to capture and store solar energy.

17. (b), (e), and (f) (these energy forms are continuously available).

19. Fermentation of corn

21. Geothermal energy will be a major energy source only in local areas that have favorable deposits.

7. Burning coal and oil to produce electricity and heat, metal smelters, petroleum refineries

9. $2SO_2 + O_2 \longrightarrow 2SO_3$
$SO_3 + H_2O \longrightarrow H_2SO_4$
$H_2SO_4 + 2NH_3 \longrightarrow (NH_4)_2SO_4$

11. Scrubbers interact pollutants such as particulates and SO_2 with materials in the smokestack to prevent the emission of the pollutants. One example of a scrubber is in Figure 12.6.

13. The brown haze comes (at least partly) from the presence of NO_2.

15. At ground level, ozone is undesirable; it causes respiratory and other damage. In the stratosphere, ozone is desirable because it shields the earth from high-energy UV radiation.

17. At high altitude, the engine would need more air (thus, a leaner fuel mixture) to provide enough oxygen to burn the fuel. This change would increase nitrogen oxide emissions and reduce hydrocarbon and carbon monoxide emissions.

19. CO

21. See Figure 12.17.

23. Antarctica

25. $(1.7) \times (1.8) = 3.1°F$

27. Pollutants from human activities often are more concentrated and in areas of high human population, so they may do more harm.

Chapter 12

1. The troposphere extends from the earth's surface upward about 8 to 12 km (5 to 7 miles). The stratosphere is 20 to 50 km (12 to 31 miles) above the earth.

3. CO

5. Mobile homes frequently have less air circulation and ventilation and more plywood and other materials from which organic compounds vaporize.

Chapter 13

1. See Figure 13.6.

3. H_2O is polar, whereas O_2 and H_2 are nonpolar. Hydrogen bonds between polar water molecules (which are absent in H_2 and O_2) make water a liquid at room temperature.

5. No. If water were linear, it would be nonpolar.

7. The six-sided (hexagonal) shapes in snowflakes reflect the pattern of hydrogen bonding in ice (see Figure 13.8).

9. Because reactions are slower at colder temperatures, colder water would require a longer waiting period before its microbes were oxidized (killed).

11. Benzene or benzenelike ring

13. Aerobic decay occurs in the presence of O_2; anaerobic decay is in the absence of O_2. Anaerobic decay is less desirable because its products have offensive odors and other undesirable qualities.

15. Heat reduces the amount of dissolved oxygen in water. Goldfish have a better chance of surviving in cool water, which has more oxygen and produces a less rapid metabolic rate.

17. Salt in water is in a high-entropy state. According to the second law, energy is required to remove the salt so that it is in a lower-entropy state.

19. Pesticides and nitrates from fertilizers

21. Precipitation, reverse osmosis, ion exchange

23. Ozone (O_3) is more soluble because it is polar; like dissolves like (Section 5.3).

Chapter 14

1. (a) A monomer is a small molecule from which a polymer can be made; examples are ethylene, isoprene, vinyl chloride, and diamines. (b) Natural polymers include rubber, DNA, proteins, and cotton. (c) Synthetic polymers include polyethylene, silicones, epoxies, nylon, and polystyrene.

3. (a) Addition polymerization occurs with no removal of atoms from the monomers; an example is polyethylene. (b) Condensation polymerization occurs with removal of atoms from the monomers to form a small molecule such as water; an example is nylon.

5. (a) and (g) (both have one or more carbon-to-carbon double bonds).

7. Nylon has many amide links, which are more reactive than the carbon–carbon and carbon–fluorine links in Teflon.

9. Polyacrylonitrile

11. (a) and (b)

13. Vulcanization

15. (c), (d), and (f)

17. Polyester means many ester ($-\overset{\overset{\displaystyle O}{\|}}{C}-O-$) linkages in the polymer. These ester linkages form by the reaction of an alcohol ($-OH$) group in one monomer with a carboxylic acid ($-\overset{\overset{\displaystyle O}{\|}}{C}-OH$) group in a different monomer with the elimination of H_2O.

19. A silicone polymer has alternating silicon and oxygen atoms for its basic structural framework. These polymers form by reactions between silicone alcohol monomers.

21. Polyvinyl chloride (PVC) and polyvinylidene chloride (Saran) could give off HCl gas when burned. Polyacrylonitrile could give off HCN gas.

23. Many exist. Two examples are

Chapter 15

1. Ethyl alcohol is polar enough to dissolve in water, a polar solvent. Cetyl alcohol contains a long, nonpolar hydrocarbon region that is insoluble in water.

3. (b) and (e) would be effective surfactants; (d) would have moderate surfactant properties; (a) and (c) would not be surfactants.

5. (b) and (c) are true; (a), (d), and (e) are false.

7. See Figure 15.7.

9. Hard water, unlike soft water, contains substantial amounts of Ca^{2+}, Mg^{2+}, or Fe^{2+} ions. Hard water is softened by removing these ions with ion-exchange resins (such as zeolite) or other methods.

11. Hard water limits the usefulness of soap for washing clothes.

13. (b) is an anionic detergent; (d) is a weak nonionic detergent; (e) is a cationic detergent.

15. Nonionic surfactants don't react with the ions in hard water and produce fewer suds to clog washing machines, but are not highly effective in keeping removed dirt suspended in the rinse water.

17. A builder is a chemical water softener that keeps mineral ions from fouling the surfactant.

19. Phosphorus is a plant nutrient that promotes eutrophication. Legislators in some states have limited the amount of phosphorus in detergents, and manufacturers are developing substitutes for phosphates as builders.

21. Optical brighteners cause additional visible blue light to be reflected from the fabric. This masks any yellow appearance and makes the clothes appear to be "whiter than white."

Chapter 16

1. $Ca_5(PO_4)_3OH \longrightarrow 5Ca^{2+} + 3PO_4^{3-} + OH^-$

3. (a) NaF, (c) PO_3F^{2-}, and (d) SnF_2

5. An emollient prevents water from evaporating from the skin; a humectant attracts water to the skin.

7. The more oil-like of the two, (b) $CH_3(CH_2)_{14}OCH_3$

9. (a) solvent, (b) solvent, (c) absorbant, (d) surfactant, (e) surfactant

11. UV light can cause sunburn, premature aging, and skin cancer.

13. Fair-skinned people, who burn easily, should use a sunscreen with a high SPF number. Olive-skinned people, who seldom burn, can choose sunscreens with lower SPF numbers. Longer exposures and greater intensity of radiation require sunscreens with higher SPF numbers.

15. Darker-tinted sunglasses may screen out more visible light, but not necessarily more UV light.

17. Perfumes have a higher content of essential oils to provide fragrance and less solvent. Colognes are essentially diluted perfumes.

19. Aluminum chlorohydrate shrinks the ducts of sweat glands, clogs those ducts, and kills bacteria.

21. Too much sebum makes your hair greasy and matted; too little makes it dry and strawlike.

23. (b), (c), (d), and (h)

Chapter 17

1. a. $6CO_2 + 6H_2O \longrightarrow C_6H_{12}O_6 + 6O_2$
   b. $C_6H_{12}O_6 + 6O_2 \longrightarrow 6CO_2 + 6H_2O$

3. Needed in large amounts: P, Ca, S, N, K; needed only in trace amounts: Mo, Zn, Mg

5. $N_2 + 3H_2 \longrightarrow 2NH_3$

7. Animal manure, plant stalks and other plant residues, and compost are forms of organic fertilizers, which are rich sources of carbon and nitrogen.

9. Add lime, $CaCO_3$, to the soil.

11. (a) chlorinated hydrocarbon, (c) and (d) organophosphates, (b) and (e) carbamates

13. (a) chloro group; (b) chloro, alkene, and phosphate groups; (c) ester and modified (with S) phosphate groups; (d) amide and ester groups that overlap

15. Contact herbicides kill by direct contact, whereas systemic herbicides travel throughout the plant and kill it.

17. Pesticides that break down rapidly in the environment do not cause damage for extended periods. But this also means they must be applied more frequently for sustained pesticide action.

19. Glyphosate inhibits a plant enzyme that normally helps synthesize needed aromatic amino acids.

21. The substance produces unusually large larvae because in its presence there is more juvenile hormone to stimulate the larvae to grow.

Chapter 18

1. Roughage or bulk is indigestible fiber in the diet. It helps your digestive system to function properly and prevents constipation. It may even absorb cancer-causing substances.

3. Fats (9 Cal/g)

5. (e) starch

7. Carbohydrates (potatoes, sugar), fats (cooking oil, butter), and proteins (meats, fish)

9. Energy is measured in calories. Food does not actually contain calories; rather, it contains chemical potential energy in the bonds of its molecules. A 245-Cal candy bar has the potential of producing that much energy when it is metabolized by the body.

11. Essential fatty acids possess multiple carbon–carbon double bonds. You must get these needed fatty acids in your diet because your body cannot synthesize them from other substances.

13. Of the twenty amino acids in our proteins, our bodies can synthesize twelve from other substances. The other eight are essential in the diet.

15. Meat, fish, poultry, eggs, and dairy products contain complete protein. A strict vegetarian must eat a variety of plant products (for example, legumes with grains) that together contain sufficient amounts of all the essential amino acids.

17. Vitamins A, D, E, and K can accumulate in the body because they are nonpolar (fat-soluble) and thus are not readily flushed from the body in urine.

19. Oxygen

21.

$$CH_3CH{=}CHCH{=}CH{-}\overset{\overset{\displaystyle O}{\|}}{C}{-}OH$$

Sorbic acid

$$\overset{\overset{\displaystyle O}{\|}}{C}{-}O^-K^+$$

potassium benzoate

23. BHA and BHT are easily oxidized molecules. Their presence in foods prevents fats and oils from becoming oxidized and rancid as easily.

Chapter 19

1. A drug is a chemical used as medicine to treat disease. (b) KI (to treat goiter) and (d) insulin (to treat diabetes) are drugs. (a) Caffeine, (c) LSD, and (e) birth control pills (to block fertility) do not clearly fit this definition of a drug.

3. Sulfa drugs are synthetic. They kill bacteria by blocking their ability to make folic acid from para-aminobenzoic acid (PABA).

5. Ampicillin has the same structure as penicillin G except that ampicillin contains an extra amino (NH_2) group, shown in color in Figure 19.7. This structural difference makes ampicillin acid-resistant, so it can be taken orally. Penicillin G must be taken by injection.

7. Excess use of antibiotics can cause (a) destruction of harmless bacteria in the body that are replaced by harmful ones, (b) allergic reactions in susceptible people, and (c) widespread development of resistant strains of bacteria.

9. (a) Benzocaine anesthetizes (numbs) the throat; (b) codeine inhibits the cough center in the brain; and (c) antihistamine blocks nasal congestion, runny noses, and other symptoms due to the release of histamine during an allergic response.

11. Thymidine

13. Drugs to treat cancer also destroy rapidly dividing normal cells, causing side effects such as decreased blood cells, nausea, vomiting, and hair loss.

15. (b) An anticoagulant is used as a rat poison. The anticoagulant is effective in rats at much lower concentrations than in people, so it is not normally toxic to humans.

17. Goiter is a disease of thyroid deficiency. It is treated by supplying the iodide ion (I^-) in the diet.

19. Oral hypoglycemic drugs lower blood glucose levels in diabetics by stimulating the pancreas to secrete more insulin.

21. (c) Cortisone and (d) estrogens (Cholesterol (a) is a steroid but is not classified as a hormone.)

23. Ketone group

Chapter 20

1. Acetylcholine (ACH) is released from the axon of a neuron and crosses a synapse to bind to a dendrite of a nearby neuron. ACH then triggers a flow of ions that allows the nerve impulse to travel across the synapse to the receiving neuron.

3. No brand-name aspirins contain 100 percent acetylsalicylic acid. All contain inert materials such as binders. Major brands of aspirin do not differ in effectiveness, and they contain similar dosages, which are listed on the label.

5. Morphine

7. Naloxone binds to the same receptor sites in the brain as does heroin, so it displaces the heroin from those sites and reduces the effects of a heroin overdose.

9. Cocaine

11. Accidental deaths occur with barbiturates because (1) the margin of safety between a lethal dose and an effective dose shrinks as tolerance develops, (2) a user may forget how much barbiturate he or she already took, and (3) barbiturates are very dangerous in combination with ethanol.

13. Cirrhosis and birth defects

15. L-Dopa is converted to dopamine in the brain. People with Parkinson's disease lack the neurotransmitter, dopamine, so L-Dopa helps relieve this deficit.

17. (a) amphetamines, (b) marijuana, (c) methylphenidate (Ritalin), and (d) various antidepressants, including lithium carbonate

19. People who have developed a tolerance for one hallucinogen frequently also have a tolerance for other hallucinogens, even though they haven't taken the substance(s).

21. See Table 20.4. Psychological dependence can develop with all these drugs. Tolerance and physical dependence occur with (a), (b), (c), (g), (h), (i), and (j). Physical dependence is not established for (d), (e), and (f).

Chapter 21

1. LD_{50} is the lethal dose for 50 percent of the test population. It is determined by finding the dose of a toxic material that kills 50 percent of the animals tested.

3. Activated charcoal (carbon) adsorbs (binds) many toxic materials to its surface and carries them into the intestine for excretion.

5. See Table 21.2.

7. Hemoglobin binds CO much more readily than it binds O_2, so 50 percent or more of a person's hemoglobin could be bound to CO even when there is more O_2 present than CO.

9. Calcium ions (Ca^{2+}) react with fluoride ions (F^-) to form calcium fluoride (CaF_2), which

is insoluble; this effectively removes fluoride ions. Using table salt (NaCl), however, would produce sodium fluoride (NaF), which is soluble; thus, fluoride ions would not be effectively removed by this treatment.

11. BAL or EDTA would *not* be effective antidotes to nitrite poisoning because BAL and EDTA bind positively charged metal ions. Nitrite (NO_2^-) has a negative charge.

13. See Table 21.4.

15. (a) has negatively charged groups that can attract and bind to positively charged metal ions.

17. An *initiator* apparently damages DNA. Subsequent, prolonged exposure to a *promoter* substance leads to the *progression* stage, in which the cancer develops.

19. No. Some mutations have essentially no effect.

Chapter 22

1. A gene is a section of DNA in a chromosome. There are many genes on each chromosome.

3. (a) Adenine (A) pairs with (d) thymine (T); (b) cytosine (C) pairs with (c) guanine (G).

5. Vaccines enable the immune system to recognize disease-causing bacteria and viruses and to produce antibodies to destroy them. Vaccines produced by genetic engineering may be safer because the patient receives bacterial or viral proteins rather than a weakened form of the actual infectious agent.

7. Diseases due to the absence of a single gene, particularly in cells (such as bone marrow) that are readily accessible, are the most likely to be treatable by genetic therapy. Other genetic diseases will be more complex to treat.

9. Some genetic diseases may be treatable by supplying enough of the right kind of cells so that the patient can produce enough of the product designated by the new gene(s). Not all cells would necessarily have to carry the new gene(s).

11. An "artificial kidney" dialysis machine removes toxic waste materials from the patient's blood. It does not work as well as a real kidney, so the best long-term solution is a kidney transplant.

13. Respirators help the patient breathe, causing the blood to become oxygenated as it passes by the lungs. Membrane oxygenators directly oxygenate the blood as it circulates out of the body.

15. Erythropoietin stimulates the production of red blood cells; this may replace blood doping, which accomplishes the same thing.

17. Carbohydrate loading increases the amount of glycogen stored in the liver and muscles. It is done by first depleting the glycogen (by exercise or diet) and then by eating carbohydrate-rich foods for three to four days before competition. Athletes competing in events of short duration are unlikely to benefit from this.

19. (a) Caffeine releases stored fats and carbohydrates; it is used prior to endurance events. (b) Beta blockers are used by shooters to slow their pulse. (c) Amphetamines are used as a stimulant for various types of competition. None of these drugs, however, is legal to use in competition.

Appendix 1

1. (b) 3.06982×10^5; (d) 1×10^{-15}, or 10^{-15}; (f) 9×10^{-9} g of mercury; (h) 1.32×10^{34}; (j) 3.060×10^{-5}

2. (b) 2.10×10^{-21}; (d) 4.2×10^{-19}; (f) 7.44×10^{18}

3. (b) 1.5×10^{-9}; (d) 6.9×10^8; (f) 8.25×10^{64}

Appendix 2

1. (b) 91.4 m; (d) 2.50×10^6 mg; (f) 1.8×10^4 cigarettes; (h) 94 L; (j) 285 mL; (l) 2.82×10^6 J

3. $-196°C$

Further Readings

Chapter 1

Akeroyd, F. Michael. "Chemistry, Biochemistry and the Growth of Scientific Knowledge," *Journal of Chemical Education*, May 1984, pp. 434–436.

Black, Bert. "Evolving Legal Standards for the Admissibility of Scientific Evidence," *Science*, 25 March 1988, pp. 1508–1512.

Bronowski, Jacob. *The Ascent of Man*. London: British Broadcasting Corporation, 1973.

Diamond, Jared. "Soft Sciences Are Often Harder Than Hard Sciences," *Discover*, August 1987, pp. 34–39.

Lenox, Ronald S. "Educating for the Serendipitous Discovery," *Journal of Chemical Education*, April 1985, pp. 282–285.

Medawar, Peter B. *The Limits of Science*. New York: Harper & Row, 1984.

Snow, Charles Percy. *The Two Cultures and the Scientific Revolution*. New York: Cambridge University Press, 1959.

"Why Scientific Fact Is Sometimes Fiction," *The Economist*, 28 February 1987, pp. 97–98.

Chapter 2

Asimov, Isaac. *The Search for the Elements*. New York: Fawcett Books, 1962.

Cole, K. C. "On Imagining the Unseeable," *Discover*, December 1982, pp. 70–72.

Fernelius, W. Conrad. "Some Reflections on the Periodic Table and Its Use," *Journal of Chemical Education*, March 1986, pp. 263–266.

Kolb, Doris. "Chemical Principles Revisited: What Is an Element?" *Journal of Chemical Education*, November 1977, pp. 696–700.

Levy, Primo. *The Periodic Table*. New York: Schocken Books, 1984.

Loening, K. L. "Recommended Format for the Periodic Table of the Elements," *Journal of Chemical Education*, February 1984, p. 136.

Treptow, Richard S. "Conservation of Mass: Fact or Fiction?" *Journal of Chemical Education*, February 1986, pp. 103–105.

Chapter 3

Foster, N. "Diagnostic Radiopharmaceuticals," *Journal of Chemical Education*, August 1984, pp. 689–693.

Kerr, Richard A. "Indoor Radon: The Deadliest Pollutant," *Science*, 29 April 1988, pp. 606–608.

Marshall, Eliot. "Nuclear Winter Debate Heats Up," *Science*, 16 January 1987, pp. 271–273.

Rhodes, Richard. *The Making of the Atomic Bomb*. New York: Simon & Schuster, 1987.

Rosenthal, Elisabeth. "The Hazards of Everyday Radiation," *Science Digest*, March 1984, pp. 38–43, 96–97.

Sime, Ruth Lewin. "Lise Meitner and the Discovery of Fission," *Journal of Chemical Education*, May 1989, pp. 373–376.

Weiss, Rick. "The Gamma-Ray Gourmet," *Science News*, 19 and 26 December 1987, pp. 398–399.

Zurer, Pamela S. "Archaeological Chemistry," *Chemical & Engineering News*, 21 February 1983, pp. 26–42.

Chapter 4

Dahl, Peter. "The Valence-Shell Electron-Pair Repulsion Theory," *Chemistry*, March 1973, pp. 17–19.

DeKock, Roger L. "The Chemical Bond," *Journal of Chemical Education*, November 1987, pp. 934–941.

Peterson, Raymond F. and David F. Treagust. "Grade-12 Students' Misconceptions of Covalent Bonding and Structure," *Journal of Chemical Education*, June 1989, pp. 459–460.

Tiernan, Natalie Foote. "Gilbert Newton Lewis and the Amazing Electron Dots," *Journal of Chemical Education*, July 1985, pp. 569–570.

Webb, Valerie J. "Hydrogen Bond 'Special Agent,'" *Chemistry*, June 1968, pp. 16–20.

Chapter 5

Arena, Blaise J. "Ammonia: Confronting a Primal Trend," *Journal of Chemical Education*, December 1986, pp. 1040–1044.

Bent, Harry A. "Carbon Dioxide: Its Principal Properties Displayed and Discussed," *Journal of Chemical Education*, February 1987, pp. 167–171.

Poole, Richard L. "Teaching Stoichiometry: A Two Cycle Approach," *Journal of Chemical Education*, January 1989, pp. 57–58.

Scott, Arthur F. "The Invention of the Balloon and the Birth of Modern Chemistry," *Scientific American*, January 1984, pp. 126–137.

Strong, L. E. "Balancing Chemical Equations," *Chemistry*, January 1974, p. 13.

Chapter 6

Ennis, John L. "Photography at Its Genesis," *Chemical & Engineering News*, 18 December 1989, pp. 26–42.

Kauffman. "The Brønsted–Lowry Acid–Base Concept," *Journal of Chemical Education*, January 1988, pp. 28–31.

Kolb, Doris. "The Chemical Equation: Part II: Oxidation–Reduction Reactions," *Journal of Chemical Education*, May 1978, pp. 326–331.

Staff Report. "Chem I Supplement: Everyday Examples of Oxidation–Reduction Processes," *Journal of Chemical Education*, May 1978, pp. 332–333.

Treptow, Richard S. "The Conjugate Acid–Base Chart," *Journal of Chemical Education*, November 1986, pp. 938–941.

Chapter 7

Feldman, Martin R. and Monica L. Tarver. "Fritz Haber," *Journal of Chemical Education*, June 1983, pp. 463–464.

Haim, Albert. "Catalysis: New Reaction Pathways, Not Just a Lowering of the Activation Energy," *Journal of Chemical Education*, November 1989, pp. 935–937.

Ihde, John. "Le Châtelier and Chemical Equilibrium," *Journal of Chemical Education*, March 1989, pp. 238–239.

Mickey, Charles D. "Chemical Equilibrium," *Journal of Chemical Education*, November 1980, pp. 801–804.

Treptow, Richard S. "Le Châtelier's Principle: A Reexamination and Method of Graphic Illustration," *Journal of Chemical Education*, June 1980, pp. 417–420.

Chapter 8

Budavari, Susan (Ed.). *The Merck Index*, 11th ed. Rahway, N.J.: Merck & Co., 1989.

Kolb, Kenneth E. and Doris Kolb. "Organic Chemicals from Carbon Monoxide," *Journal of Chemical Education*, January 1983, pp. 57–59.

Miller, Foil A. and George B. Kauffman. "Alfred Nobel and Philately: The Man, His Work, and His Prizes," *Journal of Chemical Education*, October 1988, pp. 843–846.

Schumm, Margot K. *Understanding Organic Chemistry*. New York: Macmillan, 1987.

Seymour, Raymond B. "Alkanes: Abundant, Pervasive, Important, and Essential," *Journal of Chemical Education*, January 1989, pp. 59–63.

Spitz, Peter H. *Petrochemicals: The Rise of an Industry*. New York: Wiley, 1988.

Wittcoff, Harold. "Nonleaded Gasoline: Its Impact on the Chemical Industry," *Journal of Chemical Education*, September 1987, pp. 773–776.

Chapter 9

Bohinski, Robert C. *Modern Concepts in Biochemistry*, 5th ed. Boston: Allyn and Bacon, 1987.

Cromartie, Thomas H. "The Inhibition of Enzymes by Drugs and Pesticides," *Journal of Chemical Education*, September 1986, pp. 765–768.

"Life's Recipe," *The Economist*, 30 May 1987, pp. 84–85.

Oliver, William R. and Diana Comb McGill. "Butter and Margarine: Their Chemistry, Their Conflict," *Journal of Chemical Education*, July 1987, pp. 596–598.

Stryer, Lubert. *Biochemistry*, 3rd ed. San Francisco: Freeman, 1988.

Watson, James D. *The Double Helix*. New York: Atheneum, 1968.

Chapter 10

Craig, Norman C. "Charles Martin Hall—The Young Man, His Mentor, and His Metal," *Journal of Chemical Education*, July 1986, pp. 557–559.

Fischhoff, Baruch, Sarah Lichtenstein, Paul Slovic, Stephen L. Derby, and Ralph L. Keeney. *Acceptable Risk*. Cambridge, England: Cambridge University Press, 1984.

Hanson, David J. "Hazardous Waste Management: Planning to Avoid Future Problems," *Chemical & Engineering News*, 31 July 1989, pp. 9–17.

Kumar, Vinay and Linda Milewski. "Charles Martin Hall and the Great Aluminum Revolution," *Journal of Chemical Education*, August 1987, pp. 690–691.

Miller, G. Tyler, Jr. *Living in the Environment*, 6th ed. Belmont, Calif.: Wadsworth, 1990.

Ulrich, Donald R. "Chemical Processing of Ceramics," *Chemical & Engineering News*, 1 January 1990, pp. 28–40.

Chapter 11

Anspaugh, Lynn R., Robert J. Catlin, and Marvin Goldman. "The Global Impact of the Chernobyl Reactor Accident," *Science*, 16 December 1988, pp. 1513–1519.

Burnett, W. M. and S. D. Ban. "Changing Prospects for Natural Gas in the United States," *Science*, 21 April 1989, pp. 305–310.

Hubbard, H. M. "Photovoltaics Today and Tomorrow," *Science*, 21 April 1989, pp. 297–304.

Lumpkin, Robert E. "Recent Progress in the Direct Liquefaction of Coal," *Science*, 19 February 1988, pp. 873–877.

Oliver, W. R., R. J. Kempton, and H. A. Conner. "The Production of Ethanol from Grain," *Journal of Chemical Education*, January 1982, pp. 49–52.

Raloff, Janet. "Energy Efficiency: Less Means More," *Science News*, 7 May 1988, pp. 296–298.

Raloff, Janet. "Fallout Over Nevada's Nuclear Destiny," *Science News*, 6 January 1990, pp. 11–12.

Schachter, Y. "Oil Shale—Heir to the Petroleum Kingdom?" *Journal of Chemical Education*, September 1983, pp. 750–752.

Taylor, John J. "Improved and Safer Nuclear Power," *Science*, 21 April 1989, pp. 318–325.

Chapter 12

Barnhardt, Wilton. "The Death of Ducktown," *Discover*, October 1987, pp. 34–43.

Charola, A. Elena. "Acid Rain Effects on Stone Monuments," *Journal of Chemical Education*, May 1987, pp. 436–437.

Ember, Lois R. "Survey Finds High Indoor Levels of Volatile Organic Chemicals," *Chemical & Engineering News*, 5 December 1988, pp. 23–25.

Hileman, Bette. "Global Warming," *Chemical & Engineering News*, 13 March 1989, pp. 25–44.

McElroy, Michael B. and Ross J. Salawitch. "Changing Composition of the Global Stratosphere," *Science*, 10 February 1989, pp. 763–770.

Monastersky, Richard. "Looking for Mr. Greenhouse," *Science News*, 8 April 1989, pp. 216–221.

Nero, Anthony V., Jr. "Controlling Indoor Air Pollution," *Scientific American*, May 1988, pp. 42–48.

Raloff, Janet. "CO$_2$: How Will We Spell Relief?" *Science News*, 24 and 31 December 1988, pp. 411–414.

Rowland, F. Sherwood. "Chlorofluorocarbons and the Depletion of Stratospheric Ozone," *American Scientist*, January/February 1989, pp. 36–45.

Schwartz, Stephen E. "Acid Deposition: Unraveling a Regional Phenomenon," *Science*, 10 February 1989, pp. 753–763.

Zurer, Pamela S. "Producers, Users Grapple with Realities of CFC Phaseout," *Chemical & Engineering News*, 24 July 1989, pp. 7–13.

Chapter 13

Hileman, Bette. "The Great Lakes Cleanup Effort," *Chemical & Engineering News*, 8 February 1988, pp. 22–39.

Knox, Charles E. "What's Going on Down There?" *Science News*, 3 December 1988, pp. 362–365.

Lahey, William and Michael Conner. "The Case for Ocean Waste Disposal," *Technology Review*, August/September 1983, pp. 61–68.

Roberts, Leslie. "Long, Slow Recovery Predicted for Alaska," *Science*, 7 April 1989, pp. 22–23.

Schindler, D. W. et al. "Long-Term Ecosystem Stress: The Effects of Years of Experimental Acidification on a Small Lake," *Science*, 21 June 1985, pp. 1395–1401.

Smith, Richard A., Richard B. Alexander, and M. Gordon Wolman. "Water-Quality Trends in the Nation's Rivers," *Science*, 27 March 1987, pp. 1607–1615.

Sun, Marjorie. "Ground Water Ills: Many Diagnoses, Few Remedies," *Science*, 20 June 1986, pp. 1490–1493.

Watson, Lyall. *The Water Planet*. New York: Crown Publishers, 1988.

Chapter 14

Carraher, Charles E., Jr., and Raymond B. Seymour. "Polymer Structure—Organic Aspects (Definitions)," *Journal of Chemical Education*, April 1988, pp. 314–318.

Conner, Daniel K. and Robert O'Dell. "The Tightening Net of Marine Plastics Pollution," *Environment*, January/February 1988, pp. 16–35.

Friedel, Robert. "The First Plastic," *Invention & Technology*, Summer 1987, pp. 18–23.

Greek, Bruce F. "Modest Growth Ahead for Rubber," *Chemical & Engineering News*, 21 March 1988, pp. 25–51.

Hounshell, David A. and John Kenly Smith, Jr. "The Nylon Drama," *Invention & Technology*, Fall 1988, pp. 40–55.

Rudolph, Barbara. "Second Life for Styrofoam," *Time*, 22 May 1989, p. 84.

Seymour, Raymond B. "Polymers Are Everywhere," *Journal of Chemical Education*, April 1988, pp. 327–334.

Chapter 15

Greek, Bruce F. "Detergent Components Become Increasingly Diverse, Complex," *Chemical & Engineering News*, 25 January 1988, pp. 21–53.

Greek, Bruce F. and Patricia L. Layman. "Higher Costs Spur New Detergent Formulations," *Chemical & Engineering News*, 23 January 1989, pp. 29–49.

Hammond, A. L. "Phosphate Replacements: Problems with the Washday Miracle," *Science*, 23 April 1971, pp. 361–363.

Soap and Detergent Association. "Detergents—In Depth, '84," "Soap and Detergents," "A Handbook of Industry Terms," "Measuring Your Way to a Better Wash," available free from the Soap and Detergent Association, 475 Park Avenue South, New York, NY 10016.

Staff Report. "Household Soaps and Detergents," *Journal of Chemical Education*, September 1978, pp. 596–597.

Chapter 16

Fox, Charles. "Skin Care: An Overview and Update on the State of the Art and Science," *Cosmestics & Toiletries*, March 1984, pp. 41–56.

Fox, Charles. "Shampoo Components—1985," *Cosmetics & Toiletries*, March 1985, pp. 31–46.

Layman, Patricia L. "Cosmetics: Body Care Eclipses Coverup in Some Uses," *Chemical & Engineering News*, 4 April 1988, pp. 21–41.

Pader, Morton. "Gel Toothpastes: Genesis," *Cosmetics & Toiletries*, February 1983, pp. 71–76.

Quatrale, Richard P. "The Mechanism of Antiperspirant Action," *Cosmetics & Toiletries*, December 1985, pp. 23–26.

Sbrollini, Marilyn C. "Olfactory Delights," *Journal of Chemical Education*, September 1987, pp. 799–801.

Shaath, Nadim A. "The Chemistry of Sunscreens," *Cosmetics & Toiletries*, March 1986, pp. 55–69.

Waldbott, G. L., A. W. Burstahler, and H. L. McKinney. *Fluoridation: The Great Dilemma*. Lawrence, Kan.: Coronado Press, 1978.

Chapter 17

Booth, William. "Revenge of the 'Nozzleheads,'" *Science*, 8 January 1988, pp. 135–137.

Carson, Rachel. *Silent Spring*. New York: Houghton Mifflin, 1962.

Cromartie, Thomas H. "The Inhibition of Enzymes by Drugs and Pesticides," *Journal of Chemical Education*, September 1986, pp. 765–768.

Haggin, Joseph. "Monsanto Uses Genetic Engineering to Solve Agricultural Problems," *Chemical & Engineering News*, 15 February 1988, pp. 28–36.

"How Bugs and Pests Rout the Chemists," *The Economist*, 21 March 1987, pp. 97–98.

Martin, Dean F. and Barbara B. Martin. "The Challenge of Herbicides for Aquatic Weeds," *Journal of Chemical Education*, November 1985, pp. 1006–1007.

Sanders, Howard J. "Herbicides," *Chemical & Engineering News*, 3 August 1981, pp. 20–35.

Sheldon, Richard P. "Phosphate Rock," *Scientific American*, June 1982, pp. 45–51.

Stine, William R. "Chemical Communication by Insects," *Journal of Chemical Education*, July 1986, pp. 603–606.

Chapter 18

Block, Eric. "The Chemistry of Garlic and Onions," *Scientific American*, March 1985, pp. 114–119.

Brown, J. Larry. "Hunger in the U.S.," *Scientific American*, February 1987, pp. 36–41.

Goldsmith, Robert H. "A Tale of Two Sweeteners," *Journal of Chemical Education*, November 1987, pp. 954–955.

Kuang-chih, Tseng and He Hua-zhong. "Structural Theories Applied to Taste Chemistry," *Journal of Chemical Education*, December 1987, pp. 1003–1009.

Lecos, Chris W. "Food Labels: Test Your Food Label Knowledge," *FDA Consumer*, March 1988, pp. 16–21.

Lecos, Chris W. "An Order of Fries—Hold the Sulfites," *FDA Consumer*, March 1988, pp. 8–11.

Southgate, David A. T. "Minerals, Trace Elements, and Potential Hazards," *American Journal of Clinical Nutrition*, 45 (1987): pp. 1256–1266.

Tyler, David R. "Chemical Additives in Common Table Salt," *Journal of Chemical Education*, November 1985, pp. 1016–1017.

Ziporyn, Terra. "The Food and Drug Administration: How 'Those Regulations' Came to Be," *Journal of the American Medical Association*, 18 October 1985, pp. 2037–2046.

Chapter 19

AIDS (Special Issue of *Science*, 5 February 1988).

American Medical Association Guide to Prescription and Over-the-Counter Drugs. New York: Random House, 1988.

Ares, Jeffrey J. and Duane D. Miller. "Treatment of Diabetic Complications with Aldose Reductase Inhibitors," *Journal of Chemical Education*, March 1986, pp. 243–245.

Castleton, Michael. *Cold Cures*. New York: Ballantine Books, 1987.

Cohen, Mitchell L. and Robert V. Tauxe. "Drug-Resistant *Salmonella* in the United States: An Epidemiological Perspective," *Science* 234 (1986): pp. 964–969.

Djerassi, Carl. "The Bitter Pill," *Science*, 28 July 1989, pp. 356–361.

Palca, Joseph. "AIDS Drug Trials Enter New Age," *Science*, 6 October 1989, pp. 19–21.

Riley, Thomas N. "The Prodrug Concept and New Drug Design and Development," *Journal of Chemical Education*, November 1988, pp. 947–953.

Rosenthal, Elisabeth. "Drug Designers," *Science Digest*, March 1983, pp. 22–24, 106, 113.

Saxena, Anil. "Chemistry of Birth Control Pills," *Journal of Chemical Education*, December 1984, pp. 1075–1076.

Stinson, Stephen C. "Better Understanding of Arthritis Leading to New Drugs to Treat It," *Chemical & Engineering News*, 16 October 1989, pp. 37–70.

Weiss, Rick. "Delivering the Goods," *Science News*, 4 June 1988, pp. 360–362.

Chapter 20

Ackerman, Sandra. "Drug Testing: The State of the Art," *American Scientist*, January/February 1989, pp. 19–23.

Alper, Joseph. "Depression at an Early Age." *Science 86*, May 1986, pp. 44–50.

Davis, Audrey B. "The Development of Anesthesia," *American Scientist*, September/October 1982, pp. 522–528.

Hazleton, Lesley. "Cocaine and the Chemical Brain," *Science Digest*, October 1984, pp. 58–61, 100–103.

Jacobs, Barry L. "How Hallucinogenic Drugs Work," *American Scientist*, July/August 1987, pp. 386–391.

Kozel, Nicholas J. and Edgar H. Adams. "Epidemiology of Drug Abuse: An Overview," *Science* 234 (1986): pp. 970–974.

Rosenthal, Elisabeth. "Heroin Addiction: Is It a Disease?" *Science Digest*, August 1983, pp. 76, 102–104.

Shafer, Jack. "Designer Drugs," *Science 85*, March 1985, pp. 60–67.

Snyder, Solomon H. *Drugs and the Brain*. San Francisco: Freeman, 1987.

Veca, A. and J. H. Dreisbach. "Classical Neurotransmitters and Their Significance Within the Nervous System," *Journal of Chemical Education*, February 1988, pp. 108–111.

Chapter 21

Ames, Bruce N., Renae Magaw, and Lois Swirsky Gold. "Ranking Possible Carcinogenic Hazards," *Science* 236 (1987): pp. 271–280.

Arena, Jay M. and Richard H. Drew (Eds.). *Poisoning: Toxicology, Symptoms, Treatments*, 5th ed. Springfield, Ill.: Thomas, 1986.

Beyler, Roger E. and Vera Kolb Meyers. "What Every Chemist Should Know About Teratogens— Chemicals That Cause Birth Defects," *Journal of Chemical Education*, September 1982, pp. 759–763.

Booth, William. "Voodoo Science," *Science* 240 (1988): pp. 274–277.

Graham, John D., Laura C. Green, and Marc J. Roberts. *In Search of Safety: Chemicals and Cancer Risk*. Cambridge, Mass.: Harvard University Press, 1988.

Hanson, David J. "Science Failing to Back Up Veteran Concerns About Agent Orange," *Chemical & Engineering News*, 9 November 1987, pp. 7–14.

Hart, J. Roger. "EDTA-Type Chelating Agents in Everyday Consumer Products: Some Medicinal

and Personal Care Products," *Journal of Chemical Education*, December 1984, pp. 1060–1061.

Raloff, Janet. "Detoxifying PCBs," *Science News*, 5 September 1987, pp. 154–159.

Chapter 22

Eichner, Edward R. "Blood Doping: Results and Consequences from the Laboratory and the Field," *The Physician and Sports Medicine*, January 1987, pp. 120–129.

Friedman, Theodore. "Progress Toward Human Gene Therapy, *Science*, 16 June 1989, pp. 1275–1288.

Gasser, Charles S. and Robert T. Fraley. "Genetically Engineering Plants for Crop Improvement," *Science*, 16 June 1989, pp. 1293–1307.

Hanker, Jacob S. and Beverly L. Giammara. "Biomaterials and Biomedical Devices," *Science*, 11 November 1988, pp. 885–892.

Jones, Michael D. and Jeffrey T. Fayerman. "Industrial Applications of Recombinant DNA Technology," *Journal of Chemical Education*, April 1987, pp. 337–339.

Lynch, Wilfred. *Implants: Reconstructing the Human Body*. New York: Van Nostrand Reinhold, 1982.

Reynolds, Gretchen. "Don't Drink Dry," *Runner's World*, June 1987, pp. 40–45.

Strauss, Richard H. *Drugs and Performance in Sports*. Philadelphia: Saunders, 1987.

Thornton, John I. "DNA Profiling," *Chemical & Engineering News*, 20 November 1989, pp. 18–30.

Voy, Robert. "Illicit Drugs and the Athlete," *American Pharmacy*, November 1986, pp. 39–45.

Weiss, Rick. "Sanguine Substitutes," *Science News*, 26 September 1987, pp. 200–202.

Index

Photo Credits for Part Openers
Part I, NASA; Part II, Abbott Laboratories; Part III, USDA—Soil Conservation Service; Part IV, Gunther/Photo Researchers, Inc.; Part V, Central Washington University.

Cover Photographs
Background and left: © Comstock, Inc. Right: © Bruce Hands/Comstock, Inc. Center: © Oxford Molecular Biophysics Laboratory Science Library/Photo Researchers, Inc.